Energy Studies

Third Edition

W. Shepherd

Formerly University of Bradford, UK and Ohio University, USA

D. W. Shepherd

University of Bradford, UK

Imperial College Press

ICP

Published by

Imperial College Press
57 Shelton Street
Covent Garden
London WC2H 9HE

Distributed by

World Scientific Publishing Co. Pte. Ltd.
5 Toh Tuck Link, Singapore 596224
USA office: 27 Warren Street, Suite 401-402, Hackensack, NJ 07601
UK office: 57 Shelton Street, Covent Garden, London WC2H 9HE

British Library Cataloguing-in-Publication Data
A catalogue record for this book is available from the British Library.

ENERGY STUDIES
Third Edition

ISBN 978-1-84816-850-3

Typeset by Stallion Press
Email: enquiries@stallionpress.com

Printed in Singapore by Mainland Press Pte Ltd.

PREFACE

The industrially developed countries of the world have become rich and prosperous by the profligate use of fossil fuels: coal, oil and natural gas. Countries of the developing areas of the world, mainly in the Pacific Rim and Far Fast, are starting to use fossil fuels, especially oil, at increasing rates. But both oil and natural gas reserves are fast depleting and are non-renewable. Each source has only a few tens of years of stock remaining. How is future world energy demand to be met?

To address such a fundamental problem, it is vitally important that all of the various elements comprising the problem are well understood. In the case of world energy, the problem elements are the individual energy sources, both old and new.

At least ten distinct types of energy source exist:

coal;
oil;
natural;
gas;
nuclear;
geothermal;
biological/chemical;
hydroelectric;
wind;
wave/tidal; and
solar energy.

Each of these sources is examined in Energy Studies, in an attempt to take stock of the development of each, towards either depletion or viable widespread utilization. Environmental implications, economic assessments and industrial risks are also considered.

By doing this, the authors are able to conclude with an illustrative example of an energy strategy with which to address the world energy future, so encouraging readers to weigh for themselves the complex problem that now stares mankind in the face.

Chapter 1 is written mainly for students of the physical sciences and engineering. More general readers are advised to begin reading from Chapter 2.

W. Shepherd and D. W. Shepherd

July 1997

PREFACE TO THE SECOND EDITION

In the five years that have elapsed since the original publication, the issues of energy matters and environmental concerns have become prominent. Energy supply and use is now a matter of frequent reports, not only in trade journals but also in the popular press.

Up-to-date figures are now given for items of fuel supply and also for the use of renewable sources such as wind energy and photovoltaics. The chapters on geothermal energy and nuclear energy have been extended. Increased coverage is given to waste and waste disposal, in Chapter 13.

The energy strategy proposed in the first edition is unchanged. It is the view of the authors that this remains the logical, sensible and workable way to proceed.

W. Shepherd and D. W. Shepherd
June 2002

PREFACE TO THE THIRD EDITION

Energy supply has become a dominant issue that is frequently reported and discussed in the media, both in the technical press and in the popular press.

The publication of a third edition has enabled the authors to update the figures for fossil fuel use, for nuclear energy and for renewables.

The chapters concerned with the fossil fuels — coal, oil and natural gas — have been largely rewritten and re-illustrated. Major revision has also been applied to the chapters on electricity and wind energy, and to the section of the nuclear chapter dealing with nuclear fusion. Many minor modifications have been made throughout the whole text. The end of chapter problems and answers are largely unchanged.

Complete and detailed solutions of all the end of chapter problems are given in the companion book "Problems and Solutions — Energy Studies", by the same authors, published by Imperial College Press in 2008.

The energy strategy proposed in the first edition is unchanged and is reproduced here. It remains the view of the authors that this is the logical, sensible and workable way to proceed.

W. Shepherd, D. W. Shepherd
September 2013

ACKNOWLEDGEMENTS

Much of the material in this book has been taught in undergraduate and post-graduate courses at the University of Bradford, England, and Ohio University, Athens, Ohio, USA. The authors are grateful to both universities for permission to reproduce teaching and examination materials.

The information was obtained from a vast number of sources, some original. Wherever possible the authors have attributed their sources. Thanks are due to the publishers of pre-existing material for their generous permission to reproduce previously published information. The authors apologize if any pre-existing material is not adequately attributed; this is not an attempt to deceive but due to inadvertence.

Dr James Brooks of Glasgow, Scotland, a distinguished geochemist, read the manuscript. His many helpful criticisms and suggestions have enhanced the presentation, especially the chapters on fossil fuels and on geothermal energy.

The authors' work was greatly helped by the superb facilities of the Alden Library at Ohio University. Special thanks are due to Lars Lutton, photographer, Samuel Girton and Scott Wagner, graphic artists, and especially to Peggy Sattler, graphic design manager in the Instructional Media and Technology Services Unit.

We are grateful to Mr Michael Mitchell of Bradford, England, for his valuable help with the computer-generated diagrams.

The typing of the manuscript, with its many revisions during the evolution, was largely done by Suzanne Vazzano of Athens, Ohio. Her professionalism and good nature were indispensable in its completion.

Athens, Ohio, USA
1997

ACKNOWLEDGEMENTS FOR THE SECOND EDITION

The authors would like to thank the publishers of the many new sources that are included in this second edition, in addition to re-acknowledgement of the original sources.

Once more the chief sources of information are British Petroleum plc of London, England, and the US Energy Information Administration of Washington, DC, USA.

Dr James Brooks of Glasgow, Scotland, has once again reviewed the chapters on the fossil fuels plus the work on geothermal energy. His careful scrutiny and helpful suggestions are much appreciated. Ms Ann Mandi of Brown University, USA, also reviewed the manuscript and made many helpful suggestions.

Much of the artwork is due to the staff of the Instructional Media Services Unit at the Alden Library of Ohio University. Special mention must be made of Kelly Kirves, graphic artist, and Emily Marcus, media artist. Particular thanks are due to Lara Neel, graduate assistant, who transferred the manuscript, including artwork, onto computer discs. All of this work was supervised by Peggy Sattler, the production manager of the unit. The book cover is only a small part of Peggy's significant contribution to the overall presentation.

The typing of the revised manuscript, with its many revisions, was largely done by Suzanne Vazzano, helped by Erin Dill, Tammy Jordan, Juan Echeverry and Brad Lafferty. Their professionalism and good nature were indispensable to its conclusion.

Athens, Ohio, USA
2002

ACKNOWLEDGEMENTS FOR THE THIRD EDITION

Most of the new and updated information about fuel supply and use has been obtained from literature published by British Petroleum plc of London, England, and from the US Energy Information Administration of Washington, DC, USA. The authors are grateful to these sources for their permission to publish. Wherever possible the authors have cited their quotations and references. Any reference that is not cited or is wrongly cited is due to inadvertence and not an attempt to deceive.

Once again, Dr James Brooks of Ayrshire, Scotland, undertook to review the section on fossil fuels and on geological energy. He supplied several of the new illustrations but became ill and was not able to complete the proofreading.

Much of the new typing was undertaken by Mrs Ella Chavda of Bradford and the authors are grateful for her contribution. Mr Dig Chavda, University of Bradford, was very helpful in technical aspects of the preparation. Regrettably, Mr Chavda died while the manuscript was in press.

The authors are greatly indebted to our editor Catharina Weijman of Imperial College Press. Her patience, kindness, encouragement and advice were very helpful in processing the manuscript through the editorial stages.

Yorkshire, England
2013

Contents

CHAPTER 1

ENERGY AND POWER

Energy is the capacity or capability to do work. All materials possess energy, because they can all be utilized in some form of energy conversion process. For example, most substances will burn or vaporize, and the consequent heat energy can be harnessed within mechanical energy systems that create motion against some form of mechanical resistance.

Energy can take several forms, as classified in Table 1.1. Mass or matter is a form of highly concentrated energy. Some forms of matter can be utilized in nuclear energy applications, as discussed in Chapter 8.

1.1. Energy Conversion

The many applications of the use of energy usually involve transformations between different forms of energy; a process known as energy conversion. Any conversion between different energy forms is imperfect, in that some of the energy has to be used to facilitate the conversion process. The converted energy output is lower than the energy input and this feature is usually described as the conversion efficiency. Figure 1.1 illustrates the large range of variation of energy conversion efficiencies, from very large electricity generators (mechanical to electrical converters), which can operate continuously at about 99% efficiency, to the incandescent electric lamp (electrical to radiant converter), which is only a few per cent efficient [1]. Some well-known energy conversion processes involve two successive stages. An example is the

Table 1.1. Forms of energy.

Biological	Mass
Chemical	Mechanical — kinetic
Electrical	Mechanical — potential
Gravitational	Nuclear
Heat (thermal)	Radiation
Magnetic	Sound

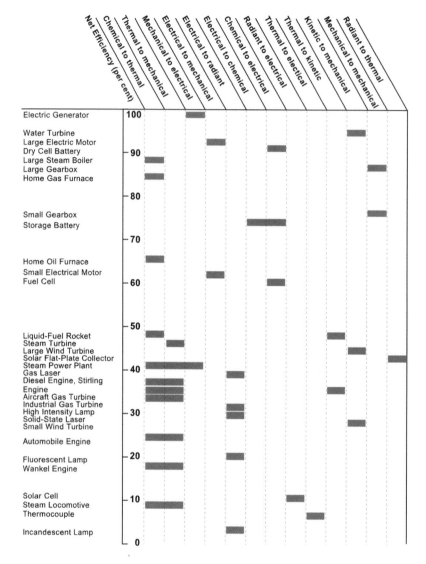

Fig. 1.1.   Efficiencies of energy converters (based on [1]).

motor car engine, in which chemical energy in the form of oil or petrol (gasoline) is converted to heat and then to rotational energy.

1.2   Mechanical Energy [2,3]

The widely used laws of motion for bodies of constant mass were developed by the English scientist Isaac Newton in the 17th century. It is now known that in extreme cases Newton's laws are insufficient: for very small masses quantum mechanics must be employed; with very high speeds, Einstein's theory of special relativity becomes relevant; with very large masses, the concepts of space and time are modified by the theory of general relativity. Nevertheless, for general conduct of life on earth using realistic sizes and time spans, the work of Newton remains valid.

1.2.1. *Linear motion*

When a constant force F is applied to an object and causes it to move through a distance x in the direction of the force, then the work done W is equal to the energy expended:

$$W = Fx \qquad (1.1)$$

In Eq. (1.1), if the force is in newtons (N) and the distance in metres (m), the work or energy (W) has the unit of joules (J) or newton-metres (Nm).

If a body of mass m moves in a straight line with a linear velocity v, which is the time rate of change of its position,

$$v = \frac{dx}{dt} \text{ for small changes of } x$$
$$v = \frac{x}{t} \text{ for large changes of } x \qquad (1.2)$$

If a body of mass m moving in a straight line is subjected to changes of velocity, the rate of change of the velocity with time is known as the acceleration a:

$$a = \frac{dv}{dt} = \frac{d}{dt}\left(\frac{dx}{dt}\right) = \frac{d^2x}{dt^2} \qquad (1.3)$$

In SI units, mostly used in this book, the velocity is measured in metres/sec (m/s) and the acceleration in metres/sec/sec or metres/sec^2(m/s^2).

When a force F is applied to a body of constant mass m and causes the linear velocity v to change, the resulting acceleration can be shown experimentally to be proportional to the applied force:

$$F = ma = m\frac{dv}{dt} = m\frac{d^2x}{dt^2} = mv\frac{dv}{dx} \qquad (1.4)$$

Equation (1.4) is sometimes referred to as Newton's Second Law of Motion.

Mass m may be combined with velocity v to define an important physical property, known as the momentum:

$$\text{Linear Momentum} = mv = m\frac{dx}{dt} \qquad (1.5)$$

A comparison of Eqs. (1.4) and (1.5) shows that

Force = time rate of change of linear momentum

$$F = m\frac{dv}{dt} = \frac{d}{dt}(mv). \tag{1.6}$$

Equation (1.6) shows that momentum has the dimension of force × time or mass × velocity.

A mass m possesses energy of two kinds, known as potential energy, associated with its position, and kinetic energy, associated with its motion. The gravitational potential energy of a body of mass m, at height h above a datum plane, is given by

$$W_{PE} = mgh \tag{1.7}$$

where g is the gravitational acceleration constant of value $g = 9.81\,\mathrm{m/s^2}$. If the mass m is in kilogrammes and height h is in metres, the potential energy W_{PE} is in joules.

While a mass m is in linear motion at a constant velocity v, the kinetic energy W_{KE} associated with the motion is

$$W_{KE} = \frac{1}{2}mv^2 \tag{1.8}$$

It can be seen from Eq. (1.8) that the derivative of kinetic energy W_{KE} with respect to velocity gives the momentum

$$\frac{dW_{KE}}{dv} = mv \tag{1.9}$$

Both kinetic energy and momentum, like mass, satisfy important conservation rules. In this book the most relevant rule is the principle of conservation of energy, which states that "in any physical system the total energy remains constant — energy may be converted to a different form, it may be wasted, but it cannot be destroyed". When a mass m in linear motion is acted upon by a force F, then, in moving between two locations:

force × distance moved = work done on or against the mass

= change of kinetic energy between the two locations

$$\tag{1.10}$$

Example 1.1.   A mass m initially rests on a ledge at height h metres above ground level, which is the datum plane. Define the conditions of velocity, kinetic energy and potential energy (i) initially, (ii) as the mass falls to ground and finally after the mass comes to rest.

(i) With the mass at rest, its initial velocity v_i is zero and therefore so are its initial momentum and kinetic energy. Its total energy is then the potential energy given by Eq. (1.7), as illustrated in Fig. 1.2.

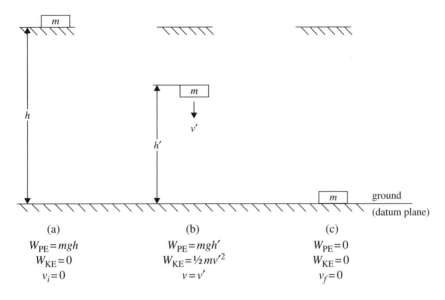

Fig. 1.2. Mass falling freely under gravity.

(ii) As the mass falls to the ground it possesses an instantaneous velocity v^t, initially zero and increasing uniformly due to gravitational acceleration. Its final velocity becomes zero on impact with the ground. At any arbitrary height h^t during the fall, the mass possesses both potential energy mgh^t and kinetic energy $\frac{1}{2}mv^{t2}$, which sum to the initial energy mgh. After striking the ground the final velocity v_f is zero, the momentum of the motion is transferred to the ground and the kinetic energy is converted to local heat and sound due to impact. Since the ground level is the datum plane, the potential energy after impact is also zero here.

1.2.2. *Rotational motion*

Most energy conversion processes involving mechanical energy incorporate rotational devices. For example, electromechanical energy converters use rotors that have the form of solid cylinders (Fig. 1.3(a)). Petrol engines and diesel engines usually incorporate flywheels (Fig. 1.3(b)). The rotor of a water or gas turbine also has the nature of a non-uniform flywheel.

To illustrate some of the principles of rotational motion, the example used is that of a concentrated mass m in circular motion at radius r about a fixed centre point (Fig. 1.4). The motion is characterized by the angular velocity ω in radians/sec (rad/s) and the instantaneous tangential velocity v of the mass in metres/sec (m/s), where

$$v = \omega r \qquad (1.11)$$

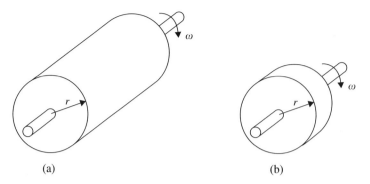

(a)   (b)

Fig. 1.3.   Structure of some mechanical energy converters. (a) Electric motor or generator. (b) Flywheel (internal combustion engine).

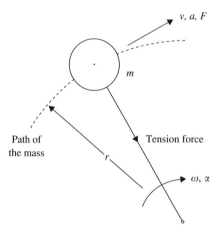

Fig. 1.4.   Mass rotating in a horizontal circle.

A centripetal force acting radially inwards is required to keep the mass moving in a circle and is provided along the tie rod. With rotational motion, the externally applied force F acting tangentially on the mass (through a rigid tie-rod; Fig. 1.4) times the radius r is called the torque T, which acts as a rotation producing force:

$$T = Fr = F\frac{v}{\omega} \tag{1.12}$$

Torque is measured in newton-metres (Nm) and is a very important property of rotating energy converters. The tangential or linear acceleration of the mass is given, from Eq. (1.11), by

$$a = \frac{dv}{dt} = r\frac{d\omega}{dt} \tag{1.13}$$

Combining Eqs. (1.12) and (1.13) leads to

$$T = Fr$$

$$= mar$$

$$= m\frac{dv}{dt}r$$

$$\therefore\ T = mr^2\frac{d\omega}{dt} = mr^2\alpha \tag{1.14}$$

In Eq. (1.14) the term α is the angular acceleration in rad/s^2. The quantity mr^2 in Eq. (1.14) is known as the polar moment of inertia J and is an important physical property in rotational structures, having the dimension kgm^2.

$$J = mr^2 = [\text{mass}]\,[\text{radius of gyration}]^2 \tag{1.15}$$

Equation (1.15) is true directly for the flywheel and cylinder of Fig. 1.3. For more complicated structures with distributed, non-uniform mass, the effective radius of gyration is more complicated but the relationship in Eq. (1.15) is still valid in principle.

The properties of Eqs. (1.14) and (1.15) can be summarised as

$$T = J\alpha = J\frac{d\omega}{dt} \tag{1.16}$$

which is directly analogous to Eq. (1.4) for linear motion.

The kinetic energy associated with rotational motion can be obtained by incorporating Eq. (1.11) into Eq. (1.8), using Eq. (1.15):

$$W_{\text{KE}} = \frac{1}{2}mv^2$$

$$= \frac{1}{2}mr^2w^2 \tag{1.17}$$

$$\therefore\ W_{\text{KE}} = \frac{1}{2}Jw^2$$

When J is in kgm^2 and ω in rad/s, the energy of motion, which is also the work done on the rotating mass, has the dimension of joules (J) or watt-seconds (Ws).

Example 1.2.   A solid mass m rotates around a fixed centre point at radius r with constant angular velocity ω. Show that the force F that impels the motion is proportional to the time rate of change of angular momentum.

In Fig. 1.4, the tangential force that causes the rotation and maintains it is given by

$$F = ma$$

$$= m\frac{dv}{dt}$$

For constant mass, therefore,

$$F = \frac{d}{dt}(mv)$$

$$= \text{time rate of change tangential momentum}, mv$$

$$= \frac{d}{dt}(mwr) = r\frac{d}{dt}(mw)$$

$$= [r] \text{ [time rate of change of angular momentum]}$$

Example 1.3.   The rotor of an electric motor has a polar moment of inertia of $10\,\text{kgm}^2$ and rotates at a steady speed of 1800 revolutions per minute (rpm). Calculate the kinetic energy of the motion.

$$J = 10\,\text{kgm}^2$$

The angular velocity is given in rpm and must be converted to the corresponding SI unit of radians/sec.

$$1800\,\text{rpm} = \frac{1800}{60}\,\text{rev/s}$$

$$= \frac{1800}{60}\,2\pi\text{rad/s}$$

$$= 60\pi\text{rad/s}$$

The kinetic energy, from Eq. (1.17), is

$$W_{\text{KE}} = \frac{1}{2}J\omega^2$$

$$= \frac{1}{2}10(6\pi)^2$$

$$= 2418\,\text{joules}$$

This value also represents the work that was done in rotating the motor from rest to its steady speed.

1.3.   Electrical Energy

Electrical energy is the universal clean form of energy that is most commonly used. It is, however, a secondary form of energy that has to be obtained by the use of a primary fuel such as coal or oil. Because of the great importance of electrical energy it is the subject of a separate section, in Chapter 3.

1.4.   Chemical Energy

Chemical energy may be associated with chemical reactions, combustion engines, rockets, electrical cells and batteries, heating from boilers, etc. The energy is usually stored within materials and is released by combustion. Some aspects of chemical energy storage are covered in Chapter 7.

1.5.   Nuclear Energy

The energy stored within an atomic nucleus is manifested, for certain chemical elements, by radioactive decay. Energy can be made available by the processes of nuclear fission and nuclear fusion, discussed in Chapter 8.

1.6.   Thermal Energy

Heat (thermal) energy is so important a feature of energy conversion systems and incorporates so many vital physical principles that it is treated in a major section immediately below.

1.7.   Thermodynamics and Heat Energy

1.7.1.   *Quantity of heat*

Heat is a form of energy. It has the capacity to do work directly as thermal warming or by conversion to other energy forms, mainly mechanical. The quantity of heat involved in a process can be measured by some effect that accompanies the process. Traditional units of measurement for the quantity of heat Q are the calorie (in cgs units) and the British thermal unit or BTU (fps units).

1 calorie is the amount of heat to raise 1 gramme (1 g) of water through 1°C.
1 BTU is the amount of heat to raise 1 pound (1 lb) of water through 1°F.

Note that heat is not the same as temperature. The quantity of heat Q is a measure of the energy capacity, whereas temperature T is a measure of the hotness.

The amount of heat energy required to raise the temperature of a particular mass of material through a specified temperature range is a characteristic property of the material. In particular, the specific heat capacity is the heat capacity per unit mass, and is measured in cals/g-centigrade or BTU/lb-fahrenheit. Water is the standard material, having a specific heat capacity of 1 cal/g°C, 1000 cals/kg°C or 1 BTU/lb°F. It is common to use a dimensionless property known as specific heat (SH),

$$SH = \frac{\text{specific heat capacity of a material}}{\text{specific heat capacity of water}} \qquad (1.18)$$

Water, therefore, has the value $SH = 1$.

A mass m of material raised through a temperature difference $(T_2 - T_1)$ possesses a quantity of heat energy Q where

$$Q = [\text{heat capacity of the body}] \times [\text{temperature difference}]$$
$$M = [\text{mass}]\,[\text{specific heat}]\,[\text{temperature difference}] \qquad (1.19)$$
$$Q = mSH(T_2 - T_1)$$

In Eq. (1.19), Q has the dimension of mass $\times$ temperature.

1.7.2. *Mechanical equivalent of heat*

Mechanical energy is expressed in units of ergs (cgs system), ft lb (fps system) or joules (SI system). The equivalence between mechanical energy units and heat energy has to be determined by experiment. This equivalence has the internationally agreed value, formerly known as the "mechanical equivalent of heat",

$$1\,\text{g-calorie} \equiv 4.186\,\text{joules}$$
$$1\,\text{BTU} \equiv 7718.26\,\text{ft lb}$$
$$\equiv 252\,\text{g-cal}$$
$$\equiv 1054.7\,\text{joules}$$

Slight approximation of the above figures is often permitted, so that the energy or work W in joules is given by

$$\begin{aligned} W &= 4.2Q \quad \text{if } Q \text{ is in calories} \\ &= 1055Q \quad \text{if } Q \text{ is in BTUs} \end{aligned} \qquad (1.20)$$

A well-known energy conversion process incorporating a heat-to-work stage is electricity generation, illustrated in Figs. 1.5 and 1.6. Chemical energy in the solid fuel is released by combustion and used to boil water in a closed-cycle system that is thermally insulated to reduce heat loss. Energy in the form of pressurized steam rotates the blades of a steam turbine. After imparting much of its energy to the turbine, the steam condenses back into hot water, which needs to be force-cooled from some large external source of cooling water such as a sea, lake or river. Rotational mechanical energy is transferred from the turbine to the electric generator. A more detailed consideration of electricity generation is given in Chapter 3.

The conversion of heat energy into mechanical work also takes place in petrol engines, diesel engines, jet engines, gas turbines and rocket motors. These may all be grouped under the title of "heat engines". Energy conversion of this form is

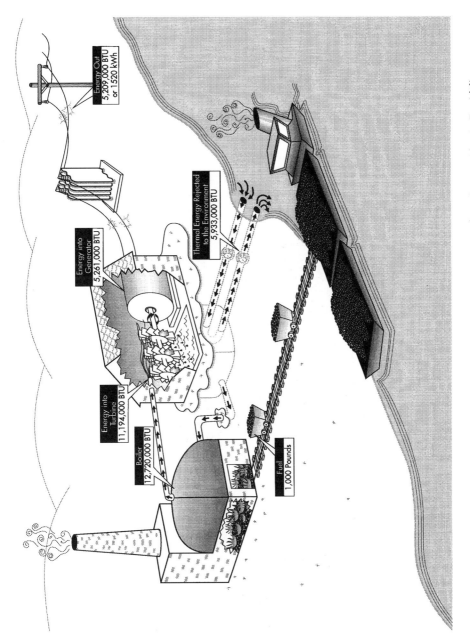

Labels within figure:

Energy Out
5,209,000 BTU
or 1,520 kWh

Energy into
Generator
5,261,000 BTU

Thermal Energy Rejected
to the Environment
5,933,000 BTU

Energy into
Turbine
11,194,000 BTU

Boiler
12,720,000 BTU

Fuel
1,000 Pounds

Fig. 1.5.   Basic form of a heat–work system for electricity generation (based on an idea in Dorf [2]).

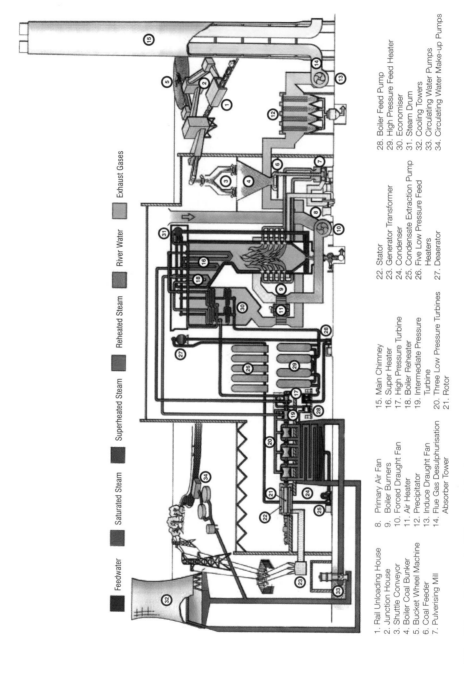

Fig. 1.6. Schematic layout of Drax coal-fired power station, Yorkshire, England (reproduced by permission of "Environmental Performance Review 2008", Drax Power Ltd.).

1. Rail Unloading House
2. Junction House
3. Shuttle Conveyor
4. Boiler Coal Bunker
5. Bucket Wheel Machine
6. Coal Feeder
7. Pulverising Mill
8. Primary Air Fan
9. Boiler Burners
10. Forced Draught Fan
11. Air Heater
12. Precipitator
13. Induce Draught Fan
14. Flue Gas Desulphurisation Absorber Tower
15. Main Chimney
16. Super Heater
17. High Pressure Turbine
18. Boiler Reheater
19. Intermediate Pressure Turbine
20. Three Low Pressure Turbines
21. Rotor
22. Stator
23. Generator Transformer
24. Condenser
25. Condensate Extraction Pump
26. Five Low Pressure Feed Heaters
27. Deaerator
28. Boiler Feed Pump
29. High Pressure Feed Heater
30. Economiser
31. Steam Drum
32. Cooling Towers
33. Circulating Water Pumps
34. Circulating Water Make-up Pumps

Feedwater Saturated Steam Superheated Steam Reheated Steam River Water Exhaust Gases

restricted in scope and efficiency, defined by certain natural laws and limitations embodied in a formulation called the laws of thermodynamics.

1.7.3.   *The first law of thermodynamics*

When the principle of conservation of energy is applied to a heat–work conversion process, it becomes known as the first law of thermodynamics and can be stated thus:

> In an isolated, enclosed heat–work system the total energy remains constant.

An expanded statement of the first law is:

> The change of internal energy of a system is equal to the net heat energy input (Q) minus the net external work done (W).

In equation form

$$\text{net heat energy supplied} - \text{net work done by the system}$$
$$= \text{change in stored energy}$$

or

$$Q - W = \text{final stored energy} - \text{initial stored energy}$$

Even if the operating level of the system changes (e.g. if it operates at, say, a higher level of heat input and consequent higher work done), the quantity $(Q - W)$ remains constant. The first law states, in effect, that not more than 4.186 joules of mechanical energy can be obtained from 1 g-calorie of suitably heated material, irrespective of the energy conversion process. It is significant that the first law does not describe any changes that may occur in the quality of the stored energy nor in its capacity to do work.

1.7.4.   *The second law of thermodynamics*

The heat energy contained within a body varies directly with its temperature, as implied in Eq. (1.19). The process of cooling implies a reduction of energy. Heat always flows spontaneously from a body of higher temperature to a body of lower temperature in an attempt to obtain a thermal energy equilibrium of uniform temperature. There can be no spontaneous flow of heat energy from a cooler body to a hotter body; this would be analogous to water spontaneously flowing uphill.

A heat transfer process is always imperfect. However well the system is designed and maintained, some heat is lost to the surroundings in the form of exhaust gases, cooling liquids or gases, conduction losses due to a temperature rise in pipes, convection to the ambient air, radiation into the surrounding space, etc.

1.7.4.1.   *Ideal heat engine*

Even in an ideal heat engine containing no losses or design imperfections, there would still remain an important natural limitation of scope. This is expressed in one of several possible formulations of the important physical principle known as the second law of thermodynamics:

> A cyclic heat–work operation needs to contain two thermal reservoirs. Even an ideal heat engine is capable of converting only part of the input heat energy into work. The remaining heat energy must be transferred to a lower temperature reservoir.

The first and second laws of thermodynamics are illustrated by application to an ideal heat engine in Fig. 1.7.

If there is no stored energy, then, from the first law statement of Eq. (1.21),

$$Q_H - Q_L = W \tag{1.22}$$

Now, the energy efficiency of any converter is the ratio of the energy output to the energy input. In the case of a heat–work converter the output energy is equal to the mechanical work done:

$$\eta = \frac{\text{work output}}{\text{work input}} = \frac{W}{Q_H} = 1 - \frac{Q_L}{Q_H} \tag{1.23}$$

It can be seen from Eq. (1.23) and Fig. 1.6(a) that the most efficient heat engine will be a form in which Q_H and T_H are as high as possible and Q_L and T_L are as low as possible. But it is important to note that Q_L and T_L cannot be reduced to

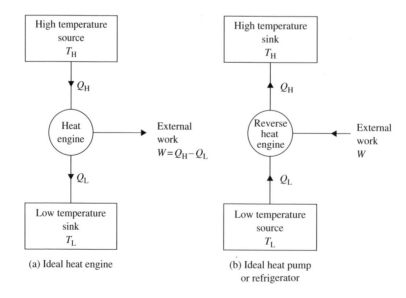

(a) Ideal heat engine

(b) Ideal heat pump or refrigerator

Fig. 1.7.   Schematic function of an ideal heat engine.

zero, which would represent 100% efficiency. Cyclic interchange between Q_H and W in Fig. 1.7(a), with $Q_L = 0$, would then constitute a form of perpetual motion machine, which the laws of thermodynamics declare to be impossible.

In 1824 a young French engineer called Sadi Carnot deduced that an ideal heat engine would have a theoretical maximum efficiency that has come to be known as the Carnot efficiency, η_{Carn},

$$\eta_{Carn} = 1 - \frac{T_L}{T_H} \tag{1.24}$$

In Eq. (1.24), T_L and T_H are absolute temperatures measured from the absolute minimum temperature $-273°C$, in units known as Kelvin (K). The two expressions, Eqs. (1.23) and (1.24), can be linked via the property of entropy, which is discussed below in Section 1.8.

1.7.4.2. *Practical heat engine*

A practical form of heat engine does not realize a working efficiency anywhere near to the level of the Carnot efficiency. The efficiency diagram of Fig. 1.1 shows that automobile petrol engines have working efficiencies of about 25%, diesel engines and gas turbines about 35% and steam turbines about 45%. This means that a lot of energy is wasted (not destroyed!) and a lot of thermal pollution is created. The most efficient engines achieve a practical efficiency of about only two-thirds that of the Carnot efficiency.

The various energy utilization features of a solid-fuel electricity generation plant, referred to in Section 1.7.2 above, are illustrated in Figs. 1.5 and 1.6. The energy into the turbine, 11,194,000 BTU in Fig. 1.5, corresponds to Q_H in Fig. 1.7(a), but only 5,261,000 BTU is converted to mechanical work W. The remaining heat, 5,933,000 BTU, corresponds to Q_L. This heat energy is dissipated into the environment, causing increased temperature of the coolant source water. In a steam engine the low temperature sink is often referred to as a condenser.

1.7.4.3. *Ideal reverse heat engine (heat pump)*

If external work is done on the engine (instead of by it), heat energy can be extracted from the low temperature source and injected into the high temperature reservoir (Fig. 1.7(b)). This process is now referred to as the use of a "heat pump". An example is the familiar air conditioner widely used in the summer of hot climates such as that of the southern states of the USA. There is seldom a need for domestic air conditioning in the UK. The best-known example of a heat pump is the domestic refrigerator (Fig. 1.8), in which heat is extracted from the food inside and expelled into the kitchen. In a refrigerator the reverse heat engine converts the electrical (or gas) input energy into thermal energy. The refrigerator mechanism of pump, coolant pipes and coolant chemical (liquid or gas) acts as a heat pump. The refrigerant fluid

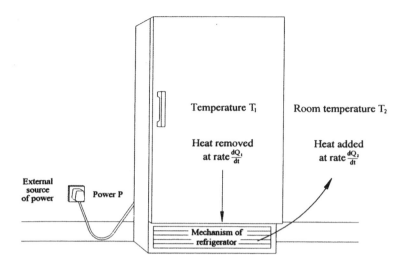

Fig. 1.8.   Thermal action of a refrigerator [3].

acts as a heat exchanger, absorbing heat as it vaporizes and releasing heat when it condenses.

In an ideal reverse heat engine (Fig. 1.7(b)),

$$Q_{\mathrm{H}} = W + Q_{\mathrm{L}} \tag{1.25}$$

The effectiveness of the performance of a heat pump may be defined in terms of a coefficient of performance (CoP), defined as

$$\text{Coefficient of performance} = \frac{\text{Heat energy generated}}{\text{Electrical energy to drive the heat pump}}$$

$$\tag{1.26}$$

For viable operation $CoP \geq 3$.

With domestic applications heat can be extracted from the ground or from the air, rendering them colder still. In pumping heat from an outside source T_{L} to a place of higher temperature T_{H} (both temperatures in Kelvin), the ideal efficiency η is

$$\eta = T_{\mathrm{H}}/(T_{\mathrm{H}} - T_{\mathrm{L}}) \tag{1.27}$$

Note that the ideal efficiency of a heat pump, Eq. (1.27), is the inverse of the Carnot efficiency for heat transfer in Eq. (1.23).

Mackay has a helpful comment about the relative merits of UK ground source and air source heat pumps:

"While in theory ground-source heat pumps might have better performance than air-source, because the ground temperature is usually closer than the air temperature to the indoor temperature, in practice an air-source heat pump might be the best and simplest choice. In cities, there may be uncertainty about the future effectiveness of ground-source heat pumps, because the more people use them in the winter, the colder the ground gets; this thermal fly-tipping problem may also show up in the summer in cities where too many buildings use ground-source (or should I say "ground-sink"?) heat pumps for air-conditioning." [4]

In the refrigerator of Fig. 1.8, heat energy is extracted from the refrigerator contents at a time rate dQ_L/dt. Additional energy enters from the electrical supply at a time rate P and energy is expelled into the surroundings at a rate dQ_H/dt. The conservation law in Eq. (1.25) is satisfied by a time rate equation

$$\frac{dQ_H}{dt} - \frac{dQ_L}{dt} = \frac{dW}{dt} = P \tag{1.28}$$

For any ideal heat engine or reverse heat engine, it can be shown that Eq. (1.32) below is true. Combining Eqs. (1.28) and (1.32) gives a relationship for the necessary input power P in terms of temperatures:

$$P \geq \frac{dQ_L}{dt}\left[\frac{T_H}{T_L} - 1\right] \tag{1.29}$$

Equation (1.29) is true for all refrigerator systems regardless of their size or mode of operation [3].

1.7.5.   *Worked examples on thermodynamics and heat energy*

Example 1.4.   A mass of material m_1 of specific heat SH_1, at temperature T_1, is mixed with a mass of fluid m_2 of specific heat SH_2 at temperature T_2. Express the final temperature of the mixture in terms of its constituent parts.

Let the final temperature of the mixture be T_f.
Assume that $T_1 > T_2$, so that $T_2 \leq T_f \leq T_1$. Heat lost by mass m_1 is

$$Q_1 = m_1 SH_1(T_1 - T_f) \tag{a}$$

Heat gained by mass m_2 is

$$Q_2 = m_2 SH_2(T_2 - T_f) \tag{b}$$

Assuming that the mixing is thermally ideal and no heat is lost, then

$$Q_1 = Q_2 \tag{c}$$

Combining (a), (b) and (c) gives

$$T_f = \frac{m_1 SH_1 T_1 + m_2 SH_2 T_2}{m_1 SH_1 + m_2 SH_2} \tag{1.30}$$

Example 1.5.   In the steam boiler electrical plant of Fig. 1.5, what are the mechanical and electrical equivalents of the thermal energy imparted to the cooling water?

$$\text{Thermal energy rejected into the coolant water} = Q = 5933000\,\text{BTU}$$

From Eq. (1.20),

$$W = 1055\,Q$$

$$\text{Mechanical energy equivalent of the thermal energy} = 1055 \times 5933000$$
$$= 6259\,\text{MJ}$$

Now 1 joule $= 1$ watt second, so that

$$\text{Electrical energy equivalent of the thermal energy} = 6259\,\text{MWs}$$
$$= \frac{6259}{3600}$$
$$= 1.739\,\text{MWh}$$

Example 1.6.   A thermal system has an initial internal energy of 50 kJ. Net work is done on the system of energy value 100 kJ, after which the final internal energy is found to be 75 kJ. Calculate the net heat transfer to the system.

From Eq. (1.21), for a lossless system,

$$Q - W = \text{final stored energy} - \text{initial stored energy}$$
$$Q - W = 75 - 50 = 25\,\text{kJ}$$

Now, in Eq. (1.21), the term W represents work done by the system. In the present case, work is done on the system (i.e. W is an energy input) and therefore

$$W = -100\,\text{kJ}$$
$$Q - (-100) = 25\,\text{kJ}$$
$$Q = 25 - 100 = -75\,\text{kJ}$$

In order to achieve a new energy balance, according to the principle of conservation of energy, the heat transfer is negative (i.e. heat energy must have been removed from the system).

Example 1.7.   A high temperature fluid at 600°C is transferred via a mechanical work converter to a lower temperature (condenser) sink at 100°C. If the fluid has an initial heat energy of 250 MJ, (a) what is the maximum theoretical efficiency of mechanical conversion, and (b) what is the maximum theoretical work output?

$$T_{\text{H}} = 600°\text{C} = 600 + 273 = 873\,\text{K}$$
$$T_{\text{L}} = 100°\text{C} = 100 + 273 = 373\,\text{K}$$

(a) The maximum or Carnot efficiency of mechanical conversion is, from Eq. (1.24),

$$\eta_{\text{Carn}} = 1 - \frac{T_{\text{L}}}{T_{\text{H}}}$$

$$= 1 - \frac{373}{873}$$

$$= 1 - 0.427 = 0.573 \quad \text{or} \quad 57.3\%$$

(b) From Eq. (1.23),

$$\eta = \frac{W}{Q_{\text{H}}}$$

so that

$$W = \eta Q_{\text{H}}$$

$$= 0.573 \times 250 = 143.25 \, \text{MJ}$$

Example 1.8. A domestic refrigerator has a freezer compartment with its temperature set at 20°F. The room temperature is 70°F. What is the minimum external energy input requirement if it takes 40 kcal of heat extraction to freeze a tray of ice cubes?

Since zero Kelvin $= -273.15°$C,

$$T_{\text{H}} = 70°\text{F} = 21.1°\text{C} = 294.25 \, \text{K}$$

$$T_{\text{L}} = 20°\text{F} = -6.7°\text{C} = 266.3 \, \text{K}$$

$$Q_{\text{L}} = 40 \, \text{kcal} \equiv 40 \times 4.18 = 167.2 \, \text{kJ}$$

Integrating both sides of Eq. (1.29),

$$\int P dt \geq \int dQ_{\text{L}} \left[\frac{T_{\text{H}}}{T_{\text{L}}} - 1 \right]$$

$$W_{\text{in}} = Pt \geq Q_{\text{L}} \left[\frac{T_{\text{H}}}{T_{\text{L}}} - 1 \right]$$

$$W_{\text{in}} \geq 167.2 \left[\frac{294.25}{266.3} - 1 \right] \text{kJ}$$

$$\geq 167.2[1.105 - 1] \, \text{kJ}$$

$$\geq 17.56 \, \text{kJ} = 17.56 \, \text{kWs}$$

The total "load" on the refrigerator, to freeze the freezer compartment and to cool the main space, is many times the above value.

The rate of heat extraction from a freezer compartment is likely to be of the order 1–3 kcal/min. If this machine extracts heat at (say) 2 kcal/min, it will take 17.52/2 or 8.76 min to freeze the water.

1.8   Entropy [5–7]

1.8.1.   *Entropy in heat–work systems*

The second law of thermodynamics has basic and profound scientific implications far beyond that of heat–work interchange in heat engines. It can be interpreted in terms of the probability of molecular arrangement in a material or in terms of the nature of spontaneous change in materials and systems [3].

In any isolated physical system the direction of spontaneous change is always from molecular order to disorder. A container of hot water, for example, undergoes spontaneous cooling as the energy of motion of its microscopic particles decreases. The molecular change is from a condition of higher order to a condition of lesser order; what mathematicians now describe as greater chaos. The same change is, conversely, from a condition of lesser probability of arrangement to a condition of greater probability of arrangement.

Entropy is a concept, not a physical property. It is a measure of the extent of disorder in a system or of the probability of the arrangement of parts of a system. Greater probability implies greater disorder and higher entropy. Lesser probability implies lesser disorder (more order) and less entropy. The entropy is usually denoted by the symbol S and was defined in 1865 by the German physicist Clausius,

$$dS = \frac{dQ}{T} \tag{1.31}$$

Equation (1.30) defines a small change of entropy of a system dS between equilibrium states as the change of heat energy dQ divided by its absolute temperature T. It is seen that entropy has the dimension heat energy/temperature with the unit of joules/Kelvin (J/K) in SI units. Because the natural tendency of change of physical systems is always towards greater disorder (more chaos), the second law of thermodynamics can be expressed in terms of entropy:

> The entropy of an isolated system spontaneously increases or remains the same.

For a heat–work system the entropy represents the amount of energy that cannot be transformed into mechanical work.

In mathematical terms the second law can be written as

$$dS \geq 0 \tag{1.32}$$

For a finite and measurable heat flow, the entropy can be obtained by integrating Eq. (1.30).

$$\int dS = \int \frac{dQ}{T} \tag{1.33}$$

Entropy increases for heat gain when dQ is positive, but decreases for heat loss when dQ is then negative.

Alternatively to Eq. (1.30), it is possible to define entropy in terms of the natural logarithm of the mathematical probability, but this aspect is not pursued here.

In the ideal heat engine of Fig. 1.6(a) the entropy flow is negative from the high temperature source $S_H = Q_H/T_H$ and positive into the low temperature sink $S_L = Q_L/T_L$. The change of entropy $\int dS$ is therefore

$$\int dS = -S_H + S_L$$

$$= \frac{Q_L}{T_L} - \frac{Q_H}{T_H} \tag{1.34}$$

Since $dS \geq 0$, Eq. (1.32) can be rearranged as

$$\frac{Q_L}{Q_H} \geq \frac{T_L}{T_H} \tag{1.35}$$

Equation (1.34) can be combined with the laws of thermodynamics for an ideal system (Eq. (1.23)) to give the thermodynamic efficiency.

$$\eta = \frac{W}{Q_H} = 1 - \frac{Q_L}{Q_H} \leq \left[1 - \frac{T_L}{T_H}\right] \tag{1.36}$$

The maximum theoretical efficiency deduced by reasoning based on entropy from Eq. (1.35) is therefore that $\eta_{max} = 1 - T_L/T_H$, which is equal to the Carnot efficiency of Eq. (1.24).

1.8.2. *Entropy on a cosmic scale*

Since energy always flows in such a direction as to make the entropy (disorder) increase, different forms of cosmological energy can be arranged in increasing entropy order. There is no associated temperature for gravitational energy or for planetary rotation and orbital motion so that the entropy is zero. Chemical reactions have entropies of the order 1–10 inverse electron volts. The cosmic microwave background radiation is the ultimate heat energy sink, with entropy 10^4 inverse electron volts; no further energy degradation or conversion is possible [1].

The structure of the universe is not inherently stable. It possesses a succession of quantitative features, such as the values of a number of physical constants, which effectively arrest the normal processes of energy degradation favoured by thermodynamics. For example, the main energy flow is associated with the gravitational contraction of very large masses, which converts the energy released into heat, light and motion. Gravitational energy remains, after 15 billion years of cosmic evolution, as predominant in quality and quantity. This is in apparent contradiction to the fact that large masses are unstable against gravitational collapse. There is no agreed scientific explanation for many paradoxical questions with regard to the structure and physical operation of the universe.

1.9. Power

Power P is the time rate of doing work or of expending energy and therefore has the dimension of energy (or work) divided by time.

$$P = \frac{dW}{dt} \quad \begin{array}{l} \text{for small increments of time} \\ P \text{ is called the instantaneous power} \end{array}$$
$$P = \frac{W}{t} \quad \begin{array}{l} \text{for larger increments of time} \\ P \text{ is called the average power} \end{array} \quad (1.37)$$

In SI units the unit of power is the joule per second (J/s), which is called the watt (W). For practical purposes it is often convenient to use the units kilowatt (kW) or megawatt (MW). Power in watts is not concerned exclusively with electrical engineering. The power ratings in watts of various animals or manufactured devices are given on a logarithmic scale in Fig. 1.9 [2], which has to be read carefully.

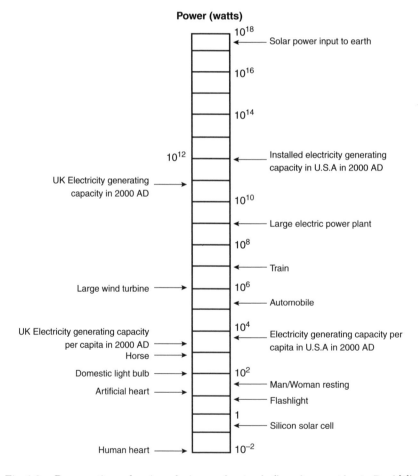

Fig. 1.9.   Power ratings of various devices and animals (based on an idea in Dorf [2]).

For example, the installed electricity generation capacities in the USA and UK are close together on the scale but differ in magnitude by a factor of ten!

In terms of human perception it may sometimes be helpful to use the old British power unit of horsepower (HP).

$$1 \text{ horsepower (HP)} \equiv 746 \text{ W} \tag{1.38}$$

Energy converters with a mechanical output, such as combustion engines, tend to be rated in horsepower and the relative power ratings can be judged from the size (and sometimes noise) of their engines. In Fig. 1.9, the automobile, for example, is shown as having a power rating of about 0.3×10^6 W, which is roughly 400 HP.

It is important to clearly distinguish between power and energy. In particular, it is essential to avoid a common lay practice of using the terms interchangeably.

Power is associated with time, whereas energy is independent of time.

$$\text{power} = \frac{\text{energy}}{\text{time}} = \frac{\text{work}}{\text{time}} \tag{1.39}$$

1.10.   Units and Conversion Factors

The most common system of units used today in science and engineering applications is the International System of Units or Systéme International d'Unités (SI). Most physical sciences and engineering textbooks are now written in SI units, although older books still on library shelves may have been written in foot-pound-second (fps) or centimetre-gramme-second (cgs) units.

Table 1.2 lists the basic physical properties of electrical and mechanical engineering systems together with their units and the symbols used in this text (which are common but not all universal). The common multiples and sub-multiples in SI units, with their names and symbols, are listed in Table 1.3.

Table 1.2.   The international system of units (SI).

Property	Unit	Symbol
Basic		
Length	metre (UK), meter (US)	m
Mass	kilogramme	kg
Time	second	s
Temperature	Kelvin	K
Electric current	ampere	A
Derived		
Velocity	metre per second	v (m/s)
Area	square metre	A (m^2)
Force	newton	F (kgm/s^2)
Energy (Work)	joule (newton-metre)	W (J or Nm)
Power	watt	P (J/s)

Table 1.3.   Multiples and sub-multiples in SI units.

Unit	Symbol	Value
pico	p	10^{-12}
nano	n	10^{-9}
micro	μ	10^{-6}
milli	m	10^{-3}
centi	c	10^{-2}
kilo	k	10^{3}
mega	M	10^{6}
giga	G	10^{9}
tera	T	10^{12}

Conversion factors between various basic SI units and their UK or US equivalents are given in Table 1.4 [5]. Many of these units are used in subsequent chapters of the book.

Table 1.5 lists some useful conversion data relating to heat energy and power and Table 1.6 represents conversion between different scales of temperature [5].

1.11.   Problems on Energy and Power

Mechanical energy

1.1. A body of constant mass m is acted on by force F which results in linear motion at constant velocity v. Show that the linear momentum mv is equal to the time integral of the force.

1.2. A body of mass 100 kg initially rests on a ledge 25 m above the ground. It then falls freely to the ground under the influence of gravity. Air friction may be neglected. Gravitational constant g $= 9.81\,\mathrm{m/s^2}$.

   (a) What are initial and final values of the potential energy?
   (b) What are the initial value and final value after impact of the kinetic energy?
   (c) Calculate the instantaneous velocity at the midheight of the fall. Hint: Use energy balance.
   (d) Calculate the values of potential energy and kinetic energy at the midheight of the fall.

1.3. A force of 100 N acts on a mass of 100 kg.

   (a) What is the resulting linear acceleration?
   (b) If the steady-state velocity is 10 m/s, what are the values of the kinetic energy and momentum?

1.4. A mass of 1 kg is rotated in a horizontal circle, at the end of a rigid tie-rod, with an angular velocity of 10 rad/s. If the radius of gyration is 0.5 m, what is the instantaneous linear velocity of the mass? Calculate the torque and angular acceleration if a force of 10 N is needed to maintain the rotation.

Table 1.4.   Conversion factors [5].

Length	
1 millimetre (mm)	0.0393701 inch (in)
1 metre (m)	3.28084 feet (ft)

Area	
1 square centimetre (cm^2)	0.155000 in^2
1 square metre (m^2)	10.7639 ft^2
1 hectare $= 10^4\, m^2$	2.4710 acres

Volume	
1 cubic centimetre (cm^3)	0.0610237 in^3
1 cubic metre (m^3)	35.31477 ft^3
1 litre (1) ($1000\, cm^3$)	1.75985 UK pints
1 imperial gallon (UK)	4.54596 litres
1 US gallon	3.78531 litres
1 barrel $= 42$ US gallons $= 34.97$ UK gallons $= 159.00$ litres	

Weight	
1 kilogramme (kg)	2.20462 lb
1 tonne ($103\, kg$)	0.9984207 ton (UK)

Force	
1 newton (N)	0.2248 lb force

Pressure	
1 pascal (Pa)	$1\, N/m^2$
1 bar $= 10^5$ Pa	14.50 lbf/in^2
1 lbf/in^2 (one pound per square inch or psi)	6.89476 kPa
Atmospheric pressure $= 14.70\, lbf\, in^{-2}$	101.325 kPa

Velocity	
1 mile per hour (mph)	0.447 m/s
1 kilometre per hour (kph)	0.278 m/s

1.5. Calculate the moment of inertia of the rotating mass in Question 1.4.

1.6. A mass of 1 kg rotates in a horizontal circle, at a radius 0.5 m about its fixed anchor point, with an angular velocity of 10 rad/s. What is the kinetic energy of the motion?

1.7. If a mass of 10 kg rotates around a circle of 1 m radius at 1800 rpm, what is its energy of motion?

Thermal energy

1.8. An imperial gallon of water is uniformly heated so that its temperature increases by 20°C. What is the rise of its heat energy content?

Table 1.5. Conversion factors in power, heat and energy [5].

	Unit	Equivalents
Power	1 watt (W)	1 Joule/sec (J/s) = 0.001341 HP
	1 kilowatt (kW)	1000 W = 1.34 HP
	1 horsepower (HP)	745.7 W = 550 ft lb/sec
Power density	1 W/m^2	3.6 kJ/m^2/h = 0.317 BTU/ft^2/h
Heat energy	1 calorie (cal)	4.1868 J
	1 British thermal unit (BTU)	1055.06 J = 778.169 ft lb = 25 × 10^{-15} mtoe
	1 therm	105 BTU = 29.3 kWh = 1.05506 × 10^8 J
	1 mtoe	0.04 quadrillion BTU = 0.04 × 10^{15} BTU
Heat energy density	1 kcal/m^2	0.3687 BTU/ft2 = 1.163 Wh/m^2
	1 BTU/ft^3	3.726 × 104 J/m^2
	1 Langley	1 cal/cm^2 = 41868 J/m^2
Energy	1 Joule	1 watt-second (Ws)
	1 electron volt (eV)	1.602 × 10^{-19} J
	1 kilowatt hour (kWh)	3.6 × 106 J = 3.412 × 10^3 BTU

Table 1.6. Scales of temperature.

The centigrade (Celsius) scale of temperature has 100 degree units between the freezing point 0°C and boiling point 100°C of water at standard pressure.

The Fahrenheit scale has 180 degree units between the freezing point 32°F and boiling point 212°F of water. Therefore

$$100^{\circ}\text{C} \equiv 180^{\circ}\text{F}$$

and

$$^{\circ}\text{C} = 5/9(^{\circ}\text{F} - 32)$$
$$^{\circ}\text{F} = 9/5\,^{\circ}\text{C} + 32$$

The Kelvin scale of temperature is measured from absolute zero -273.15°C, usually rounded to -273°C. Therefore, for temperatures greater than zero degrees centigrade,

$$\text{K} = {}^{\circ}\text{C} + 273$$

1.9. Two equal masses of water are mixed in a container. What is the final temperature of the mixture if (a) the two initial temperatures T_{in} are equal and (b) one mass has an initial temperature twice that of the other?

1.10. In the UK, the original steam engines designed by Watt and Newcomen used reservoir temperatures of 100°C and 10°C. What was the maximum theoretical efficiency?

1.11. A heat engine operates with a high temperature source of 900 K and initial heat energy of 500 MJ. Its low temperature sink operates at 300 K. The maximum realizable efficiency is one-half the value of the theoretical maximum value.

(a) Calculate the maximum working value of the efficiency.
(b) What is the maximum work output?

1.12. For the purpose of converting heat energy into useful work from an ambient temperature of 100°C, is it better to have one heat source Q of temperature 400°C or two equal sources Q of temperature 200°C?

1.13. A heating boiler has a full-load working efficiency of 65%. It is used to heat a building from the outside temperature of 35°F to 68°F. What is the total thermal efficiency?

1.14. For the steam boiler electricity generator system of Fig. 1.5:

(a) Show that the first law of thermodynamics is satisfied.
(b) Calculate the efficiency of the turbine.
(c) What information is needed in order to calculate the energy discharged through the chimney?
(d) Calculate the efficiency of the generator.

1.15. A Carnot engine has a low temperature sink of 10°C and a maximum theoretical efficiency of 38%. By how much does the temperature of the high temperature source need to increase in order to raise the efficiency to 50%?

1.16. Explain what happens to the power input to a refrigerator if its door is left open in a warm room.

1.17. A high temperature fluid contains 1000 MJ of energy at 600°C. This fluid powers a mechanical converter of Carnot efficiency 30%.

(a) What is the temperature of the sink fluid?
(b) What is the change of entropy?

Units and unit conversions

1.18. What are the centigrade (Celsius) equivalents of the following temperatures in degrees Fahrenheit? (a) 212°F, (b) 100°F, (c) 32°F and (d) 0°F.

1.19. What are the Fahrenheit equivalents of the following temperatures in degrees centigrade? (a) 212°C, (b) 100°C, (c) 32°C and (d) 0°C.

1.20. At what value of temperature is the temperature reading in degrees centigrade equal to the reading in degrees Fahrenheit?

1.21. What is the centigrade equivalent of 75.8°F?

1.22. What is the Fahrenheit equivalent of 19.6°C?

1.23. A modern electric power station has a full load of 2000 MW. What are the equivalent values in (a) horsepower, (b) joules/second, (c) kilowatts and (d) foot-pounds/second?

1.24. A solar water heating panel has a thermal energy rating of 50 MJ. What is the rating in kilowatt hours?

1.25. The large wind turbine at Burger Hill, Orkney, Scotland, is rated at 3 MW. What is the equivalent rating in horsepower?

References

[1] Summers, C.M. (1971). "The Conversion of Energy", in *Energy and Power: A Scientific American Book*, W.H. Freeman, San Francisco, Chapter 8, pp. 95–106.

[2] Dorf, R.C. (1978). *Energy, Resources and Policy*, Addison Wesley Publishing Co., Inc., Boston, MA.

[3] Ford, K.W. (1972). *Classical and Modern Physics*, Vol. 2, Xerox College Publishing, Lexington, MA.

[4] Mackay, D. (2009). *Sustainable Energy — Without the Hot Air*, Cambridge University Press, Cambridge.

[5] McVeigh, J.C. (1984). *Energy around the World*, Pergamon Press, Oxford.

[6] Sears, F.W. (1947). *Principles of Physics I*, Addison Wesley Press, Boston, MA.

[7] Kittel, C., Knight, W.D. and Ruderman, M.A. (1965). *Berkeley Physical Course*, Vol. 1, McGraw-Hill Book Co, New York, NY.

CHAPTER 2

ENERGY RESOURCES AND ENERGY USE

2.1. Energy Input to the Earth

Figure 2.1 shows the accepted energy flow pattern for the earth [1]. All of the numerical values in this figure are obtained from scientifically measured data but are subject to measurement error and to slight intrinsic variations of value.

2.1.1. *Solar radiation and annual variation*

Solar radiation flows continuously through the earth's atmosphere onto its surface. Each square metre of surface area at the atmospheric boundary intercepts a normal (i.e. perpendicular) radiation, when facing the sun, of almost constant value. The orbit of the earth around the sun is not precisely circular but is slightly elliptical. For this reason the sun–earth distance is 91.4 million miles in January and 94.5 million miles in July. This small variation accounts for an annual difference of 3% to 4% in the solar radiation rate at the edge of the atmosphere. For calculation purposes it is convenient to assume a constant radiation rate and this is embodied in the "solar constant", which has a present value 1377 W/m^2 [2].

If the earth is assumed to be perfectly spherical it has a diametric plane area πr^2, where r is the mean earth radius. The solar radiation intercepted is therefore at a rate of 1377 πr^2 watts, if radius r is in metres. A figure for the mean earth radius is 6.324 million metres, resulting in

$$\text{earth radiation reception rate} = 1377 \times \pi \times (6.324 \times 10^6)^2$$
$$= 1.73 \times 10^{17} \, \text{W}$$
$$= 1.73 \times 10^{17} \, \text{J/s}$$

In a year of 365.25 days the total input radiation is therefore

$$W_{\text{annual}} = 365.25 \times 24 \times 3600 \times 1.73 \times 10^{17}$$
$$= 5.46 \times 10^{24} \, \text{J}$$

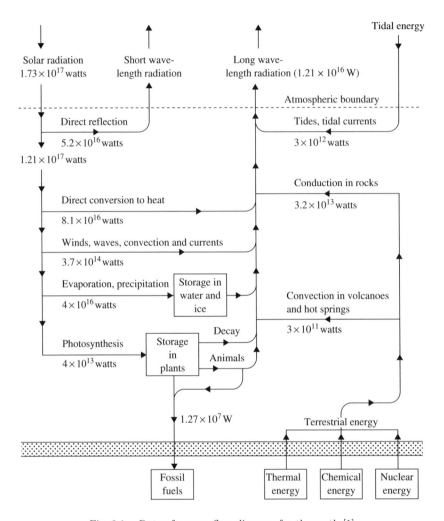

Fig. 2.1.   Rate of energy flow diagram for the earth [1].

The radiant input energy is staggeringly large. During the year 2009, for example, the total primary energy consumption in the world was 1,1164.3 million tonnes of oil equivalent (mtoe) [3]. In terms of various energy units this is

$$W_{\text{world consuption}} = 11164.3 \, \text{mtoe}$$
$$= 1164.3 \times 12 \times 10^9 \, \text{kWh}$$
$$= 11164.3 \times 12 \times 10^9 \times 3.6 \times 10^6 \, \text{J}$$
$$= 4.82 \times 10^{20} \, \text{J}$$

The input solar radiation for the year 2009 was therefore about $(5.46 \times 10^{24}/4.82 \times 10^{20})$, or 11,300 times the total world primary energy consumption. Most of the input radiant energy is used for the thermal heating of land and water masses and to power the hydrological cycle.

2.1.2.   *Terrestrial energy from inside the earth*

Energy flows outwardly from the interior of the earth to its surface at an estimated rate $0.063 \, W/m^2$, being a resultant of thermal, chemical and nuclear sources (Fig. 2.1). If the earth is assumed to be a perfect sphere, the outward flow of power is then $0.063 \times 4\pi r^2$ watts, where r is the earth's radius in metres. The total heat flow rate is about $3.2 \times 10^{12} \, W$ from rocks plus $3 \times 10^{11} \, W$ due to convection by hot springs and volcanoes. It is thus seen that about 99% of the outward-flowing terrestrial energy is due to conduction in rocks. This is discussed further in Chapter 7.

2.1.3.   *Tidal (gravitational) input energy*

Gravitational energy derived from the locations and motions of the sun, earth and moon is responsible for the ocean tides and currents. The use of this energy is discussed in Chapter 9. Tidal energy is estimated to represent an energy input at the rate $3 \times 10^{12} \, W$, or about one-tenth of the energy flow rate outward from the earth's interior.

A comparison of the three sources of input energy in Fig. 2.1 shows that the solar input is about 5,000 times the sum of the other sources. It is clear that the earth does not suffer from an energy shortage. There is natural energy in abundance. But serious technical problems arise with regard to the distribution, location, collection, conversion and storage of energy in its various forms. Even more serious are the human issues concerning the ownership and stewardship of the resources.

2.2.   Energy Flow upon Earth from Natural Sources

About 47% of the incoming solar radiation, i.e. $8.1 \times 10^{16} \, W$, is absorbed by the oceans, landmass and atmosphere. This is converted directly to heat energy at the ambient surface temperature.

The earth's hydrological cycle consists of evaporation, precipitation and surface run-off of water. As water vaporizes, it absorbs heat — sometimes referred to as the latent heat of evaporation — and rises into the air and atmosphere. On precipitation, most of this heat is released, together with the potential energy of its location and the kinetic energy of its pre-impact motion, as low temperature heat. Some portion of this hydrological component of energy is stored in rivers and lakes or in masses of ice. About 23% ($4 \times 10^{16} \, W$) of the solar input radiation is routed via the hydrological cycle.

A small fraction, about 0.21% ($3.7 \times 10^{14} \, W$), of the input radiation drives the ocean and atmospheric convections and circulations. This energy appears in the form of wind, wave and ocean current motions, which are all dissipated as low temperature heat due to friction.

The photosynthesis of plants is the process whereby some of the energy of solar radiation is captured by the chlorophyll of plant leaves. This is discussed in detail in Chapter 13. Solar energy combines with carbon dioxide (CO_2) and water (H_2O) to produce chemical carbohydrates plus oxygen. The stored energy in the leaves and plants is dissipated when the vegetation is consumed or decays (Fig. 2.1). Only a small proportion, 4×10^{12} W or 0.0023%, of the solar input power is involved in photosynthesis. Nevertheless, the annual energy of photosynthesis is of the same order of magnitude (10^{20} J) as the present world commercial energy consumption. It is the process of plant decay that is largely the source of the fossil fuel deposits [1, 2].

2.3. Energy Outflow from the Earth

About 30% of the incoming solar radiation is immediately reflected back into space in the form of short-wave radiation, at a rate of 5.2×10^{16} W. But the earth has to maintain an energy balance in order to avoid cumulative rises or falls of overall temperature. For energy balance, the total input energy must equal the sum of the component directly re-radiated plus the sum of the components stored or converted to other forms.

For example, 47% (8.1×10^{16} W) of the incident radiation is converted to low-grade heat (Fig. 2.1) and then re-radiated as long-wavelength radiation. Also re-radiated is most of the energy from the other input routes, including the tidal and terrestrial input components. It can be seen in Fig. 2.1 that the net input solar radiation, 1.21×10^{17} W, is almost equal to the sum of the direct conversion (to heat) plus the hydrological components. The amount of input energy being stored is of a much lower order of magnitude.

2.4. Energy Stored Within the Fossil Fuels

The burial of vegetation under successive layers of sedimentary sands and rocks has been going on for an estimated 600 million years. Each major climatic or geological change causes further changes in the earth's crust. Physical compressions and chemical decomposition have, during the course of time, resulted in deposits of the so-called fossil fuels: coal, oil-shale, petroleum and natural gas. These constitute sites of extremely concentrated, stored solar energy. The time scale of the laying down of the fossil fuel deposits is such that, in terms of the human lifespan, they can be considered non-renewable. Once they are extracted and used, they are gone forever.

The main change of human lifestyle and conduct in Western countries over the past 300 years has been to convert from agriculture-based societies and economies into industry-based societies and economies. Many of the key events and features of the profound historical change, known as the Industrial Revolution, started in Britain, which was a dominant world power during the 18th and 19th centuries. The Industrial Revolution was initially fuelled by coal and continues to be fuelled by coal, oil and natural gas. These fuels, together with nuclear and natural-flow hydro, are often called primary fuels. They are usually converted to other forms of energy. Great inroads have been made into these initially vast stores of concentrated energy, so that they are now significantly depleted. Given that the world demand for energy still continues to rise, it is important to assess the remaining reserve stocks of the fossil fuels. Present rates of fuel consumption can be measured accurately but future demand has to be estimated and is therefore subject to uncertainties. Further uncertainty arises with regard to the possible discovery of new and accessible sources of fossil fuels. It is certain that the passage of time reduces both the number of undiscovered fuel deposits and the reserves available.

Estimates of the rates of use and the years of fossil-fuel reserves remaining are given in Figs. 2.2 and 2.3 [3]. Some sources of information quote the reserves data in terms of the reserve/production (R/P) ratio. If the R/P ratio increases, this implies that the new discoveries (or revised estimates) have exceeded the energy consumption within the particular time period. Figure 2.3 includes information concerning oil supplies, which is the most critical from the viewpoint of remaining years of reserves. Although the R/P ratio for oil has been steadily growing since the mid-1970s, the critical feature is that the stock of natural crude oil reserves remaining is just over 40 years for the world. Most of the accessible oil reserves

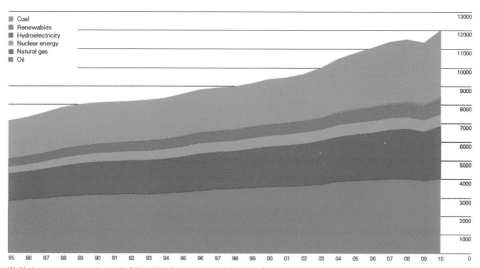

World primary energy consumption grew by 5.6% in 2010, the strongest growth since 1973. Growth was above average for oil, natural gas, coal, nuclear, hydroelectricity, as well as for renewables in power generation. Oil remains the dominant fuel (33.6% of the global total) but has lost share for 11 consecutive years. The share of coal in total energy consumption continues to rise, and the share of natural gas was the highest on record.

Fig. 2.2.   World consumption of prime fuels 1985–2010 [3] (2011).

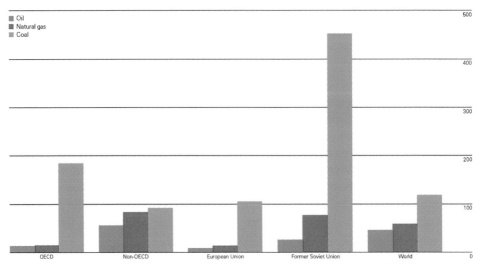

■ Oil
■ Natural gas
■ Coal

Coal remains the most abundant fossil fuel by global R/P ratios, though oil and natural gas proved reserves have generally risen over time. Non-OECD countries account for 93.4% of the world's proved oil reserves; 90.9% of natural gas reserves, and 56% of coal reserves. The Middle East holds the largest share of proved oil and natural gas reserves; Europe and Eurasia hold a significant share of the world's natural gas and the largest coal reserves. Asia and North America also hold substantial coal reserves.

Fig. 2.3.   Fossil fuel R/P ratio (years) at the end of 2009 [3] (2010).

remain in the Middle East, which continues to be the main provider of oil for the rest of the world. This is discussed in detail in Chapter 5.

Numerical information for the remaining world stock of natural gas is also shown in Fig. 2.3. As in the case of oil, the R/P value is slowly increasing, but the remaining world reserve stock is estimated at between 60 and 70 years. It would be foolish to assume that the R/P values for oil and for gas will continue to rise so as to always maintain two or three human generations of reserves.

The situation with regard to coal reserves is less serious because the estimated stock will outlast both oil and gas by about 200 years. Moreover, the distribution of the coal reserves is geographically very widespread, such that it is likely to remain available all over the world, as discussed in Chapter 4.

Figure 2.3 summarizes the year 2009 world fossil fuel reserve situation, divided between three groupings. The Organisation for Economic Co-operation and Development (OECD) group (see Table 2.1), which includes the UK and the USA, is well off for coal, mainly due to vast deposits in the USA. For both oil and natural gas, however, the long-term situation for OECD is bleak.

The data of Figs. 2.2 and 2.3 have the most profound implications. Will the large reserve stocks of oil and gas in the Middle East continue to be available for purchase to the rest of the world? In the future, will there be major dislocations of the world economy similar to that of 1973, when the price of oil quadrupled within a year? Will the need for oil create political, perhaps military, crises between nations or groups of nations?

Table 2.1.   Country groupings on the "energy" map (Fig. 2.4). Taken from Reference [4].

- **OECD** (18 percent of the 2010 world population):

  North America — United States, Canada, and Mexico; **OECD Europe** — Austria, Belgium, Czech Republic, Denmark, Finland, France, Germany, Greece, Hungary, Iceland, Ireland, Italy, Luxembourg, the Netherlands, Norway, Poland, Portugal, Slovakia, Spain, Sweden, Switzerland, Turkey, and the United Kingdom. **OECD Asia** — Japan, South Korea, Australia, and New Zealand.

- **Non-OECD** (82 percent of the 2010 world population):

  — **Non-OECD Europe and Eurasia** (5 percent of the 2010 world population) — Albania, Armenia, Azerbaijan, Belarus, Bosnia and Herzegovina, Bulgaria, Croatia, Cyprus, Estonia, Georgia, Kazakhstan, Kyrgyzstan, Latvia, Lithuania, Macedonia, Malta, Moldova, Montenegro, Romania, Russia, Serbia, Slovenia, Tajikistan, Turkmenistan, Ukraine, and Uzbekistan.

  — **Non-OECD Asia** (53 percent of the 2010 world population) — Afghanistan, American Samoa, Bangladesh, Bhutan, Brunei, Cambodia (Kampuchea), China, Cook Islands, Fiji, French Polynesia, Guam, Hong Kong, India, Indonesia, Kiribati, Laos, Macau, Malaysia, Maldives, Mongolia, Myanmar (Burma), Nauru, Nepal, New Caledonia, Niue, North Korea, Pakistan, Papua New Guinea, Philippines, Samoa, Singapore, Solomon Islands, Sri Lanka, Taiwan, Thailand, Timor-Leste (East Timor), Tonga, U.S. Pacific Islands, Vanuatu, Vietnam, and Wake Islands.

  — **Middle East** (3 percent of the 2010 world population) — Bahrain, Iran, Iraq, Israel, Jordan, Kuwait, Lebanon, Oman, Qatar, Saudi Arabia, Syria, the United Arab Emirates, and Yemen.

  — **Africa** (14 percent of the 2010 world population) — Algeria, Angola, Benin, Botswana, Burkina Faso, Burundi, Cameroon, Cape Verde, Central African Republic, Chad, Comoros, Congo (Brazzaville), Congo (Kinshasa), Côte d'Ivoire, Djibouti, Egypt, Equatorial Guinea, Eritrea, Ethiopia, Gabon, Gambia, Ghana, Guinea, Guinea-Bissau, Kenya, Lesotho, Liberia, Libya, Madagascar, Malawi, Mali, Mauritania, Mauritius, Morocco, Mozambique, Namibia, Niger, Nigeria, Reunion, Rwanda, Sao Tome and Principe, Senegal, Seychelles, Sierra Leone, Somalia, South Africa, St. Helena, Sudan, Swaziland, Tanzania, Togo, Tunisia, Uganda, Western Sahara, Zambia, and Zimbabwe.

  — **Central and South America** (7 percent of the 2010 world population) — Antarctica, Antigua and Barbuda, Argentina, Aruba, the Bahamas, Barbados, Belize, Bolivia, Brazil, British Virgin Islands, Cayman Islands, Chile, Colombia, Costa Rica, Cuba, Dominica, Dominican Republic, Ecuador, El Salvador, Falkland Islands, French Guiana, Grenada, Guadeloupe, Guatemala, Guyana, Haiti, Honduras, Jamaica, Martinique, Montserrat, Netherlands Antilles, Nicaragua, Panama, Paraguay, Peru, Puerto Rico, St. Kitts-Nevis, St. Lucia, St. Vincent/Grenadines, Suriname, Trinidad and Tobago, Turks and Caicos Islands, Uruguay, U.S. Virgin Islands, and Venezuela.

In addition, the following commonly used country groupings are referenced with regard to energy matters.

- **European Union (EU):** Austria, Belgium, Bulgaria, Cyprus, Czech Republic, Denmark, Estonia, Finland, France, Germany, Greece, Hungary, Ireland, Italy, Latvia, Lithuania, Luxembourg, Malta, the Netherlands, Poland, Portugal, Romania, Slovakia, Slovenia, Spain, Sweden, and the United Kingdom.

- **Organization of the Petroleum Exporting Countries (OPEC):** Algeria, Angola, Ecuador, Iran, Iraq, Kuwait, Libya, Nigeria, Qatar, Saudi Arabia, the United Arab Emirates, and Venezuela.

(Continued)

Table 2.1.   (*Continued*)

- **Persian Gulf Countries:** Bahrain, Iran, Iraq, Kuwait, Oman, Qatar, Saudi Arabia, and the United Arab Emirates.

- **Arabian natural gas producers:** Bahrain, Kuwait, Oman, the United Arab Emirates, and Yemen.

- **Non-OECD Developed Asia:** Hong Kong, Macau, Singapore, and Taiwan.

- **Non-OECD Asia LNG exporters:** Brunei, Indonesia, Malaysia, and Papua New Guinea.

- **Central and South America northern producers:** Colombia, Ecuador, Trinidad and Tobago, and Venezuela.

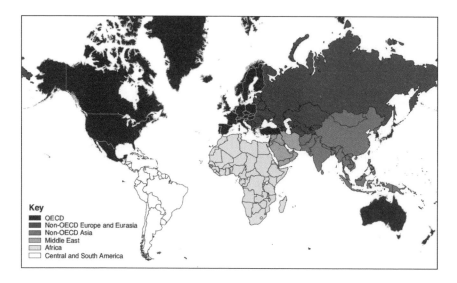

Fig. 2.4.   Basic country energy groupings [4].

Source: Energy Information Administration. Office of Integrated Analysis and Forecasting.

2.5.   Energy Production and Consumption

2.5.1.   *Energy consumption in the world*

For the purpose of energy production and consumption, the countries of the world can be divided into the six groupings shown in Fig. 2.4. Member countries of the various groups are listed in Table 2.1 [4]. Details of the past production and consumption of separate energy forms, country by country, are given and discussed in later sections of the book.

The world situation, as at the end of 2009, is shown in Fig. 2.2 [3]. Energy consumption, in almost all its forms, continues to increase. From 1970 to 2000 the primary energy consumption increased from 4900 to 8762 mtoe, which represents an increase of 78.6% over the original value. There is no evidence that this trend will reverse, although the rate of increase has slowed down since 2007. If figures are taken for the 11-year period 1999–2009, the world consumption of primary energy increased by an average of (1164.3 − 9030)/11, which is 194 mtoe/year, or about 2% a year.

Some country-by-country details of primary energy consumption in the period 1990–2009 are given in Table 2.2 [3]. It can be seen that in 2009 the USA was, by far, the largest consumer, followed by China, Russia, India and Japan. The countries of North America and Western Europe consume just about one-half of the total world primary energy.

Various predictions have been made about future world energy consumption. The figures differ slightly according to the assumed rates of growth, but all surveys indicate continued increase. Figure 2.5, quoted from Gruenspecht [5], uses the energy unit quadrillions of British thermal units ($1\,\text{mtoe} = 0.04 \times 10^{15}\,\text{BTU} = 0.04$ quadrillion BTU). Conversion factors between fuel values are subject to slight inconsistencies due to variations of fuel quality, but the data of Fig. 2.5 agree with the value of Fig. 2.2 for 2007 (the only common date).

Recent consumption figures and future projections for different primary fuels are given in Fig. 2.6 [5] and Table 2.3 [3, 5]. The primary fuel most used is oil, which is needed in all regions. It is clear that oil and natural gas account for about 64% of present world fuel demand. This makes even more urgent the need to address the remaining reserves situation indicated in Figs. 2.2 and 2.3. Country-by-country details of remaining individual fossil fuel stocks of coal, oil and natural gas are given as appropriate tables in Chapters 4, 5 and 6, respectively.

World energy consumption in different nations occurs in patterns and is not only unequal but inequitable. Rather than considering only the energy consumption figures, it is instructive to consider the energy consumption per capita (i.e. per person of the population). Table 2.4 shows the situation in 2009 with regard to different regions of the world. Canada and the USA consume twice as much energy per person as the Western Europeans, who, in turn, consume energy at about twice the average value for the world.

The use of energy is related directly to industrial productivity and to the gross domestic product (GDP) of a country. The GDP is the market value of all final goods and services produced in a year. It is often positively correlated with the standard of living. Other factors, such as the climate and the industry/agriculture ratio, are relevant, but the GDP is a useful indicator of national prosperity.

A further indicator is greenhouse gasses/capita data, which usually represents the degree of industrialization. This, in turn, indicates the level of GDP and national prosperity. Figure 2.7 shows the greenhouse gas emissions per capita versus the energy use per capita for many countries.

The countries of the EU achieve a high per capita prosperity with about one-half the per capita energy consumption of the North Americans. The former Soviet

Table 2.2.   Primary energy consumption [3].*

Million tonnes oil equivalent	2000	2001	2002	2003	2004	2005	2006	2007	2008	2009	2010	Change 2010 over 2009	2010 share of total
US	2313.7	2259.7	2295.5	2302.3	2348.8	2351.2	2332.7	2372.7	2320.2	2204.1	**2285.7**	3.7%	19.0%
Canada	302.3	298.2	303.1	312.3	315.2	325.3	323.6	329.0	326.6	312.5	**316.7**	1.3%	2.6%
Mexico	141.2	140.7	140.3	147.1	153.0	159.0	164.9	168.2	171.2	167.1	**169.1**	1.2%	1.4%
Total North America	2757.2	2698.5	2738.9	2761.8	2817.1	2835.4	2821.2	2870.0	2818.0	2683.7	**2771.5**	3.3%	23.1%
Argentina	59.3	58.6	55.3	59.8	63.1	67.6	71.3	74.6	77.5	75.1	**77.1**	2.6%	0.6%
Brazil	185.2	182.9	187.1	192.5	199.9	207.2	212.7	225.4	235.1	234.1	**253.9**	8.5%	2.1%
Chile	24.2	24.8	25.1	25.4	27.0	28.3	29.4	29.9	29.3	28.5	**28.4**	−0.4%	0.2%
Colombia	25.8	25.3	25.2	25.8	26.8	28.3	29.6	29.3	30.8	31.5	**32.2**	2.2%	0.3%
Ecuador	7.8	7.8	7.8	8.0	8.3	9.5	10.5	11.5	12.4	12.6	**13.0**	2.8%	0.1%
Peru	11.9	11.7	11.9	11.7	12.6	13.1	13.4	14.6	15.9	16.3	**18.3**	12.2%	0.2%
Trinidad & Tobago	11.4	11.9	13.1	14.6	13.9	15.8	20.6	20.6	21.9	20.8	**22.0**	5.4%	0.2%
Venezuela	65.2	69.2	69.4	61.4	68.6	71.3	77.3	77.0	80.9	80.6	**80.3**	−0.4%	0.7%
Other S. & Cent. America	73.7	75.7	77.1	80.2	80.2	81.3	83.4	86.7	86.0	85.4	**86.8**	1.6%	0.7%
Total S. & Cent. America	464.4	467.8	472.0	479.4	500.2	522.3	548.1	569.6	589.9	585.0	**611.9**	4.6%	5.1%
Austria	31.8	33.5	33.2	34.6	34.0	34.8	34.1	33.4	34.1	33.1	**33.3**	0.7%	0.3%
Azerbaijan	11.4	11.1	10.9	11.8	12.7	13.8	13.6	12.2	12.3	10.7	**10.0**	−6.9%	0.1%
Belarus	21.2	21.5	21.7	21.5	23.7	23.7	25.2	24.3	25.5	23.9	**24.4**	2.0%	0.2%
Belgium & Luxembourg	66.8	64.5	65.4	69.0	70.4	70.6	70.4	70.3	72.5	65.7	**69.8**	6.1%	0.6%
Bulgaria	17.9	18.5	18.4	19.8	19.2	19.9	20.7	19.8	20.1	18.3	**18.0**	−1.8%	0.2%
Czech Republic	40.1	41.6	41.6	43.7	44.9	44.8	44.4	43.6	42.0	40.6	**41.3**	1.7%	0.3%
Denmark	20.1	20.0	19.9	21.4	20.5	19.8	21.9	20.9	20.1	18.8	**19.5**	3.8%	0.2%
Finland	28.0	28.3	28.8	31.0	30.9	28.4	30.0	29.8	29.2	27.1	**29.1**	7.6%	0.2%

(Continued)

Table 2.2.   (Continued)

Million tonnes oil equivalent	2000	2001	2002	2003	2004	2005	2006	2007	2008	2009	2010	Change 2010 over 2009	2010 share of total
France	254.2	258.4	255.4	259.3	263.6	261.2	259.2	256.7	257.8	244.0	**252.4**	3.4%	2.1%
Germany	332.3	338.8	334.0	337.1	337.3	333.2	339.5	324.2	326.8	307.4	**319.5**	3.9%	2.7%
Greece	32.0	32.1	33.0	32.7	34.3	34.0	35.0	34.9	34.7	33.3	**32.5**	−2.3%	0.3%
Hungary	23.0	24.1	23.4	24.1	24.2	25.8	25.4	25.0	24.8	22.8	**23.4**	2.7%	0.2%
Republic of Ireland	13.7	14.7	14.5	14.1	14.6	15.0	15.5	15.8	15.7	14.6	**14.6**	†	0.1%
Italy	176.5	177.5	176.3	182.3	185.8	186.2	185.4	182.4	180.7	168.3	**172.0**	2.3%	1.4%
Kazakhstan	41.3	42.1	47.4	53.1	61.6	64.5	66.7	68.2	73.0	67.5	**72.8**	7.9%	0.6%
Lithuania	7.0	8.1	8.6	9.1	9.2	8.4	8.1	8.7	8.7	8.0	**6.1**	−24.3%	0.1%
Netherlands	87.6	90.6	91.0	91.8	95.0	97.1	97.2	98.4	98.1	95.6	**100.1**	4.8%	0.8%
Norway	46.1	41.2	43.1	38.8	39.5	45.6	42.0	45.7	46.7	43.4	**41.8**	−3.7%	0.3%
Poland	88.5	88.9	87.7	90.1	91.7	91.5	94.9	95.8	95.4	91.3	**95.8**	−4.9%	0.8%
Portugal	25.2	25.4	25.5	25.8	25.4	25.6	25.5	25.4	24.3	24.6	**27.1**	10.0%	0.2%
Romania	37.0	37.3	38.6	37.8	39.0	39.8	40.6	37.5	38.5	34.0	**34.5**	1.5%	0.3%
Russian Federation	620.4	631.4	633.1	649.9	657.8	657.4	675.3	685.8	691.0	654.7	**690.9**	5.5%	5.8%
Slovakia	18.1	18.6	18.7	18.1	17.6	18.8	17.8	17.2	17.7	16.0	**16.2**	0.8%	0.1%
Spain	130.2	134.0	135.9	144.4	151.1	153.4	154.1	158.6	157.1	146.1	**149.7**	2.5%	1.2%
Sweden	50.6	54.4	51.1	48.8	52.6	54.3	51.4	52.0	51.7	47.4	**50.7**	6.8%	0.4%
Switzerland	29.3	31.3	29.3	29.1	28.8	27.7	28.9	28.6	29.7	29.7	**29.0**	−2.4%	0.2%
Turkey	76.7	71.6	75.2	79.7	84.4	89.5	95.8	102.2	103.8	101.0	**110.9**	9.8%	0.9%
Turkmenistan	14.5	14.9	15.4	17.0	17.9	19.0	21.2	24.2	23.8	23.4	**26.0**	11.2%	0.2%
Ukraine	134.7	134.7	132.3	135.2	137.4	136.1	137.5	135.2	131.9	112.0	**118.0**	5.4%	1.0%
United Kingdom	224.1	226.7	221.9	225.6	227.4	228.3	225.6	218.4	214.9	203.6	**209.1**	2.7%	1.7%
Uzbekistan	50.9	54.0	55.5	51.4	48.8	46.2	45.6	49.0	52.6	48.0	**49.8**	3.9%	0.4%
Other Europe & Eurasia	70.0	71.4	73.8	77.3	81.4	81.0	79.2	80.9	82.4	79.0	**83.4**	5.6%	0.7%
Total Europe & Eurasia	2821.4	2861.3	2860.7	2925.4	2982.7	2995.4	3027.6	3025.6	3037.4	2853.8	**2971.5**	4.1%	24.8%

(Continued)

Table 2.2. (*Continued*)

Million tonnes oil equivalent	2000	2001	2002	2003	2004	2005	2006	2007	2008	2009	2010	Change 2010 over 2009	2010 share of total
Iran	121.2	128.0	141.8	149.3	156.1	177.0	185.7	189.6	197.4	205.9	**212.5**	3.2%	1.8%
Israel	19.7	19.7	20.1	20.7	21.0	21.6	21.8	22.9	23.8	23.3	**23.7**	1.9%	0.2%
Kuwait	20.0	21.0	21.0	23.5	25.9	27.7	26.5	26.2	27.7	28.0	**30.6**	9.1%	0.3%
Qatar	10.7	12.3	13.0	14.1	17.0	20.8	22.3	22.8	23.6	24.2	**25.7**	6.1%	0.2%
Saudi Arabia	117.9	123.0	127.6	135.8	147.4	152.3	158.5	165.2	179.6	187.8	**201.0**	7.0%	1.7%
United Arab Emirates	48.4	54.0	53.5	57.3	61.0	64.6	67.4	74.3	85.5	83.0	**86.8**	4.6%	0.7%
Other Middle East	78.3	81.5	83.0	80.9	86.1	89.0	97.0	100.3	107.3	112.7	**120.7**	7.1%	1.0%
Total Middle East	416.2	439.5	460.0	481.5	514.5	553.0	579.1	601.4	644.9	664.9	**701.1**	5.4%	5.8%
Algeria	26.8	27.9	28.9	30.2	31.3	32.7	33.7	35.5	37.5	39.7	**41.1**	3.6%	0.3%
Egypt	49.8	52.6	53.6	57.0	59.6	62.5	65.8	69.9	74.1	76.5	**81.0**	6.0%	0.7%
South Africa	101.2	100.1	103.4	109.5	115.7	113.5	115.4	118.0	116.3	118.8	**120.9**	1.7%	1.0%
Other Africa	94.4	97.8	100.8	103.6	111.4	115.9	114.2	120.1	126.2	125.2	**129.5**	3.4%	1.1%
Total Africa	272.1	278.4	286.7	300.4	318.0	324.6	329.0	343.6	354.1	360.1	**372.6**	3.4%	3.1%
Australia	106.7	109.7	112.8	112.0	116.9	117.7	124.3	125.2	124.3	125.6	**118.2**	-5.8%	1.0%
Bangladesh	12.7	14.1	14.8	15.7	16.6	18.3	19.1	20.0	21.8	23.4	**23.6**	1.2%	0.2%
China	1038.2	1072.5	1140.4	1313.5	1531.3	1691.5	1858.1	1996.8	2079.9	2187.7	**2432.2**	11.2%	20.3%
China Hong Kong SAR	16.1	19.4	20.8	21.2	24.3	22.9	24.7	26.2	24.5	24.3	**25.9**	6.4%	0.2%
India	295.8	297.4	308.7	317.2	345.5	364.0	381.4	414.5	444.6	480.0	**524.2**	9.2%	4.4%
Indonesia	98.4	104.0	108.7	117.6	116.9	120.5	122.0	129.6	123.6	132.2	**140.0**	5.9%	1.2%
Japan	514.1	512.9	510.3	511.0	522.1	527.2	528.3	523.6	516.2	473.0	**500.9**	5.9%	4.2%
Malaysia	46.5	48.6	52.2	53.6	53.8	59.6	62.6	63.4	62.2	60.8	**62.9**	3.6%	0.5%
New Zealand	18.6	18.5	19.1	18.7	19.3	18.8	19.1	19.0	19.1	19.0	**18.9**	-0.5%	0.2%
Pakistan	44.3	45.5	47.4	52.4	56.8	58.9	61.8	65.1	64.8	66.9	**67.6**	1.1%	0.6%

(*Continued*)

Table 2.2. (*Continued*)

Million tonnes oil equivalent	2000	2001	2002	2003	2004	2005	2006	2007	2008	2009	2010	Change 2010 over 2009	2010 share of total
Philippines	25.2	25.1	25.8	26.6	27.5	27.7	25.9	27.1	26.9	27.3	**27.6**	1.3%	0.2%
Singapore	33.4	37.2	38.7	37.5	42.7	48.4	51.5	56.8	59.4	63.4	**69.8**	10.2%	0.6%
South Korea	188.9	193.5	202.4	209.0	213.2	220.6	222.7	231.3	235.3	236.7	**255.0**	7.7%	2.1%
Taiwan	87.5	90.9	94.9	99.3	104.8	107.1	108.7	113.7	106.7	104.2	**110.5**	6.0%	0.9%
Thailand	68.0	70.9	76.3	81.3	87.7	93.1	95.0	97.9	100.6	102.4	**107.9**	5.4%	0.9%
Vietnam	17.7	19.9	21.4	22.5	28.7	29.8	32.3	34.9	36.7	42.1	**44.0**	4.5%	0.4%
Other Asia Pacific	39.0	39.9	38.8	40.2	41.5	44.0	45.4	43.2	44.6	46.9	**44.6**	−5.0%	0.4%
Total Asia Pacific	2651.2	2720.1	2833.5	3049.3	3349.5	3570.2	3782.8	3988.3	4091.5	4215.6	**4573.8**	8.5%	38.1%
Total World	**9382.4**	**9465.6**	**9651.8**	**9997.8**	**10482.0**	**10800.9**	**11087.8**	**11398.4**	**11535.8**	**11363.2**	**12002.4**	**5.6%**	**100.0%**
of which: OECD	5435.4	5405.0	5445.0	5511.4	5622.5	5667.3	5673.2	5714.6	5658.7	5378.4	**5568.3**	3.5%	46.4%
Non-OECD	3947.0	4060.6	4206.8	4486.4	4859.5	5133.7	5414.6	5683.8	5877.1	5984.8	**6434.1**	7.5%	53.6%
European Union	1720.4	1751.8	1738.9	1777.0	1805.9	1808.2	1814.0	1786.9	1783.9	1678.6	**1732.9**	3.2%	14.4%
Former Soviet Union	922.3	939.0	945.7	971.2	992.6	993.9	1018.2	1034.5	1044.1	972.1	**1023.3**	5.3%	8.5%

*In this Review, primary energy comprises commercially traded fuels only, including modern renewables used to generate electricity.
†Less than 0.05%.
Notes: Oil consumption is measured in million tonnes; other fuels in million tonnes of oil equivalent.

energy consumption
quadrillion Btu

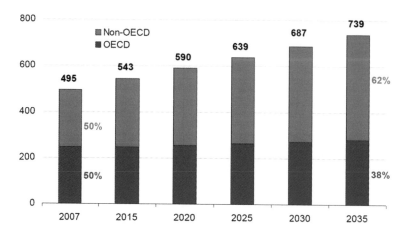

Fig. 2.5.   World energy demand 2007–2035 [5].

world primary energy consumption
quadrillion Btu

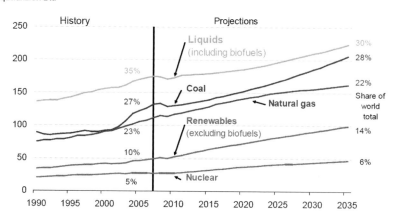

Fig. 2.6.   World energy consumption by fuel type, 1990–2035 [5].

Union and its former Eastern European satellites expend a lot of per capita energy to achieve relatively poor economic performance. China, a large and rapidly developing country, uses energy/capita at about the world average. Although the large Asian country of India remains low on energy consumption/capita its level of GDP is growing rapidly due to increasing industrialization.

It can be seen in Fig. 2.7 that the great economic success of the USA and Canada is realized with the expenditure of large amounts of per capita energy. The information in Fig. 2.7 should not be used to invoke uninformed criticism of

Table 2.3.   Primary energy consumption (mtoe) by fuel [3].*

Million tonnes oil equivalent	2009							2010						
	Oil	Natural gas	Coal	Nuclear energy	Hydro-electricity	Renew-ables	Total	Oil	Natural gas	Coal	Nuclear energy	Hydro-electricity	Renew-ables	Total
US	833.2	588.3	496.2	190.3	62.5	33.6	2204.1	850.0	621.0	524.6	192.2	58.8	39.1	2285.7
Canada	97.1	85.0	23.3	20.2	83.6	3.3	312.5	102.3	84.5	23.4	20.3	82.9	3.3	316.7
Mexico	88.5	59.9	8.6	2.4	6.0	1.8	167.1	87.4	62.0	8.4	1.3	8.3	1.7	169.1
Total North America	1018.8	733.1	528.1	212.9	152.1	38.7	2683.7	1039.7	767.4	556.3	213.8	149.9	44.2	2771.5
Argentina	23.7	38.8	1.2	1.8	9.2	0.4	75.1	25.7	39.0	1.2	1.6	9.2	0.4	77.1
Brazil	107.0	17.8	11.7	2.9	88.5	6.2	234.1	116.9	23.8	12.4	3.3	89.6	7.9	253.9
Chile	15.6	2.8	3.7	—	5.6	0.7	28.5	14.7	4.2	3.7	—	4.9	0.8	28.4
Colombia	10.5	7.8	3.7	—	9.3	0.2	31.5	11.0	8.2	3.8	—	9.1	0.2	32.2
Ecuador	10.1	0.4	—	—	2.1	†	12.6	10.6	0.4	—	—	2.0	0.1	13.0
Peru	8.1	3.1	0.5	—	4.5	0.1	16.3	8.4	4.9	0.5	—	4.4	0.1	18.3
Trinidad & Tobago	2.1	18.8	—	—	—	†	20.8	2.1	19.8	—	—	—	†	22.0
Venezuela	33.7	27.5	†	—	19.5	—	80.6	35.2	27.6	†	—	17.4	—	80.3
Other S. & Cent. America	57.9	4.5	2.0	—	19.2	1.7	85.4	57.3	5.0	2.1	—	20.7	1.7	86.8
Total S. & Cent. America	268.6	121.6	22.9	4.7	157.9	9.3	585.0	282.0	132.9	23.8	4.9	157.2	11.1	611.9
Austria	13.0	8.4	2.2	—	8.2	1.4	33.1	13.0	9.1	2.0	—	7.8	1.4	33.3
Azerbaijan	3.2	7.0	†	—	0.5	†	10.7	3.3	5.9	†	—	0.8	†	10.0
Belarus	9.3	14.5	†	—	†	†	23.9	6.6	17.7	†	—	†	†	24.4
Belgium & Luxembourg	33.4	15.7	4.6	10.7	0.1	1.3	65.7	35.0	17.4	4.9	10.9	0.1	1.5	69.8
Bulgaria	5.6	2.1	6.3	3.4	0.8	0.1	18.3	4.2	2.3	6.6	3.5	1.3	0.2	18.0
Czech Republic	9.7	7.4	16.2	6.2	0.7	0.5	40.6	9.2	8.4	16.0	6.3	0.8	0.6	41.3
Denmark	8.5	4.0	4.0	—	†	2.3	18.8	8.7	4.5	3.8	—	†	2.5	19.5
Finland	9.9	3.2	3.7	5.4	2.9	2.0	27.1	10.4	3.5	4.6	5.2	3.2	2.2	29.1

(*Continued*)

Table 2.3.   (*Continued*)

Million tonnes oil equivalent	2009							2010						
	Oil	Natural gas	Coal	Nuclear energy	Hydro-electricity	Renew-ables	Total	Oil	Natural gas	Coal	Nuclear energy	Hydro-electricity	Renew-ables	Total
France	87.5	38.0	9.9	92.8	13.1	2.8	244.0	83.4	42.2	12.1	96.9	14.3	3.4	252.4
Germany	113.9	70.2	71.7	30.5	4.2	16.9	307.4	115.1	73.2	76.5	31.8	4.3	18.6	319.5
Greece	20.2	3.0	8.1	—	1.3	0.6	33.3	18.5	3.3	8.5	—	1.7	0.6	32.5
Hungary	7.1	9.1	2.5	3.5	0.1	0.6	22.8	6.7	9.8	2.6	3.6	†	0.7	23.4
Republic of Ireland	8.0	4.3	1.3	—	0.2	0.7	14.6	7.6	4.8	1.4	—	0.1	0.7	14.6
Italy	75.1	64.4	13.1	—	11.1	4.6	168.3	73.1	68.5	13.7	—	11.2	5.6	172.0
Kazakhstan	12.1	22.1	31.7	—	1.6	—	67.5	12.5	22.7	36.1	—	1.5	—	72.8
Lithuania	2.6	2.5	0.1	2.5	0.3	0.1	8.0	2.7	2.8	0.2	—	0.3	0.1	6.1
Netherlands	49.4	35.0	7.9	1.0	†	2.3	95.6	49.8	39.2	7.9	0.9	†	2.2	100.1
Norway	10.3	3.7	0.3	—	28.8	0.3	43.4	10.7	3.7	0.5	—	26.7	0.3	41.8
Poland	25.3	12.0	51.9	—	0.7	1.4	91.3	26.3	12.9	54.0	—	0.8	1.9	95.8
Portugal	12.8	4.2	3.3	—	2.0	2.3	24.6	12.6	4.5	3.4	—	3.8	2.8	27.1
Romania	9.2	11.9	6.6	2.7	3.6	†	34.0	9.1	12.0	6.2	2.6	4.6	0.1	34.5
Russian Federation	135.2	350.7	91.9	37.0	39.9	0.1	654.7	147.6	372.7	93.8	38.5	38.1	0.1	690.9
Slovakia	3.7	4.4	3.5	3.2	1.0	0.1	16.0	3.7	5.1	2.7	3.3	1.3	0.1	16.2
Spain	75.7	31.1	10.5	11.9	6.0	10.9	146.1	74.5	31.0	8.3	13.9	9.6	12.4	149.7
Sweden	14.6	1.0	1.6	11.9	14.9	3.4	47.4	14.5	1.4	2.0	13.2	15.1	4.3	50.7
Switzerland	12.3	2.7	0.1	6.2	8.1	0.3	29.7	11.4	3.0	0.1	6.0	8.2	0.3	29.0
Turkey	28.2	32.1	32.0	—	8.1	0.5	101.0	28.7	35.1	34.4	—	11.7	1.0	110.9
Turkmenistan	5.4	17.9	—	—	—	—	23.4	5.6	20.4	—	—	—	—	26.0
Ukraine	13.3	42.3	35.0	18.8	2.7	†	112.0	11.6	46.9	36.4	20.2	2.9	†	118.0
United Kingdom	74.4	78.0	29.6	15.6	1.2	4.7	203.6	73.7	84.5	31.2	14.1	0.8	4.9	209.1
Uzbekistan	4.8	39.2	1.4	—	2.6	—	48.0	5.0	41.0	1.3	—	2.5	—	49.8
Other Europe & Eurasia	28.5	12.3	15.2	1.9	19.6	1.5	79.0	28.3	14.1	15.7	1.8	22.3	1.2	83.4
Total Europe & Eurasia	922.2	954.5	466.4	265.1	184.0	61.6	2853.8	922.9	1023.5	486.8	272.8	195.9	69.6	2971.5

(*Continued*)

Table 2.3. (*Continued*)

Million tonnes oil equivalent	2009							2010						
	Oil	Natural gas	Coal	Nuclear energy	Hydro-electricity	Renew-ables	Total	Oil	Natural gas	Coal	Nuclear energy	Hydro-electricity	Renew-ables	Total
Iran	85.1	118.2	1.1	—	1.5	†	205.9	86.0	123.2	1.1	—	2.2	0.1	212.5
Israel	11.5	4.1	7.7	—	—	†	23.3	11.2	4.8	7.7	—	—	†	23.7
Kuwait	17.2	10.9	—	—	—	—	28.0	17.7	12.9	—	—	—	—	30.6
Qatar	6.2	18.0	—	—	—	—	24.2	7.4	18.4	—	—	—	—	25.7
Saudi Arabia	117.2	70.6	—	—	—	—	187.8	125.5	75.5	—	—	—	—	201.0
United Arab Emirates	29.8	53.2	—	—	—	†	83.0	32.3	54.5	—	—	—	†	86.8
Other Middle East	77.3	34.7	—	—	0.7	†	112.7	80.2	39.6	—	—	0.9	†	120.7
Total Middle East	344.3	309.7	8.8	—	2.1	0.1	664.9	360.2	329.0	8.8	—	3.0	0.1	701.1
Algeria	14.9	24.5	0.2	—	0.1	—	39.7	14.9	26.0	0.3	—	†	—	41.1
Egypt	34.4	38.3	0.6	—	2.9	0.2	76.5	36.3	40.6	0.7	—	3.2	0.3	81.0
South Africa	24.7	3.0	87.7	3.1	0.2	0.1	118.8	25.3	3.4	88.7	3.1	0.3	0.1	120.9
Other Africa	77.0	23.2	5.5	—	18.9	0.6	125.2	79.0	24.4	5.7	—	19.6	0.7	129.5
Total Africa	150.9	89.0	94.1	3.1	22.2	0.9	360.1	155.5	94.5	95.3	3.1	23.2	1.1	372.6
Australia	42.2	27.7	51.7	—	2.6	1.3	125.6	42.6	27.3	43.4	—	3.4	1.5	118.2
Bangladesh	4.8	17.8	0.4	—	0.3	—	23.4	4.8	18.0	0.5	—	0.3	—	23.6
China	388.2	80.6	1556.8	15.9	139.3	6.9	2187.7	428.6	98.1	1713.5	16.7	163.1	12.1	2432.2
China Hong Kong SAR	14.0	2.8	7.6	—	—	†	24.3	16.1	3.4	6.3	—	—	†	25.9
India	151.0	45.9	250.6	3.8	24.0	4.6	480.0	155.5	55.7	277.6	5.2	25.2	5.0	524.2
Indonesia	59.2	33.6	34.6	—	2.6	2.1	132.2	59.6	36.3	39.4	—	2.6	2.1	140.0
Japan	198.7	78.7	108.8	65.0	16.5	5.4	473.0	201.6	85.1	123.7	66.2	19.3	5.1	500.9
Malaysia	24.5	30.3	4.0	—	2.0	†	60.8	25.3	32.2	3.4	—	2.1	†	62.9

(*Continued*)

Table 2.3. (*Continued*)

Million tonnes oil equivalent	2009							2010						
	Oil	Natural gas	Coal	Nuclear energy	Hydro-electricity	Renew-ables	Total	Oil	Natural gas	Coal	Nuclear energy	Hydro-electricity	Renew-ables	Total
New Zealand	6.8	3.5	1.6	—	5.5	1.6	19.0	6.9	3.7	1.0	—	5.5	1.8	18.9
Pakistan	20.6	34.6	4.7	0.6	6.4	—	66.9	20.5	35.5	4.6	0.6	6.4	—	67.6
Philippines	13.1	3.0	6.7	—	2.2	2.4	27.3	13.1	2.8	7.7	—	1.8	2.3	27.6
Singapore	56.1	7.3	—	—	—	—	63.4	62.2	7.6	—	—	—	—	69.8
South Korea	103.0	30.5	68.6	33.4	0.6	0.4	236.7	105.6	38.6	76.0	33.4	0.8	0.5	255.0
Taiwan	44.1	10.2	38.7	9.4	0.8	1.0	104.2	46.2	12.7	40.3	9.4	0.9	1.0	110.5
Thailand	49.9	35.3	14.5	—	1.6	1.1	102.4	50.2	40.6	14.8	—	1.2	1.1	107.9
Vietnam	14.1	7.2	14.0	—	6.8	—	42.1	15.6	8.4	13.7	—	6.3	—	44.0
Other Asia Pacific	13.4	4.6	22.1	—	6.8	†	46.9	13.5	4.8	18.9	—	7.4	†	44.6
Total Asia Pacific	1203.8	453.5	2185.3	128.2	218.0	26.7	4215.6	1267.8	510.8	2384.7	131.6	246.4	32.6	4573.8
Total World	**3908.7**	**2661.4**	**3305.6**	**614.0**	**736.3**	**137.4**	**11363.2**	**4028.1**	**2858.1**	**3555.8**	**626.2**	**775.6**	**158.6**	**12002.4**
of which: OECD	2094.8	1313.9	1049.5	511.5	299.4	109.5	5378.4	2113.8	1397.6	1103.6	520.9	309.5	123.0	5568.3
Non-OECD	1813.9	1347.5	2256.1	102.5	436.9	27.9	5984.8	1914.3	1460.5	2452.2	105.3	466.1	35.6	6434.1
European Union	670.2	412.6	259.9	202.5	74.1	59.3	1678.6	662.5	443.3	269.7	207.5	83.0	66.9	1732.9
Former Soviet Union	192.7	503.0	161.1	58.8	56.2	0.3	972.1	201.5	537.1	169.1	59.3	55.9	0.3	1023.3

*In this Review, primary energy comprises commercially traded fuels, including modern renewables used to generate electricity.
†Less than 0.05%.
Note: Oil consumption is measured in million tonnes; other fuels in million tonnes of oil equivalent.

Table 2.4.   Primary energy consumption, population and GDP (2009).

Country	Population (millions)[1]	Energy consumption 2009			GDP/capita (US dollars)[3]
		Mtoe[2]	% world Total	Per capita Toe/year	
Algeria	34.9	39.7	0.4	1.14	3996
Australia	22.5	119.2	1.1	5.3	45285
Argentina	40.13	73.3	0.7	1.83	7725
Belgium/Lux	10.83	69.4	0.6	6.41	43794
Brazil	193.79	225.7	2.0	1.16	8220
Canada	33.7	319.2	2.9	9.47	39658
China	1340.5	2177	19.5	1.62	3735
Denmark	5.53	16.1	0.1	2.91	56263
Egypt	78.99	76.3	0.7	0.97	2450
France	62.79	241.9	2.2	3.85	42413
Germany	81.76	289.8	2.6	3.54	40832
Greece	11.3	32.7	0.3	2.89	29635
Hungary	10.013	22.4	0.2	2.24	12914
India	1189.9	468.9	4.2	0.394	1032
Indonesia	231.4	128.2	1.1	0.55	2329
Iran	74.196	204.8	1.8	2.76	4399
Ireland	4.45	13.9	0.1	3.12	49863
Italy	60.2	163.4	1.5	2.71	35435
Japan	127.38	463.9	4.2	3.64	39740
Malaysia	28.3	55.7	0.5	1.97	6950
Mexico	107.55	163.2	1.5	1.52	8134
Netherlands	16.63	93.3	0.8	5.61	48209
New Zealand	4.32	17.6	0.2	4.07	27259
Norway	4.9	42.5	0.4	8.67	78178
Pakistan	171	65.8	0.6	0.385	989
Poland	38.16	92.3	0.8	2.42	11302
Portugal	10.63	22.3	0.2	2.1	21970
Romania	21.47	34.6	0.3	1.61	7523
Russian Fed.	141.9	635.3	5.7	4.48	8681
Saudi Arabia	28.15	191.5	1.7	6.8	14745
Singapore	4.988	60.8	0.5	12.2	36379
S.Africa	49.32	126.8	1.1	2.57	5824
S.Korea	48.46	237.5	2.1	4.9	17074
Spain	46.09	132.6	1.2	2.88	32030
Sweden	9.37	43.2	0.4	4.61	43668
Switzerland	7.76	29.4	0.3	3.79	63536
Taiwan	23.07	105.7	0.9	4.58	16372
Thailand	64.2	95.1	0.9	1.48	3894
Turkey	72.56	93	0.8	1.28	8711
United Arab Emirates	4.6	75	0.7	16.3	45615
UK	62	198.9	1.8	3.21	35257
Ukraine	46.94	112.5	1.0	2.4	2569
USA	310.67	2182	19.5	7.02	45934
Venezuela	29	73.6	0.7	2.54	11383
World	6880.3	11164.3	100	1.62	8587

(1) Wikipedia, 2010 (IMF).
(2) BP Statistical Review of World Energy, June 2010.
(3) World Development Indicators, World Bank 2010.

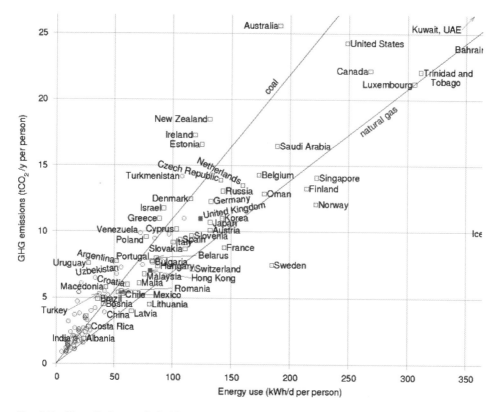

Fig. 2.7.   From Reference [20]: "Greenhouse-gas emissions per capita, versus power consumption percapita. The lines show the emission-intensities of coal and naturalgas. Squares show countries having 'high human development;' circles, 'medium' or 'low.'[...].

Source: UNDP Human Development Report, 2007." Reproduced, by permission [20].

the North Americans. Canada has lavish hydro sources that provide 82.9/775.6 (10.7%) of the world primary energy from this renewable source. The USA, with 4.5% of the world population, uses one-quarter of all the world's energy, but is generously endowed with abundant coal, natural gas and oil. One could argue that the North Americans are burning their own fuel. But are they wise to burn it at the present rates of consumption? In the event, the USA generated (in 2009) a national productivity at 4.3 times the world average value by using energy at 5.35 times the world average value [4].

One of the challenges of the energy future is that many of the world's poorest countries are not on the energy–GDP grid at all (Fig. 2.7). Moreover, China and India, with 37% of the world population between them, have so far achieved low percapita economic performance. All the recent signs indicate, however, that both China and India, with abundant coal supplies, have begun to make a great surge forward.

Available forecasts of the future indicate that the world population, energy consumption and GDP will all continue to increase. It is likely that GDP/capita will rise faster than the growth of population as prosperity increases. Most of the countries that have achieved high economic prosperity have done so through the extravagant use of industrial energy, especially fossil fuels. Fast-developing countries like Malaysia and Singapore are attempting to do the same. The poor and underdeveloped countries of the world may never have this opportunity because there is not enough primary fossil energy. Also, oil-rich countries, which could fund energy development in the Third World, are reluctant to lend sums of money on a long-term, speculative basis; they prefer safe and short-term investments in the USA and Europe. Nevertheless, there is abundant renewable energy for all and forever. The world problem is not an energy supply problem but a human behaviour problem.

2.5.2. *Energy production and use in the UK*

Recent changes in primary fuel production in the UK are summarized in Table 2.5 and illustrated by the graph of Fig. 2.8 [6]. Coal production has slowly declined since the early 1970s. Corresponding information about fuel consumption since 1999 is given in Table 2.6 [3, 7, 8] and Fig. 2.9 [6]. Slight differences in the figures for some fuels between the tabled data and the diagram arise because the data is taken from different sources at different times.

After the peak figures of the early 1970s, oil consumption declined for about ten years but has now (2013) levelled out to about 45% of the total demand. Nuclear-generated electricity now accounts for about 11% of the total energy consumption.

A feature that has recently occurred is that electricity is now imported from France using undersea cables beneath the English Channel. Like all generated electricity, it is a secondary fuel and in this case is probably obtained from nuclear generation.

Table 2.5. UK primary energy production (mtoe) [3].

Year	Coal	Oil	Natural Gas	Nuclear	Hydro	Total
1999	22.5	137.4	89.2	21.5	1.2	271.8
2000	19	126.2	97.5	19.3	1.2	26.3
2001	19.4	116.7	95.2	20.4	0.9	253
2002	18.2	115.9	93.2	19.9	1.1	248
2003	17.2	106.1	92.6	20.1	0.7	237
2004	15.3	95.4	86.7	18.1	1.1	217
2005	12.5	84.7	79.4	18.5	1.1	196
2006	11.3	76.6	72	17.1	1.0	178
2007	10.3	76.8	64.9	14.3	1.2	168
2008	11	71.7	62.7	11.9	1.2	159
2009	10.9	68	53.7	15.7	1.2	150
Share of 2009 total	7.27%	45.3%	35.8%	10.5%	0.08%	100%

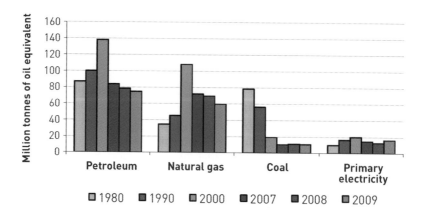

Fig. 2.8.   UK Primary fuel production, 1980–2009 [6].

Table 2.6.   UK primary energy consumption (mtoes) [3].

Year	Coal	Oil	Natural Gas	Nuclear	Hydro	Total
1999	34.3	79.4	84.2	21.5	1.2	221
2000	36.7	78.6	87.2	19.3	1.2	223
2001	38.9	78.4	86.7	20.4	0.9	225
2002	35.7	78	85.6	19.9	1.1	220
2003	38.1	79	85.8	20.1	0.7	224
2004	36.6	81.7	87.7	18.1	1.1	225
2005	37.4	83	85.2	18.5	1.1	225
2006	40.8	82.3	80.9	17.1	1.0	222
2007	38.2	79.2	81.9	14.3	1.2	215
2008	35.5	77.9	84.4	11.9	1.2	210
2009	29.7	74.4	77.9	15.7	1.2	199
Share of 2009 Total	14.9%	37.4%	39.1%	7.89%	0.6%	100%

The contribution of natural-flow hydropower increases slowly as wind energy stations come on-stream. All of the available natural hydro sites have been exploited. Vast schemes, such as the possible Severn Barrage hydro scheme (see Chapter 9), would make a big contribution but would be classed as man-made hydropower.

The flow of UK energy from source to end user is illustrated in the revised Fig. 2.10 [7]. Most of the primary fuel input is converted to heat and then to mechanical or/and electrical energy. Iron and steel, once a UK base industry and massive energy user, has shrunk to a fraction of its size 30 years ago. The UK industry sector is now smaller than the domestic sector, while transport accounts for about one-quarter of all the end user energy.

The energy industries make a major contribution to the UK economy. In 2009, 49.6% of industrial investment and 2.1% of annual business expenditure on research and development came from the energy industries. Overall, the energy industries

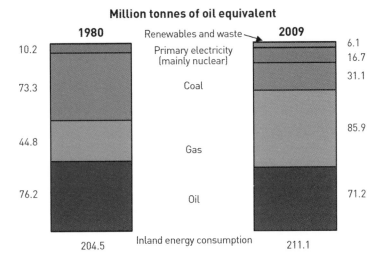

Fig. 2.9. UK Primary fuel consumption, 1980 and 2009 [6].

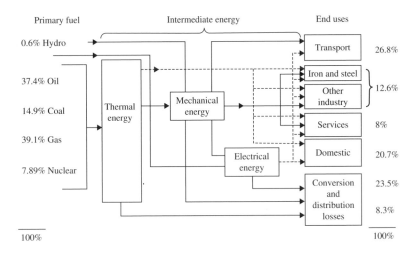

Fig. 2.10. UK Energy Flows, 2009 (adapted from Ref. 7, using data from Ref. 3).

contribute 3.7% of the GDP [6]. Of this total, 1.1% is due to the electricity industry and just under 2% due to the oil and gas industries [8].

2.5.3. *Energy production and use in the USA*

Energy consumption in the USA for 2009 reduced by 5% compared with 2008. These figures are shown in Table 2.3, in mtoe, broken down by different sources [3]. The fossil fuels coal, oil and natural gas account for almost 90% of the total.

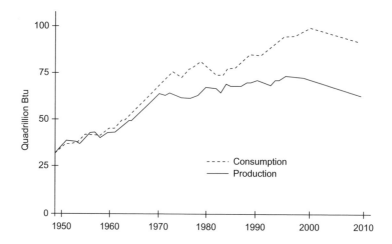

Fig. 2.11.   Prime energy production and consumption in the USA (updated from Ref. 9).

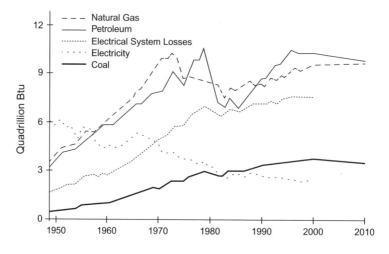

Fig. 2.12.   USA industrial energy consumption [9].

Since the late 1950s, energy consumption in the USA has exceeded production, mainly driven by the demand for petroleum [9]. In 2009 the USA produced
1659.4 mtoe of energy but consumed 2182 mtoe, resulting in a deficit of 523 mtoe,
or 24% of the consumption (Fig. 2.11).

In the USA the industrial sector is the largest sector consumer, but all sectors
continue to demand more energy. Industrial demand continues to be dominated by
oil and natural gas (Fig. 2.12) [8]. The use of coal, once the leading source, grows
more slowly, as it also does in the residential and commercial sectors.

About three-fifths of the energy consumed in the industrial sector is used for
manufacturing. The remainder goes to mining, construction, agriculture, fisheries

and forestry. Within manufacturing, large consumers of energy are the petroleum and coal products, chemicals and allied products and primary metal industries. Natural gas is the most commonly-consumed energy resource in manufacturing. The predominant end-use activity is process heating, followed by machine drives and then facility heating, ventilation and air conditioning combined.

Just under 7% of all energy consumed in the USA is used for non-fuel purposes, such as asphalt and roads, oil for roofing products and road building, conditioning liquefied petroleum gases for feed stocks and petrochemical plants, waxes for packaging, cosmetics, pharmaceuticals, inks, adhesives and still gas for chemical and rubber manufacture.

While variety and change in energy sources are the hallmarks of the industrial sector and the residential and commercial sectors, transportation's reliance on petroleum has been nearly total since 1949 [9]. The enormous appetite for energy of the USA is illustrated in the energy per capita chart of Fig. 2.7. Americans use energy at about twice the rate of Europeans.

2.5.4. *World fossil fuel production and consumption*

Production figures for the world's leading fossil fuel producers in 2009 are listed in Table 2.7. The USA produces more oil than any country other than Russia and Saudi Arabia and dominates the world energy market, both as a producer and a

Table 2.7.   Fossil fuel production in 2009.

Country	Oil (mto)	Natural Gas (mtoe)	Coal (mtoe)	Total (mtoe)
China	189	76.7	1552.9	1820
USA	325.3	541.8	539.9	1410
Russian Fed.	494.2	474.8	140.7	1110
Saudi Arabia	459.5	69.7	—	565
Canada	155.7	145.3	32.8	334
Iran	202.4	118.1	—	321
Australia	23.6	38.1	228	290
India	35.4	35.3	211.5	282
Indonesia	49	64.7	155.3	269
Mexico	147.5	52.4	5.3	205
Norway	108.3	93.1	—	201
United Arab Emirates	120.6	44	—	164.6
Venezuela	124.8	25.1	3.6	154
Algeria	77.6	73.3	—	151
South Africa	—	—	140.9	140.9
Qatar	57.9	80.4	—	138
UK	68	53.7	10.9	133
Kuwait	121.3	11.3	—	132.6
Brazil	100.4	10.7	19	123
Nigeria	99.1	22.4	—	122
Iraq	121.8	—	—	121.8
Libya	77.1	13.8	—	90.9

Table 2.8.   Fossil fuel consumption 2009.

Country	Oil (mto)	Natural Gas (mtoe)	Coal (mtoe)	Total (mtoe)
China	404.6	79.8	1537.4	2020
USA	842.9	588.7	498	1930
Russian Fed.	124.9	350.7	82.9	559
India	148.5	46.7	245.8	441
Japan	197.6	78.7	108.8	385
Germany	113.9	70.2	71	255
Canada	97	85.2	26.5	209
Iran	83.6	118.4	1.4	204
South Korea	104.3	30.4	68.6	203
UK	74.4	77.9	29.7	182
Mexico	85.6	62.7	6.8	155
Italy	75.1	64.5	13.4	153
France	87.5	38.4	10.1	136
Brazil	104.3	18.3	11.7	134
Indonesia	62	33	30.5	126
South Africa	24.3	—	99.4	124
Australia	42.7	23.1	50.8	117
Spain	72.9	31.1	10.6	115
Taiwan	46.6	10.2	38.7	95.5
Netherlands	49.4	35	7.9	92.3
Poland	25.5	12.3	53.9	91.7
Ukraine	14.1	42.3	35	91.7

consumer. Figures for the consumption of fossil fuels in 2009 are shown in Table 2.8. The production and consumption figures are calculated as a surplus or deficit in Fig. 2.13. Most of the OECD countries consume more primary energy than they produce and are net importers of energy, as shown in Fig. 2.13.

The OECD countries are largely dependent on fossil fuels for their energy. Of the present 33 OECD countries (2013), only Iceland and Norway supply more than 50% of their primary energy consumption from non-fossil sources. The oil producer countries of the Organization of Petroleum Exporting Countries (OPEC) are also net exporters.

A historical perspective of the fossil fuel era is demonstrated in the time scale shown in Fig. 2.14. The beginning of the Christian era is denoted by the Star of Bethlehem in the middle of the time axis. The present time is now slightly to the right of the apex of the fossil fuel "blip" [7].

The mass use of fossil fuels began in Europe during the 18th century. There was no mass fossil fuel use before the Industrial Revolution in Britain or elsewhere. Once the fossil fuels have been largely depleted there can never be another fossil fuel era. The present period of fuel history is unique. It is conjectural to put a firm date to the end of the fossil fuel era, but it is likely that virtually all of the usable oil and natural gas will be finished in less than 100 years. Much of the coal will be finished in less than 200 years from now. Is the end of the fossil fuel era, especially

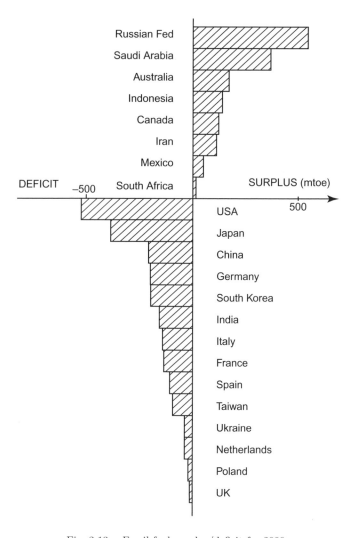

Fig. 2.13.   Fossil fuel surplus/deficit for 2009.

the ending of natural crude oil supplies, sufficiently near to engage the attention and action of the nations and peoples of the world?

2.6.   Risks Associated with Energy Systems

All energy systems entail risks. Some risks affect public health and welfare directly.

Other risks affect the environment and have direct and indirect ecological impacts. The extraction, transportation, distribution and use of prime fuels, especially fossil fuels, create some environmental and ecological side effects that may be not only undesirable but dangerous to human life and welfare. Risk assessment, as

Fig. 2.14.   Time scale of the fossil fuel era.

applied to energy systems, is unreliable, subjective and controversial. No attempt is made here to apportion different risks in a quantitative or even relative manner. The approach is to note the main areas for concern. Different observers may reach very different conclusions as to the relative seriousness of the various features delineated.

Features of the risks associated with energy production and use are discussed under the headings listed below [10].

2.6.1 Industrial accidents and industrial diseases
2.6.2 Large-scale accidents and sabotage
2.6.3 Management of energy waste
2.6.4 Ecosystem effects
2.6.5 Water supply problems
2.6.6 Emissions

2.6.1.   *Industrial accidents and industrial diseases*

Industrial accidents and industry-induced diseases are well recorded and can therefore be assessed fairly accurately. From the accidental death statistics, coal mining is the most dangerous form of fuel procurement, resulting in more fatalities than oil, gas and nuclear combined.

In the UK before the Second World War (i.e. in the 1930s), the toll in the deep coal mines was about 50 fatalities per year. This appalling record was greatly reduced by post-war modernization of the coal industry and is now further reduced by the drastic contraction of the industry. Corresponding reductions have occurred in the incidence of industrial injuries and lung diseases caused by mining gases and coal dust.

2.6.2.   *Large-scale accidents and sabotage*

There are risks of low probability, high consequence accidents associated with nuclear reactors, hydroelectric dams and the transportation and storage of liquefied natural gas (LNG). Between 1918 and 1958 there were an average of 40 deaths per year from dam failures in the USA, including some high casualty individual failures. These figures are comparable with the fatality figures from the British coal industry in the same period, and both have been significantly reduced in the past 30 years. Worst case scenarios for both dams and LNG facilities suggest numbers of casualties comparable with those anticipated from severe nuclear accidents. An important difference is that casualties from dam failures, oilfield fires and explosions and LNG accidents are immediate and obvious. By comparison, nuclear accidents may cause delayed effects that affect a large number of people and therefore engender a greater public apprehension.

In 1988 an explosion and fire on an offshore oil platform, the Piper Alpha, in the UK North Sea killed 167 people. This was the worst single energy-industry incident in recent times, emphasizing the fact that even industries considered low-risk by the public exact a death toll on their workers.

Industrial plants are always vulnerable to sabotage. Nuclear plants tend to be better-guarded than dams or LNG refineries. A discussion of this issue is included in Chapter 8.

2.6.3.   *Management of energy waste*

All energy systems produce waste. The management of industrial wastes involves risks to health and poses environmental concerns. In the developed industrial countries, the transportation, processing and disposal of industrial wastes is highly developed and the subject of safe and reliable control procedures. Waste management relevant to individual fossil fuels and biofuels is discussed in appropriate chapters below. The issue of radioactive waste management is treated as a special case in Chapter 8.

2.6.4.   *Ecosystem effects*

Energy production and use has some adverse ecological effects, including the loss of arable land, water resources, natural beauty, open space, habitat, wildlife preservation and wilderness areas. The relative importance that might be attached to these various criteria, in comparison with the universal need for energy, is very much a feature of individual opinion. The loss of ecological diversity has long-range human consequences that are not well understood compared with the more immediate effects of energy development.

It is difficult to assign quantitative levels to the destructive side-effects of different energy systems. From the point of view of ecosystems, it is possible that hydroelectric power development, with its "clean" public image, is more destructive per unit of energy output than fossil fuel plants because of the large land area involved. Similarly, the land-based production of biomass can have very serious destructive side-effects that are less obvious than those of fossil or nuclear systems.

The ecological side-effects of nuclear power are smaller than those for any other established energy source. Only if it becomes necessary to deep-mine for low-grade uranium ores will the effects become comparable with coal mining. The widespread use of breeder reactors would eliminate that necessity.

Fossil fuel mining, processing and usage have some very evident ecological effects, depending mainly on the locale. The location of offshore oil rigs, for example, can have significant effects on local marine life. Synthetic oil production from oil shale or coal products has major ecological impact in the pollution and contamination of local groundwater flow.

2.6.5.   *Water supply problems*

Fossil fuel and nuclear systems for generating electricity require the availability of large masses of water. Any heat–work energy conversion system, such as a steam turbine or a nuclear reactor, involves a necessary cooling process, for which water is most likely used as the coolant fluid. Consequent heating and discharge of the coolant water can have ecological impact. In the choice of site for an electricity-generating station, the availability of an adequate cooling water supply is a vital consideration. When a station is located on a seashore or lakeshore, the operation of a station may affect the immediate and downstream water turbidity and temperature. If groundwater has to be used at an inland site, this may affect the local hydrological cycle, irrigation, water table levels, water run-off streams and the availability and cost of water for other consumers.

2.6.6.   *Emissions*

The combustion of fossil fuels emits a great variety of airborne pollutants. Gaseous pollutants include sulphur oxides, nitrogen oxides, carbon monoxide, carbon dioxide and various hydrocarbons (mainly methane). The gaseous emissions are considered in Section 2.6.6.2, below. In addition to gaseous emissions, fossil fuel burning gives off solid particles of carbon, some less than 1 mm in size, plus trace amounts of heavy metals. A detailed technical analysis of the effects of carbon particulates is included in Section 4.4.2.4 of Chapter 4 (coal). A more general summary of gaseous emissions follows here.

Table 2.9. World carbon dioxide emissions, 2007 [11].

Rank	Country	Annual CO_2 emissions (in thousands of metric tons)	Percentage	Per Capita (metric tonnes)
	World	29,321,302	100%	
1	China	6,538,367	22.30%	4.9
2	United States	5,838,381	19.91%	19.9
—	European Union	4,177,817.86	14.04%	—
3	India	1,612,362	5.50%	1.4
4	Russia	1,537,357	5.24%	10.8
5	Japan	1,254,543	4.28%	9.8
6	Germany	787,936	2.69%	9.6
7	Canada	557,340	1.90%	16.9
8	United Kingdom	539,617	1.84%	8.9
9	South Korea	503,321	1.72%	10.5
10	Iran	495,987	1.69%	6.8
11	Mexico	471,459	1.61%	4.4
12	Italy	456,428	1.56%	7.7
13	South Africa	433,527	1.48%	8.8
14	Saudi Arabia	402,450	1.37%	16.3
15	Indonesia	397,143	1.35%	1.8
16	Australia	374,045	1.28%	17.9
17	France	371,757	1.27%	6.0
18	Brazil	368,317	1.26%	1.9
19	Spain	359,260	1.23%	8.2
20	Ukraine	317,537	1.08%	6.9
21	Poland	317,379	1.08%	8.3
22	Turkey	288,681	0.98%	4.0
23	Thailand	277,511	0.95%	4.1
24	Taiwan	275,577	0.94%	12
25	Kazakhstan	227,394	0.78%	14.8
26	Malaysia	194,476	0.66%	7.3
27	Egypt	184,659	0.63%	2.3
28	Argentina	183,728	0.63%	4.7
29	Netherlands	173,244	0.59%	32.5
30	Venezuela	165,550	0.56%	6.0
31	Pakistan	156,394	0.53%	0.9
32	Algeria	140,120	0.48%	4.1

2.6.6.1. *Carbon dioxide emissions*

World carbon dioxide emissions for 2007 (the latest year for which figures are available) are given in Table 2.9 [11]. The main producers are, in order, China, the USA, India and Russia, followed by Japan and Germany. Between them the above countries generate almost 60% of the total world sum, very largely due to fossil fuel burning. There is a close correlation between the amount of primary fuel consumption in Table 2.2 and the carbon dioxide emissions in Table 2.9. The six countries listed above as the biggest emitters account for 56% of the prime energy consumption.

China and the USA consume 39% of the world prime energy (Table 2.2) and emit 42.2% of the carbon dioxide (Table 2.9). The figures are not precise because they refer to different years of data but the picture that they indicate is correct.

The non-industrial countries that are high emitters of carbon dioxide (Table 2.9) include Indonesia and Brazil. In both cases it is likely that part of the reason is the widespread deforestation that now occurs due to logging of the rainforests. Environmental objections to the reduction of the rainforests include not only reduction of the carbon dioxide capture capability but severe reduction of habitat for some of the remaining wild animals.

Projections of future fuel use (Fig. 2.6) suggest that consumption of the fossil fuels is likely to continue to rise until 2035 and beyond. If there are no agreed policy changes the consequent production of greenhouse gases, particularly carbon dioxide, will rise correspondingly, as forecast in Fig. 2.15 [5]. Further discussion about gaseous emissions, greenhouse gases and carbon dioxide trading is given in Section 2.6.6.3 below, in the following chapters on fossil fuels and in Chapter 13, concerning biological and chemical energy.

International concern about the levels of carbon dioxide emissions and the possible effects on climate resulted in the Kyoto Protocol agreement in 1998. This agreement calls for quantifiable goals for emission reductions from participating countries [12]. As of March 1999, 83 countries had signed the Protocol, indicating their agreement in principle. But ratification and action on the agreement would involve major reductions of energy use and major changes of industrial organization and employment patterns. In 2001 the USA formally rejected the Protocol and declined to sign because of possible damage to its energy industries and also because of scientific reservations about the global warming levels suggested in the

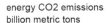

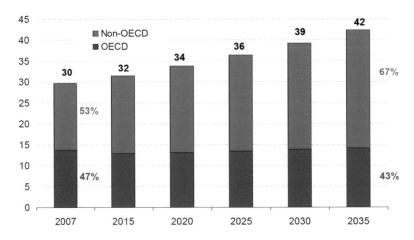

Fig. 2.15.   Possible growth of carbon dioxide emissions [5].

calculations. Everyone is agreed on the need for action but none of the major players have yet been able to implement it.

A further international conference on energy matters, including pollution and global warming, took place in Cancun, Mexico, in 2010.

2.6.6.2. *Gaseous emissions and the "greenhouse" effect*

Fossil fuel burning plants emit a variety of flue gases, including oxides of sulphur, oxides of nitrogen and oxides of carbon plus methane and chlorofluorocarbons (CFCs). In addition, a major source of air pollution is automobile exhausts, which are particularly responsible for carbon monoxide, nitrogen dioxide and hydrocarbons. These automobile pollutants may interact biologically and chemically with those from electricity-generating plants. Carbon monoxide (CO) is a dangerous gas to animals and humans because it is an active oxygen seeker. It readily combines with haemoglobin, which is the oxygen carrier in the bloodstream [13].

Carcinogens are also present in fossil fuel emissions, particularly those from coal. Various environmental effects, such as acid rain, are discussed in Chapter 4.

The so-called "greenhouse" effect, applied to the earth's climate, is a scientific theory that was first proposed about 100 years ago but is still the object of some dispute. The earth's gaseous atmosphere permits the easy transmission of incoming ultraviolet, visible and infrared solar radiation, which are mostly reflected back from the earth's surface. The net radiation effect on the atmosphere is to create a thermal barrier around the earth, resulting in a possible slow increase of the earth's ground temperature. There is uncertainty over the precise rate and ultimate magnitude of any global temperature rise. A 1983 report by the US Environmental Protection Agency (EPA) stated a possible increase of $2°C$ by the middle of the 21st century and a $5°C$ increase by 2100. This would result in "a rise in global average sea level and dramatic changes in precipitation and storm levels" [14].

A more recent investigation, involving 28 experts from 12 countries plus 6 modelling teams from the USA, studied 40 different emissions scenarios. A consensual view is that global temperatures will increase in the range of $1.4°C$–$5.8°C$ during the 21st century, with rises in sea levels in the range 8 cm–99 cm [15].

The principal greenhouse gas emissions carbon dioxide (CO_2), methane (CH_4) and nitrous oxide (N_2O) are combined to show results in "equivalent carbon dioxide" in Table 2.10 [16]. The equivalences were obtained using the global warming potentials reported by the Intergovernmental Panel on Climate Change, 1999 [17]. Because some countries did not provide data under certain headings, or did not report at all, the information in Table 2.10 should be regarded as partial and subject to revision.

In the column labelled "Forestry and Change of Land Use" in Table 2.10, most of the figures are negative, implying the absorption of greenhouse gases. This is a result of forestation and agricultural planting, with the resultant increased digestion of carbon dioxide.

Table 2.10.   Greenhouse gas emissions/capita [16, 17] (1995 figures, mtoe).

Country	Fuel combustion	Industry	Agriculture	Waste	Forestry and change of land use	Other	Total emissions	Population	Per capita emissions
Australia	291.77	7.45	8.36	16.36	+51.87	27.13	481.94	18.07	26.67
Belgium	112.83	14.27	11.52	4.99	-2.06	1.01	142.56	10.14	14.06
Canada	478.96	36.34	25.04	19.47	0	51.51	611.32	29.62	20.64
Czech. Rep.	130.37	5.22	3.45	3.02	-5.45	8.81	145.42	10.33	14.08
Denmark	58.91	1.31	16.17	1.55	-0.96	1.17	78.14	5.23	14.94
France	365.79	40.79	48.88	19.15	-46.8	24.2	452.06	58.14	7.78
Germany	885.13	50.31	61.52	39.9	-30	24.57	1031.43	81.66	12.63
Greece	84.79	8.33	8.37	2.77	0	1.03	105.29	10.45	10.08
Ireland	33.27	2.58	19.28	2.95	-6.23	0.98	52.83	3.6	14.68
Italy	425.2	29.31	41.84	21.65	-24.51	22.46	515.95	57.3	9.0
Japan	1162.1	68.65	20.65	28.54	-94.62	5.06	1190.38	125.57	9.48
Luxembourg	9.16	0.41	0.51	0.08	-0.3	0.05	9.92	0.41	24.19
Netherlands	183.66	7.61	18.31	9.13	-1.7	5.05	222.06	15.46	14.36
New Zealand	24.95	2.74	44.33	2.77	-13.49	1.39	62.69	3.66	17.13
Norway	29.89	8.52	3.88	6.78	-13.64	2.69	38.12	4.36	8.74
Poland	365.18	13.76	22.87	17.96	-41.95	19.13	396.94	38.59	10.29
Portugal	47.92	4.01	6.33	13.77	-1.15	0.53	71.41	9.92	7.2
Russian Fed.	1607.27	24.37	114.53	41.04	-568	307.2	1526.37	148.2	10.3
Spain	221.62	18.85	37.64	15.3	-28.97	13.45	277.88	39.21	7.09
Sweden	56.29	5.17	4.2	1.28	-30	0.27	37.21	8.83	4.21
Switzerland	40.95	2.71	5.84	2.85	-5.1	0.46	47.72	7.08	6.74
Ukraine	671.17	33.7	50.5	19.68	-51.98	138.06	861.12	51.55	16.7
UK	533.77	28.93	26.19	38.44	+9.95	25.47	662.75	58.61	11.31
USA	5206.4	96.43	268.23	236.44	-428	202.49	5581.99	263.17	21.21

Australia had the world's highest greenhouse gas emissions per person at 26.7 tonnes in 1995, which was twice the average level for industrialized countries (13.4 tonnes) and 25% higher than the figure for the USA (21.21 tonnes).

It is seen from Table 2.10 that the nations with the highest per capita emissions were, in order, Australia, the USA and Canada. Luxembourg's very high per capita emissions are due to the small population and the presence of a large steel plant. New Zealand has low emissions from energy use, due to the predominance of hydroelectricity, but very high emissions from agriculture due to the large number of sheep!

More up-to-date figures of the per capita emissions of carbon dioxide are included in Table 2.9. France has a low emissions figure, 6.0, due to a combination of energy efficiency, industrial structure and the use of nuclear power (which does not create significant greenhouse gases) [16, 17].

A major increase of CO_2 emission from earth-bound activities would cause the level of atmospheric CO_2 to build up, possibly increasing the thermal insulation effect. Any additional heat produced by the greenhouse effect would first be absorbed and distributed by the oceans, so introducing a time lag of several decades before the air temperatures increased.

Increasing the level of atmospheric CO_2 would, via any consequent temperature changes, have the effect of redistributing agricultural productivity across the world and possibly lead to mutant changes in plant and animal life. In some regions there could be ecological disaster. It remains conjectural as to whether the danger posed by the greenhouse effect is a reality. The attitude "wait and see" is potentially as dangerous as the threat itself. Due to various positive feedback effects within the earth's complicated climate cycle, by the time that climatic changes due to excessive CO_2 were detected and attributed, they would be irreversible and progressively cumulative.

Nuclear reactors have smaller effects on climate than fossil fuel systems because they do not emit CO_2. Hydroelectric and geothermal sources also do not contribute significantly to greenhouse gases. The manufacture of solar photovoltaic panels is an energy use that might require fossil-fuel generated electricity but the operation of the panels themselves is benign.

Will the use of fossil fuels be controlled to limit the "greenhouse" danger, by the application of the Kyoto Protocol or otherwise? Will the oil and natural gas run out before their use has irrevocably changed the atmospheric CO_2 levels? Will nature find a way to automatically redress any temperature rise? Is it already too late because the initiating changes are irreversible and are already in place? Have the scientists misread or misinterpreted the data and, in fact, there is no danger?

2.6.6.3.   *Carbon trading*

An attempt to reduce greenhouse gas emissions by the business community is referred to as carbon trading. This is a market-based strategy in which the carbon market trades emissions under "cap-and-trade" schemes that pay for or offset gaseous emissions.

> "The scheme governing body sets a cap on allowable emissions and then auctions off 'emissions allowances' up to the cap limit. Members that do not have enough allowances to cover their emissions must either make reductions or buy another member's spare credits. Members with extra allowances can sell them or bank them for future use. Cap and trade schemes can be either mandatory or voluntary.
>
> "A successful cap and trade scheme relies on a strict but feasible cap that decreases emissions over time. If the cap is set too high, an excess of emissions will enter the atmosphere and the scheme will have no effect on the environment. A high cap can also drive down the value of allowances, causing losses in firms that have reduced their emissions and blanked credits. If the cap is set too low allowances are scarce and overpriced. Some cap and trade schemes have safety valves to keep the value of allowances within a certain range. If the price of allowances gets too high, the scheme's governing body will release additional credits to stabilize the price. The price of allowances is usually a function of supply and demand." [18]

Obviously the usefulness and effectiveness of a cap-and-trade scheme depends on the determination of the members to make it work. At the United Nations climate summit in Cancun, Mexico, in 2010, an attempt was made to secure agreement on a global scheme concerning deforestation. The objective was to reduce the global deforestation rate by rich countries investing money into a scheme to compensate countries that refrain from or limit their deforestation. The biggest obstacle to the scheme is that, with the possible exception of Brazil, the countries with the highest rates of deforestation have the worst governance, the weakest enforcement of laws and inadequate records of land use [19].

2.7. Summary: Where Do We Go from Here?

The sociology and politics of energy planning are enormously complicated. Should a particular country have an overall energy policy or is this an unwarranted interference in the freedom of its citizens? Are there overall considerations regarding the welfare of the country, in energy matters, that transcend the right of individuals? Is energy use and conservation important enough to justify mandated government control?

For the OECD countries the critical near-term factor in energy supplies is the availability of liquid fuels. Modifications of petroleum usage and energy conservation measures of various kinds may delay the final scene but the end point is inevitable. The oil is running out. Energy planning on the scale that is needed involves a lead time of several decades.

For the industrialized and developed countries of the OECD, two principal energy problems exist:

Problem 1

A need for continuing sources of gasoline and diesel fuel for motor vehicles and aircraft.

Problem 2

A need for continuing sources of prime fuel for use in the generation of electricity (on the assumption that oil and natural gas will ultimately be unavailable).

2.7.1. *An energy strategy*

Problem 1, concerning oil supply, can be addressed in terms of:

- new oilfield discoveries;
- enhanced recovery rates of natural crude oil;
- development of a synthetic fuels industry;
- major social adjustments in the pattern of private motoring;
- some alternative form of road transportation (such as the electric car); and
- further development of mass transportation systems using electricity-powered rail vehicles.

Problem 2, concerning electricity generation, can be addressed simultaneously on both short-term and long-term levels.

In the short term:

- new natural gas field discoveries and enhanced recovery rates;
- coal and coal products, such as coal-bed methane; and
- reinstatement and increased use of nuclear fission power, using breeder reactors.

In the long term, the future seems to lie with renewable energy sources:

- solar energy:

  — photovoltaic conversion;
  — solar-thermal systems;
  — wind turbine systems;
  — ocean thermal currents;
  — water wave energy;
  — hydropower; and
  — biomass and photosynthesis;

- geothermal energy;
- gravitational energy;
- tidal energy; and
- thermonuclear fusion.

World prime energy consumption grew by 24% from 1999 to 2009. Consumption in all of the world's geographical areas showed increases.

A recent forecast of future energy demand is given in Fig. 14.3 of Chapter 14. This puts forward the view that in the next 50 years there will be a decreasing use of fossil fuels, especially oil, and an increasing use of renewable energy sources.

2.8.   Problems and Review Questions

2.1. If the solar total radiation intercepted by the earth has the value equivalent to 1.73×10^{17} J/s, calculate the diametric area if the solar constant is $1377\,\mathrm{W/m^2}$.

2.2. The outward flow of power from the earth's interior proceeds at an estimated rate of $0.063\,\mathrm{W/m^2}$. If the total flow rate is 3.203×10^{32} W, calculate an approximate value for the earth's radius.

2.3. The daily output of energy from the sun is 3×10^{32} J. What fraction of this is intercepted by the earth?

2.4. The solar constant for the planet earth is $1377\,\mathrm{W/m^2}$. If the effective diametric radius is 6.36×10^6 m, what is the radiation rate?

2.5. Compile a table of data, based on Fig. 2.1, showing the proportions of input solar energy for the earth that are used for (i) direct conversion to heat, (ii) winds, waves, water currents and convection, (iii) the evaporation, precipitation and drainage cycle and (iv) photosynthesis.

2.6. What were the principal reasons for the sudden mass use of fossil fuels in around 1700 AD?

2.7. Use the data from Fig. 2.3 to roughly estimate the number of years of crude oil supply remaining, based on present predictions of use and present reserves for (a) the world, (b) the OECD countries and (c) Russia.

2.8. Repeat Problem 2.7 for natural gas supply.

2.9. Repeat Problem 2.7 for coal supply.

2.10. Why does Norway occupy so favourable a position in Fig. 2.7?

2.11. What should be the ambition of a country with regard to its Fig. 2.7 coordinate position?

2.12. Japan has achieved very successful economic performance in spite of poor indigenous energy supplies. What are some of the implications of this for Third World developing countries?

2.13. Why are the agriculture-based economies of Canada, Australia and New Zealand favourably located in Fig. 2.7?

2.14. What are the main reasons why North Americans consume such a high per capita amount of energy?

2.15. In Fig. 2.7 the countries in the bottom left-hand corner, including Brazil, China, India and Indonesia, total 44% of the world population. What are some of the implications of these countries seeking to obtain a "bigger share of the cake"?

2.16. The data of UK primary fuel consumption and production since 1999 are given in Tables 2.5 and 2.6. What do you deduce about the place, in the UK industrial scene, of the coal, gas and oil industries?

2.17. Which countries of the OECD group supply more than 50% of their energy needs from non-fossil sources?

2.18. What measures would you advocate to reduce the amount of carbon dioxide produced by automobiles?

2.19. Should there be an "emissions tax" to penalize the parts of industry that generate large volumes of greenhouse gases and/or particulates?

2.20. Is there any way in which implementation of the Kyoto Protocol agreement can be enforced?

2.21. Use the data in Table 2.9 to list, in order, (a) the six largest emitters and (b) the six largest per capita emitters.

2.22. Why are the agriculture-based economies of countries such as Australia, Ireland and New Zealand such large per capita producers of greenhouse gases?

2.23. Why is Japan, for example, with a large population and highly industrialized economy, such as a low per capita contributor to the production of greenhouse gases?

References

[1] King Hubbert, M. (1971). "The Energy Resources of the Earth", in W.H. Freeman (ed.), *Energy and Power: A Scientific American Book*, San Francisco, Chapter 3, pp. 30–40.

[2] McVeigh, J.C. (1984). *Energy around the World*, Pergamon Press, Oxford.

[3] "BP Statistical Review of World Energy", British Petroleum plc, London, June 2010 and 2011.

[4] "International Energy Outlook 2010", US Dept. of Energy, Washington, DC, Report DOE/EIA-0484, March 2010.

[5] Presentation by Howard Gruenspecht, Center for Strategic and International Studies, US Energy Information Administration (EIA), Washington, DC, May 2010.

[6] "UK Energy in Brief 2010", Dept. of Energy and Climate Change, London, July 2010.

[7] Shepherd, W. and Shepherd, D.W. (2003). *Energy Studies*, 2nd Edition, Imperial College Press, London.

[8] "UK Energy Sector Indicators 2010", Dept. of Energy and Climate Change, London, July 2010.

[9] "Monthly Energy Review", US Energy Information Administration, Dept. of Energy, Washington, DC, November 2010.

[10] "Energy in Transition 1985–2010", National Research Council, National Academy of Sciences, Washington, DC, 1979.

[11] "UN Statistics Division, Millenium Development Goals Indicators: Carbon Dioxide Emissions", collected by the US Dept. of Energy's Carbon Dioxide Information Analysis Center (CDIAC), 2008.

[12] "Impacts of the Kyoto Protocol on US Energy Markets and Economic Activity", US Dept. of Energy, Washington, DC, Report DOE/EIA-SR/OIAF/98-03, October 1998.

[13] "Greenhouse Gas Emissions", International Energy Agency, OECD, Paris, 1991.

[14] Seidal, S. and Keyes, D. (1983). "Can We Delay a Greenhouse Warning?", US Environmental Protection Agency, Washington, DC, September 1983.

[15] Nakicenovic, N. and Swart, R. (2000). *Emissions Scenarios: Intergovernment Panel on Climate Change*, Cambridge University Press, Cambridge.

[16] Hamilton, C. and Turton, H. (1999). "Greenhouse Gas Emissions Per Capita of Annexe B Parties to the Kyoto Protocol", Australian National University ACT 2602, Australia, November 1999, https://www.tai.org.au/file.php?file=WP19.pdf.

[17] National Greenhouse Gas Inventory Committee (NGGIC) (1999). *National Greenhouse Gas Inventory 1997*, Australian Greenhouse Office, Canberra.

[18] Dowdey, S. (2010). "How Carbon Trading Works", http://science.howstuffworks.com/environmental/green-science/carbon-trading.htm.

[19] Webster, B. (2010). "A Forest Clearing", *Eureka*, Issue 15, The Times, London.

[20] Mackay, D.J.C (2009). *Sustainable Energy — Without the Hot Air*, UIT, Cambridge, page 337.

CHAPTER 3

ELECTRICITY

3.1. Introduction

Electricity is the universally popular form of energy and is always in demand. It is invisible, odourless, available at the flick of a switch, easy to measure and meter and has very wide and varied applications. It continues (in 2013) to be the most rapidly growing form of energy consumption. When electricity has to be manufactured from some source of fossil fuel it can be regarded as a secondary form of energy. Electricity generated from nuclear or hydro sources can be regarded as primary energy.

3.2. Some Basic Electrical Relationships

3.2.1. *Voltage, current and power*

The potential energy of an electricity supply system is measured in terms of electrical "pressure" or voltage, named after the Italian scientist Alessandro Volta. The unit of potential difference between two points in a circuit is the volt (V), which represents an energy of one joule per coulomb of charge. Electricity supplies for domestic consumers are usually rated at 240 V in the UK and 120 V in the USA. Silicon solar cells are rated at about 0.5 V, small dry batteries are rated at 1.5 V, while car batteries are usually 12 V. For the transmission of electricity by overland line or underground cable, the transmission voltages range up to several hundred thousand volts (i.e. several hundred kV).

The application of an electrical voltage to a circuit results in the flow of electric current, measured in amperes (A), named after the French scientist Andre Marie Ampère. Small values of current are measured in microamperes (μA) or milliamperes (mA). Domestic appliances such as washing machines operate with currents of the order 3 A–5 A. Large industrial machines or processes take currents of tens (or even hundreds) of amperes. A lightning stroke is an electrical discharge that might involve millions of amperes, for its few seconds of duration. It is the current rating of an electrical load that determines the size of connecting wires or conductors.

Electrical current flow in a circuit is opposed by the circuit impedance. With a direct current (DC) supply the circuit impedance is entirely resistive and the load is a resistance R measured in ohms, after the German scientist Georg Simon Ohm. The relationship between voltage, current and resistance (or impedance) is called Ohm's Law.

$$R = \frac{V}{I} \tag{3.1}$$

When V is in volts and I is the current in amperes, then resistance R has the unit of ohms.

The time rate of energy dissipation in a DC circuit, or circuit power rating, is given by

$$P = VI \tag{3.2}$$

When V is in volts and I is in amperes, the unit of power is the watt (W), named after Scottish engineer James Watt. One watt is an energy dissipation rate of one joule per second (J/s).

Equations (3.1) and (3.2) can be combined to give the power in terms of load resistance R:

$$P = I^2 R = \frac{V^2}{R} \tag{3.3}$$

Most electricity supply systems generate alternating current (AC) rather than direct current (DC), for economic reasons. In an AC system the voltages and currents vary sinusoidally (usually) in magnitude and direction at a time rate defined as the frequency of operation. Frequency is measured in units of cycles per second or hertz (Hz), named after the German scientist Heinrich Hertz. In the USA the domestic and most industrial electricity supplies operate at 60 Hz, whereas in the UK the preferred frequency is 50 Hz. Radio transmission usually operates at frequencies of hundreds of thousands of Hz (i.e. several hundred kHz). Television picture signals are in the order of hundreds of millions of Hz (i.e. hundreds of MHz). The visible spectrum of solar radiation is approximately of frequency 6×10^{14} Hz.

Because of the time variation of the voltages and currents, the circuit properties of AC quantities that are usually used are the root mean square (rms) values. The rms value of an AC variable is sometimes called the effective value because it is effectively equal to the DC value that would dissipate the same power. When a circuit is rated at (say) 240 V it means that the applied voltage has an rms value of 240 V. Most electricity supply systems operate with constant rms voltage.

With an AC supply the load may be resistive, as with electric lamp and heating loads. For industrial loads, however, where the load is often due to three-phase AC induction motors, the load is partly inductive in nature. The combination of resistance R and inductance L effects is conjoined in the AC impedance Z, which depends on the (usually constant) frequency of the AC supply.

For AC circuits Eq. (3.1) is modified to

$$Z = \frac{V}{I} = \sqrt{R^2 + \omega^2 L^2} \tag{3.4}$$

Equation (3.3) is still valid for the resistance components in an AC circuit provided that I and V are the rms values.

In an AC circuit the power dissipation in watts is no longer generally described by Eq. (3.2). Because of time-phase effects caused by magnetic field and electric field properties, the power is reduced and is now given by

$$P = VI \cos \psi \tag{3.5}$$

In Eq. (3.5) the angle ψ depends on the amount of inductance or capacitance in the circuit load. With purely resistive loads $\psi = 0°$, $\cos \psi = 1$, and Eq. (3.5) becomes equal to Eq. (3.2). The property $\cos \psi$ in Eq. (3.5) is often referred to as the power factor (PF). It is a constant aim of electrical engineers to operate circuits and systems at the highest possible value of power factor. When PF $= 1$, its highest possible value, it means that the specified load power P is being supplied using the minimum possible value of supply current I and this is the most economic condition of operation. In some electrical engineering books the term "power factor" is alternatively designated as "energy efficiency".

The most effective and economic form of electricity generation and distribution is to use three-phase systems. With a three-phase AC generator supplying a balanced load it is possible to supply three separate one-phase loads using four (sometimes three) conductors. With three-phase loads Eqs. (3.1)–(3.5) remain valid but refer to per phase quantities.

One of the great advantages of using AC supply systems is that the voltage levels can be easily changed using electrical transformers. In order to use small area conductors, thereby saving on the cost and weight of copper, a transmission circuit is made to operate with low values of current. But with low current it is necessary to use very high transmission voltages in order to transmit a lot of power, as implied in Eqs. (3.2) and (3.5). Figure 3.1 shows a typical arrangement of three-phase, high-voltage towers or pylons, rated at several hundred kV [1]. At the receiving end of the transmission lines the high voltages are transformed to much lower values for distribution and use, while the low transmission line currents correspondingly increase. The large sizes of apparatus needed are illustrated by the high-voltage transformer and switchgear of Fig. 3.2 [2].

The efficiency (η) of operation of an electrical circuit or system is defined as the ratio of the output power to the input power:

$$\eta = \frac{P_{out}}{P_{in}} \tag{3.6}$$

Fig. 3.1.   Three-phase, high-voltage transmission line [1].

Fig. 3.2.   High-voltage transformer and switchgear (reproduced by permission of "Energy", by Marshall *et al.*, Cavendish Books Ltd., London, England, 1978).

For a resistive circuit such as an electric heater the efficiency is high, probably well over 90%. With an electric motor load the full-load efficiencies vary from about 60% for small motors to over 90% for large motors.

It should be noted that efficiency is not the same as power factor (i.e. energy efficiency). For a full assessment of an AC circuit performance both the efficiency and the power factor need to be considered. The total tariff (i.e. electricity bill) paid by an industrial consumer depends not only on the kilowatt hours of electrical energy consumed but also on the maximum demand figure and the operating power factor.

3.2.2. *Worked examples on electrical circuits*

Example 3.1. A 120-V, 60 Hz single-phase electric supply is connected to a load of resistance 24 ohms. Calculate the current and the power dissipation.

From Eq. (3.1)

$$I = \frac{V}{R} = \frac{120}{24} = 5 \, \text{A}$$

From Eq. (3.3)

$$P = I^2 R = (5)^2 . 24 = 600 \, \text{W}$$

Example 3.2. For a single-phase 240-V, 50-Hz supply the instrument readings with a certain electric motor load are $I = 12 \, \text{A}$, $P = 1500 \, \text{W}$. Calculate the input impedance of the motor and its operating power factor.

From Eq. (3.4)

$$Z = \frac{V}{I} = \frac{120}{24} = 20 \, \text{ohms}$$

From Eq. (3.5)

$$PF = \cos \psi = \frac{P}{VI}$$

$$= \frac{1500}{240 \times 12} = 0.52$$

Note that this is a low (poor) value of PF and represents a poor form of utilization.

Example 3.3. An AC electric motor has a nameplate rating of 1 kW. When it delivers its rated load from a 240-V, 50-Hz supply, the input power is 1500 W. What is the motor efficiency?

A motor nameplate rating represents its output power at full load. Therefore,

$$P_{\text{out}} = 1000 \, \text{W}$$
$$P_{\text{in}} = 1500 \, \text{W}$$

From Eq. (3.6) the efficiency is

$$\eta = \frac{P_{\text{out}}}{P_{\text{in}}} = \frac{1000}{1500} = 0.667$$
$$= 66.7\%$$

3.3.   The Generation of Electricity

An electricity generator is a machine for converting mechanical energy into electrical energy. When coal, natural gas, oil or nuclear fuels are used the fuel is first
converted into heat energy and then into mechanical energy by a turbine. The
mechanical energy is usually rotational in form and the so-called prime mover that
rotates the generator shaft is usually a steam turbine, diesel engine or gas turbine.
Generation plants that operate via a heat–work process are called thermal stations
to distinguish them from more direct systems such as wind turbines and water
turbines. An illustration showing the essential features of a coal-fired electricity
generation system is given as Fig. 1.5 in Chapter 1.

Electrical generators work on the principle of electromagnetic induction, discovered by the English scientist Michael Faraday in the early 1800s. An arrangement of
coils on the rotating part (rotor) causes a magnetic field to rotate at the shaft speed.
This induces electromotive forces (voltages) in the conductors of the armature coils
mounted on the machine frame (stator). Both the magnitudes and frequencies of the
induced or generated voltages are proportional to the speed of shaft rotation. Generators and transformers are usually rated in terms of voltamperes (volts × amps
at full load) rather than in terms of watts, to accommodate power factor effects.
The largest individual generators, often called turbo-alternators, are rated at about
600 MW. The sizes of comparable machines are illustrated in Fig. 3.3. Generators
as small as 1 kVA rating are used in some applications. The 400-Hz generators used
on board large jet airliners, driven by the high-speed jet engines, are rated at about
40 kVA.

3.4.   The Siting of Electrical Power Plants

The precise choice of location for an electric power station or power plant involves
several technical and non-technical considerations [2]:

   (i)   fuel supply,
   (ii) water supply,
   (iii) land elevation,
   (iv) road and rail access,
   (v)   height of the structure,
   (vi) disposal of waste products,

Fig. 3.3. Four 500-MW turbo-alternators (reproduced by permission of "Energy" by Marshall *et al.*, Cavendish Books Ltd., London, England, 1978).

(vii) proximity to populated areas, and

(viii) environmental implications.

3.4.1. *Fuel supply*

There must be proximity to economic sources of fuel. For coal-fired stations there must be rail access. For oil-fired stations the site must be within pipeline range of a refinery or have access to a deep-water anchorage. For nuclear stations the source and distance of the fuel supplies is not a critical issue.

3.4.2. *Water supply*

There must be an adequate and secure supply of cold water to cool the condensed (liquefied) steam of thermal electrical power plants. This is illustrated in Fig. 1.5 by the closed water pipe that transfers water to and from the river to the turbine-generator house. A 2000-MW plant requires up to 60 cubic metres/second (m^3/s). The piping of an adequate water supply over large distances would be prohibitively expensive, so power plants are often built on the sea coast or alongside rivers or lakes. When the local water supply has an insufficient flow rate, cooling towers must be used. These tend to be large, unsightly, concrete and steel structures (Fig. 3.4), with their effluent steam being visible for several miles around. At a 2000-MW station the cooling towers evaporate water at the rate of 50,000 cubic metres per day. This is about one hundredth of the natural water content of the ambient air passing over on an average day. The cooling towers themselves require, during their

Fig. 3.4.   Cooling towers.

periods of operation, about $2\,\text{m}^3/\text{s}$ of water to make good the evaporation losses and to prevent the build-up of dissolved impurities in the cooling water.

3.4.3.   *Land elevation*

A power station requires a fairly level area of ground of the range 120–370 acres. It must be sited above the flood-water level but not so far above that the necessary pumping of cooling water becomes too expensive. Power plant buildings and equipment are large and heavy, requiring ground areas that have solid foundations of rock.

3.4.4.   *Road and rail access*

The construction phase of a power plant requires a heavy flow of site traffic, which in turn requires access to the existing road network. Rail access is necessary to transport the fuel for coal-fired stations.

3.4.5. *Height of the structures*

Chimneys and cooling towers are tall structures, sometimes being more than 100 metres high, to elevate the plumes of smoke and steam. The heights of the structures must comply with local environmental planning rules and also with airport safety regulations if the site is in the vicinity of an airport.

3.4.6. *Disposal of waste products*

For coal-fired stations it is necessary to have nearby facilities for dumping large quantities of pulverized fuel ash. This can subsequently be grassed and landscaped. The disposal of waste from nuclear stations is a special case that is dealt with in Chapter 8.

3.4.7. *Proximity to populated areas*

In the UK it is necessary to site nuclear stations remotely from large population areas for legal reasons. A correspondingly remote location of fossil-fuel burning stations is quite impossible in a small and densely populated country like Britain.

3.4.8. *Environmental implications*

The environmental implications of burning coal, oil, natural gas and nuclear fuels to generate electricity are discussed in Chapters 4, 5, 6 and 8, respectively. For example, electricity generation accounts for about one-third of total carbon dioxide emissions in Western Europe and North America. Carbon dioxide is believed to be the main cause of global warming due to human activity [3].

3.5. World Electricity Consumption

A listing of countries, in order, representing total electricity consumption in MWh/year is given in Table 3.1 [4]. The two world leaders, the USA and China, now consume roughly equal electricity demand. The right-hand column of the table gives information of the electricity consumption/capita. This is represented in bar chart form in Fig. 3.5, for the leading consumers. The leading countries of Fig. 3.5, being Norway, Finland, Canada, Sweden and Kuwait, have small populations compared with the industrial countries of Western Europe.

Coal is, by far, the largest prime fuel source for electricity generation, and this is expected to continue until the year 2035, as shown in Fig. 3.6 [5]. The proportional share of prime fuels and hydroelectricity is expected to remain about the same over the period but the total amount of each fuel used will increase as the demand for electrical power increases. In the 28-year period from 2007 to 2035, the total world

Table 3.1.   World electricity consumption [4].

Rank	Country	Electricity consumption (MW·h/yr)	Year of Data	Source	Population	As of	Electricity consumption per Capita MWh/yr/ person
	World	17,109,665,000	2007	EIA	6,464,750,000	2005	2.65
1	United States	3,872,598,000	2008	EIA	298,213,000	2005	13
	European Union	2,950,297,000	2007	EIA	459,387,000	2005	6.42
2	China	3,650,600,000	2009	WSJ	1,315,844,000	2009	2.77
3	Japan	1,007,067,000	2007	EIA	128,085,000	2005	7.86
4	Russia	840,380,000	2007	EIA	141,927,297	2010	5.92
5	Brazil	600,029,000	2009	EIA	186,405,000	2005	3.22
6	India	568,000,000	2007	EIA	1,103,371,000	2005	0.515
7	Germany	547,326,000	2007	EIA	82,329,758	2009 (CIA Est.)	6.65
8	Canada	536,054,000	2007	EIA	32,268,000	2005	16.6
9	France	447,223,000	2007	EIA	60,496,000	2005	7.39
10	South Korea	386,169,000	2007	EIA	47,817,000	2005	8.08
11	United Kingdom	345,798,000	2007	EIA	59,668,000	2005	5.8
12	Italy	307,100,000	2005	CIA	58,093,000	2005	5.29
13	Spain	243,000,000	2005	CIA	43,064,000	2005	5.64
14	South Africa	241,400,000	2007	CIA	47,432,000	2005	5.09
15	Taiwan (Republic of China)	221,000,000	2006	CIA	22,894,384	2005	9.65
16	Australia	219,800,000	2005	CIA	20,155,000	2005	10.9
17	Mexico	183,300,000	2005	CIA	107,029,000	2005	1.71
18	Ukraine	181,900,000	2006	CIA	46,481,000	2005	3.91
19	Saudi Arabia	146,900,000	2005	CIA	24,573,000	2005	5.98
20	Iran	136,200,000	2005	CIA	69,515,000	2005	1.96
21	Sweden	134,100,000	2005	CIA	9,041,000	2005	14.8
22	Turkey	129,000,000	2005	CIA	73,193,000	2005	1.76
23	Poland	140,400,000	2005	CIA	38,530,000	2005	3.12
24	Thailand	117,700,000	2005	CIA	64,233,000	2005	1.83
25	Norway	113,900,000	2005	CIA	4,620,000	2005	24.7
26	Netherlands	108,200,000	2005	CIA	16,299,000	2005	6.64
27	Indonesia	108,000,000	2006	CIA Est.	222,781,000	2005	0.485
28	Argentina	88,980,000	2005	CIA	38,747,000	2005	2.3
29	Finland	88,270,000	2007	CIA Est.	5,249,000	2005	16.8
30	Egypt	84,490,000	2005	CIA	74,033,000	2005	1.14
31	Belgium	82,990,000	2005	CIA	10,419,000	2005	7.97
32	Malaysia	78,720,000	2005	CIA	25,347,000	2005	3.11
33	Kazakhstan	76,430,000	2007	CIA	14,825,000	2005	5.16
34	Venezuela	73,360,000	2005	CIA	26,749,000	2005	2.74
35	Pakistan	67,060,000	2005	CIA	157,935,000	2005	0.425

(*Continued*)

Table 3.1. (*Continued*)

Rank	Country	Electricity consumption (MW·h/yr)	Year of Data	Source	Population	As of	Electricity consumption per Capita MWh/yr/ person
36	Austria	60,250,000	2005	CIA Est.	8,189,000	2005	7.36
37	Czech Republic	59,720,000	2005	CIA	10,220,000	2005	5.84
38	Romania	58,490,000	2007	CIA	21,711,000	2005	2.69
39	Switzerland	58,260,000	2005	CIA	7,252,000	2005	8.03
40	Greece	54,310,000	2005	CIA	11,120,000	2005	4.88
41	United Arab Emirates	52,620,000	2005	CIA	4,496,000	2005	11.7
42	Vietnam	51,350,000	2005	CIA	84,238,000	2005	6.1
43	Portugal	48,550,000	2006	CIA	10,495,000	2005	4.63
44	Chile	48,310,000	2005	CIA	16,295,000	2005	2.96
45	Uzbekistan	47,000,000	2006	CIA Est.	26,593,000	2005	1.77
46	Philippines	46,860,000	2005	CIA	83,054,000	2005	0.564
47	Israel	43,280,000	2005	CIA	6,725,000	2005	6.44
	Hong Kong	40,300,000	2006	CIA	7,041,000	2005	5.72
48	Colombia	38,910,000	2005	CIA	45,600,000	2005	4.56
49	Bulgaria	37,400,000	2006	CIA	7,726,000	2005	4.84
50	New Zealand	37,390,000	2006	CIA Est.	4,028,000	2005	9.28
51	Kuwait	36,280,000	2005	CIA	2,687,000	2005	13.5
52	Hungary	35,980,000	2005	CIA	10,098,000	2005	3.56
53	Singapore	35,920,000	2006	CIA	4,326,000	2005	8.3
54	Iraq	35,840,000	2007	CIA Est.	28,807,000	2005	1.24
55	Denmark	34,020,000	2005	CIA	5,431,000	2005	6.26

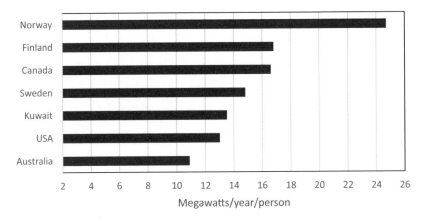

Fig. 3.5. Electricity consumption/capita.

world electricity generation
trillion kilowatt hours

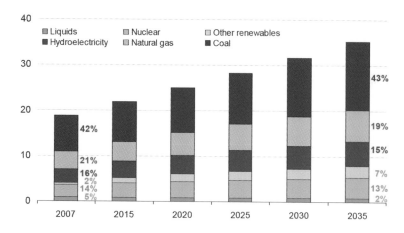

Fig. 3.6.   World electricity consumption, by fuel, up to 2035 [5].

demand is expected to increase from 18 to 35 million kilowatt hours, which is almost a doubling.

3.6.   UK Electricity

3.6.1.   *Organization in 2010*

The UK has a privatized electricity sector, where generators and distributors trade electricity on a wholesale market. The largest power producer in the country is Electricité de France (EDF) Energy, which controls most of the nuclear power capacity and generates one-sixth of the total electricity supply. Other important generating companies include E.ON UK, RWE-npower, Scottish and Southern Energy (SSE) and Scottish Power (SP). National Grid owns and operates the national transmission system in England and Wales, whereas SSE and SP operate the grid in Scotland and Northern Ireland Electricity (NIE) operates the grid in Northern Ireland [6].

3.6.2.   *Consumption and supply*

In the UK the daily variation of electricity demand retains the same form year-by-year. Typical characteristics are given in Fig. 3.7. The weekday load greatly exceeds the Sunday load because of the industrial demand. The nature of the daily variation, shown for 1993, is still valid in midwinter 2013, except that the maximum demand has now risen to 60 GW and the average daily demand is about 48 GW. This represents an increase of about 37% in 17 years, or more than 2% per year. The total generation capacity is about 85 GW.

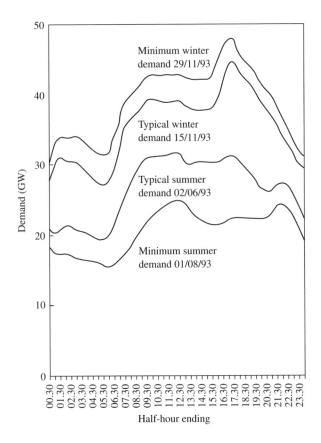

Fig. 3.7.   UK daily electricity demand, 1993.

On a typical day the big load demand starts between 7 am and 9 am and lasts until 5 pm to 6 pm. A typical winter load is about 30% higher than the corresponding summer load (Fig. 3.7). It is a feature of electricity supply systems that they must be capable of supplying the maximum demand even if, for a lot of the time, the expensive generation equipment is underused or is completely idle.

The mix of fuels used to generate electricity continues to change (see Fig. 3.8 and Table 3.2). Since 1990, the decline of coal and oil and the rise of gas have been the most distinct features. Gas rose quite markedly from 1.6 TWh to 153.7 TWh in 2004, before falling back to 137.8 TWh in 2006 and then rising again to new peaks in both 2007 and 2008. Gas then fell in 2009 reflecting lower demand for electricity, but retained the same share of supply as in 2008. Nuclear reached its peak in 1998, before bottoming during 2006–2008, as station closures and maintenance outages reduced supply, but it went on to recover again in 2009. Coal made up for the reduced availability of nuclear stations and, as a substitute for high priced gas, was recorded at its highest level of use for a decade in 2006. Between 2006 and 2009 its use fell. Wind power, however, has been on an upward trend since 2000. Electricity

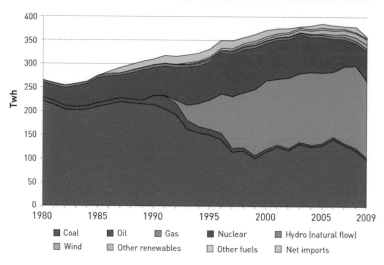

Fig. 3.8.   UK electricity supply, by fuel type, 1980–2009 [6].

Table 3.2.   UK electricity supply by fuel type [6].

	1980	1990	2000	2007	2008	2009
Coal	220.8	213.4	114.7	129.6	119.0	99.3
Oil	8.1	20.0	5.9	3.9	5.0	3.8
Gas	—	0.4	144.9	162.4	173.0	162.5
Nuclear	32.3	58.7	78.3	57.2	47.7	62.8
Hydro^2	3.9	5.2	4.2	3.8	3.8	4.1
Wind	—	—	0.9	5.3	7.1	9.3
Other fuels	—	—	8.3	11.8	11.5	12.6
Net Imports	—	11.9	14.2	5.2	11.0	2.9
Total electricity available for supply	**264.9**	**309.4**	**371.4**	**379.2**	**378.1**	**357.2**

available for supply rose continuously from 1997 to reach a peak in 2005. It has subsequently fallen every year, and in 2009 was 357.2 TWh [6].

In 1999 natural gas supplied more electricity than coal for the first time. Coal's share of the market was $99.3/357.2 = 27.8\%$ in 2009 whereas natural gas had grown to $162.5/357.2 = 45.5\%$. Nuclear's share peaked at 26% in 1997 but in 2009 fell back to $62.8/357.2$, which is 17.6%.

In the UK the consumption of electricity is shared roughly equally between the industry, domestic and services sectors. Consumption grew by 19% and 29.5%, respectively, in the domestic and services sectors in the ten years to 2000. Industrial consumption also grew to its highest level ever of 115.3 TWh in 2000 but fell back to 98.7 TWh in 2009 [6].

In 2010 the cost of electrical energy to UK domestic consumers was complicated because most suppliers used a two-part tariff. The first (approximately) 120 kWh of electrical energy was charged at (say) 25 p/kWh. Higher amounts of electrical energy were charged at (say) 11 p/kWh. The average tariff paid therefore depends on the total energy used, with the overall average price reducing slightly as more energy is used.

3.6.3.   *Comparative costs of electrical generation in the UK*

A study by PB Power was commissioned by the UK Royal Academy of Engineering in 2005. This study compared the various generation technologies listed below:

- Coal plant:

  - pulverized fuel (PF) steam plant
  - circulating fluidized-bed combustion (CFBC) plant
  - integrated gasification combined-cycle (IGCC) plant

- Gas plant:

  - open-cycle gas turbine (OCGT) plant
  - combined-cycle gas turbine (CCGT) plant

- Nuclear fission plant
- biomass (poultry litter):

  - bubbling fluidized-bed combustion (BFBC) plant

- Wind turbines:

  - onshore
  - offshore

- Wave and marine.

The cost of generating electricity, as defined within the scope of this study, is expressed in terms of a unit cost (pence per kWh) delivered at the boundary of the power station site. This cost value, therefore, includes the capital cost of the generating plant and equipment; the cost of fuel burned (if applicable); and the cost of operating and maintaining the plant in keeping with UK best practices. Within the study, however, the "cost of generating electricity" is deemed to refer to that of providing a dependable (or "firm") supply. For intermittent sources of generation, such as wind, an additional amount has been included for the provision of adequate standby generation. The findings of this study are summarized in Fig. 3.9, which illustrates the present-day costs of generating electricity from different types of technology appropriate to the UK [7]. Corresponding comparative tables for other

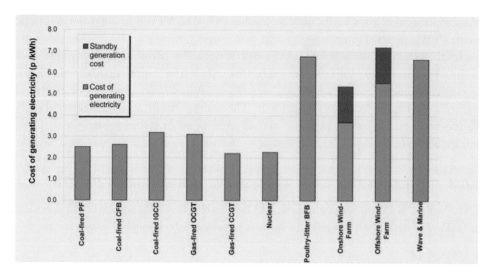

Fig. 3.9. Cost of generating UK electricity (pence per kWh) with no cost of CO_2 emissions included [7].

Table 3.3.   Comparative cost of generating base-load electricity in the UK [7].

Gas-fired CCGT	2.2
Nuclear fission plant	2.3
Coal-fired pulverized-fuel (PF) steam plant	2.5
Coal-fired circulating fluidized bed (CFB) steam plant	2.6
Coal-fired integrated gasification combined cycle (IGCC)	3.2

countries may be significantly different dependent on their fuel resources, subsidies of the different technologies and their attitudes to carbon dioxide emissions.

The UK costs are dependent, to some extent, on whether the plant is designed for base-load operation (i.e. continuous operation) or to provide standby for limited periods of high demand.

For base-load operation the cheapest way to generate electricity in the future from new plants, i.e. ignoring rehabilitation of existing plants, is by constructing CCGT plants designed to burn natural gas.

Table 3.3 summarizes the comparative cost of generating electricity for the different "base-load" plants considered by this study [7].

For peaking operations, i.e. generating for limited periods of high demand and providing standby capacity, OCGT fired on natural gas are the most appropriate new plant candidates. OCGT is ideally suited for the role of peaking duty, which requires flexibility, reliability and can be started quickly should the need arise [7].

Renewables are generally more expensive than conventional generation technologies. This is due to the immaturity of the technology and the more limited opportunity to take advantage of cost savings brought about by economies of scale usually associated with more traditional fossil-fuel types of generation. In addition, fluctuations in the energy source itself may limit the output of generation available from these technologies and, thus, raise the unit costs of the generator on two counts:

1. As capacity factor falls, unit costs of generation rise.
2. Additional, fast response, standby generating plants may have to be provided to maintain system security as the energy source fluctuates.

The "capacity factor" is an operational term to represent the extent to which the generator is producing electricity over a period, e.g. a year. Wind turbines have typical capacity factors of 25–45% whilst large coal or nuclear plants may have capacity factors in excess of 90% when operating on base-load [7].

3.7. US Electricity Consumption and Production

The USA is the world's largest producer and consumer of electricity, followed by China and Japan. In January 2010 the USA had a total electricity generation of 3992 billion kWh while China had 3715 billion kWh. A pie chart showing the US electricity generation fuel shares in 2009 is shown in Fig. 3.10 [8]. Electricity demand is expected to increase annually so that total demand increases by 41% by 2030.

Coal is the fuel that is cheapest with regard to operating costs and is the biggest fuel source of electricity generation. Nuclear plants provided about 20% of demand. The use of natural gas supplied 23.3% of demand in 2009, while hydropower and renewables supplied 11%.

Information concerning the predictions up to 2035, concerning electricity supply and fuel use, is given in Table 3.4 [9]. The section regarding "Total Electricity Generation by Fuel" indicates that natural gas and renewables are expected to show the greatest increases of use, with coal and nuclear increasing more slowly.

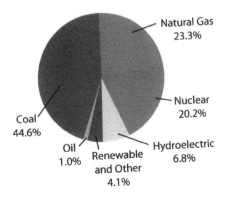

Fig. 3.10.   US electricity generation fuel shares, 2009 [8].

Table 3.4.   US electricity supply and use (billion kWh). Modified from [9].

Supply, disposition, and prices	Reference case							Annual growth 2009–2035 (percent)
	2008	2009	2015	2020	2025	2030	2035	
Generation by Fuel Type Electric Power Sector[1] Power Only[2]								
Coal	1932	1719	1746	1849	1987	2028	2076	0.7%
Petroleum	39	32	37	39	39	40	41	1.0%
Natural Gas[3]	683	722	729	716	701	817	921	0.9%
Nuclear Power	806	799	839	877	877	877	874	0.3%
Pumped Storage/Other[4]	0	2	−0	−0	−0	−0	−0	—
Renewable Sources[5]	347	380	491	521	541	554	569	1.6%
Distributed Generation (Natural Gas)	0	0	1	2	3	4	5	—
Total	**3807**	**3653**	**3843**	**4004**	**4148**	**4319**	**4485**	**0.8%**
Combined Heat and Power[6]								
Coal	37	30	23	26	29	30	31	0.1%
Petroleum	4	4	0	0	0	0	0	−9.1%
Natural Gas	119	119	129	125	119	120	113	−0.2%
Renewable Sources	4	4	3	4	4	3	3	−1.0%
Total	**167**	**161**	**155**	**155**	**153**	**153**	**148**	**−0.3%**
Total Net Generation	**3974**	**3814**	**3998**	**4158**	**4300**	**4472**	**4633**	**0.8%**
Less Direct Use	35	35	33	33	33	33	33	−0.2%
Net Available to the Grid	**3939**	**3779**	**3965**	**4125**	**4267**	**4439**	**4600**	**0.8%**
End-Use Generation[7]								
Coal	19	23	30	32	52	79	111	6.3%
Petroleum	3	5	5	5	5	5	5	−0.3%
Natural Gas	80	90	141	160	180	211	250	4.0%
Other Gaseous Fuels[8]	11	11	15	15	15	15	15	1.4%
Renewable Sources[9]	34	36	63	82	128	147	152	5.7%
Other[10]	2	2	1	1	1	1	1	−1.8%
Total	**149**	**167**	**255**	**295**	**381**	**458**	**533**	**4.6%**
Less Direct Use	120	135	205	230	276	334	392	4.2%
Total Sales to the Grid	**29**	**31**	**49**	**65**	**105**	**124**	**142**	**6.0%**
Total Electricity Generation by Fuel								
Coal	1987	1772	1799	1907	2069	2137	2218	0.9%
Petroleum	46	41	43	44	44	45	46	0.5%
Natural Gas	882	931	1000	1002	1003	1152	1288	1.3%
Nuclear Power	806	799	839	877	877	877	874	0.3%
Renewable Sources[5,9]	385	420	556	608	673	703	724	2.1%
Other[11]	16	18	16	16	16	16	16	−0.3%
Total Electricity Generation	**4123**	**3981**	**4253**	**4453**	**4682**	**4930**	**5167**	**1.0%**
Total Net Generation to the Grid	**3968**	**3810**	**4014**	**4190**	**4372**	**4563**	**4742**	**0.8%**
Net Imports	**33**	**34**	**33**	**27**	**22**	**13**	**14**	**−3.4%**

(*Continued*)

Table 3.4. (*Continued*)

Supply, disposition, and prices	Reference case							Annual growth 2009–2035 (percent)
	2008	2009	2015	2020	2025	2030	2035	
Electricity Sales by Sector								
Residential	1380	1363	1348	1394	1461	1538	1613	0.7%
Commercial	1336	1323	1416	1526	1636	1761	1886	1.4%
Industrial	1009	882	1038	1046	1031	997	962	0.3%
Transportation	7	7	8	10	13	18	22	4.6%
Total	**3732**	**3574**	**3811**	**3976**	**4142**	**4314**	**4483**	**0.9%**
Direct Use	154	170	239	263	309	367	425	3.6%
Total Electricity Use	**3886**	**3745**	**4049**	**4240**	**4451**	**4681**	**4908**	**1.0%**
Electric Power Sector Emissions[1]								
Sulphur Dioxide (million tons)	7.62	5.72	3.77	3.68	4.09	3.97	3.94	−1.4%
Nitrogen Oxide (million tons)	3.01	1.99	1.99	1.98	2.00	2.03	2.05	0.1%
Mercury (tons)	45.27	40.66	26.88	26.82	28.21	29.08	29.91	−1.2%

[1]Includes electricity-only and combined heat and power plants whose primary business is to sell electricity, or electricity and heat, to the public.

[2]Includes plants that only produce electricity.

[3]Includes electricity generation from fuel cells.

[4]Includes non-biogenic municipal waste. The U.S. Energy Information Administration estimates approximately 7 billion kilowatt hours of electricity were generated from a municipal waste stream containing petroleum-derived plastics and other non-renewable sources. See U.S. Energy Information Administration, *Methodology for Allocating Municipal Solid Waste to Biogenic and Non-Biogenic Energy* (Washington. DC, May 2007).

[5]Includes conventional hydroelectric, geothermal, wood, wood waste, biogenic municipal waste, landfill gas, other biomass, solar, and wind power.

[6]Includes combined heat and power plants whose primary business is to sell electricity and heat to the public (i.e., those that report North American Industry Classification System code 22).

[7]Includes combined heat and power plants and electricity-only plants in the commercial and industrial sectors; and small on-site generating systems in the residential, commercial, and industrial sectors used primarily for own-use generation, but which may also sell some power to the grid.

[8]Includes refinery gas and still gas.

[9]Includes conventional hydroelectric, geothermal, wood, wood waste, all municipal waste, landfill gas, other biomass, solar, and wind power.

[10]Includes batteries, chemicals, hydrogen, pitch, purchased steam, sulphur, and miscellaneous technologies.

[11]Includes pumped storage, non-biogenic municipal waste, refinery gas, still gas, batteries, chemicals, hydrogen, pitch, purchased steam, sulphur, and miscellaneous technologies.

− = Not applicable.

Note: Totals may not equal sum of components due to independent rounding. Data for 2008 and 2009 are model results and may differ slightly from official EIA data reports.

Sources: 2008 and 2009 electric power sector generation; sales to utilities; net imports; electricity sales; electricity end-use prices; and emissions: U.S. Energy Information Administration (EIA), *Annual Energy Review 2009*, DOE/EIA-0384 (2009) (Washington, DC, August 2010), and supporting databases. 2008 and 2009 prices: EIA, AEO2011 National Energy Modeling System run REF2011.D020911A. **Projections**: EIA, AEO2011 National Energy Modeling System run REF2011.D020911A.

Total electricity generation is anticipated to increase by 25% at the rate of 1% per year.

The "Electricity Sales by Sector" section of Table 3.4 shows that the residential and commercial sectors were about equal consumers in 2009, both being about 50% greater than the industrial sector. This trend is expected to continue into the next 25 years of the forecast [9].

The price of domestic electricity in the USA was about one-half the cost in the UK during 2010.

3.8.   Combined Heat and Power (CHP)

A combined heat and power plant (CHP) is an installation in which there is simultaneous generation of usable heat and power (usually electricity) in a single process. High temperature heat from fuel combustion is used to generate electricity while lower temperature exhaust heat can be used for industrial process heat, district heating and space heating.

In conventional electricity generation, by the use of steam turbines, only 30–35% of the energy available in the fuel is converted to electricity. Even the most modern combined cycle plants achieve efficiencies of only 50–60%. Most of the energy that is wasted in this conversion process is released to the environment as waste heat; power station cooling towers are a very recognizable sight. Each year, UK power stations typically reject more energy as waste heat than is consumed by the entire domestic sector.

The principle of CHP, also known as co-generation, is to recover and make beneficial use of this heat, significantly raising the overall efficiency of the conversion process. The very best CHP schemes can achieve fuel conversion efficiencies of the order of 90%. Most of the heat wasted in electricity generation is carbon-based and so if CHP could be more widely deployed there are potentially significant energy, environmental and economic benefits that could be realized [10].

3.8.1.   *Technologies used in current UK CHP schemes* [10]

The CHP principle can be realized in many ways, using a range of fuels and prime movers and implemented in capacities from 1 kWe to hundreds of MW$_e$. CHP schemes operate for decades and so current UK CHP capacity comprises schemes based on old and new forms of CHP technology.

The main technologies used in current UK CHP schemes are:

- Gas Engines: reciprocating engines, similar to those used in road vehicles, using natural gas as fuel, driving a generator and recovering heat from the engine exhaust and cooling jacket. Sizes range from 10 kWe to several MW$_e$. This form of CHP is widely used in buildings, where the use of low-grade heat in space and water heating systems enables a high level of waste recovery.

- Gas Turbines: derived from aerospace and industrial designs, these produce exhaust gas at high temperatures, which is ideal for generating steam or direct use in drying processes. Sizes range from below 100 kWe (e.g. microturbines) to over 100 MW$_e$. This form of CHP is commonly found in industry, e.g. the process industries such as the chemicals or paper industries, which have high demands for steam.
- Steam Turbines: this technology was the mainstay of industrial CHP capacity before the introduction of gas turbines and is based on the designs found in coal fired power stations. Sizes range from around 100 kWe to many MW$_e$. As the heat output is in the form of steam the main applications are also in the process industries.
- Combined Cycle Gas Turbines: these combine a gas turbine feeding a heat recovery steam generator (HRSG), which in turn drives a steam turbine. Sizes range from 20 MW$_e$ to many hundreds of MW$_e$. This system offers a high power efficiency and is also normally found in the process industries.
- Absorption Chilling: this technology uses heat (as hot water or steam) to produce chilled water or chilled glycol. In combination with one of the other CHP technologies, this can be used to provide power, heat and chilling. Such "tri-generation" systems are found in buildings and on some industrial sites.

Whilst the most common fuel since 1990 has been natural gas, the range of fuels that can be used by these technologies is also wide and includes:

- natural gas;
- landfill and sewage gas;
- fuel and gas oils;
- coal, lignite and coke;
- biomass and biogas;
- solid waste, e.g. refuse, tyres;
- waste gases, e.g. refinery off gas; and
- waste process heat.

Schematic diagrams of two forms of CHP systems are shown in Fig. 3.11.

3.8.2.   *CHP in the UK*

Some statistics of CHP plant ratings in the UK are shown in Table 3.5 [6]. Most plants use fossil fuels but the use of biofuels such as sewage gases and municipal waste incinerator gases is increasing. Each of the five forms of CHP plants described above contribute significantly to overall efficiency improvement and energy saving. The main feature of the larger scale CHP market in the past ten years has been the increasing use of gas turbines. For steam-turbine-based plants the major fuel is coal. It is seen from Table 3.5 that in 2009 the UK had 1,465 CHP plants with a

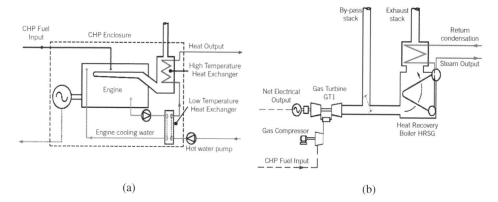

(a)   (b)

Fig. 3.11.   Examples of CHP systems [10]. (a) Gas engine combined with single grade hot water output. (b) Gas turbine with heat recovery steam generator.

Table 3.5.   CHP plants in the UK [6].

	1995	2000	2007	2008	2009
CHP electrical capacity [MWe]	3354	4451	5438	5494	5569
CHP electrical generation [GWh]	14778	25246	27846	27901	27777
CHPheat generation [GWh]	56,833	54,877	51,314	52,778	50,721
Number of CHP sites					
Less than 100 kWe	617	556	456	457	454
100 kWe to 999 kWe	396	532	687	706	738
1 MWe to 9.9 MWe	139	182	203	202	201
10 MWe and greater	68	70	70	72	72
Total	**1,220**	**1,340**	**1,415**	**1,437**	**1,465**

total generating capacity of 5569 MW$_e$ (i.e. megawatts electrical). This represented about 6.5% of the total capacity.

CHP electrical capacity and generation has remained broadly unchanged over the last three years. Despite the adverse economic climate, electricity generation in 2009 was only slightly lower than in 2008 (0.4%). A bigger fall was recorded for CHP heat generation, which was 3.9% lower in 2009 than 2008. A third of the CHP installations in the UK are small schemes with an electrical capacity of less than 100 kWe, however schemes larger than 10 MW$_e$ account for 84% of the total CHP installed electrical capacity. In 2009, just over 7% of the total electricity generated in the UK came from CHP plants [6].

The realizable efficiency of the CHP option in terms of primary energy input to electricity/heat output is often claimed to be 70% or more; double the value from a conventional coal-fired plant. In addition to the more efficient use of fuels, CHP systems produce fewer air pollutants and lower thermal discharges than equivalent single-purpose systems. Initial cost is a disincentive to their introduction, but as fuel prices rise the savings due to CHP increasingly offset the first costs.

Many district heating schemes, where the heat is centrally generated and piped to a distribution of local users, are combined with CHP schemes in the UK [11, 12].

3.8.3. *CHP in the USA* [13]

The first commercial electricity power plant in the USA — Thomas Edison's Pearl Street station in Manhattan, New York, in 1881 — was a CHP facility. Many of the early US electricity generation schemes were industrial facilities that added generators to existing steam systems. Gradually the use of CHP facilities declined as the use of large power plants developed at sites remote from their customers.

In the year 2009 the total electric power capacity in the USA was 1121 GW with about 8% of this due to CHP facilities. A comparison of the USA's position with some other industrial countries is shown in Fig. 3.12. It is seen that the greatest proportion of CHP is undertaken in Denmark, the Netherlands and Germany.

3.9. Efficient Utilization of Electrical Energy

It is in the interest of everyone (except, perhaps the vendors of prime fuels) that electricity should be used efficiently. This accomplishes the twin objectives of saving money and conserving prime energy.

Among the many features that might be addressed with regard to efficient utilization are [14]:

(a) avoiding waste;
(b) monitoring and control;
(c) redesigning to reduce energy costs;
(d) maintenance of equipment;
(e) power factor correction;

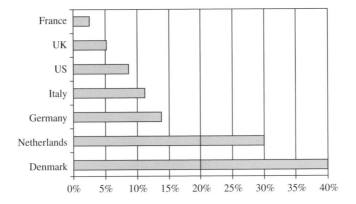

Fig. 3.12. CHP electrical power production [13] (% of total generation).

(f) maintenance of supply current waveform;
(g) choice and use of electric motors;
(h) load factor; and
 (i) choice of lighting systems.

3.9.1. *Avoiding waste*

It is a waste of fuel and money to leave electric lights or machines switched on unnecessarily, to overheat buildings, to have too much ventilation, etc.

3.9.2. *Monitoring and control*

Continuous human or automatic monitoring of an environment will eliminate the input of excess energy and reduce the need for emergency intervention.

3.9.3. *Redesigning to reduce energy costs*

It is good engineering practice to consider the energy costs of a machine or process. When different options are available it may become economical to replace an energy-inefficient machine with a more costly but efficient one.

3.9.4. *Maintenance of equipment*

Carrying out regular, planned routine maintenance is essential to efficient operation. Records of the cycles, performances and replacements of components need to be known.

3.9.5. *Power factor correction*

The power factor of a load usually indicates the ratio of power to current. For economic operation this ratio should be as near as possible to its maximum realizable per-unit value of unity. Continuous operation with a poor (i.e. low) power factor is extremely wasteful and expensive. Small, fully loaded AC induction motors can have power factors below 0.6 and, at low torque load, the power factor can drop to 0.1, especially for low speeds of operation. Power factor correction can be obtained by connecting suitable capacitors in parallel with the motor load. If the compensating capacitors are connected across the motor terminals, the wiring all the way back to the generating station is relieved. The reduction of supply currents causes reduced voltage (IR) drop and reduced resistance loss (I^2R) in the line conductors. Also, the reduction of maximum demand of voltamperes may be an economic advantage to the customer in some electrical utility tariff structures. Power factor correction equipment is expensive and careful calculation is required to determine if it would be an economical investment.

3.9.6. *Maintenance of supply current waveform*

Many modern electrical AC loads cause deformation from the ideal sinusoidal current waveform. Large industrial loads such as arc furnaces and arc welders draw distorted current but so do television sets and personal computers. The combined effects of very many small distorting appliances can be worse than the effects of one very large industrial process. Distortion of the current waveform causes wastage of electrical energy and other serious problems [15].

3.9.7. *Choice and use of electric motors*

The highest efficiency is achieved when an electric motor is the right size and operates at its rated load and speed. A motor purchase price varies directly with its rating (and size) so that oversizing is uneconomical in first cost and in running costs. Operation of a motor at reduced speed or reduced torque load is accompanied by loss of efficiency.

3.9.8. *Load factor*

Load factor is concerned with variations of electrical load on the supply network, usually on an hourly, daily or weekly basis. It can be defined as the ratio of average demand to maximum demand. In all types of operation high load factors are economically desirable, which implies steady and predictable load demand.

3.9.9. *Choice of lighting systems*

Different forms of electric light bulbs of the same rating give different amounts and qualities of illumination. The effectiveness of a lamp is sometimes called its efficacy and is a measure of the illumination in lumens per watt of electrical power input. Also relevant is the life expectancy of a typical lamp. Some typical values are given in Table 3.6 [16]. An installation of electric lighting is typical of engineering systems in that careful calculation has to be made of the initial costs, running costs, anticipated lifetime and replacement costs. A judgement of the overall quality of a

Table 3.6.   Properties of types of electric lamps [16].

Type	Illuminance (lm)	Power (W)	Efficacy (lm/W)	Life expectancy (h)
Tungsten GLS	1200	100	12	1000
Tungsten halogen	50000	2000	25	4000
Compact fluorescent	1200	28	43	8000
Fluorescent tube	4500	70	64	10000
High-frequency tube	5000	62	80	10000
High-pressure sodium	25000	280	90	12000

lighting system is not only technical but psychological. The colour and shadowing effects are important factors in its acceptability.

Many countries have regulations specifying the minimum illumination, in lumens/unit area, to be used for various grades of accommodation, such as factories, offices and classrooms. Appropriate figures for (as an example) the USA are given in the *Lighting Handbook* [17].

Example 3.4.   A factory building of area $6000\,\mathrm{m}^2$ is lighted by 200 twin-tube fluorescent fixtures. Each tube is rated at 80 W, has an efficacy of 60 lumens/watt and a life expectancy of 8,000 hours. Each fluorescent tube costs £4. It is intended that the lighting system be replaced to reduce running costs. The proposed new system would use 40 high-pressure sodium lamps rated at 550 W, efficacy 100 lumens/watt, with a life expectancy of 10,000 hours. Each sodium lamp costs £5. The factory is lighted for 12 hours/day, 5 days/week, 50 weeks/year. The labour charge for the replacement of a tube or lamp at any fixture is £5. The relevant electricity tariff is 6.3 p/unit (i.e. 6.3 p/kWh).

(a) Calculate the annual electricity bill for the two systems and the total cost saving due to changing to the sodium system.
(b) If the installation of the sodium lighting system would cost £3,000, evaluate the break-even time. Neglect the effects of inflation and depreciation and write off any scrap value of the fluorescent system.
(c) Calculate the percentage increase in illumination in the factory.

The lights are in use for $12 \times 5 \times 50 = 3,000$ hours/year.

(a) For ease of comparison the various stages of the solution are evaluated in parallel vertical columns.

	Flourescent system	Sodium system
Units of electricity used/year	$200 \times 2 \times 80 \times 3000$ $= 96000000$ Wh $= 96000$ kWh	$40 \times 550 \times 3000$ $= 66000000$ Wh $= 66000$ kWh
Annual electricity bill	$96000 \times 6.3/100$ $= £6048$	$66000 \times 6.3/100$ $= £4158$
Lamp/tube replacement period = life expectancy annual "on" time	$8000/3000 = 2\frac{2}{3}$ years	$10000/3000 = 3\frac{1}{3}$ years
Labor cost of lamp/tube replacement	$\dfrac{200 \times 5}{2\frac{2}{3}} = £375/\text{year}$	$\dfrac{40 \times 5}{3\frac{1}{3}} = £\,60/\text{year}$
Equipment replacement costs/year	$\dfrac{200 \times 2 \times 4}{2\frac{2}{3}} = £600$	$\dfrac{40 \times 25}{3\frac{1}{3}} = £300$

The annual cost saving due to changing to sodium lights:

electricity bill	$6048 - 4158 = 1890$
labour charge	$375 - 60 = 315$
equipment replacement	$600 - 300 = 300$
Total Saving	$= £2505$

(b) The capital cost of installing a new system is £3,000, so that the payback for this investment is $\frac{3000}{2505} = 1.2$ years.

Such a short payback period represents a terrific bargain in commercial terms!

(c)

	Flourescent	Sodium
Level of illumination per unit area		
$= \dfrac{\text{lamps} \times \text{rating} \times \text{efficacy}}{\text{area}}$	$\dfrac{200 \times 2 \times 80 \times 60}{6000}$	$\dfrac{40 \times 550 \times 100}{6000}$
	$= 320 \text{ Im/m}^2$	$= 366.7 \text{ Im/m}^2$

The proposed change to sodium lighting therefore represents energy saving, annual cost saving, 14.6% increased illumination, and would repay the investment cost in 1.2 years.

3.10.   Problems and Review Questions

Problems on electrical circuits and systems

3.1. A 200-V DC supply is applied to a resistor of value 1000 ohms. Calculate the current and power dissipation.

3.2. An electrical supply of 240 V is applied to a resistor, resulting in a current of 12 A. Calculate the value of the resistor and the power dissipation.

3.3. When an electrical supply of 240 V, 50 Hz was applied to an electrical load the ammeter and wattmeter gave readings of 10 A and 750 W respectively. Calculate the circuit impedance and the power factor.

3.4. A DC electric motor rated at 2 HP is connected to a 200-V DC supply. If the input current at full load is 10 A, what is the motor efficiency?

3.5. A 120-V, 60-Hz electrical supply is applied to a single-phase AC motor rated at 2 HP. If the full-load current is 20.7 A and power input is 2000 W, calculate the motor efficiency and the power factor.

3.6. If the power factor for Problem 3.5 is improved to unity by the connection of parallel capacitors, what will be the new value of the current?

3.7. Why are AC generators and power transmission systems usually three-phase in nature?

3.8. Why do long-distance electrical transmission lines usually operate at high voltages?

3.9. Briefly discuss some of the main considerations in choosing a suitable location and site for an electrical power generation plant.

3.10. Which countries of the world consume the most electricity at present?

3.11. List the countries of regions that have the highest consumption of electricity per capita. Why does Canada have such a high per capita consumption?

3.12. Compare the world prime fuels used for electricity generation in 2007 with the projected figures for 2035. What are the trends in prime fuel use?

3.13. Compare the proportions of prime fuels used for electricity generation in the UK between 1980 and 2009. What are the present trends?

3.14. Compare the proportions of prime fuels used for electricity generation in the USA between 2008 and 2035.

3.15. What proportion of the total electricity consumption in the UK is now attributable to imported electricity? Where is the electricity imported from?

3.16. What proportion of the total electricity generation in the UK was attributable to combined heat and power schemes (a) in 2000 (b) in 2009?

3.17. What proportion of the total electricity generation in the USA was contributed by CHP schemes in the past year?

3.18. Explain, using a diagram, the principle of combined heat and power schemes. What improvement of overall efficiency can be obtained compared with a conventional coal-fired electricity generation plant?

3.19. Why is operating an unnecessarily oversized electric motor uneconomical?

3.20. The data of Table 3.6 suggests that high-pressure sodium lighting has the highest efficacy and illuminance. Does it have any disadvantages?

3.21. A classroom of area $150 \, \text{m}^2$ is illuminated by 20 standard incandescent lamps rated at 100 W with an illuminance of 1200 lumens and an efficacy of 12 lm/W. Each lamp costs 50p and is switched on for 8 hours/day, 5 days/week, 40 weeks/year.

   (a) If electricity costs 7 p/kWh, calculate the annual running cost.
   (b) If the life expectancy of a lamp is 1,000 hours and the replacement labour cost is £5 per item, calculate the annual replacement cost.
   (c) Calculate the total annual electricity costs.

3.22. An alternative plan to light the classroom of Problem 3.21 is to use 30 fluorescent tubes, rated at 70 W with an illuminance of 4500 lumens and an efficacy of 64 lumens/watt. Each fluorescent tube costs £4 and has a life expectancy of 10,000 hours.

   (a) Calculate the annual running costs if the electricity tariff is 7 p/kWh.
   (b) Calculate the annual replacement cost if the replacement labour charge remains at £5 per item.
   (c) Calculate the total annual electricity costs.

3.23. Compare the performance of the incandescent lamp system of Problem 3.21 with the corresponding fluorescent lamp system of Problem 3.22.

(a) What would be the annual cost saving in moving to the fluorescent system?

(b) The cost of the modified installation would be £660. Neglecting the effects of inflation and depreciation and ignoring any scrap value of the incandescent system, what would be the payback period of the modification?

(c) Compare the illuminance of the classroom using the two systems.

References

[1] "Tomorrow's Energy Today", Office of Energy Efficiency and Renewable Energy, US Department of Energy, Washington, DC, November 1993.

[2] Clarke, A.J. (ed.) (1980). *Electricity Supply and the Environment*, Central Electricity Generating Board, London.

[3] Sweet, W. (1996). Power and Energy, *IEEE Spectrum*, **33**, 70–75.

[4] "Electricity Consumption", The World Factbook, Central Intelligence Agency, Washington, DC, 2010.

[5] Gruenspecht, H. (2010). Presentation by Howard Gruenspecht, Centre for Strategic and International Studies, US Energy Information Administration (EIA), Washington.

[6] "UK Energy in Brief 2010", Dept. of Energy and Climate Change, London, July 2010.

[7] "The Cost of Generating Electricity", Commentary on a study carried out for The Royal Academy of Engineering, London, 2005.

[8] "US Electricity Generation Fuel Shares 2009", US Energy Information Administration, Washington, DC, http://www.nei.org/resourcesandstats/documentlibrary/reliableandaffordableenergy/graphicsandcharts/uselectricitygenerationfuelshares/.

[9] "Electricity Supply Disposition, Prices and Emissions", Annual Energy Outlook 2011, Energy Information Administration, Washington, DC, DOE/EIA-0383, April 2011, http://www.electricdrive.org/index.php?ht=a/GetDocumentAction/id/27843.

[10] "Combined Heat and Power (CHP)" (2008). Factfile, The Institution of Engineering and Technology, London.

[11] "Digest of UK Energy Statistics 2000", Annex C — Combined Heat and Power, Department of Trade and Industry, London, 2000.

[12] Woods, P.S and Zdaniuk, G. (2011). "CHP and District heading — How Efficient are these Technologies?" CIBSE Technical Symposium, De Montfort University, Leicester, UK.

[13] "Combined Heat and Power (CHP)", USCHP Association, CHP Vision Workshop, Washington, DC, September 1999.

[14] "Economic Use of Electricity", Fuel Efficiency, No. 9, Energy Efficiency Office, Dept. of Energy, London, 1991.

[15] Shepherd, W., Hulley, L.N. and Liang, D.T.W. (1995). *Power Electronics and Motor Control*, 2nd Edition, Cambridge University Press, Cambridge.

[16] Personal communication to the authors from Dr M.J. West, Coventry University, 1995.

[17] Rea, M.S. (ed.) (1993). *Lighting Handbook*, Illuminating Engineering Society of North America, New York.

CHAPTER 4

COAL

4.1. The History and Status of Coal

4.1.1. *Fossil fuel resources*

The dominance of coal in the world stock of fossil fuel reserves is illustrated in Fig. 4.1, for most areas of the world. This is discussed in detail in various sections of this chapter.

It should be noted that Fig. 4.1 does not include figures for various unconventional fossil fuels that are now increasing in importance and also in market penetration. Unconventional fossil fuels are either liquid or gaseous and are discussed in Chapter 5, on oil, and in Chapter 6, on natural gas.

4.1.2. *Composition and ranking of coal*

Coal is a hard, combustible, sedimentary rock. In spite of centuries of use its chemical nature is still not fully understood. It is a very complex and varied substance with a quality determined by two classes of material: (a) the organic remains of plants solidified by the combined action of heat and pressure; and (b) inorganic substances contributed by plants, water seepage and surrounding geological mineral matter. The organic content consists mostly of carbon plus smaller amounts of nitrogen, hydrogen and oxygen. The inorganic or mineral content, which is of the order 9–30% of the coal by weight, contains sulphur plus small but potentially toxic amounts of antimony, arsenic, beryllium, cadmium, mercury, lead, selenium, zinc, heavy radionuclides and asbestos [1].

When coal is heated many products arise, some of which are useful but some are potentially hazardous. Gases such as carbon monoxide, carbon dioxide, methane and water vapour are formed as the hydrogen and oxygen are driven off. The solid combustible residue remaining after the gases are burned off is called "fixed carbon" [2].

During coal formation as the materials undergo evolutionary change from dead vegetation to coal, they pass through various stages of "coalification". This becomes

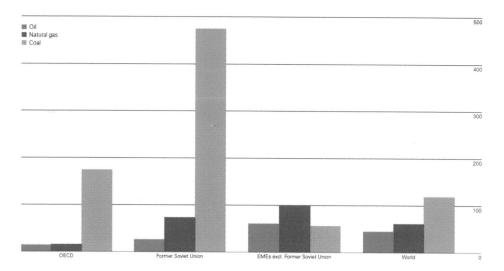

Fig. 4.1.   World conventional fossil fuel reserves [5].

the basis on which various categories of coal can be ranked, according to their calorific (i.e. heating) value and carbon–hydrogen ratios. Table 4.1 gives a ranking from the oldest formation, anthracite, with its low volatility and high carbon content, to the youngest formation, lignites [2]. Anthracite is the slowest- and cleanest-burning coal. Low volatile bituminous coal creates less ash than anthracite and is the best coal for making coke, which is a solid fuel that remains when coal is heated to a high temperature out of contact with air, as in the manufacture of coal gas. Coke is widely used in the manufacture of iron and steel in coke-oven plants.

It can be seen in Table 4.1 that there can be a factor of 2.1 in calorific value between the best and worst of the fuel coals. Anthracite is classed as a "hard" coal, whereas lignites are classed as a "soft" coal.

4.1.3.   *Coal mining*

Coal is extracted from the ground by (a) deep mining or (b) surface (open-cast) mining. In the deep mining process vertical shafts are sunk to the level of the coal

Table 4.1.   Classification of coal by rank [2].

Class	Carbon %	Hydrogen %	Fixed carbon %	Calorific value MJ kg^{-1}
Anthracite	95–98	2.9–3.8	91–95	>32.5
Low volatile bituminous	91–92	4.2–4.6	80–85	>32.5
Medium volatile bituminous	87–92	4.6–5.2	70–80	>32.5
High volatile bituminous	82.5–87	5.0–5.6	60–70	26.7–32.5
Sub-bituminous	78–82.5	5.2–5.6	55–60	19.3–26.7
Lignites	73–78	5.2–5.6	50–55	<19.3

Fig. 4.2.   Classical image of winding gear at the head of a deep mine [3].

seams, sometimes more than 4,000 feet deep, and horizontal bores are then made along the seams. The shaft of an older deep mine is identifiable by the surface winding-gear structure (Fig. 4.2) [3]. It is necessary to support the roofs of the diggings by pit-props to prevent collapse. The extracted rock and coal is transported by conveyer belts to the shafts and then taken to the surface. About 50–60% of a deep-mine seam content can be extracted.

The area around a developed deep mine is often identifiable by huge mounds of diggings, known in the UK as slag heaps. Nowadays these are covered with soil and grass in an attempt to minimise the visual environmental impact. Working in a deep coal mine is not only a very dirty occupation but can be unhealthy and dangerous. Deep mining involves the release of various gases, including carbon monoxide and methane, which represent serious explosion and underground fire risks.

The rates of fatalities and serious accidents in the coal industry are among the worst of all industries. In the USA, about ten times as many accidental deaths occur in the coal energy cycle, from mine to power plant, as in the production of the equivalent amount of power from oil, gas and nuclear fuels together. A methane explosion at the Millfield mine in Southern Ohio, USA, in 1930 killed 82 people, including 73 miners, five mine officials and four visitors.

Until the period of the Second World War (1939–1945), there were about 50 fatalities each year in the UK coal industry. On several occasions serious fires and underground explosions caused the deaths of dozens of miners. The UK coal industry passed from private to public ownership in 1947 and thereafter the safety record improved. Working conditions in the mines have also improved drastically with increasing mechanization (Fig. 4.3) and better underground ventilation. There is

Fig. 4.3.   Mechanized coal extraction in a deep mine [4].

also now a much lower incidence of crippling accidents and industrial lung diseases such as silicosis and pneumoconiosis (black lung disease).

Coal seams are commonly 5–50 feet in thickness but may extend for miles horizontally.

When the coal seams are within about 200 feet of the surface it is possible to mine it by open-cast or surface methods. This involves the despoliation of large areas of ground that may not be reclaimable after the seams are exhausted. The process of extraction involves conventional earth-moving equipment and "draglines": enormous shovels with a boom 180–375 feet long and a bucket that will hold up to 5,400 cubic feet of material. Open-cast mined coal is cheaper to obtain than deep-mined coal, even after factoring in the reclamation costs. In the UK about 55% of the home-produced coal in 2009 was obtained by open-cast methods [4]. Surface mining has a bigger extraction yield than deep mining and up to 90% of a seam's content may be removed. In the former Soviet Union, the USA and Australia, surface mine production now greatly exceeds deep mine production.

4.2.   World Reserves, Production and Consumption of Coal

4.2.1.   *World coal reserves*

World coal reserves are widely distributed and are accessible by existing mining methods. The reserve/production (R/P) figures in years of reserves remaining, at the end of 2010, are included in Table 4.2 for the main "coal" countries. Some of this information is also illustrated in Fig. 2.3 of Chapter 2. The biggest proportions of the world reserve are held by the USA (27.6%), Russia (18.2%), China (13.2%), Australia (8.9%) and India (7.0%).

Table 4.2. World coal reserves (years), 2010 [5].

Proved reserves at end 2010

Million tonnes	Anthracite and bituminous	Sub-bituminous and lignite	**Total**	Share of total	R/P ratio
US	108501	128794	**237295**	27.6%	241
Canada	3474	3108	**6582**	0.8%	97
Mexico	860	351	**1211**	0.1%	130
Total North America	112835	132253	**245088**	28.5%	231
Brazil	—	4559	**4559**	0.5%	*
Colombia	6366	380	**6746**	0.8%	91
Venezuela	479	—	**479**	0.1%	120
Other S. & Cent. America	45	679	**724**	0.1%	*
Total S. & Cent. America	6890	5618	**12508**	1.5%	148
Bulgaria	2	2364	**2366**	0.3%	82
Czech Republic	192	908	**1100**	0.1%	22
Germany	99	40600	**40699**	4.7%	223
Greece	—	3020	**3020**	0.4%	44
Hungary	13	1647	**1660**	0.2%	183
Kazakhstan	21500	12100	**33600**	3.9%	303
Poland	4338	1371	**5709**	0.7%	43
Romania	10	281	**291**	♦	9
Russian Federation	49088	107922	**157010**	18.2%	495
Spain	200	330	**530**	0.1%	73
Turkey	529	1814	**2343**	0.3%	27
Ukraine	15351	18522	**33873**	3.9%	462
United Kingdom	228	—	**228**	♦	13
Other Europe & Eurasia	1440	20735	**22175**	2.6%	317
Total Europe & Eurasia	92990	211614	**304604**	35.4%	257
South Africa	30156	—	**30156**	3.5%	119
Zimbabwe	502	—	**502**	0.1%	301
Other Africa	860	174	**1034**	0.1%	*
Middle East	1203	—	**1203**	0.1%	*
Total Middle East & Africa	32721	174	**32895**	3.8%	127
Australia	37100	39300	**76400**	8.9%	180
China	62200	52300	**114500**	13.3%	35
India	56100	4500	**60600**	7.0%	106
Indonesia	1520	4009	**5529**	0.6%	18
Japan	340	10	**350**	♦	382
New Zealand	33	538	**571**	0.1%	107
North Korea	300	300	**600**	0.1%	16
Pakistan	—	2070	**2070**	0.2%	*
South Korea	—	126	**126**	♦	60
Thailand	—	1239	**1239**	0.1%	69
Vietnam	150	—	**150**	♦	3
Other Asia Pacific	1582	2125	**3707**	0.4%	114
Total Asia Pacific	159326	106517	**265843**	30.9%	57

(*Continued*)

Table 4.2.   (*Continued*)

Million tonnes	Anthracite and bituminous	Sub-bituminous and lignite	Total	Share of total	R/P ratio
Total World	**404762**	**456176**	**860938**	**100.0%**	**118**
of which: OECD	155926	222603	**378529**	44.0%	184
Non–OECD	248836	233573	**482409**	56.0%	92
European Union	5101	51047	**56148**	6.5%	105
Former Soviet Union	86725	141309	**228034**	26.5%	452

*More than 500 years.
♦Less than 0.05%.
Source of reserves data: Survey of Energy Resources, World Energy Council 2010.
Notes: Proved reserves of coal — Generally taken to be those quantities that geological and engineering information indicates with reasonable certainty can be recovered in the future from known deposits under existing economic and operating conditions. Reserves-to-production (R/P) ratio — If the reserves remaining at the end of the year are divided by the production in that year, the result is the length of time that those remaining reserves would last if production were to continue at that rate.

Anthracite and bituminous coal, with its high heat energy value, is widely traded around the world. Lignite or "brown coal" is not traded to any significant extent in world markets because of its relatively low heat content. The present (2013) figures of reserves suggest that coal will far outlast both oil and natural gas as a primary fuel source. The widespread nature of the deposits will hopefully prevent any violent political or military action to gain control over the remaining coal. Nevertheless, there are wide disparities between countries. If the indigenous reserves remaining are divided by the current production figures to give the present R/P ratios, these indicate the needs/opportunities to participate in trading in the world coal market. For example, the European countries of Greece, Poland, Turkey and the UK have less than 50 years of reserves remaining (Table 4.2). On the other hand, there are many countries, including the USA, that have more than 200 years of reserves at present rates of usage. It is possible that estimates of coal reserves might reduce in the future, as coal is used to offset the demand for depleting reserves of oil and natural gas. Although the coal will outlast the fluid fuels, it too will eventually run out. A need exists for the conservation of coal supplies and also for the continual exploration of more efficient methods of coal use.

4.2.2.   *World coal production*

Details of country-by-country coal production between the years 2000 and 2010 are given in Table 4.3 [5], in million tonnes of oil equivalent (mtoe). Over that 11-year period world coal production increased 53%, with the main increase of production due to China, which increased by 240% over the period. North American production reduced by about 10% and Europe by 1%, in spite of the large increase by Russia. In Europe, Germany, Poland and the UK all made significant reductions due to increased use of natural gas.

Table 4.3. World coal production* (mtoe), 2000–2010 [5].

Million tonnes oil equivalent	2000	2001	2002	2003	2004	2005	2006	2007	2008	2009	2010	Change 2010 over 2009	2010 share of total
US	570.1	590.3	570.1	553.6	572.4	580.2	595.1	587.7	596.7	540.9	**552.2**	2.1%	14.8%
Canada	36.1	36.6	34.2	31.7	33.8	33.5	33.9	36.0	35.6	32.5	**34.9**	7.2%	0.9%
Mexico	5.4	5.3	5.3	4.6	4.7	5.2	5.5	6.0	5.5	5.1	**4.5**	−11.4%	0.1%
Total North America	**611.6**	632.2	609.5	589.9	610.9	618.8	634.5	629.7	637.8	578.5	**591.6**	2.3%	15.9%
Brazil	2.9	2.1	1.9	1.8	2.0	2.4	2.2	2.3	2.5	1.9	**2.1**	8.2%	0.1%
Colombia	24.9	28.5	25.7	32.5	34.9	38.4	42.6	45.4	47.8	47.3	**48.3**	2.1%	1.3%
Venezuela	5.8	5.6	5.9	5.1	5.9	5.3	5.4	5.6	4.5	2.7	**2.9**	8.1%	0.1%
Other S. & Cent. America	0.4	0.5	0.4	0.5	0.2	0.3	0.6	0.3	0.4	0.5	**0.5**	−7.0%	◆
Total S. & Cent. America	33.9	36.8	33.9	39.9	43.0	46.3	50.8	53.6	55.2	52.4	**53.8**	2.6%	1.4%
Bulgaria	4.4	4.4	4.4	4.6	4.5	4.1	4.2	4.7	4.8	4.6	**4.8**	5.8%	0.1%
Czech Republic	25.0	25.4	24.3	24.2	23.5	23.5	23.7	23.3	21.1	19.5	**19.4**	−0.7%	0.5%
France	2.3	1.5	1.1	1.3	0.4	0.2	0.2	0.2	0.1	†	**†**	—	◆
Germany	56.5	54.1	55.0	54.1	54.7	53.2	50.3	51.5	47.7	44.4	**43.7**	−1.5%	1.2%
Greece	8.2	8.5	9.1	9.0	9.1	9.0	8.3	8.6	8.3	8.4	**8.8**	5.0%	0.2%
Hungary	2.9	2.9	2.7	2.8	2.4	2.0	2.1	2.0	1.9	1.9	**1.9**	1.0%	0.1%
Kazakhstan	38.5	40.7	37.8	43.3	44.4	44.2	49.1	50.0	56.8	51.5	**56.2**	9.2%	1.5%
Poland	71.3	71.7	71.3	71.4	70.5	68.7	67.0	62.3	60.5	56.4	**55.5**	−1.6%	1.5%
Romania	6.4	7.1	6.6	7.0	6.7	6.6	6.5	6.7	6.7	6.4	**5.8**	−9.2%	0.2%
Russian Federation	116.0	122.6	117.3	127.1	131.7	139.2	145.1	148.0	153.4	142.1	**148.8**	4.7%	4.0%
Spain	8.0	7.6	7.2	6.8	6.7	6.4	6.2	6.0	3.7	3.5	**3.3**	−6.3%	0.1%
Turkey	13.9	14.2	11.5	10.5	10.5	12.8	13.4	15.8	17.2	17.4	**17.4**	◆	0.5%
Ukraine	42.0	43.5	42.8	41.6	42.2	41.0	41.7	39.9	41.3	38.4	**38.1**	−0.8%	1.0%

(*Continued*)

Table 4.3.   (*Continued*)

Million tonnes oil equivalent	2000	2001	2002	2003	2004	2005	2006	2007	2008	2009	2010	Change 2010 over 2009	2010 share of total
United Kingdom	19.0	19.4	18.2	17.2	15.3	12.5	11.3	10.3	11.0	10.9	**11.0**	1.6%	0.3%
Other Europe & Eurasia	14.0	14.4	15.3	15.8	15.6	14.7	15.7	16.7	17.3	16.9	**16.1**	-4.3%	0.4%
Total Europe & Eurasia	428.6	438.2	424.7	436.7	438.2	438.1	444.9	446.1	452.0	422.1	**430.9**	2.1%	11.5%
Total Middle East	0.8	0.7	0.8	0.7	0.8	0.8	0.9	1.0	1.0	1.0	**1.0**	—	♦
South Africa	126.6	126.1	124.1	134.1	137.2	137.7	138.0	139.6	142.4	141.2	**143.0**	1.3%	3.8%
Zimbabwe	2.8	2.9	2.5	1.8	2.4	2.2	1.4	1.3	1.0	1.1	**1.1**	—	♦
Other Africa	1.2	1.2	1.3	1.5	1.2	1.1	1.0	0.9	0.9	0.8	**0.8**	—	♦
Total Africa	130.6	130.1	127.9	137.4	140.8	141.0	140.3	141.8	144.2	143.1	**144.9**	1.3%	3.9%
Australia	166.5	180.2	184.3	189.9	198.5	205.8	210.3	217.2	220.7	228.8	**235.4**	2.9%	6.3%
China	762.5	809.5	853.8	1013.4	1174.1	1302.2	1406.4	1501.1	1557.1	1652.1	**1800.4**	9.0%	48.3%
India	132.2	133.6	138.5	144.4	155.7	162.1	170.2	181.0	195.6	210.8	**216.1**	2.5%	5.8%
Indonesia	47.4	56.9	63.5	70.3	81.4	93.9	119.2	133.4	147.8	157.6	**188.1**	19.4%	5.0%
Japan	1.7	1.8	0.8	0.7	0.7	0.6	0.7	0.8	0.7	0.7	**0.5**	-28.4%	♦
New Zealand	2.2	2.4	2.7	3.2	3.2	3.2	3.5	3.0	3.0	2.8	**3.3**	16.8%	0.1%
Pakistan	1.4	1.5	1.6	1.5	1.5	1.6	1.7	1.6	1.8	1.6	**1.5**	-5.2%	♦
South Korea	1.9	1.7	1.5	1.5	1.4	1.3	1.3	1.3	1.2	1.1	**0.9**	-17.3%	♦
Thailand	5.1	5.6	5.7	5.3	5.6	5.8	5.3	5.1	5.0	5.0	**5.0**	0.5%	0.1%
Vietnam	6.5	7.5	9.2	10.8	14.7	18.3	21.8	22.4	23.0	25.2	**24.7**	-2.0%	0.7%
Other Asia Pacific	19.6	20.1	19.6	20.6	22.4	24.7	25.1	23.3	24.3	29.2	**33.4**	14.5%	0.9%
Total Asia Pacific	1147.1	1220.8	1281.0	1461.5	1659.2	1819.4	1965.6	2090.2	2180.1	2314.8	**2509.4**	8.4%	67.2%

(*Continued*)

Table 4.3. (*Continued*)

Million tonnes oil equivalent	2000	2001	2002	2003	2004	2005	2006	2007	2008	2009	2010	Change 2010 over 2009	2010 share of total
Total World	**2352.5**	**2458.9**	**2477.7**	**2666.1**	**2893.0**	**3064.4**	**3237.1**	**3362.4**	**3470.3**	**3511.8**	**3731.4**	**6.3%**	**100.0%**
of which: OECD	994.2	1027.6	1003.5	987.3	1012.0	1021.0	1036.6	1036.6	1039.3	978.2	**996.0**	1.8%	26.7%
Non-OECD	1358.3	1431.3	1474.2	1678.8	1881.0	2043.4	2200.5	2325.8	2431.1	2533.7	**2735.5**	8.0%	73.3%
European Union	206.6	205.1	202.5	200.7	195.8	188.1	181.5	177.4	167.7	157.7	**156.0**	−1.1%	4.2%
Former Soviet Union	197.4	207.8	198.9	212.8	219.4	225.5	237.0	239.0	252.9	233.2	**244.4**	4.8%	6.5%

*Commercial solid fuels only, i.e. bituminous coal and anthracite (hard coal), and lignite and brown (sub-bituminous) coal.
†Less than 0.05.
♦Less than 0.05%.
Notes: Coal production data expressed in million tonnes is available at www.bp.com/statisticalreview.

Coal production in the Pacific region increased greatly due to large increases in coal use in the major countries of China, India, Indonesia and Australia.

4.2.3.   *World coal consumption*

World coal consumption represents a source of heat energy plus chemical products. Nation-by-nation details for the period 2000–2010 are given in Table 4.4. Over this period the world consumption increased by 59%, mainly due to increases by the Asian countries of China, India and Japan. Consumption in the North American countries and in Europe decreased by about 9%.

Coal remains a vital part of the world fuels scene and in 2010 accounted for 524.6/2285.7 mtoe, which was 23% of the prime energy consumed.

Comparison of the data for coal production in Table 4.3 and coal consumption in Table 4.4 indicates which countries now operate as exporters or importers of coal. The main exporter countries are listed in the left-hand column of Table 4.5, with the total export, in millions of short tons, shown in the right-hand column. The estimation of the US Energy Information Administration for the year 2010, shown in Table 4.5, was that Australia is likely to remain the principal exporter.

The term "steam" coal refers to the high grade coal used mainly for electricity generation, whereas "coking" coal is used extensively for steel making. Australia, the USA and Canada have substantial reserves of the premium coals that can be used to manufacture coke.

During the 1980s, Australia became the leading coal exporter in the world, primarily by meeting increased demand for steam coal in Asia. Some growth in exports of coking coal also occurred as countries such as Japan began using some of Australia's semi-soft or weak coking coals in their coke blends. As a result, imports of hard coking coals from other countries, including the USA, were displaced. Australia's share of total world coal trade, which increased from 17% in 1980 to 33% in 1997, is projected to reach 38% in 2020 [6]. Australia should continue as the major exporter to Asia, continuing to meet approximately one-half of the region's total coal import demand.

For the nations of Western Europe, future coal consumption is expected to decline. Western European countries are relying on increasingly available natural gas supplies for future growth in electricity production. The elimination of subsidies in the UK was largely responsible for a 50% drop in the nation's coal production between 1989 and 1997 and a greatly reduced role for coal in electricity generation.

With the exception of Germany, coal imports to Western Europe are not expected to increase to compensate for a reduction in indigenous coal production. Rather, increased use of natural gas, renewable energy and nuclear power (primarily in France) is expected to fill the gap in energy supply left by the continuing reductions in the region's indigenous coal production [6].

Table 4.4. World coal consumption* (mtoe), 2000–2010 [5].

Million tonnes oil equivalent	2000	2001	2002	2003	2004	2005	2006	2007	2008	2009	2010	Change 2010 over 2009	2010 share of total
US	569.0	552.2	552.0	562.5	566.1	574.2	565.7	573.3	564.1	496.2	**524.6**	5.7%	14.8%
Canada	31.8	34.0	31.6	33.4	29.9	31.7	31.0	32.3	28.9	23.3	**23.4**	0.4%	0.7%
Mexico	6.2	6.8	7.6	8.6	7.0	9.1	9.4	9.1	6.9	8.6	**8.4**	−2.4%	0.2%
Total North America	606.9	593.0	591.1	604.5	603.0	614.9	606.1	614.7	599.9	528.1	**556.3**	5.3%	15.6%
Argentina	0.8	0.6	0.5	0.7	0.8	0.9	0.3	0.4	1.1	1.2	**1.2**	—	◆
Brazil	12.5	12.2	11.5	11.8	12.8	12.7	12.5	13.4	13.5	11.7	**12.4**	6.0%	0.3%
Chile	3.0	2.3	2.4	2.3	2.6	2.6	3.2	3.8	4.1	3.7	**3.7**	1.0%	0.1%
Colombia	2.7	2.7	2.2	2.4	2.0	2.7	2.4	2.4	2.8	3.7	**3.8**	2.1%	0.1%
Ecuador	—	—	—	—	—	—	—	—	—	—	—	—	—
Peru	0.5	0.4	0.4	0.4	0.5	0.5	0.4	0.5	0.5	0.5	**0.5**	6.3%	◆
Trinidad & Tobago	†	†	†	†	—	—	†	†	†	†	†	—	—
Venezuela	†	†	†	†	†	†	†	†	†	†	†	−1.9%	◆
Other S. & Cent. America	0.6	0.7	1.0	2.1	1.9	1.8	2.1	2.1	2.2	2.0	**2.1**	1.9%	0.1%
Total S. & Cent. America	20.1	19.0	18.0	19.6	20.5	21.2	21.0	22.6	24.2	22.9	**23.8**	3.9%	0.7%
Austria	2.8	3.1	3.0	3.3	3.3	3.1	3.1	2.9	2.7	2.2	**2.0**	−10.5%	0.1%
Azerbaijan	—	†	†	†	†	†	†	†	†	†	†	−7.8%	◆
Belarus	0.1	0.1	0.1	0.1	0.1	0.1	0.1	†	†	†	†	−23.0%	◆
Belgium & Luxembourg	7.6	7.6	6.7	6.5	6.4	6.1	6.1	5.5	4.8	4.6	**4.9**	6.2%	0.1%
Bulgaria	6.3	6.9	6.5	7.1	6.9	6.9	7.1	7.8	7.5	6.3	**6.6**	4.8%	0.2%

(Continued)

Table 4.4.   (*Continued*)

Million tonnes oil equivalent	2000	2001	2002	2003	2004	2005	2006	2007	2008	2009	2010	Change 2010 over 2009	2010 share of total
Czech Republic	21.0	21.2	20.6	20.8	20.5	19.8	19.4	19.3	17.4	16.2	**16.0**	−1.3%	0.5%
Denmark	4.0	4.2	4.2	5.7	4.6	3.7	5.6	4.7	4.1	4.0	**3.8**	−4.9%	0.1%
Finland	3.5	4.0	4.4	5.8	5.3	3.1	5.2	4.6	3.4	3.7	**4.6**	24.9%	0.1%
France	13.9	12.1	12.4	13.3	12.8	13.3	12.1	12.3	11.9	9.9	**12.1**	23.2%	0.3%
Germany	84.9	85.0	84.6	87.2	85.4	82.1	83.5	85.7	80.1	71.7	**76.5**	6.7%	2.2%
Greece	9.2	9.3	9.8	9.4	9.0	8.8	8.1	8.5	8.1	8.1	**8.5**	4.6%	0.2%
Hungary	3.2	3.4	3.1	3.4	3.1	2.7	2.9	2.9	2.8	2.5	**2.6**	6.0%	0.1%
Republic of Ireland	1.8	1.9	1.8	1.7	1.8	1.8	1.6	1.5	1.4	1.3	**1.4**	3.8%	◆
Italy	13.0	13.7	14.2	15.3	17.1	17.0	17.2	17.2	16.7	13.1	**13.7**	5.0%	0.4%
Kazakhstan	23.2	22.5	22.8	25.2	26.5	27.2	28.1	30.8	34.0	31.7	**36.1**	13.8%	1.0%
Lithuania	0.1	0.1	0.1	0.2	0.2	0.2	0.2	0.2	0.2	0.1	**0.2**	19.6%	◆
Netherlands	8.6	8.5	8.9	9.1	9.1	8.7	8.5	9.0	8.5	7.9	**7.9**	−0.1%	0.2%
Norway	0.7	0.6	0.5	0.5	0.6	0.5	0.4	0.4	0.5	0.3	**0.5**	32.0%	◆
Poland	57.6	58.0	56.7	57.7	57.3	55.7	58.0	57.9	56.0	51.9	**54.0**	3.9%	1.5%
Portugal	4.5	3.7	4.1	3.8	3.7	3.8	3.8	3.3	3.2	3.3	**3.4**	3.4%	0.1%
Romania	7.0	7.2	7.6	7.8	7.4	7.6	8.5	7.4	7.4	6.6	**6.2**	−6.6%	0.2%
Russian Federation	105.2	102.4	103.0	104.0	99.5	94.2	96.7	93.5	100.4	91.9	**93.8**	2.1%	2.6%
Slovakia	4.0	4.1	4.0	4.2	4.1	3.9	3.8	3.8	3.7	3.5	**2.7**	−24.3%	0.1%
Spain	21.6	19.5	21.9	20.5	21.0	21.2	18.5	20.2	15.6	10.5	**8.3**	−21.3%	0.2%
Sweden	1.9	2.0	2.2	2.2	2.3	2.2	2.3	2.2	2.0	1.6	**2.0**	23.6%	0.1%
Switzerland	0.1	0.1	0.1	0.1	0.1	0.1	0.1	0.1	0.1	0.1	**0.1**	5.7%	◆
Turkey	25.5	21.8	21.2	21.8	23.0	26.1	28.8	31.0	31.3	32.0	**34.4**	7.4%	1.0%
Turkmenistan	—	—	—	—	—	—	—	—	—	—	—	—	—
Ukraine	38.8	39.4	38.3	39.0	39.1	37.5	39.8	39.7	40.3	35.0	**36.4**	4.2%	1.0%
United Kingdom	36.7	38.9	35.7	38.1	36.6	37.4	40.9	38.4	35.6	29.6	**31.2**	5.2%	0.9%

(*Continued*)

Table 4.4. (*Continued*)

Million tonnes oil equivalent	2000	2001	2002	2003	2004	2005	2006	2007	2008	2009	2010	Change 2010 over 2009	2010 share of total
Uzbekistan	1.0	1.1	1.0	1.0	1.2	1.1	1.1	1.4	1.4	1.4	**1.3**	−3.3%	◆
Other Europe & Eurasia	17.4	16.6	18.8	19.7	20.2	18.0	15.8	16.0	16.8	15.2	**15.7**	3.1%	0.4%
Total Europe & Eurasia	525.2	518.9	518.5	534.3	528.1	513.9	527.2	528.3	517.8	466.4	**486.8**	4.4%	13.7%
Iran	1.1	1.1	1.1	1.1	1.0	1.2	1.2	1.3	0.9	1.1	**1.1**	1.0%	◆
Israel	6.2	7.2	7.6	7.9	8.0	7.9	7.8	8.0	7.9	7.7	**7.7**	◆	0.2%
Kuwait	—	—	—	—	—	—	—	—	—	—	—	—	—
Qatar	—	—	—	—	—	—	—	—	—	—	—	—	—
Saudi Arabia	—	—	—	—	—	—	—	—	—	—	—	—	—
United Arab Emirates	—	—	—	—	—	—	—	—	—	—	—	—	—
Other Middle East	—	—	—	—	—	—	—	—	—	—	—	—	—
Total Middle East	7.3	8.3	8.7	9.0	9.0	9.1	9.1	9.3	8.7	8.8	**8.8**	0.1%	0.2%
Algeria	0.5	0.6	0.9	0.8	0.8	0.6	0.7	0.7	0.6	0.2	**0.3**	6.1%	◆
Egypt	1.4	1.2	1.3	1.4	1.3	1.3	1.2	1.2	1.2	0.6	**0.7**	5.1%	◆
South Africa	74.6	73.4	75.9	81.4	85.4	82.9	84.0	85.1	84.7	87.7	**88.7**	1.1%	2.5%
Other Africa	6.5	7.3	7.2	6.4	7.1	7.3	6.7	6.0	6.2	5.5	**5.7**	3.8%	0.2%
Total Africa	82.9	82.6	85.4	90.0	94.7	92.1	92.6	93.1	92.7	94.1	**95.3**	1.3%	2.7%
Australia	46.7	48.2	51.0	49.8	52.7	53.6	55.6	54.2	51.8	51.7	**43.4**	−16.1%	◆
Bangladesh	0.3	0.4	0.4	0.4	0.4	0.4	0.5	0.4	0.6	0.4	**0.5**	6.0%	1.2%
China	737.1	751.9	794.9	936.3	1084.3	1218.7	1343.9	1438.4	1479.3	1556.8	**1713.5**	10.1%	48.2%

(*Continued*)

Table 4.4.   (Continued)

Million tonnes oil equivalent	2000	2001	2002	2003	2004	2005	2006	2007	2008	2009	2010	Change 2010 over 2009	2010 share of total
China Hong Kong SAR	3.7	4.9	5.4	6.6	6.6	6.7	7.0	7.5	7.0	7.6	**6.3**	−16.3%	0.2%
India	144.2	145.2	151.8	156.8	172.3	184.4	195.4	210.3	230.4	250.6	**277.6**	10.8%	7.8%
Indonesia	13.7	16.8	18.0	24.2	22.2	25.4	30.1	37.8	30.1	34.6	**39.4**	13.7%	1.1%
Japan	98.9	103.0	106.6	112.2	120.8	121.3	119.1	125.3	128.7	108.8	**123.7**	13.7%	3.5%
Malaysia	1.9	2.6	3.6	4.2	5.7	6.3	7.3	7.1	5.0	4.0	**3.4**	−16.1%	◆
New Zealand	1.1	1.3	1.3	1.9	2.0	2.2	2.2	1.6	2.0	1.6	**1.0**	−37.3%	0.1%
Pakistan	2.0	2.1	2.4	2.9	3.8	4.1	4.2	5.1	5.3	4.7	**4.6**	−2.3%	0.2%
Philippines	4.3	4.5	4.7	4.7	5.0	5.7	5.5	5.9	7.0	6.7	**7.7**	15.8%	—
Singapore	—	—	—	—	—	—	—	—	—	—	—	—	
South Korea	43.0	45.7	49.1	51.1	53.1	54.8	54.8	59.7	66.1	68.6	**76.0**	10.8%	2.1%
Taiwan	28.7	30.6	32.7	35.1	36.6	38.1	39.6	41.8	40.2	38.7	**40.3**	4.0%	1.1%
Thailand	7.8	8.8	9.2	9.4	10.4	11.2	12.4	14.1	15.3	14.5	**14.8**	2.6%	0.4%
Vietnam	4.7	5.0	5.3	5.5	8.2	8.0	9.5	10.1	10.0	14.0	**13.7**	−2.0%	0.4%
Other Asia Pacific	18.9	19.5	18.6	18.9	19.2	20.7	21.4	18.0	19.7	22.1	**18.9**	−14.5%	0.5%
Total Asia Pacific	1157.1	1190.7	1254.9	1419.9	1603.1	1761.6	1908.6	2037.5	2098.4	2185.3	**2384.7**	9.1%	67.1%
Total World	**2399.7**	**2412.4**	**2476.7**	**2677.3**	**2858.4**	**3012.9**	**3164.5**	**3305.6**	**3341.7**	**3305.6**	**3555.8**	**7.6%**	**100.0%**
of which: OECD	1333.6	1124.9	1130.7	1161.4	1170.5	1179.7	1179.8	1200.2	1171.5	1049.5	**1103.6**	5.2%	31.0%
Non-OECD	1266.1	1287.5	1345.9	1515.9	1687.9	1833.2	1984.7	2105.4	2170.2	2256.1	**2452.2**	8.7%	69.0%
European Union	314.9	315.7	314.0	324.3	319.1	310.4	317.7	316.7	294.4	259.9	**269.7**	3.8%	7.6%
Former Soviet Union	169.1	166.1	166.0	170.3	167.6	161.1	166.8	166.4	177.2	161.1	**169.1**	5.0%	4.8%

*Commercial solid fuels only, i.e. bituminous coal and anthracite (hard coal), and lignite and brown (sub-bituminous) coal.

†Less than 0.05.

◆ Less than 0.05%.

Table 4.5. World coal flows by importing and exporting regions, by reference case, 2000, 2010 and 2020 [7]. In million short tons.

	Importers											
	Steam[a]				Coking				Total			
Exporters	Europe[b]	Asia	America	Total[c]	Europe[b]	Asia[d]	America	Total[c]	Europe[b]	Asia	America	Total[c]
2000												
Australia	13.8	83.2	2.3	96.7	25.7	76.2	6.6	109.1	39.5	159.4	8.9	205.8
United States	5.8	4.3	15.4	25.6	21.6	2.3	8.9	32.8	27.4	6.6	24.3	58.4
South Africa	55.6	14.4	1.3	74.3	0.4	0.3	1.0	2.8	56.0	14.7	2.3	77.1
Former Soviet Union	18.4	6.0	0.1	23.3	3.1	3.7	0.0	8.0	21.5	9.7	0.1	31.3
Poland	15.7	0.0	0.0	14.6	3.3	0.0	0.1	3.0	19.0	0.0	0.1	17.6
Canada	0.3	3.3	0.7	5.1	8.2	19.3	3.6	32.8	8.5	22.6	4.3	37.9
China	3.2	53.8	0.2	53.0	0.3	7.1	0.0	7.4	3.5	60.9	0.2	60.4
South America[e]	30.4	0.0	15.0	46.4	0.4	0.1	0.1	0.7	30.8	0.1	15.1	47.1
Indonesia[f]	4.5	46.5	2.4	59.8	0.5	10.6	0.0	11.2	5.0	57.1	2.4	71.0
Total	**147.7**	**211.5**	**37.5**	**398.8**	**63.5**	**119.6**	**20.4**	**204.7**	**211.2**	**331.1**	**57.9**	**603.5**
2010												
Australia	10.0	108.2	0.7	118.8	35.6	85.5	8.0	129.1	45.6	193.7	8.7	247.9
United States	3.1	6.7	8.6	18.4	13.4	1.3	15.5	30.2	16.5	8.0	24.2	48.7
South Africa	70.5	8.2	4.4	83.0	1.1	0.5	0.0	1.7	71.6	8.7	4.4	84.7
Former Soviet Union	19.6	6.1	0.0	25.6	3.0	4.3	0.0	7.3	22.5	10.4	0.0	32.9
Poland	8.0	0.0	0.0	8.0	1.1	0.0	0.0	1.1	9.1	0.0	0.0	9.1
Canada	5.0	0.0	0.0	5.0	6.9	13.8	3.3	24.0	11.9	13.8	3.3	29.0
China	0.0	113.5	0.0	113.5	0.0	12.4	0.0	12.4	0.0	125.9	0.0	125.9
South America[e]	36.4	0.0	34.8	71.2	0.0	0.0	0.0	0.0	36.4	0.0	34.8	71.2
Indonesia[f]	7.6	65.9	0.0	73.5	0.5	9.1	0.0	9.6	8.1	75.0	0.0	83.1
Total	**160.3**	**308.5**	**48.4**	**517.2**	**61.6**	**126.9**	**26.8**	**215.3**	**221.9**	**435.5**	**75.2**	**732.6**

(Continued)

Table 4.5.   (*Continued*)

Exporters	Steam[a]				Coking				Total			
	Europe[b]	Asia	America	Total[c]	Europe[b]	Asia[d]	America	Total[c]	Europe[b]	Asia	America	Total[c]
						2020						
Australia	9.3	112.7	0.7	**122.7**	35.8	89.7	12.4	**137.9**	45.1	202.4	13.1	**260.6**
United States	1.9	7.5	7.2	**16.6**	12.1	1.4	18.1	**31.7**	14.1	8.9	25.3	**48.3**
South Africa	67.7	17.0	4.3	**89.0**	0.9	0.6	0.0	**1.5**	68.6	17.6	4.3	**90.5**
Former Soviet Union	16.1	7.2	0.0	**23.3**	3.0	4.7	0.0	**7.7**	19.1	11.9	0.0	**31.0**
Poland	5.5	0.0	0.0	**5.5**	1.1	0.0	0.0	**1.1**	6.6	0.0	0.0	**6.6**
Canada	2.9	0.0	0.0	**2.9**	6.8	14.0	1.7	**22.5**	9.7	14.0	1.7	**25.4**
China	0.0	121.3	0.0	**121.3**	0.0	12.4	0.0	**12.4**	0.0	133.6	0.0	**133.6**
South America[e]	50.0	0.0	36.1	**86.1**	0.0	0.0	0.0	**0.0**	50.0	0.0	36.1	**86.1**
Indonesia[f]	0.0	83.8	0.0	**83.8**	0.4	9.2	0.0	**9.6**	0.4	93.0	0.0	**93.4**
Total	**153.4**	**349.4**	**48.3**	**551.1**	**60.2**	**132.1**	**32.2**	**224.4**	**213.6**	**481.4**	**80.5**	**775.5**

[a]Reported data for 2000 are consistent with data published by the International Energy Agency (IEA). The standard IEA definition for "steam coal" includes coal used for pulverized coal injection (PCI) at steel mills; however, some PCI coal is reported by the IEA as "coking coal."

[b]Coal flows to Europe include shipments to the Middle East and Africa.

[c]In 2000, total world coal flows include a balancing item used by the International Energy Agency to reconcile discrepancies between reported exports and imports. The 2000 balancing items by coal type were 2.1 million tons (steam coal), 1.2 million tons (coking coal), and 3.3 million tons (total).

[d]Includes 14.4 million tons of coal for pulverized coal injection at blast furnaces shipped to Japanese steelmakers in 2000.

[e]Coal exports from South America are projected to originate from mines in Colombia and Venezuela.

[f]In 2000, coal exports from Indonesia include shipments from other countries not modeled for the forecast period. The 2000 non-Indonesian exports by coal type were 6.2 million tons (steam coal), 1.5 million tons (coking coal), and 7.7 million tons (total).

Notes: Data exclude non-seaborne shipments of coal to Europe and Asia. Totals may not equal sum of components due to independent rounding. The sum of the columns may not equal the total, because the total includes a balancing item between importers' and exporters' data.

Sources: **2000:** International Energy Agency, *Coal Information 2001* (Paris, France, September 2001); Energy Information Administration, *Quarterly Coal Report, October–December 2000*, DOE/EIA-0121(2000/4Q) (Washington, DC, May 2001). **Projections:** Energy Information Administration, National Energy Modeling System run IEO2002.D011402A (January 2002).

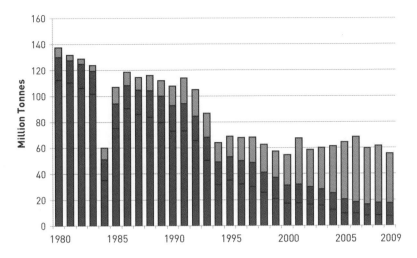

Fig. 4.4.   UK coal production and imports, 1980–2009 [4].

Table 4.6.   UK coal production, 1980–2009 [4].

Million tonnes

	1980	1990	2000	2007	2008	2009
Deep mined	112.4	72.9	17.2	7.7	8.1	7.5
Opencast	15.8	18.1	13.4	8.9	9.5	9.9
Total (including slurry)	**130.1**	**92.8**	**31.2**	**17.0**	**18.1**	**17.9**
Coal imports	7.3	14.8	23.4	43.4	43.9	38.2

4.2.4.   *UK coal production and consumption*

UK coal production in the past 30 years is illustrated by the bar chart of Fig. 4.4 [4]. Further specific numerical information is given in Table 4.6. Coal production was 0.3% lower in 2010 than in 2009, deep-mined production fell by 7%, while open-cast production increased by 4%. Imports, initially of coal types in short supply in the UK, started in 1970 and then grew steadily to reach around 20 million tonnes a year by the late 1990s. The very rapid expansion of imports in 2001 meant that imports exceeded the level of UK production for the first time. Since 2002 imports have been rising at 15% a year on average and in 2006 imports were at a record 50 million tonnes to meet strong demand from generators and the steel industry. However, since the end of 2008, levels have started to decrease and in 2009 UK imports (38.2 million tonnes) were 13% lower than 2008. Despite this fall imports still account for more than two-thirds of UK coal supply.

The proportion of coal consumed by electricity power stations increased steadily from the 1970s to reach 86% in 2006, before falling back to 83% in 2008 (Table 4.7). The decline in coal consumption at power stations reached a low of 41.8 million

Table 4.7. UK coal consumption, 1980–2009 [4].

Million tonnes

	1980	1990	2000	2007	2008	2009
Power stations	89.6	84.0	46.8	53.0	48.3	40.1
Domestic	8.9	4.2	1.9	0.6	0.7	0.7
Industry	7.9	6.3	1.9	1.9	1.9	1.8
Services	1.8	1.2	0.1	<0.1	<0.1	<0.1
Other energy industries	15.3	12.5	9.2	7.4	7.4	6.1
Total consumption	**123.5**	**108.3**	**59.9**	**63.0**	**58.4**	**48.8**

tonnes in 1999, before climbing to 57.9 million tonnes in 2006. Since then it has declined and in 2009 it stood at 40 million tonnes, the lowest level on record. But the figure means that 40/48.8 million tonnes, or 83%, of the UK coal consumption in 2009 was used for electricity generation [4].

The figures for coal production (Table 4.3) and coal consumption (Table 4.4) show that in 2009 the UK produced 11/31.2 mtoe, or 35.3%, of its coal consumption. Coal imports for 2009 represented the remainder of the consumption. The number of working deep-coal mines in the UK reduced from more than 100 in 1970, to 10 in 2009, to only two in 2013. This is due to a combination of financial and domestic political reasons. This spectacular decline of what was once one of Britain's base industries may yet prove to have been a big mistake [6]. Coal is still a major factor in the UK energy situation.

The price of European coal (see Table 4.8) has been more expensive than surface-mined American coal over most of the period since 1990. In 1999 the prices were nearing parity but during 2000 European coal again became significantly more expensive than American coal. This compares with the import price of Japanese coal, where steam coal (for making electricity) is significantly cheaper than coking coal (for making steel).

4.2.5. *US coal production and consumption*

The USA contributed 14.8% of the world total coal production in 2010 (Table 4.3) and accounted for 14.8% of the world consumption (Table 4.4). Both production and consumption reduced over the period since 2000 [5]. Coal continues to be the most commonly used fuel for electricity generation. In 2009, 44.5% of US electricity generation was provided by coal, but electricity producers are now increasingly turning to natural gas as the fuel source for the new generation [6–9]. Coal deposits are widespread throughout the USA (Fig. 4.5) and active mining, mainly surface mining, takes place in about 20 states [10]. The use of coal deposits as sources of coal-bed methane — a form of natural gas — is discussed in Chapter 6.

Table 4.8.   International coal prices, 1990–2010 [5].

US dollars per tonne	Northwest Europe marker price[†]	US Central Appalachian coal spot price index[‡]	Japan coking coal import cif price	Japan steam coal import cif price
1990	43.48	31.59	60.54	50.81
1991	42.80	29.01	60.45	50.30
1992	38.53	28.53	57.82	48.45
1993	33.68	29.85	55.26	45.71
1994	37.18	31.72	51.77	43.66
1995	44.50	27.01	54.47	47.58
1996	41.25	29.86	56.68	49.54
1997	38.92	29.76	55.51	45.53
1998	32.00	31.00	50.76	40.51
1999	28.79	31.29	42.83	35.74
2000	35.99	29.90	39.69	34.58
2001	39.03	50.15	41.33	37.96
2002	31.65	33.20	42.01	36.90
2003	43.60	38.52	41.57	34.74
2004	72.08	64.90	60.96	51.34
2005	60.54	70.12	89.33	62.91
2006	64.11	62.96	93.46	63.04
2007	88.79	51.16	88.24	69.86
2008	147.67	118.79	179.03	122.81
2009	70.66	68.08	167.82	110.11
2010	92.50	71.63	158.95	105.19

[†] *Source*: McCloskey Coal Information Service. Prices for 1990–2000 are the average of the monthly marker, 2001–2010 the average of weekly prices.
[‡] *Source*: Platts. Prices are for CAPP 12,500 Btu, 1.2 SO_2 coal, fob. Prices for 1990–2000 are by coal price publication date, 2001–2010 by coal price assessment date.
Note: CAPP = Central Appalachian; cif = cost + insurance + freight (average prices); fob = free on board.

Other coal-producing countries, including Australia, South Africa, Colombia and Venezuela, have increased production and by aggressive pricing, coupled with favourable currency exchange rates, have achieved a growing share of traditional US export markets, such as the European countries of Italy, Portugal, Spain and the UK. In addition, the competition has gained footholds within the USA. These factors have tended to cap the expansion of markets for US coal and placed strong competitive pressures among domestic producers to keep coal prices low, relative to other fuels, in order to maintain existing sales quantities and market share. Concerns related to the ultimate impacts of the Kyoto Protocol and subsequent measures regarding greenhouse gases also loom over future decisions that will affect coal use [8].

In 2001 the USA announced that because of fears of economic recession it would not accept the Kyoto Protocol in its present form. This matter is still at the time of writing in abeyance.

Fig. 4.5.   Principal US coal basins and estimated in-place coal-bed methane sources.

4.3.   Coal Transportation

4.3.1.   *Surface transportation*

In Europe and North America the common method of transporting coal is by rail. The great rivers of the USA are used as waterways to transport convoys of coal-laden barges often hundreds of miles, and about 20% of US coal is shipped in that way [11]. As in Europe, however, more than one-half the coal travels by rail. Barge transportation is cheaper than rail but is limited by the geography of the waterways and the sizes of river and canal locks. Transporting coal by road is several times more expensive than the use of rail.

In Britain, the transportation of coal is part of a complex, interactive energy-related process. The transportation of coal is by far the largest freight section operation of the railway system and its biggest revenue earner. There are urgent needs to limit road traffic and also to keep the railway system economically viable. But the amount of home-produced coal is declining and traditional coal-rail routes have been abandoned in favour of routes needed for imported coal that is being transported from the seaports.

It would be an obvious advantage to build electricity generation stations nearer to the supplies of coal. A successful example of this in the UK is the Drax power station, rated at 2000 MW$_e$, built in the Selby coalfield, Yorkshire.

4.3.2.   *Coal slurry pipelines*

It is technically feasible to pulverize coal and to mix it, in equal amounts, with water to form a slurry. This can be transported to the user site by pipeline.

At the user location the coal particles can be extracted from the slurry by centrifuges, creating a pulverized fuel that is ideal for big boilers [12]. An 8-inch diameter pipeline for slurry was operational in London in 1914. Existing pipelines operate over the distance range 10–300 miles. Operation of a slurry pipeline depends on the availability of the necessary amount of water. There are cost and efficiency advantages to slurry pipelines compared with the rail transportation of coal.

The long distance slurry pipeline is a reality in the USA, where a 273-mile pipeline carries 5 million tons of coal annually, through an 18-inch pipe, from Arizona to south-western Nevada. This is equivalent to the rail transportation of about 150 rail freight cars per day. In the UK, the widespread use of coal slurry pipeline would reduce the delivered cost of coal but would also jeopardize the income of the railway companies. A typical UK pipeline distance would need to be of the order of tens of miles rather than hundreds of miles.

4.4.   Emissions and Effluents from Coal

In Western Europe and North America most of the coal consumption is now used for generating electricity. Large power stations burn pulverized coal in a suspension of fine particles. A 1000 MW_e coal-fired plant burns about 2.5 million tons of coal and generates roughly 6.5 billion kWh of electrical energy per year [1].

4.4.1.   *Open coal fires*

Until the 1950s coal was the prime domestic and industrial fuel in the UK and was mostly burned on open fires. The smoke and gaseous effluent from these millions of fires was expelled directly into the atmosphere above residential areas. Some degree of effluent cleansing or filtration was carried out by some industries, including the electricity-generating industry, but the overall effect of open-fire coal burning, mostly residential, was massive air pollution. This was evidenced in the smoke blackening of building stone and brick and in the incidence of thick fogs in the industrial cities. A great fog in London in 1954 was a mixture of natural fog and man-made smoke and was called "smog". This lasted several days, brought transportation to a standstill and is thought to have been directly responsible for hundreds of accelerated deaths due to lung diseases [13].

Apart from the domestic inconvenience, the use of coal-burning open fires or coke-burning stoves is dirty, unhealthy and grossly inefficient. Most of the heat goes up the chimney and is wasted. The number of victims, including fatalities, of lung diseases created by or compounded by 200 years of open coal fires cannot be calculated but is probably enormous in the UK alone. A succession of legislation, the "Clean Air Acts", has largely banned open-fire coal burning in UK cities, which are now described as "smokeless".

4.4.2.   *Effluents due to coal burning*

The three main pollutants from coal-burning plant flue gases are sulphur dioxide and nitrogen dioxide, which produce "acid rain", plus particulates.

4.4.2.1.   *Sulphur oxides (SO_x)*

The most important pollutants released by coal burning are sulphur products. "Inorganic" sulphur (sometimes called "mineral" sulphur) is physically distinct from but attached to the coal and can be largely removed by washing prior to combustion. "Organic" sulphur is chemically combined with the coal material and cannot be washed out. Emissions of sulphur dioxide (SO_2) are measured in terms of pounds (or kilogrammes) per million BTU burned.

Sulphur may be removed from the combustion gases by chemical "scrubbing". The sulphur-oxide-laden exhaust gases are brought into contact with a scrubbing agent, such as lime or limestone, to produce a chemical reaction that removes the sulphur. The stream of effluent from the absorbent scrubbers has its water evaporated off, producing a sludge that has to be disposed of as solid waste. For example, a unit burning 2% sulphur coal produces about 200 lb of sludge (dry weight) per ton of coal burned. A power plant of (say) 500 MW$_e$ would create a 560-acre sludge disposal area 40 ft deep over its lifetime [1]. More modern scrubbing agents such as sodium or recyclable metal hydroxides neutralize the sulphur oxides and also reduce the scaling of the flue pipes.

4.4.2.2.   *Nitrogen oxides (NO_x)*

During the combustion of coal, nitrogen is drawn both from the coal and from the air in which it burns. Nitrogen oxide formation includes nitrogen monoxide (NO), nitrogen dioxide (NO_2) and nitrous oxide (N_2O). The particular mix of gases depends on the flame temperature, the time of combustion, the excess air present and the rate of cooling. To lower nitrogen oxide emissions it is necessary to modify the combustion processes. Design arrangements are effective if they involve lower combustion temperature, restrictions of air intake, recirculation of flue gases and injection of water into the fire-box [1]. Motor vehicle exhaust gases are a bigger source of nitrogen oxides than power station flue gases in the UK. For this reason there is now UK legislation requiring all new cars to be fitted with exhaust gas filters. In the USA about one-half the nitrous oxide emissions in 1999 were due to the nitrogen fertilization of soils in the agricultural sector [14].

Acid precipitation or "acid rain" is caused by sulphur dioxide and nitrogen oxides mixing chemically with water vapour in the atmosphere. These emissions are mainly derived from coal and oil combustion, which is dispersed through tall chimneys to prevent high concentrations at ground level. Due to air motion, increases in the acidity of the local rainfall may occur hundreds of miles downwind from the site of the pollution. The emitted gases may be oxidized to sulphates and other chemical changes may occur, influenced by varying conditions of temperature, humidity and solar irradiation [1]. Also, the chemical nature of the gases may be affected by substances originating in the territories over which they travel.

There is evidence that pollution from power plants in the UK is blown from west to east and deposits, in the form of acid rain, onto locations in Norway and Sweden. The rain acidity can accelerate the leaching of ground toxic materials such as aluminium and mercury into water courses. There is also evidence of damage to forests and fish stocks, especially in areas with acid soils [2].

The problem of acid rain has to be approached on the basis of international collaboration. Technical aspects of the acid rain problem can only be tackled at the source. There is no form of prevention once the acidity is created.

4.4.2.3. *Particulates*

Flue gases from coal-burning operations contain small particles of solid materials, mainly carbon. There is a range of particle sizes down to less than 1 mm, and the suspensions of smaller particles can be ingested or inhaled by respiration. Particulates may travel hundreds of miles from their point of origin, accompanying emissions of gas from the same source. The chemical nature of particulates can change in transit. Both increases and decreases of toxicity have been noted [15]. Also, during transit, the emitted gases may form aerosols (fine suspensions) through condensation and coagulation. These may react with other fine suspensions, leading to increases of particulate size with changed physiological effects if ingested or inhaled into human bodies.

In order to satisfy stringent environmental regulations, chimney effluents are controlled by one or more of four filtration processes: mechanical collectors, electrostatic precipitators, scrubbers and fabric baghouses. All the methods are more effective in capturing the larger size particles. Sometimes two types are used in series, with a cheap and relatively inefficient first stage acting to reduce the loading on a more efficient and more expensive second stage.

Mechanical collectors use gravity, inertia or centrifugal forces to separate (mainly) heavier particles from the gas. The simplest form consists of enlarged chambers in the gas stream that slows down the flow rate, enabling the heavier particles to settle. Other methods use centrifugal swirling of the gases, which causes heavier particles to travel to the outer walls and drop to the bottom of the chamber [1].

Electrostatic precipitators are used by the UK electricity generation industry. The flue gases are passed between a high-voltage electrode and a grounded collection plate. The ionized particles move towards the grounded plate, where they are collected and removed. As much as 99.9% (by weight) of the particles can be removed, but the precipitation works best on the heavier particles. With low sulphur coals the particles tend to be electrically more resistive. This makes electrostatic site precipitation more expensive because the precipitator may then need to operate in a hotter part of the flue gas, where the higher temperature reduces the resistivity of the carbon particles. Alternatively, bigger precipitators may be required.

Wet scrubbers are sometimes used to wash solid particles from the gas stream using water. This is essentially a physical scrubbing process that is different from the chemical scrubbing of sulphur oxides described in Section 4.4.2.1.

For the filtration of the finer particles, the most effective method is to use fabric filter baghouses. The flue gas is forced through fine filters, effecting filtration but causing a pressure drop. This increases the operating costs. The method is widely used for general purpose industrial applications. High temperature, corrosive gases due to coal combustion in power utility boilers pose particular problems for filter methods [1].

4.4.2.4. *Carbon dioxide*

The burning of coal releases carbon dioxide (CO_2) gas and thereby contributes to the possibility of global warming due to the accumulation of so-called "greenhouse" gases. This is discussed in Sections 2.6.6.1 and 2.6.6.2 of Chapter 2.

4.4.2.5. *Carbon dioxide emissions due to coal*

Table 4.9 gives figures for past and projected future emissions of carbon dioxide due to the use of coal. Comparable information is given in Table 5.9 of Chapter 5 for oil and in Table 6.6 of Chapter 6 for natural gas. Carbon dioxide emissions due to coal are seen to be smaller than those due to oil in the OECD countries and the Middle East but much larger in China. There is a close correlation between the amount of coal consumed (Table 4.4) and the consequent amount of carbon dioxide emitted (Table 4.9).

The total carbon dioxide emitted due to fossil fuel burning in 2010, in million metric tons carbon equivalent, was:

2814 due to coal (Table 4.9)

1774 due to natural gas (Table 6.6)

3863 due to oil (Table 5.9)

<u>8451</u> Total

Table 4.9. World carbon dioxide emissions from coal use, by region, high economic growth case, 1990–2020 [7]. In million metric tons carbon equivalent.

Region/Country	History			Projections				Average Annual Percent Change, 1999–2020
	1990	1998	1999	2005	2010	2015	2020	
Industrialized Countries								
North America	**520**	**595**	**592**	**676**	**713**	**737**	**794**	**1.4**
United States[a]	485	550	549	630	665	685	737	1.4
Canada	31	38	36	36	37	41	44	0.9
Mexico	4	7	6	9	10	11	13	3.6
Western Europe	**315**	**225**	**216**	**200**	**195**	**191**	**178**	**−0.9**
United Kingdom	68	42	39	37	36	34	28	−1.5
France	20	17	15	12	7	8	5	−4.8
Germany	137	87	83	79	81	80	77	−0.4
Italy	14	11	12	11	11	11	10	−0.5
Netherlands	11	13	11	10	8	8	7	−2.4
Other Western Europe	66	56	56	51	51	51	50	−0.5
Industrialized Asia	**104**	**130**	**135**	**149**	**156**	**166**	**172**	**1.2**
Japan	66	78	81	93	98	106	111	1.5
Australasia	38	52	53	56	58	60	62	0.7
Total Industrialized	**939**	**951**	**943**	**1,026**	**1,064**	**1,094**	**1,144**	**0.9**
EE/FSU								
Former Soviet Union	333	160	168	176	176	166	158	−0.3
Eastern Europe	189	127	113	112	98	78	66	−2.5
Total EE/FSU	**522**	**287**	**280**	**288**	**274**	**245**	**224**	**−1.1**
Developing Countries								
Developing Asia	**704**	**870**	**773**	**1,035**	**1,302**	**1,594**	**1,920**	**4.4**
China	514	600	495	702	927	1,175	1,457	5.3
India	101	148	156	189	212	233	253	2.3
South Korea	21	33	36	42	48	54	57	2.3
Other Asia	67	88	87	103	115	132	152	2.7
Middle East	**20**	**30**	**29**	**32**	**39**	**42**	**44**	**2.1**
Turkey	16	23	21	23	27	29	31	1.8
Other Middle East	4	7	7	9	12	13	13	2.7
Africa	**74**	**93**	**90**	**102**	**109**	**123**	**135**	**2.0**
Central and South America	**15**	**23**	**22**	**24**	**26**	**29**	**33**	**1.9**
Brazil	9	14	13	15	17	18	21	2.1
Other Central/South America	5	9	9	9	9	10	12	1.6
Total Developing	**812**	**1,017**	**914**	**1,194**	**1,476**	**1,788**	**2,132**	**4.1**
Total World	**2,274**	**2,254**	**2,137**	**2,507**	**2,814**	**3,127**	**3,500**	**2.4**

[a]Includes the 50 States and the District of Columbia. U.S. Territories are included in Australasia.
Notes: EE/FSU = Eastern Europe/Former Soviet Union.
Sources: **History:** Energy Information Administration (EIA), *International Energy Annual 1999*, DOE/EIA-0219(99) (Washington, DC, January 2001). **Projections:** EIA, *Annual Energy Outlook 2001*, DOE/EIA-0383 (2001) (Washington, DC, December 2000), Table B19; and World Energy Projection System (2001).

If the figures for carbon dioxide emissions due to coal, oil and natural gas in the three tables are compared with the total it is found that, from the viewpoint of carbon dioxide emission, natural gas is the cleanest fuel. Natural gas contributed $\frac{2653,1}{11164.3} = 23.8\%$ of world primary energy consumption in 2009 (Table 2.3), and $\frac{3314.4}{12476.6} = 26.6\%$ in 2012. Natural gas is projected to account for 30% of the world fossil fuel use in 2040. But because of its relatively low carbon intensity, the natural gas share of energy-related carbon dioxide emissions is then anticipated to be only 22% [25]. Oil use contributed to carbon emissions pro rata to its world consumption of 40%. Coal was relatively "dirty" in generating $\frac{2814}{2451} = 33.3\%$ of the total fossil fuel carbon dioxide emissions in 2009 (Table 4.9), while contributing $\frac{3278.3}{1164.3} = 29.4\%$ of the world primary energy consumption (Table 2.3).

The dominance of carbon dioxide in world greenhouse gas emissions is illustrated in Fig. 4.6 [16].

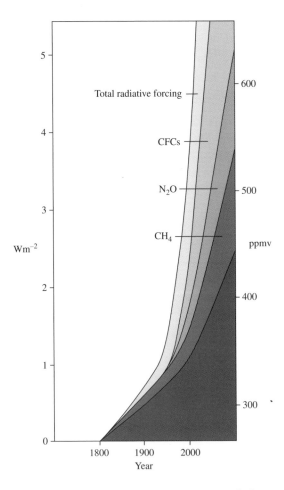

Fig. 4.6.   World greenhouse gas emissions [16].

4.4.2.6.   *An American case study. The side effects of a coal plant* [16]

A 500-MW coal plant produces 3.5 billion kilowatt-hours per year, enough to power a city of about 140,000 people. It burns 1,430,000 tons of coal, uses 2.2 billion gallons of water and 146,000 tons of limestone.

It also puts out the following, each year.

- *10,000 tons of sulphur dioxide.* Sulphur dioxide (SO_x) is the main cause of acid rain, which damages forests, lakes and buildings.
- *10,200 tons of nitrogen oxide.* Nitrogen oxide (NO_x) is a major cause of smog, and also a cause of acid rain.
- *3.7 million tons of carbon dioxide.* Carbon dioxide (CO_2) is the main greenhouse gas, and is the leading cause of global warming. There are no regulations limiting carbon dioxide emissions in the US.
- *500 tons of small particles.* Small particulates are a health hazard, causing lung damage. Particulates smaller than 10 microns are not regulated, but may be soon.
- *220 tons of hydrocarbons.* Fossil fuels are made of hydrocarbons; when they don't burn completely, they are released into the air. They are a cause of smog.
- *720 tons of carbon monoxide.* Carbon monoxide (CO) is a poisonous gas and contributor to global warming.
- *125,000 tons of ash and 193,000 tons of sludge from the smokestack scrubber.* A scrubber uses powdered limestone and water to remove pollution from the plant's exhaust. Instead of going into the air, the pollution goes into a landfill or into products like concrete and drywall. This ash and sludge consists of coal, ash, limestone and many pollutants, such as toxic metals like lead and mercury.
- *225 pounds of arsenic, 114 pounds of lead, 4 pounds of cadmium and many other toxic heavy metals.* Mercury emissions from coal plants are suspected of contaminating lakes and rivers in northern and north-eastern states of the USA and Canada. In Wisconsin alone, more than 200 lakes and rivers are contaminated with mercury. Health officials warn against eating fish caught in these waters, since mercury can cause birth defects, brain damage and other ailments. Acid rain also causes mercury poisoning by leaching mercury from rocks and making it available in a form that can be taken up by organisms.
- *Trace elements of uranium.* All but 16 of the 92 naturally-occurring elements have been detected in coal, mostly as trace elements below 0.1% (1,000 parts per million, or ppm). A study by the US DOE's Oak Ridge National Lab found that radioactive emissions from coal combustion are greater than those from nuclear power production.

The 2.2 billion gallons of water it uses for cooling is raised by 16°F on average before being discharged into a lake or river. By warming the water year-round it changes the habitat of that body of water.

Coal mining creates tons of hazardous and acidic waste which can contaminate groundwater. Strip (surface) mining also destroys habitats and can affect water tables. Underground mining is a hazard to water quality and to coal miners. In the mid-1970s, the fatality rate for underground miners was 0.4 per million tons of coal: one miner would be killed every two years to supply our 500-MW plant. The disabling injury rate was 38 people per million tons: 106 miners would be disabled every two years to supply this plant. Since coal mining is much more automated now, there are far fewer coal miners, and thus far fewer deaths and injuries.

Transportation of coal is typically by rail and barge; much coal now comes from the coal basins of Wyoming and the western USA. Injuries from coal transportation (such as at train crossing accidents) are estimated to cause 450 deaths and 6,800 injuries per year. Transporting enough coal to supply just this one 500-MW plant requires 14,300 train cars. That's 40 cars of coal per day.

4.5. Carbon Capture and Storage (CCS)

Carbon dioxide gas is released in the effluent from many fossil fuel combustion facilities. The biggest contributor in the USA and Western Europe is the generation of electricity from coal-fired power generating plants, illustrated in Fig. 4.7.

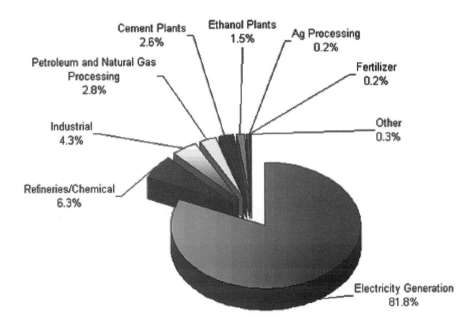

Fig. 4.7.   Carbon dioxide emissions from industrial sources [18].

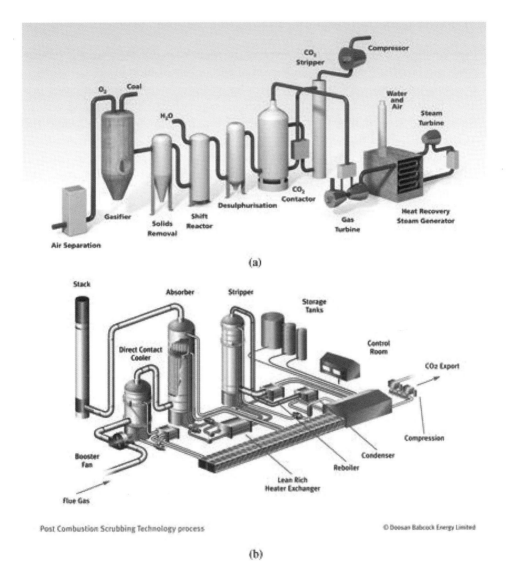

Fig. 4.8.   The principle of carbon capture and storage [19]. (a) Pre-combustion capture. (b) Post-combustion capture.

In order to reduce the amount of carbon dioxide emitted from fossil fuel burning entering the atmosphere it is proposed that a system of carbon capture and storage (sometimes known as "carbon sequestration and storage" or CCS) is used. The principle is illustrated in Fig. 4.8 [19].

Carbon dioxide gas is captured from the gaseous effluent arising from coal burning in electricity-generating stations. The process of separation and extraction is expensive and still releases 10–30% of the carbon dioxide. But it can combine with other processes to remove carbon monoxide and methane from the effluent.

The three methods for capturing CO_2 may be classified as follows.

- Post-combustion capture of CO_2 involves the separation of CO_2 from flue (stack) gases produced through the combustion of carbon-based fuels in air.
- Pre-combustion capture of CO_2 involves the pre-processing of carbon-based primary fuel to produce a synthesis gas (syngas) consisting of hydrogen and various carbon oxides, including CO_2, that can be separated and captured while the hydrogen is burned as fuel.
- Oxy-fuel combustion involves the burning of carbon-based fuel in pure oxygen, instead of air, to produce a flue gas consisting mainly of water and CO_2 that can be separated and captured.

The captured gas is compressed into pressurized gas or liquid form and transported by pipeline. Pressurization is an expensive process, especially if the fluid is compressed to a supercritical state to reduce its volume.

Pipeline technology is a mature process that has been widely used all over the world. Existing pipeline standards and regulations will generally dictate the composition of the fluid to be transported. In any pipeline technology there is a danger of leakage. The leakage of carbon dioxide could be dangerous to people and animals due to the possibility of asphyxiation.

There are a number of options for storing CO_2.

- Abandoned deep coal mines.
- Saline aquifers (rock formations which hold water): the underground store with the largest potential capacity.
- Depleted oil and gas fields: an alternative underground store with a potential capacity several times smaller than aquifers.
- Enhanced oil recovery sites: a comparatively smaller-capacity underground store likely to be developed first since the CO_2 injection improves the economics of the oilfield; however, this benefit only lasts until the oilfield is depleted, and some of the CO_2 is extracted with the oil [17].

For all of the above options there are questions about the reliability of the storage: how long will pressurized CO_2 stay underground? The public needs to be assured that the storage facilities are safe and reliable. Since this is a new technology, there is no body of experience to accurately predict the leakage rate. Perhaps the biggest challenge in CCS is finding storage caverns with an absorbent surrounding layer and an impermeable surface layer.

It should be noted that CCS refers only to carbon dioxide and does not affect emissions of sulphur and nitrogen oxides (i.e. acid rain).

There are a number of experimental and prototype plants in operation. The largest working project in Norway in 2010 pumped 1 million tons of CO_2/year into a natural gas basin. But it would take 5–10 of such projects to store the emissions of a single large coal-fired plant.

Carbon capture and storage is an excellent idea in principle. Economically, however, it is very expensive. One has to question if it could be made viable on a large commercial scale.

4.6.   Advanced Coal Technologies

Research and development of the use of solid fuels has two main aims: to improve the efficiency of energy conversion and to reduce the adverse environmental effects.

4.6.1.   *Fluidized-bed combustion*

Fluidization is a technology for causing small solid particles to behave as if they were a liquid. Coal is burned in a mixture with limestone (to absorb sulphur) suspended in a stream of combustion air rising from beneath the heated bed (Fig. 4.9) [2]. Optimum combustion and heat transfer to the heating coils requires the appropriate design of the air flow and coal configuration. Compared with the combustion of pulverized coal, the sulphur dioxide is largely absorbed within the high-pressure gas circuit and nitrogen oxide emissions are substantially reduced due to lower working temperatures. The combustion techniques can also utilize lower grades of coal [20].

Pressurized fluidized-bed systems are more efficient and result in lower levels of emissions than atmospheric level fluidized-bed systems.

4.6.2.   *Combined-cycle generation*

Combustion systems that combine a gas-turbine-driven generator with a separate steam-turbine-driven generator are called "combined-cycle" systems. Figure 4.9

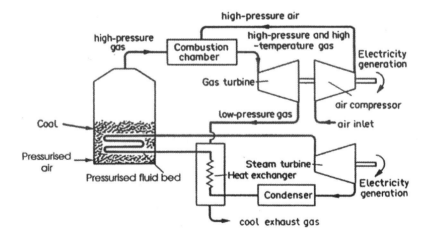

Fig. 4.9.   Combined-cycle integrated gas and steam turbines [2].

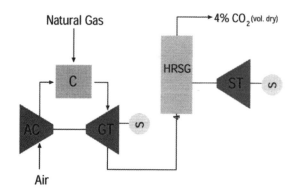

Fig. 4.10.   A schematic of the gas turbine combined-cycle power generation system [21]. AC = air conditioner; HRSG = heat recovery steam generator; GT = gas turbine; ST = steam engine.

shows an integrated system in which a fluidized-bed heater simultaneously heats the fluid in the hot water/steam loop feeding the steam turbine and also heats the coal gas feeding the gas turbine (the hotter the gas, the more efficient the gas turbine operation). Exhaust heat from the gas turbine, in the form of low-pressure gas, is used to reinforce the heat transferred from the fluidized-bed steam circuit. This is inherently the most efficient power cycle available and is gradually overtaking steam-only generation. A steam-only station, most of which are coal-fired, has a full-load efficiency of about 30%. Combined-cycle plants now have full-load efficiencies greater than 50%, although the fuel costs are greater. The modern practice is to combine a coal-fired steam plant with a natural gas system (Fig. 4.10) [21]. It is also possible to combine the steam cycle with heat energy from nuclear, wind or solar generation systems [22]. A relevant solar system is described in Section 11.5.3 of Chapter 11.

Large scale combined-cycle plants would be rated with a 400-MW gas turbine coupled to a 200-MW steam turbine.

In 1999 the USA had 199 combined-cycle units with a capacity of 16817 MW, which was only 2% of the total electric power capability of 785990 MW [23]. It was planned to add an additional 71 combined-cycle units, with nameplate capacity of 17288 MW, through to the year 2004. In the UK the use of gas turbine combined-cycle stations increased from zero in 1973 to 11.6% of the total generation in 1994.

The use of natural gas is now increasing to the extent that one-third of the total UK electricity generation is expected to come from combined-cycle plants within the next few years.

The chief advantage of natural-gas-steam combined-cycle generation compared with steam only is the great increase of overall efficiency. In addition, the use of combined-cycle systems results in a much lower coal use and a significant reduction of greenhouse gases. There are big reductions of sulphur dioxide, nitrogen oxides, carbon dioxide, particulates and fly ash.

It is important to note that combined-cycle operation, where two fuel sources are used, is not the same as cogeneration or combined heat and power (CHP), reported in Section 3.8 of Chapter 3. The two technologies are quite different, although sometimes they are incorporated on the same plant site.

4.7. Liquid Fuels from Coal

The extraction of liquid fuels from coal has been the subject of much investigation for over 100 years. It is not an active process at present.

Four main approaches are used:

1. indirect liquefaction;
2. pyrolysis;
3. solvent extraction; and
4. direct hydrogenation (catalytic liquefaction).

4.7.1. *Indirect liquefaction*

In this process the feed coal is completely gasified to produce a "synthesis" gas. Purification methods are then used to remove particulates and sulphur compounds. Liquid products, usually rich in aromatic compounds ranging from gasoline to heavy oils and waxes or methanol, are obtained by high temperature and high-pressure reactions in the presence of appropriate catalysts. Indirect liquefaction processes have low thermal efficiencies, of the order 40–45%.

The best-known method is the Fischer–Tropsch, developed in Germany in the 1930s. A large-scale commercial operation is the SASOL plants in South Africa, which initially produced petrochemicals and fertilizers. By the mid-1980s three plants used 30 million tonnes of coal per year to produce gasoline, diesel fuel, LNG, kerosene and other products [20].

A great advantage of indirect liquefaction, compared with the three direct methods described below, is that it does not require the heavy use of hydrogen in refining the heavier liquids. This may be a significant cost advantage.

4.7.2. *Pyrolysis*

In pyrolysis the coal is heated in the absence of air or oxygen, which breaks down the coal molecules to form ethane and methane, leaving a remnant called "char". Hydrogenation can be achieved using the intrinsic hydrogen from the coal or by applying a stream of hydrogen, which improves the liquid yield. Crude benzoil refining using pyrolysis began about a hundred years ago. Although the thermal efficiency can exceed 80%, the overall production cost is not, at present (2013), competitive with natural crude oil.

4.7.3.   *Solvent extraction*

In the solvent extraction processes, finely crushed and dried coal is treated with a hot, liquid, hydrogenated solvent oil, derived from coal.

The hydrogenation reaction is catalysed to some extent by inorganic substances in the coal. After processing it is possible to generate hydrogen plus a range of hydrocarbon products that can be upgraded to gasoline or diesel fuel. One of the technical problems in the solvent extraction is how to separate the undigested coal and ash from the liquid products [1]. The heavier residual oils can be recycled as the solvent oil [2, 24]. Thermal efficiencies of 60–65% have been realized in experimental process reactors. Solvent extraction methods can typically yield 2.5–3 barrels of liquid per ton of coal.

4.7.4.   *Direct hydrogenation (catalytic liquefaction)*

With the direct hydrogenation process a reducing gas such as high-purity hydrogen at high temperature and pressure reacts with the coal in the presence of a catalyst. Some catalytic processes use a suspension of coal in heavy oil and pass this over a catalyst bed to avoid intimate contact between the coal and the catalyst. This eliminates the need for the large-scale replacement of catalysts that would arise where the catalyst is in direct contact with the coal. The coal products are mainly aromatic. In the UK the Imperial Chemical Industries plant at Billingham had a capacity of 100,000 tonnes per year of liquid fuel in the 1930s.

The commercial viability of extracting liquid fuel from coal depends on the supply and price of oil. At the present time (2013) the price of natural crude oil is high enough for synthetic liquid fuels from coal to represent serious competition. It is not clear if the use of coal liquids is environmentally "cleaner" than the burning of coal itself. The combustion processes in coal liquefaction may be cleaner but the overall processes themselves may represent a greater industrial hazard because of the distillation of impurities. There is a concentration of polycyclic organic matter in the heavy fractions of coal liquids [1].

4.8.   Problems and Review Questions

4.1. Use the information in Tables 4.3 and 4.4 to list the world's ten largest producers and consumers of coal at the present time.

4.2. Which countries of the world were big exporters or importers of coal during the year 2009?

4.3. In which countries of the world did coal production (a) increase and (b) decrease most significantly between 2000 and 2010?

4.4. In which countries of the world did consumption of coal (a) increase and (b) decrease most significantly between 2000 and 2010?

4.5. Use the data of Table 4.6 and Fig. 4.3 to discuss the figures for coal mined and coal used in the UK from 1980 to 2009.

4.6. Use the data of Tables 4.3 and 4.4 to discuss the figures for coal production and consumption in the USA between 2000 and 2010.

4.7. Compare the UK's coal production since 1980 due to (a) deep mining and (b) open-cast (surface) mining.

4.8. How has the proportion of coal used in the UK domestic sector changed from 1980 onwards? What were the reasons for the change?

4.9. How has the proportion of coal used in electricity generation in the UK changed since 1980?

4.10. Despite high labour costs, the price of US coal remained lower than those of European and Japanese coal during the 1990s. Why was this so?

4.11. The USA accounts for more than one-quarter of world coal consumption. What is the coal used for?

4.12. What are the (a) advantages and (b) disadvantages of burning domestic coal in open fire places?

4.13. What are the (a) advantages and (b) disadvantages of using coal as the prime fuel for electricity generation?

4.14. How is coal transported from the pit-head to the user sites?

4.15. Explain the nature and purpose of coal slurries.

4.16. What are the main pollutants due to coal burning?

4.17. Explain what is meant by the term "acid rain". How does this occur and what are its effects?

4.18. Briefly describe the nature of the electrostatic precipitation operations used in power utility boiler emission systems.

4.19. Briefly explain four methods of obtaining synthetic liquid fuels from coal. Why are such methods not widely used to supplement the world's dwindling oil supplies?

4.20. In the USA, as in Western Europe, the coal mining industry has passed through a deep recession in the past 20 years. Why is this so?

References

[1] "Energy in Transition 1985–2010", Final Report of the Committee on Nuclear and Alternative Energy Systems, Chapter 4, National Research Council, Washington, DC, 1980.

[2] McVeigh, J.C. (1984). *Energy around the World*, Pergamon Press, Oxford.

[3] Arnold, G. (1985). *Coal, Energy Today Series*, Gloucester Press, New York.

[4] "UK Energy in Brief 2010", Dept. of Energy and Climate Change, London, July 2010.

[5] "BP Statistical Review of World Energy", British Petroleum plc, London, June 2011.

[6] Shepherd, W. and Shepherd, D.W. (2003). *Energy Studies*, 2nd Edition, Imperial College Press, London.

[7] "International Energy Outlook 2001", US Energy Information Administration (EIA), Document DOE/EIA-0484, Washington, DC, March 2001.

[8] "US Coal Industry Annual 1998 Executive Summary", Energy Information Administration (EIA), Electric Power Monthly, Document DOA/EIA-0226(99/03), Washington, DC, March 1999.

[9] Flynn, E.J. (2000). "Impact of Technological Change and Productivity on the Coal Market", in *Issues in Midterm Analysis and Forecasting 2000*, Energy Information Administration, Washington, DC.

[10] "Methane from Coal Seams Technology", US Gas Research Institute, Washington, DC, March 1990.

[11] Harvard Business School (1983). *Energy Future*, 3rd Edition, Vintage Books, New York.

[12] Dorf, R.C. (1978). *Energy, Resources and Policy*, Addison-Wesley Publishing Company, Reading, MA.

[13] Author's (WS) personal reminiscence.

[14] "Emissions of Greenhouse Gases in the USA, 1999 — Executive Summary", EIA Report EIAIDOE 0573(99), US Dept. Of Energy, Washington, DC, October 2000.

[15] Amdur, M (1976). "Toxicological Guidelines for Research on Sulphur Oxides and Particulates", Proceedings of the Fourth Symposium of Statistics and the Environment, National Academy of Sciences, Washington, DC.

[16] "Coal Information 1997", International Energy Agency, Paris, September 1998.

[17] "The Future for Coal", *New Scientist*, January 23 1993, pp. 20–41.

[18] "Carbon Capture and Geologic Sequestration" (2008). Candace Cady, Utah Division of Water Quality, Utah, http://www.climatechange.utah.gov/capture_ sequestration.htm.

[19] Carbon capture and Storage Association (CSSA) London, England, 2013 (a) http://www.ccsassociation.org/what-is-ccs/capture/pre-combustion-capture/ (b)http://www. ccsassociation.org/index.php/what-is-ccs/capture/post-combustion-capture/.

[20] Myers, R.A. (ed.) (1981). *Coal Handbook*, Marcel Dekker Inc., New York.

[21] "Natural Gas Fired Combined Cycle", from *Greenhouse Gas Emissions from Power Stations*, London, 2000, http://www.ieagreen.org.uk/emis5.htm.

[22] "Feasibility of a Solar-Driven Combined-Cycle", Section 3 — System Design and Performance, ConSolar Project, February 2001, http://magnet.consortia.org.il/ConSolar//Tintin/stepp4.htm.

[23] "Annual Electric Generator Report Utility", Form EIA-860A, Energy Information Administration, Washington, DC, December 2000.

[24] Ezra, D. (1981). "A Review Plan for Coal", The Mining Engineer, UK, January 1981.

[25] "International Energy Outlook 2013", US Energy Information Administration (EIA), Document DOE/EIA-0484(2013), Washington, DC, July 2013.

CHAPTER 5

OIL

5.1. Geological Formation

The sequence of geological formation of oil and natural gas is illustrated in Fig. 5.1. Possible products from each stage are also shown and include several of the fuels that we now regard as unconventional fuel sources.

5.2. Origin and Extraction of Oil

In the energy industries the word "petroleum" is usually understood to include both "oil" and "natural gas". But in this chapter the word refers only to oil.

It is believed that oil was formed principally from the remains of marine plants and animals buried in sedimentary rocks. The earliest oils have been found in Precambrian rock formations (i.e. more than 1,000 million years old) and predate the formation of coal by hundreds of millions of years; long before vegetation appeared on the land masses [2,3]. Because of the marine origin, oil deposits are widely distributed on the earth. They frequently occur in present coastal areas, beneath continental shelves and under inland areas that were once ocean beds. Large accumulations occur in lacustrine rocks (i.e. rocks associated with lakes or wetlands). After extraction from sediments, naturally-occurring liquid oil is known as "crude oil". It then needs to be refined for most of its applications.

Most of the world's oilfields extend only a few square miles, although some of the largest fields cover several hundreds of square miles. Some deposits contain only natural gas and these are often found at great depths. Where a deposit contains oil it also contains natural gas, in layers above the oil or dissolved into the oil (like carbon dioxide gas dissolved into lemonade to give it "fizz"). For high-pressure deposits the oil and gas, combined in the reservoir as "condensate", may separate naturally before they reach the surface. This separation can be reinforced by pressurized recovery techniques. Modern oil exploration is now very sophisticated at predicting

135

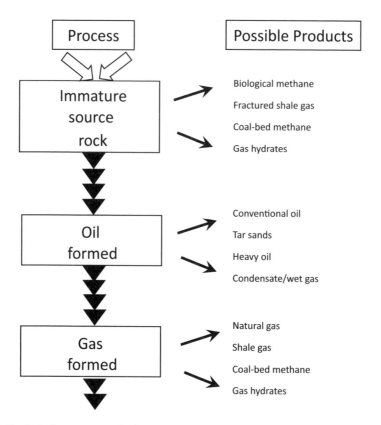

Fig. 5.1.   Geological sequence in the formation of conventional and unconventional fossil fuels [1].

the locations of oil deposits (occurrences), using remote sensing and geophysical techniques, but such predictions can only be confirmed by drilling. Not all oil well drilling strikes oil. In spite of the reasonable scientific expectation the success ratio is only about 1 in 5. Drilling and exploration costs for new well sites constitute over 40% of oil company capital expenditures.

When an oilfield is successfully located, not all of the oil is recoverable. If the natural oil pressure in the deposit is sufficient to drive the liquid crude oil upwards to the ground or ocean bed surface, then about 30% of the oil reservoir can be economically recovered (primary recovery). If the natural pressure of an oil deposit is enhanced by injecting pressurized air or water or by reinjecting natural gas, a higher proportion of the oil deposit can be recovered but at consequent greater expense (secondary recovery). Any reinjected gas can be eventually recovered. A still greater yield can be obtained by other methods of recovery (tertiary recovery), such as chemical injection or microbial injection. This is one of the reasons why estimates of oil deposits (occurrences) and recoverable reserves are rather speculative and dependent on economic as well as technical considerations. The extraction of oil results in high revenue for the oil companies, compared with the extraction of natural gas only.

The general term "petroleum" is now used to cover a wide range of hydrocarbons, including natural gas. Various "cuts" of hydrocarbons include bitumen and solid paraffin waxes. Petrol (or gasoline) and diesel fuel, used as engine fuels, have to be obtained by the "refining" of crude oil.

Oil is measured by volume, but characterized by weight. A barrel of oil is 159 litres, which is 42 US gallons or 34.97 imperial gallons. In terms of weight a barrel of oil is 0.136 tonnes and has an equivalent heat energy of 5.694×10^9 J. For the large-scale use of oil it is customary to use MMBL (million barrels of oil) or the quantity million tonnes (mto). To afford a direct comparison of the energy values, other fuels such as coal or natural gas use the quantity MMBL oil equivalent or million tonnes of oil equivalent (mtoe).

The capacity of an oilfield is usually estimated in millions of barrels. For example, the Prudhoe Bay deposit in Alaska has been estimated at 12 billion (12×10^9) barrels. From British Petroleum's offshore fields in the North Sea, between Scotland and Norway, the production has been about a million (10^6) barrels per day. This compares with the 2010 world production figure of 87,439 thousand barrels daily [4, 5].

5.3. History and Development of the Petroleum Industry

Oil and its products have been used in some of the earliest civilizations, such as those of Mesopotamia (now Iraq) and Egypt (i.e. 3000–2000 BC). This was usually obtained from surface seepages. In the form of asphalt it was used for caulking boats, setting jewels into frames, preserving mummies and for building construction and waterproofing.

Forms of light oil produced by elementary methods of distillation were used by the Arabs as incendiary weapons of war as early as the 7th century [6]. In the 18th century, oil produced from oil shale was used for street lighting in Modena, Italy, and for paraffin wax candles in Scotland.

Modern oil exploration in the form of the familiar drilling rig began in Pennsylvania, USA, in 1859 with Edwin L. Drake's well, which was about 20 metres deep. By the end of the 19th century in the USA, Standard Oil of New Jersey — owned by John D. Rockefeller, who consequently became the richest man in the world — held a virtual monopoly over oil production. The Standard Oil industrial cartel was broken up into smaller units by US anti-trust legislation in 1911. The international oil industry was dominated by the USA, because of its vast natural resources and business acumen, until the period of the Second World War (1939–1945). The great oilfields of the Middle East, particularly Saudi Arabia, now a repository of much of the world's reserves, were discovered and developed during the 1920s and 1930s by American and European explorers and engineers.

5.3.1. *The Seven Sisters* [7]

Until 1973 the production and price of crude oil were mostly controlled by seven great international companies, known in the oil trade as the "Seven Sisters". In descending order of assets, these were: Exxon (USA), Royal Dutch Shell (Netherlands/Britain), Texaco (USA), Gulf Oil (USA), Mobil (USA), British Petroleum (BP; Britain) and Socal (now Chevron; USA). In the UK Exxon used to operate under the trade name of "Esso". A 1999 merger between BP and the American Oil Company (Amoco), and later with the Atlantic Richfield Company (ARCO), moved BP/Amoco into third place in the Seven Sisters list. Exxon and Mobil joined forces recently to become the world's largest oil company.

In 2011 the great oil companies of the world, in alphabetical order, were
British Petroleum (BP),
Chevron,
Conoco Phillips,
Exxon Mobil,
Royal Dutch Shell and
Total.

The huge financial profits made by the Seven Sisters from Middle Eastern oil lured other American companies into the Middle Eastern market during the 1950s. Companies such as Standard Oil of Indiana (now Amoco), Standard Oil of Ohio (SOHIO), Continental (CONOCO) and ARCO also became large, powerful and wealthy. Various individual business entrepreneurs, such as John Paul Getty and Harold Layfayette Hunt, operated independently to their immense profit. Armand Hammer turned Occidental into a large and profitable company. Relations between the independents and the Seven Sisters were often conducted in terms of bitter rivalry and contributed to the eventual downfall of the Western oil monopoly.

Two major oil discoveries occurred in the 1960s and 1970s, both outside the Middle East. In 1959 a joint Exxon–Shell team found the giant Groningen natural gas field offshore of Holland. Also, in 1965 BP found natural gas off the coast of England and later discovered oil in the northern North Sea. Meanwhile, in 1968, one of the American independents (Arco) discovered oil in Prudhoe Bay, Alaska. For financial reasons Arco was obliged to bring in the giant Exxon, which soon became the dominant partner. Nine months later BP also struck oil in Alaska. The Alaskan fields proved to be massive, exceeding in size the deposits in western Texas.

5.3.2. *European oilfields*

In 1970 BP discovered the giant Forties oilfield under the North Sea, north of Aberdeen, Scotland. One year later, in 1971, Shell/Exxon found the equally large Brent oilfield off the Shetland Islands, north of Scotland. During the 1970s and 1980s a significant number of large and medium-sized oilfields were discovered in the North

Sea. Having been a consumer nation throughout its history, the UK was then faced with the challenges of becoming a major producer. Initially the British government was lacking in both the expertise and the will to become an oil producer. Huge areas of the sea for exploration were leased to US companies at ridiculously low rates, which excluded them as sources of UK tax revenue. By 1973 BP controlled 20% and Shell 15% of the North Sea oil but the majority was still controlled by the American sisters. It was not until seven years after the first exploration concessions that the UK parliament eventually caught up with the extent of the oil companies bonanza. A government report in 1973 was a devastating criticism of the UK civil service, in that Britain would receive a smaller share of oil revenues than other countries and that British taxes were being pre-empted by the tax demands of administrations abroad [7]. A new government, elected in 1974, passed legislation to resolve the balance in favour of the UK taxpayers.

The UK government was a significant shareholder in BP from 1914 to 1987. For example, in 1974, 49% of BP stock was held by the government, although BP was never state-owned. This asset was largely sold off by privatization in 1987, leaving only a 1.8% shareholding. The small remainder was fully liquidated in December 1995 and raised more than £500 million [1].

5.3.3.   *OPEC*

In 1960 an organization was formed by the major oil producer countries (excluding the USA) to counter the influence of the Seven Sisters. This became known (and remains known) as the Organization of Petroleum Exporting Countries (OPEC). The 12 OPEC countries, listed in Table 2.1 of Chapter 2, led by Saudi Arabia, contain more than 65% of the world's oil reserves. In 1973, after much dispute between the supplier countries of OPEC and the Western oil companies, the OPEC suppliers acted unilaterally and suddenly increased the price of crude oil. The result was widespread disruption in the industries of North America and Western Europe. The price of oil quadrupled in a year, leading to what Western politicians described as the "oil crisis". In effect, control over the production and price of oil had passed from the historical masters to the producers. The price and production of oil are now fixed by negotiations between OPEC and the oil producers, but subject to consultation with the governments of various concerned countries. There are a number of major producers (including Russia, the USA, the UK and Norway) who are not members of OPEC.

The historical price of crude oil is shown in Fig. 5.2 [4]. Characteristics are shown for both the historical price at the time and the price adjusted to the 2009 value of the US dollar. For 100 years the price of crude oil was uniformly low: of the order of a few US dollars/barrel. This increased dramatically during the 1970s but fell again during the 1980s. Table 5.1 shows that, in 2000, Middle Eastern oil rose to $26.24/barrel compared with $1.90/barrel in 1972; a 14-fold increase. The price of

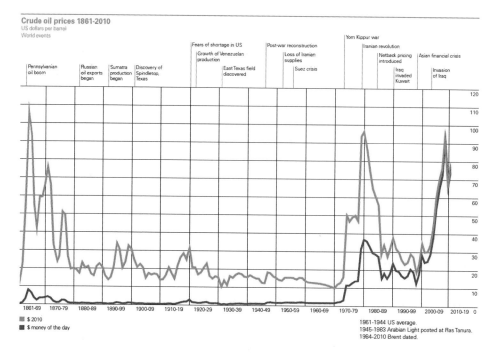

Fig. 5.2.   Crude oil prices, 1861–2010 [4].

oil slumped to about \$18/barrel in 2001, so that the OPEC producers and some major non-OPEC producers introduced cutbacks of production to maintain the oil price. In 2011 the price of oil reached more than \$100/barrel.

Figure 5.2 cites certain "world events", demonstrating the volatility of the price of crude oil, linked particularly to major political events in the Middle East in the past 30 years. The years 2005–2010 were particularly volatile in the oil business due to warfare in the Middle East and to major oil discoveries in Russia and West Africa.

5.3.4.   *Major oil areas*

The major oil producing areas of the world in 2011 were:

Russia, Kazakhstan, Azerbaijan,
East Texas,
Persian Gulf,
Western Canada,
Gulf of Mexico,
Venezuela,
Niger Delta and
North Sea (Norway and the UK).

Further detail concerning particular countries is included in Section 5.5, below.

Table 5.1. Crude oil prices, 1973–2010.

Spot crude prices

US dollars per barrel	Dubai $/bbI*	Brent $/bbI[†]	Nigerian forcados $/bbI	West Texas intermediate $/bbI[‡]
1973	2.83	—	—	—
1974	10.41	—	—	—
1975	10.70	—	—	—
1976	11.63	12.80	12.87	12.23
1977	12.38	13.92	14.21	14.22
1978	13.03	14.02	13.65	14.55
1979	29.75	31.61	29.25	25.08
1980	35.69	36.83	36.98	37.96
1981	34.32	35.93	36.18	36.08
1982	31.80	32.97	33.29	33.65
1983	28.78	29.55	29.54	30.30
1984	28.06	28.78	28.14	29.39
1985	27.53	27.56	27.75	27.98
1986	13.10	14.43	14.46	15.10
1987	16.95	18.44	18.39	19.18
1988	13.27	14.92	15.00	15.97
1989	15.62	18.23	18.30	19.68
1990	20.45	23.73	23.85	24.50
1991	16.63	20.00	20.11	21.54
1992	17.17	19.32	19.61	20.57
1993	14.93	16.97	17.41	18.45
1994	14.74	15.82	16.25	17.21
1995	16.10	17.02	17.26	18.42
1996	18.52	20.67	21.16	22.16
1997	18.23	19.09	19.33	20.61
1998	12.21	12.72	12.62	14.39
1999	17.25	17.97	18.00	19.31
2000	26.20	28.50	28.42	30.37
2001	22.81	24.44	24.23	25.93
2002	23.74	25.02	25.04	26.16
2003	26.78	28.83	28.66	31.07
2004	33.64	38.27	38.13	41.49
2005	49.35	54.52	55.69	56.59
2006	61.50	65.14	67.07	66.02
2007	68.19	72.39	74.48	72.20
2008	94.34	97.26	101.43	100.06
2009	61.39	61.67	63.35	61.92
2010	78.06	79.50	81.05	79.45

*1972–1985 Arabian Light, 1986–2010 Dubai dated.
[†]1976–1983 Forties, 1984–2010 Brent dated.
[‡]1976–1983 Posted WTI prices, 1984–2010 Spot WTI (Cushing) prices.

5.4.   World Oil Reserves

Estimates of proved oil reserves tend to be revised upwards each year as more sophisticated methods of evaluation are employed and recovery rates increase due to improved technology. The upward revisions are, however, marginal and at present are roughly keeping pace with increased oil consumption. Depletion of the world's oil reserves is largely balanced by additions from new discoveries and higher recovery rates from existing fields [4].

Numerical values of reserves since 1990 are given in Table 5.2 [4]. The right-hand column of this table gives the amount of oil remaining as the reserve/production ratio (R/P) in years. The world total shows that there was at the end of 2010 only 46.7 years oil remaining. Of the world reserves, 54.4% is in the Middle East and a total of 77.2 % in the OPEC countries.

The geographical distribution of the presently known crude oil deposits is shown in Fig. 5.3, which correlates with Table 5.2. Increased discoveries in Venezuela now surpass the Middle Eastern reserves. Russia, Kazakhstan and Azerbaijan contained 9.0% of the world reserves at the end of 2010. It is obvious that urgent planning is needed on the part of the Western countries, especially Western Europe, in anticipation of their own oil supplies running out. Figure 5.3 shows that the world R/P ratio has been roughly constant at 40+ years since the late 1980s, due to revised upward reserve estimates. But it would be unwise to assume that this trend will always continue.

When world oil consumption, which is rising, starts to outstrip the revised reserve estimates, the years of remaining world oil reserves will inevitably decline. Only the discovery of new massive, extractable, economic crude oil deposits would cause the years of reserves to rise dramatically. Improved methods of primary recovery (see Section 5.1 above) and re-entry to oilfields that had stopped producing are major new ways of increasing production and would increase available world reserves. Some authors would dispute the accuracy of the estimate of 40 years. Their argument is based on the assumption that political and (especially) economic factors will reduce the rate of oil usage in future years, thereby prolonging the life of the oil supply [8].

There has to be great concern over the possible effects of severe oil shortages. In 1992 Iraq invaded the adjacent small country of Kuwait and rapidly conquered it, so posing a threat to other oil-producing countries, notably Saudi Arabia. This provoked an immediate response from the Western allies, led by the USA. A short and brilliantly successful military campaign, known in the West as the first Gulf War, resulted in an Iraqi withdrawal from Kuwait. A second military campaign, beginning in 2003 with the invasion of Iraq, led to massive dislocation of the oil industry. The Middle East remains a region of the world subject to political uncertainties. There can be no guarantee that Middle Eastern oil will continue to be available to the rest of the world in sufficient quantity and at an acceptable price. This will

Table 5.2. World reserves (years) of natural crude oil [4].

	At end 1990 Thousand million barrels	At end 2000 Thousand million barrels	At end 2009 Thousand million barrels	At end 2010 Thousand million tonnes	At end 2010 Thousand million barrels	At end 2010 Share of total	R/P ratio
US	33.8	30.4	30.9	**3.7**	**30.9**	2.2%	11.3
Canada	11.2	18.3	32.1	**5.0**	**32.1**	2.3%	26.3
Mexico	51.3	20.2	11.7	**1.6**	**11.4**	0.8%	10.6
Total North America	96.3	68.9	74.6	**10.3**	**74.3**	5.4%	14.8
Argentina	1.6	3.0	2.5	**0.3**	**2.5**	0.2%	10.6
Brazil	4.5	8.5	12.9	**2.0**	**14.2**	1.0%	18.3
Colombia	2.0	2.0	1.4	**0.3**	**1.9**	0.1%	6.5
Ecuador	1.4	4.6	6.3	**0.9**	**6.2**	0.4%	34.1
Peru	0.8	0.9	1.1	**0.2**	**1.2**	0.1%	21.6
Trinidad & Tobago	0.6	0.9	0.8	**0.1**	**0.8**	0.1%	15.6
Venezuela	60.1	76.8	211.2	**30.4**	**211.2**	15.3%	*
Other S. & Cent. America	0.6	1.3	1.4	**0.2**	**1.4**	0.1%	28.9
Total S. & Cent. America	71.5	97.9	237.6	**34.3**	**239.4**	17.3%	93.9
Azerbaijan	n/a	1.2	7.0	**1.0**	**7.0**	0.5%	18.5
Denmark	0.6	1.1	0.9	**0.1**	**0.9**	0.1%	9.9
Italy	0.8	0.9	1.0	**0.1**	**1.0**	0.1%	25.0
Kazakhstan	n/a	25.0	39.8	**5.5**	**39.8**	2.9%	62.1
Norway	8.6	11.4	7.1	**0.8**	**6.7**	0.5%	8.5
Romania	1.5	1.2	0.5	**0.1**	**0.5**	◆	14.8
Russian Federation	n/a	59.0	76.7	**10.6**	**77.4**	5.6%	20.6
Turkmenistan	n/a	0.5	0.6	**0.1**	**0.6**	◆	7.6

(Continued)

Table 5.2. (*Continued*)

	At end 1990 Thousand million barrels	At end 2000 Thousand million barrels	At end 2009 Thousand million barrels	At end 2010 Thousand million tonnes	At end 2010 Thousand million barrels	At end 2010 Share of total	R/P ratio
United Kingdom	4.0	4.7	2.8	**0.4**	**2.8**	0.2%	5.8
Uzbekistan	n/a	0.6	0.6	**0.1**	**0.6**	♦	18.7
Other Europe & Eurasia	65.3	2.3	2.3	**0.3**	**2.4**	0.2%	17.5
Total Europe & Eurasia	80.8	107.9	139.2	**19.0**	**139.7**	10.1%	21.7
Iran	92.9	99.5	137.0	**18.8**	**137.0**	9.9%	88.4
Iraq	100.0	112.5	115.0	**15.5**	**115.0**	8.3%	*
Kuwait	97.0	96.5	101.5	**14.0**	**101.5**	7.3%	*
Oman	4.4	5.8	5.5	**0.7**	**5.5**	0.4%	17.4
Qatar	3.0	16.9	25.9	**2.7**	**25.9**	1.9%	45.2
Saudi Arabia	260.3	262.8	264.6	**36.3**	**264.5**	19.1%	72.4
Syria	1.9	2.3	2.5	**0.3**	**2.5**	0.2%	17.8
United Arab Emirates	98.1	97.8	97.8	**13.0**	**97.8**	7.1%	94.1
Yemen	2.0	2.4	2.7	**0.3**	**2.7**	0.2%	27.7
Other Middle East	0.1	0.2	0.1	**†**	**0.1**	♦	9.3
Total Middle East	659.6	696.7	752.6	**101.8**	**752.5**	54.4%	81.9
Algeria	9.2	11.3	12.2	**1.5**	**12.2**	0.9%	18.5
Angola	1.6	6.0	13.5	**1.8**	**13.5**	1.0%	20.0
Chad	—	0.9	1.5	**0.2**	**1.5**	0.1%	33.7
Republic of Congo (Brazzaville)	0.8	1.7	1.9	**0.3**	**1.9**	0.1%	18.2
Egypt	3.5	3.6	4.4	**0.6**	**4.5**	0.3%	16.7
Equatorial Guinea	—	0.8	1.7	**0.2**	**1.7**	0.1%	17.1

(*Continued*)

Table 5.2. (*Continued*)

	At end 1990 Thousand million barrels	At end 2000 Thousand million barrels	At end 2009 Thousand million barrels	At end 2010 Thousand million tonnes	At end 2010 Thousand million barrels	At end 2010 Share of total	R/P ratio
Gabon	0.9	2.4	3.7	**0.5**	**3.7**	0.3%	41.2
Libya	22.8	36.0	46.4	**6.0**	**46.4**	3.4%	76.7
Nigeria	17.1	29.0	37.2	**5.0**	**37.2**	2.7%	42.4
Sudan	0.3	0.6	6.7	**0.9**	**6.7**	0.5%	37.8
Tunisia	1.7	0.4	0.4	**0.1**	**0.4**	◆	14.6
Other Africa	0.9	0.7	0.7	**0.2**	**2.3**	0.2%	44.2
Total Africa	58.7	93.4	130.3	**17.4**	**132.1**	9.5%	35.8
Australia	3.2	4.9	4.1	**0.4**	**4.1**	0.3%	19.9
Brunei	1.1	1.2	1.1	**0.1**	**1.1**	0.1%	17.5
China	16.0	15.2	14.8	**2.0**	**14.8**	1.1%	9.9
India	5.6	5.3	5.8	**1.2**	**9.0**	0.7%	30.0
Indonesia	5.4	5.1	4.3	**0.6**	**4.2**	0.3%	11.8
Malaysia	3.6	4.5	5.8	**0.8**	**5.8**	0.4%	22.2
Thailand	0.3	0.5	0.4	**0.1**	**0.4**	◆	3.6
Vietnam	0.2	2.0	4.5	**0.6**	**4.4**	0.3%	32.6
Other Asia Pacific	1.0	1.3	1.3	**0.2**	**1.3**	0.1%	11.3
Total Asia Pacific	36.3	40.1	42.2	**6.0**	**45.2**	3.3%	14.8
Total World	**1003.2**	**1104.9**	**1376.6**	**188.8**	**1383.2**	**100.0%**	**46.2**
of which: OECD	115.4	93.3	92.0	**12.4**	**91.4**	6.6%	13.5
OPEC	763.4	849.7	1068.6	**146.0**	**1068.4**	77.2%	85.3
Non-OPEC‡	176.5	168.2	182.6	**25.5**	**188.7**	13.6%	15.1

(*Continued*)

Table 5.2.   (*Continued*)

	At end 1990 Thousand million barrels	At end 2000 Thousand million barrels	At end 2009 Thousand million barrels	At end 2010 Thousand million tonnes	At end 2010 Thousand million barrels	At end 2010 Share of total	At end 2010 R/P ratio
European Union #	8.1	8.8	6.2	**0.8**	**6.3**	0.5%	8.8
Former Soviet Union	63.3	87.1	125.4	**17.3**	**126.1**	9.1%	25.6
Canadian oil sands•	n/a	163.3	143.1	**23.3**	**143.1**		
Proved reserves and oil sands	n/a	1268.2	1519.6	**212.0**	**1526.3**		

*More than 100 years.

◆Less than 0.05%.

†Less than 0.05.

‡Excludes Former Soviet Union.

#Excludes Lithuania and Slovenia in 1990.

•"Remaining established reserves", less reserves "under active development".

Notes: Proved reserves of oil — Generally taken to be those quantities that geological and engineering information indicates with reasonable certainty can be recovered in the future from known reservoirs under existing economic and operating conditions. **Reserves-to-production (R/P) ratio** — If the reserves remaining at the end of any year are divided by the production in that year, the result is the length of time that those remaining reserves would last if production were to continue at that rate. **Source of data** — The estimates in this table have been compiled using a combination of primary official sources, third-party data from the OPEC Secretariat, *Oil & Gas Journal* and an independent estimate of Russian reserves based on information in the public domain. Canadian proved reserves include an official 26.5 billion barrels for oil sands "under active development". Venezuelan reserves are taken from the *OPEC Annual Statistical Bulletin*, that noted in 2008 that the figure included "proven reserves of the Magna Reserve Project in the Orinoco Belt, which amounted to 94,168 mb".

Reserves include gas condensate and natural gas liquids (NGLs) as well as crude oil.

Annual changes and shares of total are calculated using thousand million barrels figures.

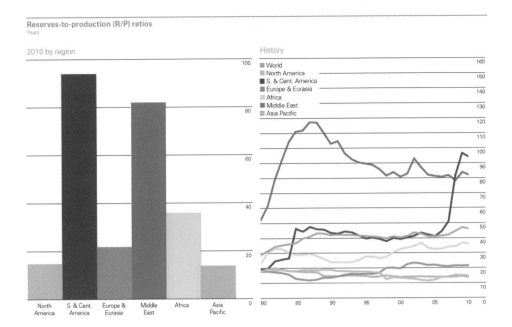

Fig. 5.3.   Oil reserve/production (R/P) ratios (years) [4].

become of critical importance to the Western OECD countries in less than 20 years' time. It is likely that the large reserves in Russia, Azerbaijan and Kazakhstan will become more significant.

There is currently (2013) a resurgence of interest in deep water oil exploration. The shallower fields of the continental shelf of the Gulf of Mexico, off the coast of Texas and Louisiana, have been the site of 3,000 drilling platforms in the past 50 years. These fields are now largely depleting beyond their economic recovery levels. Several major oil companies are now drilling in the deeper waters of the Mexican Gulf, offshore of West Africa and the continental shelf, west of Scotland. Huge drilling platforms are being designed for flotation rather than for resting on the sea bed. Deep water drilling can cost up to five times as much as shallow water exploration but the world need for oil is enormous. New technologies such as the "tension leg platform" are being used to drill to ocean depths of 4,000 feet in areas 150 miles offshore. Exploration is now being developed in the Gulf of Mexico for ocean depths of 8,000 feet at 200 miles offshore. It is not known how far and to what depths this technology can proceed.

Several countries along the Niger Delta coast of West Africa are anticipated sites of major deep sea exploration. These include Nigeria, Equatorial Guinea, Gabon, the Republic of Congo and, most notably, Angola. In terms of reserves the deep water field discoveries to date have an average field size of 535 million barrels of oil equivalent. The new West African fields have a potential in excess of the deep sea fields in North America [9].

5.5.   World Production and Consumption of Crude Oil

5.5.1.   *World oil production*

Oil is the world's most important primary fuel, accounting for 4028.1/12002.4 mtos, or 33.6%, of the total prime fuel consumption in 2010, as shown in Table 2.3 of Chapter 2. Data for world oil production for the 11-year period 2000–2010 are given in Table 5.3 [4]. The Middle East contributed the greatest proportion, 30.3%, followed by 22.4% from Europe and Eurasia, due to the large production in Russia, Kazakhstan and Azerbaijan. The decline of production from the North Sea fields is evidenced by the falling production figures of Norway and the UK.

Certain countries that are heavy users of oil, notably Japan, have no indigenous oil at all and are not listed in the producer data of Table 5.3. The worldwide trend of increasing oil production is illustrated in Fig. 5.4 [4].

5.5.2.   *World oil consumption*

World oil consumption since 2000 is given in Table 5.4. Oil consumption continues to increase significantly in most areas of the world, except Russia. In the 11-year period 2000–2010, world oil consumption increased from 3571.6 mtos to 4028.1 mtos, a rise of 11% from the 2000 figure. World oil consumption per capita, in tonnes, is shown in Fig. 5.5 [3].

The oil trade patterns in 2010 are illustrated in Fig. 5.6. Oil moves from the Middle East to Japan, South East Asia, Western Europe and North America (mainly the USA). In 2010 the USA had a net inflow (consumption minus production) of $850 - 339.1 = 511$ mtos, which was 60.1% of its consumption. The US oil "deficit" in 2010 was bigger than the consumption of China. The biggest oil importers in 2010 were the USA, Western Europe and Japan, while the big exporters were the Middle East, West Africa, Russia and Venezuela. More detailed information is available by comparing the production figures of Table 5.3 with the consumption figures of Table 5.4. Oil consumption is growing fastest in China and India as industrialization moves forward, accompanied by an increase in the general standard of living.

5.5.3.   *UK oil production and consumption* [10–12]

North Sea oil and gas reserves were first discovered in the 1960s. The North Sea did not emerge immediately as a key non-OPEC oil-producing area, but North Sea production grew as major discoveries continued throughout the 1980s and into the 1990s. Although the region is a relatively high cost producer, its high quality crude oil, political stability, and proximity to major European consumer markets allowed it to play a major role in world oil and gas markets.

Table 5.3. World oil production*, 2000–2010 [4].

Million tonnes	2000	2001	2002	2003	2004	2005	2006	2007	2008	2009	2010	Change 2010 over 2009	2010 share of total
US	352.6	349.2	346.8	338.4	329.2	313.3	310.2	309.8	304.9	328.6	**339.1**	3.2%	8.7%
Canada	126.9	126.1	135.0	142.6	147.6	144.9	153.4	158.3	156.8	156.1	**162.8**	4.3%	4.2%
Mexico	171.2	176.6	178.4	188.8	190.7	187.1	183.1	172.7	157.7	147.5	**146.3**	−0.8%	3.7%
Total North America	650.8	651.8	660.2	669.8	667.4	645.3	646.7	640.8	619.5	632.2	**648.2**	2.5%	16.6%
Argentina	40.4	41.5	40.9	40.2	37.8	36.2	35.8	34.9	34.1	33.8	**32.5**	−3.8%	0.8%
Brazil	63.2	66.3	74.4	77.0	76.5	84.6	89.2	90.4	93.9	100.4	**105.7**	5.3%	2.7%
Colombia	35.3	31.0	29.7	27.9	27.3	27.3	27.5	27.6	30.5	34.1	**39.9**	16.9%	1.0%
Ecuador	20.9	21.2	20.4	21.7	27.3	27.6	27.7	26.5	26.2	25.2	**25.2**	◆	0.6%
Peru	4.9	4.8	4.8	4.5	4.4	5.0	5.1	5.1	5.3	6.4	**6.9**	8.2%	0.2%
Trinidad & Tobago	6.8	6.5	7.5	7.9	7.3	8.3	8.3	7.2	6.9	6.8	**6.5**	−4.3%	0.2%
Venezuela	167.3	161.6	148.8	131.4	150.0	151.0	144.2	133.9	131.5	124.8	**126.6**	1.4%	3.2%
Other S. & Cent. America	6.6	6.9	7.8	7.8	7.3	7.2	7.0	7.1	7.0	6.7	**6.6**	−1.6%	0.2%
Total S. & Cent. America	345.3	339.9	334.2	318.3	337.9	347.1	344.9	332.7	335.5	338.2	**350.0**	3.5%	8.9%
Azerbaijan	14.1	15.0	15.4	15.5	15.6	22.4	32.5	42.8	44.7	50.6	**50.9**	0.5%	1.3%
Denmark	17.7	17.0	18.1	17.9	19.1	18.4	16.7	15.2	14.0	12.9	**12.2**	−5.8%	0.3%
Italy	4.6	4.1	5.5	5.6	5.5	6.1	5.8	5.9	5.2	4.6	**5.1**	11.7%	0.1%
Kazakhstan	35.3	40.1	48.2	52.4	60.6	62.6	66.1	68.4	72.0	78.2	**81.6**	4.4%	2.1%
Norway	160.2	162.0	157.3	153.0	149.9	138.2	128.7	118.6	114.2	108.8	**98.6**	−9.4%	2.5%
Romania	6.3	6.2	6.1	5.9	5.7	5.4	5.0	4.7	4.7	4.5	**4.3**	−4.7%	0.1%
Russian Federation	323.3	348.1	379.6	421.4	458.8	470.0	480.5	491.3	488.5	494.2	**505.1**	2.2%	12.9%
Turkmenistan	7.2	8.0	9.0	10.0	9.6	9.5	9.2	9.8	10.3	10.4	**10.7**	2.8%	0.3%
United Kingdom	126.2	116.7	115.9	106.1	95.4	84.7	76.6	76.8	71.7	68.2	**63.0**	−7.7%	1.6%
Uzbekistan	7.5	7.2	7.2	7.1	6.6	5.4	5.4	4.9	4.8	4.5	**3.7**	−17.8%	0.1%

(*Continued*)

Table 5.3. (*Continued*)

Million tonnes	2000	2001	2002	2003	2004	2005	2006	2007	2008	2009	**2010**	Change 2010 over 2009	2010 share of total
Other Europe & Eurasia	22.4	22.2	23.6	24.0	23.5	22.0	21.7	21.6	20.6	19.6	**18.2**	−7.0%	0.5%
Total Europe & Eurasia	724.7	746.6	785.9	818.9	850.2	844.8	848.1	860.0	850.8	856.5	**853.3**	−0.4%	21.8%
Iran	191.3	191.4	180.9	203.7	207.8	206.3	208.2	209.7	209.9	201.5	**203.2**	0.9%	5.2%
Iraq	128.8	123.9	104.0	66.1	100.0	90.0	98.1	105.2	119.5	119.8	**120.4**	0.6%	3.1%
Kuwait	109.1	105.8	98.2	114.8	122.3	129.3	132.7	129.9	137.2	121.7	**120.5**	0.6%	3.1%
Oman	46.4	46.1	43.4	39.6	38.1	37.4	35.7	34.5	35.9	38.7	**41.0**	5.9%	1.0%
Qatar	36.1	35.7	35.2	40.8	46.0	47.3	50.9	53.6	60.8	57.9	**65.7**	13.5%	1.7%
Saudi Arabia	456.3	440.6	425.3	485.1	506.0	526.8	514.3	494.2	515.3	464.7	**467.8**	0.7%	12.0%
Syria	27.3	28.9	27.2	26.2	24.7	22.4	21.6	20.6	19.8	18.6	**19.1**	2.7%	0.5%
United Arab Emirates	122.1	118.0	110.2	124.5	131.7	137.3	145.5	140.7	142.9	126.3	**130.8**	3.5%	3.3%
Yemen	21.3	21.5	21.5	21.1	19.9	19.6	17.9	16.3	14.4	13.5	**12.5**	−7.9%	0.3%
Other Middle East	2.2	2.2	2.2	2.2	2.2	1.6	1.4	1.6	1.5	1.7	**1.7**	0.6%	◆
Total Middle East	1140.9	1114.1	1048.3	1124.3	1198.9	1217.9	1226.4	1206.4	1257.2	1164.4	**1184.6**	1.7%	30.3%
Algeria	66.8	65.8	70.9	79.0	83.6	86.4	86.2	86.5	85.6	77.9	**77.7**	−0.3%	2.0%
Angola	36.9	36.6	44.6	42.8	54.5	69.0	69.6	82.5	92.2	87.4	**90.7**	3.8%	2.3%
Chad	—	—	—	1.2	8.8	9.1	8.0	7.5	6.7	6.2	**6.4**	3.5%	0.2%
Republic of Congo (Brazzaville)	13.1	12.1	12.3	11.2	11.6	12.6	14.3	11.7	12.4	13.9	**15.1**	8.1%	0.4%
Egypt	38.8	37.3	37.0	36.8	35.4	33.9	33.7	34.1	34.6	35.3	**35.0**	−0.6%	0.9%
Equatorial Guinea	4.5	8.8	11.4	13.2	17.4	17.7	16.9	17.3	17.2	15.2	**13.6**	−10.8%	0.3%
Gabon	16.4	15.0	14.7	12.0	11.8	11.7	11.7	11.5	11.8	11.5	**12.2**	6.5%	0.3%
Libya	69.5	67.1	64.6	69.8	76.5	81.9	84.9	85.0	85.3	77.1	**77.5**	0.5%	2.0%
Nigeria	105.4	110.8	102.3	109.3	119.0	122.1	117.8	112.1	103.0	99.1	**115.2**	16.2%	2.9%

(*Continued*)

Table 5.3. (*Continued*)

Million tonnes	2000	2001	2002	2003	2004	2005	2006	2007	2008	2009	2010	Change 2010 over 2009	2010 share of total
Sudan	8.6	10.7	11.9	13.1	14.9	15.0	16.3	23.1	23.7	23.6	**23.9**	1.5%	0.6%
Tunisia	3.7	3.4	3.5	3.2	3.4	3.4	3.3	4.6	4.2	4.0	**3.8**	−4.7%	0.1%
Other Africa	7.2	6.6	6.7	6.8	8.1	7.7	7.6	8.3	8.1	7.7	**7.1**	−8.0%	0.2%
Total Africa	370.9	374.1	379.8	398.4	444.9	470.7	470.4	484.4	484.9	458.9	**478.2**	4.2%	12.2%
Australia	35.3	31.8	31.5	26.6	24.8	24.5	23.2	23.5	23.7	21.9	**23.8**	8.9%	0.6%
Brunei	9.4	9.9	10.2	10.5	10.3	10.1	10.8	9.5	8.5	8.2	**8.4**	2.5%	0.2%
China	162.6	164.8	166.9	169.6	174.1	181.4	184.8	186.3	190.4	189.5	**203.0**	7.1%	5.2%
India	34.2	34.1	35.2	35.4	36.3	34.6	35.8	36.1	36.1	35.4	**38.9**	9.8%	1.0%
Indonesia	71.5	67.9	63.0	57.3	55.2	53.1	48.9	47.5	49.0	47.9	**47.8**	−0.3%	1.2%
Malaysia	33.7	32.9	34.5	35.6	36.5	34.4	33.5	34.2	34.6	33.1	**32.1**	−3.1%	0.8%
Thailand	7.0	7.5	8.2	9.6	9.1	10.8	11.8	12.5	13.3	13.7	**13.8**	0.9%	0.4%
Vietnam	16.2	17.1	17.3	17.7	20.8	19.4	17.8	16.4	15.4	16.8	**18.0**	6.9%	0.5%
Other Asia Pacific	9.4	9.1	9.0	9.1	10.5	12.5	13.2	13.9	14.7	14.3	**13.6**	−4.7%	0.3%
Total Asia Pacific	379.2	375.1	375.8	371.4	377.7	380.8	379.7	380.1	385.9	380.8	**399.4**	4.9%	10.2%
Total World	**3611.8**	**3601.6**	**3584.2**	**3701.1**	**3877.0**	**3906.6**	**3916.2**	**3904.3**	**3933.7**	**3831.0**	**3913.7**	**2.2%**	**100.0%**
of which: OECD	1011.5	1000.0	1005.8	996.0	978.2	932.2	912.2	896.2	864.0	863.3	**864.7**	0.2%	22.1%
Non-OECD	2600.3	2601.6	2578.5	2705.1	2898.8	2974.4	3004.1	3008.0	3069.8	2967.7	**3049.0**	2.7%	77.9%
OPEC	1510.3	1478.3	1405.4	1489.1	1624.9	1675.0	1680.2	1660.0	1709.4	1583.5	**1623.3**	2.5%	41.5%
Non-OPEC‡	1708.0	1698.7	1712.6	1698.4	1693.6	1654.4	1635.3	1620.2	1597.3	1603.2	**1632.9**	1.9%	41.7%
European Union	166.3	155.6	158.2	148.2	137.7	125.7	114.6	113.1	105.4	99.0	**92.6**	−6.5%	2.4%
Former Soviet Union	393.4	424.6	466.2	513.6	558.5	577.1	600.7	624.1	627.1	644.3	**657.5**	2.0%	16.8%

*Includes crude oil, shale oil, oil sands and NGLs (the liquid content of natural gas where this is recovered separately). Excludes liquid fuels from other sources such as biomass and coal derivatives.

◆ Less than 0.05%.

‡ Excludes Former Soviet Union.

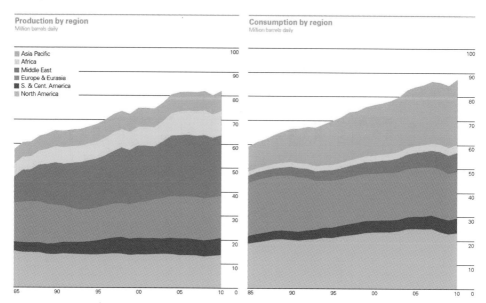

Fig. 5.4.   World oil production/consumption, 1985–2010 [4].

Fig. 5.5.   World oil consumption per capita (tonnes) [3].

Table 5.4. World oil consumption*, 2000–2010 [4].

Million tonnes	2000	2001	2002	2003	2004	2005	2006	2007	2008	2009	2010	Change 2010 over 2009	2010 share of total
US	884.1	884.1	884.9	900.7	936.5	939.8	930.7	928.8	875.8	833.2	**850.0**	2.0%	21.1%
Canada	88.1	90.5	92.2	95.9	100.6	100.3	100.5	103.8	102.5	97.1	**102.3**	5.4%	2.5%
Mexico	87.3	86.6	82.8	85.0	88.6	90.9	89.8	92.1	91.7	88.5	**87.4**	−1.2%	2.2%
Total North America	1059.5	1061.2	1059.9	1081.6	1125.8	1131.0	1120.9	1124.7	1070.0	1018.8	**1039.7**	2.1%	25.8%
Argentina	20.5	19.8	17.9	18.4	19.3	20.6	21.5	24.2	25.9	23.7	**25.7**	8.5%	0.6%
Brazil	91.5	93.3	92.0	90.9	91.3	94.0	95.1	100.6	107.1	107.0	**116.9**	9.3%	2.9%
Chile	10.8	10.5	10.4	10.5	11.1	11.8	12.3	16.2	16.8	15.6	**14.7**	−6.0%	0.4%
Colombia	10.7	9.9	9.7	9.7	9.9	10.5	11.0	10.7	10.6	10.5	**11.0**	4.1%	0.3%
Ecuador	5.8	5.9	5.9	6.2	6.3	7.5	8.2	8.9	9.4	10.1	**10.6**	5.0%	0.3%
Peru	7.4	7.0	6.9	6.5	7.3	7.1	6.9	7.1	8.0	8.1	**8.4**	3.6%	0.2%
Trinidad & Tobago	1.8	1.4	1.7	1.6	1.8	2.1	2.4	2.4	2.2	2.1	**2.1**	4.4%	0.1%
Venezuela	25.8	28.8	30.3	24.9	27.2	29.1	30.5	31.5	32.9	33.7	**35.2**	4.7%	0.9%
Other S. & Cent. America	52.3	55.0	55.1	56.2	56.8	57.1	58.1	59.9	58.6	57.9	**57.3**	−1.0%	1.4%
Total S. & Cent. America	226.6	231.7	229.9	224.9	231.1	239.9	246.0	261.5	271.4	268.6	**282.0**	5.0%	7.0%
Austria	11.8	12.8	13.1	14.2	13.8	14.2	14.2	13.4	13.5	13.0	**13.0**	0.2%	0.3%
Azerbaijan	6.3	4.0	3.7	4.3	4.6	5.3	4.8	4.5	3.5	3.2	**3.3**	4.0%	0.1%
Belarus	7.0	7.3	7.1	7.2	7.4	7.1	8.0	7.3	8.1	9.3	**6.6**	−29.3%	0.2%
Belgium & Luxembourg	33.9	32.2	33.5	36.4	37.6	37.7	37.4	37.7	40.8	33.4	**35.0**	4.8%	0.9%

(*Continued*)

Table 5.4.   (*Continued*)

Million tonnes	2000	2001	2002	2003	2004	2005	2006	2007	2008	2009	**2010**	Change 2010 over 2009	2010 share of total
Bulgaria	4.1	4.2	4.4	5.1	4.7	5.0	5.3	5.1	5.4	5.6	**4.2**	−25.6%	0.1%
Czech Republic	7.9	8.4	8.1	8.7	9.5	9.9	9.8	9.7	9.9	9.7	**9.2**	−5.0%	0.2%
Denmark	10.4	9.8	9.6	9.2	9.1	9.3	9.6	9.7	9.5	8.5	**8.7**	2.0%	0.2%
Finland	10.7	10.5	10.9	11.4	10.6	11.0	10.6	10.6	10.5	9.9	**10.4**	4.9%	0.3%
France	94.9	95.5	92.9	93.1	94.0	93.1	93.0	91.4	90.8	87.5	**83.4**	−4.7%	2.1%
Germany	129.8	131.6	127.4	125.1	124.0	122.4	123.6	112.5	118.9	113.9	**115.1**	1.1%	2.9%
Greece	19.9	20.1	20.3	19.7	21.4	21.2	22.2	21.7	21.4	20.2	**18.5**	−8.7%	0.5%
Hungary	6.8	6.7	6.4	6.3	6.5	7.5	7.8	7.7	7.5	7.1	**6.7**	−5.2%	0.2%
Republic of Ireland	8.2	9.0	8.8	8.5	8.9	9.3	9.3	9.4	9.0	8.0	**7.6**	−5.0%	0.2%
Italy	93.5	92.8	92.9	92.1	89.7	86.7	86.7	84.0	80.4	75.1	**73.1**	−2.7%	1.8%
Kazakhstan	7.8	8.7	9.3	10.1	10.7	11.3	11.6	11.8	12.8	12.1	**12.5**	3.2%	0.3%
Lithuania	2.4	2.7	2.5	2.4	2.6	2.8	2.8	2.8	3.1	2.6	**2.7**	3.0%	0.1%
Netherlands	42.5	44.6	44.6	44.9	47.1	50.6	52.0	53.5	51.1	49.4	**49.8**	0.9%	1.2%
Norway	9.5	9.8	9.6	10.2	9.9	10.0	10.3	10.5	10.2	10.3	**10.7**	3.5%	0.3%
Poland	20.0	19.5	19.9	20.2	21.6	22.4	23.3	24.2	25.3	25.3	**26.3**	3.9%	0.7%
Portugal	15.5	15.8	16.2	15.2	15.4	16.0	14.4	14.4	13.6	12.8	**12.6**	−1.6%	0.3%
Romania	10.0	10.6	10.6	9.4	10.9	10.5	10.3	10.3	10.4	9.2	**9.1**	−1.4%	0.2%
Russian Federation	129.7	128.6	129.9	130.1	130.6	129.9	135.8	135.7	141.4	135.2	**147.6**	9.2%	3.7%
Slovakia	3.4	3.2	3.5	3.3	3.2	3.8	3.4	3.6	3.9	3.7	**3.7**	−0.3%	0.1%
Spain	70.0	72.4	72.8	75.8	79.3	80.4	79.7	80.7	79.0	75.7	**74.5**	−1.6%	1.8%
Sweden	16.2	16.5	16.7	17.2	16.5	16.5	16.9	16.1	15.7	14.6	**14.5**	−0.1%	0.4%
Switzerland	12.2	13.1	12.4	12.1	12.0	12.2	12.6	11.3	12.1	12.3	**11.4**	−7.1%	0.3%
Turkey	31.1	29.9	30.6	31.0	31.0	30.2	29.5	30.5	30.9	28.2	**28.7**	1.7%	0.7%
Turkmenistan	3.6	3.7	3.8	4.2	4.3	4.5	4.7	5.1	5.3	5.4	**5.6**	3.6%	0.1%
Ukraine	12.0	13.4	13.2	13.5	14.2	13.5	14.1	15.5	14.8	13.3	**11.6**	−13.2%	0.3%
United Kingdom	78.6	78.4	78.0	79.0	81.7	83.0	82.3	79.2	77.9	74.4	**73.7**	−1.0%	1.8%
Uzbekistan	7.5	7.1	7.0	7.5	7.0	5.3	5.3	4.9	4.8	4.8	**5.0**	2.8%	0.1%

(*Continued*)

Table 5.4. (*Continued*)

Million tonnes	2000	2001	2002	2003	2004	2005	2006	2007	2008	2009	**2010**	Change 2010 over 2009	2010 share of total
Other Europe & Eurasia	21.4	22.9	23.9	25.2	26.1	27.3	27.8	29.6	29.9	28.5	**28.3**	−0.7%	0.7%
Total Europe & Eurasia	938.6	945.6	943.8	952.6	966.0	970.1	978.9	964.4	971.5	922.2	**922.9**	0.1%	22.9%
Iran	62.7	63.0	67.5	71.4	74.5	78.4	82.4	82.5	87.4	85.1	**86.0**	1.0%	2.1%
Israel	13.5	12.5	12.4	12.8	12.0	12.2	11.9	12.4	12.2	11.5	**11.2**	−2.2%	0.3%
Kuwait	11.3	11.5	12.5	13.6	15.2	16.7	15.2	15.3	16.3	17.2	**17.7**	2.8%	0.4%
Qatar	2.0	2.4	3.0	3.1	3.4	4.0	4.6	5.4	6.2	6.2	**7.4**	18.1%	0.2%
Saudi Arabia	73.0	74.7	76.6	81.7	88.3	88.1	92.3	98.2	107.2	117.2	**125.5**	7.1%	3.1%
United Arab Emirates	20.1	19.9	20.7	23.2	24.8	26.7	28.3	30.0	32.0	29.8	**32.3**	8.4%	0.8%
Other Middle East	56.3	59.2	59.7	57.3	60.9	62.3	67.4	70.2	73.6	77.3	**80.2**	3.8%	2.0%
Total Middle East	239.0	243.2	252.5	263.1	279.0	288.5	302.3	314.1	334.9	344.3	**360.2**	4.6%	8.9%
Algeria	8.5	8.8	9.7	10.1	10.6	11.0	11.5	12.9	14.0	14.9	**14.9**	−0.1%	0.4%
Egypt	27.2	26.1	25.2	25.9	26.8	29.8	28.7	30.6	32.6	34.4	**36.3**	5.4%	0.9%
South Africa	22.0	22.5	23.1	23.9	24.7	24.6	25.3	26.2	25.3	24.7	**25.3**	2.7%	0.6%
Other Africa	59.2	60.9	62.6	63.7	66.4	69.1	68.2	71.1	74.8	77.0	**79.0**	2.6%	2.0%
Total Africa	116.9	118.2	120.6	123.6	128.5	134.5	133.7	140.8	146.8	150.9	**155.5**	3.0%	3.9%
Australia	37.7	38.1	38.0	38.3	39.1	40.2	41.5	41.8	42.5	42.2	**42.6**	0.8%	1.1%
Bangladesh	3.2	3.9	3.9	4.0	4.0	4.6	4.5	4.6	4.7	4.8	**4.8**	0.4%	0.1%
China	224.2	228.4	247.5	271.7	318.9	327.8	351.2	369.3	376.0	388.2	**428.6**	10.4%	10.6%
China Hong Kong SAR	9.7	11.8	12.9	13.0	15.4	13.8	15.0	16.1	14.6	14.0	**16.1**	15.2%	0.4%
India	106.1	107.0	111.3	113.1	120.2	119.6	120.4	133.4	144.1	151.0	**155.5**	2.9%	3.9%
Indonesia	54.5	55.3	57.5	58.5	62.0	61.2	58.3	59.5	59.1	59.2	**59.6**	0.7%	1.5%
Japan	255.0	247.4	243.5	248.7	241.0	244.8	238.0	229.7	222.1	198.7	**201.6**	1.5%	5.0%

(*Continued*)

Table 5.4.   (*Continued*)

Million tonnes	2000	2001	2002	2003	2004	2005	2006	2007	2008	2009	2010	Change 2010 over 2009	2010 share of total
Malaysia	21.3	22.0	23.9	23.6	24.5	23.9	23.4	24.8	24.8	24.5	**25.3**	3.3%	0.6%
New Zealand	6.1	6.1	6.4	6.9	6.9	7.1	7.2	7.2	7.3	6.8	**6.9**	0.1%	0.2%
Pakistan	18.8	18.3	17.9	15.8	16.0	15.3	17.6	19.2	19.3	20.6	**20.5**	−0.6%	0.5%
Philippines	16.6	16.5	15.5	15.5	16.0	14.9	13.3	14.0	12.3	13.1	**13.1**	0.1%	0.3%
Singapore	33.4	36.4	35.5	33.9	38.1	42.3	45.1	49.0	52.0	56.1	**62.2**	10.9%	1.5%
South Korea	103.2	103.1	104.7	105.6	103.9	104.4	104.5	107.1	101.9	103.0	**105.6**	2.5%	2.6%
Taiwan	42.6	44.0	44.8	46.5	48.7	49.1	48.4	50.2	45.0	44.1	**46.2**	4.7%	1.1%
Thailand	38.7	38.0	40.8	43.9	48.4	50.6	50.1	49.2	49.0	49.9	**50.2**	0.5%	1.2%
Vietnam	8.3	9.0	9.8	10.5	12.5	12.2	12.0	13.3	14.1	14.1	**15.6**	10.4%	0.4%
Other Asia Pacific	11.6	12.0	12.0	12.1	12.7	12.8	12.8	13.5	13.0	13.4	**13.5**	0.9%	0.3%
Total Asia Pacific	991.1	997.3	1025.5	1061.6	1128.3	1144.5	1163.5	1201.9	1201.9	1203.8	**1267.8**	5.3%	31.5%
Total World	**3571.6**	**3597.2**	**3632.3**	**3707.4**	**3858.7**	**3908.5**	**3945.3**	**4007.3**	**3996.5**	**3908.7**	**4028.1**	**3.1%**	**100.0%**
of which: OECD	2217.1	2215.7	2207.9	2242.2	2287.2	2303.6	2289.7	2276.3	2210.5	2094.8	**2113.8**	0.9%	52.5%
Non-OECD	1354.5	1381.5	1424.4	1465.3	1571.6	1604.9	1655.6	1731.0	1786.0	1813.9	**1914.3**	5.5%	47.5%
European Union	699.3	706.2	702.3	706.6	717.7	723.1	724.6	708.4	709.0	670.2	**662.5**	−1.1%	16.4%
Former Soviet Union	180.4	179.9	181.2	184.1	186.6	185.4	193.0	194.3	200.8	192.7	**201.5**	4.6%	5.0%

*Inland demand plus international aviation and marine bunkers and refinery fuel and loss. Consumption of fuel ethanol and biodiesel is also included.
Note: Differences between these world consumption figures and world production statistics are accounted for by stock changes, consumption of non-petroleum additives and substitute fuels, and unavoidable disparities in the definition, measurement or conversion of oil supply and demand data.

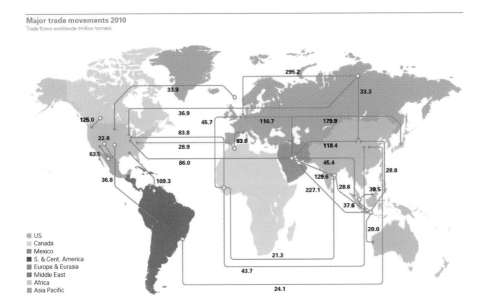

Major trade movements 2010
Trade flows worldwide (million tonnes)

US
Canada
Mexico
S. & Cent. America
Europe & Eurasia
Middle East
Africa
Asia Pacific

Fig. 5.6.   Major trade movements in oil, 2010 [4].

Many of the world's major crude oil prices (Table 5.1) are linked to the price of the North Sea's Brent crude oil. (Brent crude is a blend of North Sea crude oils and does not come exclusively from the Brent field.) Because Brent crude is traded on the International Petroleum Exchange in London, fluctuations in the market are reflected in the price of Brent. Therefore, all other crude oils linked to Brent can be priced according to the latest market conditions.

The North Sea is considered a "mature" area, with few large new discoveries likely to be made. BP (Forties Field and Brut) has sold these assets and is "moving out" of the North Sea. In both of the major North Sea producing nations, Norway and the UK, government and industry are taking steps to restructure their oil and gas sectors to make them more internationally competitive.

The UK held just over 2.8 billion barrels of proven oil reserves in 2010, almost all of which is located in the North Sea. Most of the country's production comes from basins east of Scotland in the northern and central North Sea. The northern North Sea also holds considerable reserves, and smaller deposits are located offshore in the North Atlantic Ocean, west of the Shetland Islands. There are over 100 oil and gas fields currently on-stream, and several hundred companies are active in the area. In 1999, the UK produced the highest volume of oil ever, at 2.95 million barrels per day. Most of the UK's crude oil production ranges in gravity from 30° to 40° API (American Petroleum Institute Gravity). This means that the oil is a "good crude", with ranges in viscosity from "normal oil" to "light oil".

Foreign trade in crude oil and petroleum products is shown in Fig. 5.7 and in Table 5.5 [12]. Most high quality crude oil (e.g. North Sea production) is exported, while cheaper, lower quality (mainly from the Middle East) crude oils are imported

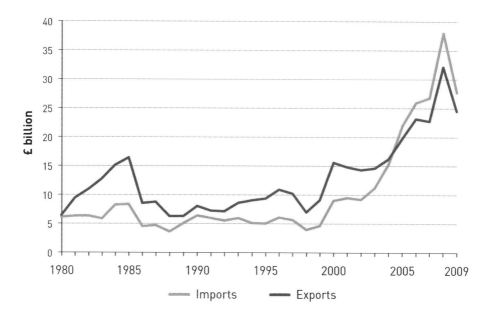

Fig. 5.7.   UK trade in crude oil and petroleum products, 1980–2009 [12].

Table 5.5.   UK trade in crude oil and petroleum products, 1980–2009 [12].

Crude oil and petroleum products   £billion

	1980	1990	2000	2007	2008	2009
Exports	6.5	8.1	15.6	22.8	32.1	24.5
Imports	6.2	6.4	9.0	26.8	38.0	27.8
Net imports	**−0.3**	**−1.6**	**−6.5**	**4.0**	**5.9**	**3.2**

Source: Office for National Statistics

Crude oil and petroleum products   Million tonnes

	1980	1990	2000	2007	2008	2009
Exports	54.3	73.9	113.6	81.0	77.2	70.9
Imports	56.0	63.7	68.6	82.4	84.0	76.8
Net imports	**1.7**	**−10.2**	**−45.0**	**1.4**	**6.8**	**5.9**

into the UK for refining. Since the first "surplus" on oil trade (£0.3 billion), which occurred in 1980, oil trade contributed £77 billion to the UK balance of payments. The largest "surplus" (£8 billion) in 1985 reflected high crude oil production and prices. In 1990 the "surplus" fell from this peak due to lower prices but managed to peak again in 2000 (£6.5 billion). Since 2000 the surplus has steadily declined and in 2005 the UK became a net importer of oil (−£2.2 billion). In 2009 the deficit was £3.2 billion, a reduction of £2.6 billion from the previous year [12].

The annual estimate of UK oil reserves remaining in known fields has been fairly constant over the last 20 years, despite a large increase in the amount of oil extracted annually. This is due to new discoveries being made and new technology allowing the exploitation of discoveries that were previously regarded as not viable.

An area with potential for new production is a remote area of the North Atlantic between the Shetland Islands and the Faroe Islands. A territorial dispute between the government of the Faroe Islands (the Faroe Islands are a self-governed part of Denmark) and the UK prevented exploration until an agreement was reached in the summer of 1999. A licensing round for Faroese development was completed in late May of 2000. A UK licensing round for the Atlantic Ocean, north of Scotland and west of the Shetland Islands, was postponed in May 2000 to allow time to implement the European Commission Habitats Directive, which requires governments to assess potential environmental damage before allowing drilling [10].

The distribution of applications of the use of oil in the UK is shown in Table 5.6. In the period since 1980 the use of fuel oil has declined, representing a move towards gas as the preferred source of energy by electricity generators and by industry. Transport fuel growth has slowed in recent years, with a switch being seen from motor spirit to DERV (i.e. diesel) fuel, but with continued growth in aviation turbine fuel use.

Comparisons of the pump prices of gasoline in various countries are listed in Table 5.7 [13]. Prices are standardized in terms of US$/litre, since the litre is the metric standard (1 litre $= 1/3.785 = 0.264$ US gallon). Some countries subsidize the price of gasoline, including Saudi Arabia, Malaysia, Kuwait, China, Nigeria and Bolivia. The actual price of the gasoline is fairly standard across the world, but all

Table 5.6.   UK petroleum use, 1980–2009 [12].

	Million tonnes					
	1980	1990	2000	2007	2008	2009
Energy uses[1]						
Motor spirit (Petrol)	19.2	24.3	21.4	17.6	16.7	15.8
DERV fuel	5.9	10.7	15.6	21.1	20.6	20.1
Aviation turbine fuel	4.7	6.6	10.8	12.6	12.1	11.5
Burning oil	2.1	2.1	3.8	3.6	3.7	3.7
Gas oil	11.6	8.0	6.8	5.9	5.8	5.2
Fuel oil	22.7	14.0	3.3	3.2	3.3	2.7
Other	4.3	4.9	5.3	5.3	5.5	5.3
Total energy uses	**70.5**	**70.6**	**67.1**	**69.3**	**67.7**	**64.2**
Of which:						
Transport fuels	31.9	43.5	49.5	53.5	51.9	49.6
Industry	14.9	7.2	5.5	5.9	5.4	4.9
Energy Industry use	6.3	5.1	5.3	4.7	4.5	4.5
Non-energy uses	**7.0**	**9.2**	**10.1**	**8.0**	**7.9**	**7.3**
Total deliveries	**77.5**	**79.8**	**77.2**	**77.9**	**76.2**	**72.0**

[1]Energy uses includes uses for transformation (e.g. electricity generation) and energy industry own use (e.g. refinery fuels).

Table 5.7.   Pump price of gasoline, 2009 [13].

Country	Pump price of gasoline US$/litre
Turkey	2.53
Norway	2.31
Netherlands	2.18
Greece	2.09
Denmark	2.07
UK	1.98
Israel	1.98
Germany	1.96
Belgium	1.95
Portugal	1.89
Italy	1.84
France	1.82
Finland	1.8
Ireland	1.75
Sweden	1.69
Austria	1.68
Spain	1.59
New Zealand	1.55
Japan	1.37
India	1.3
Canada	1.01
USA	0.734

the European countries impose a significant taxation levy on their gasoline. It is notable that the pump price of gasoline in the USA is smaller than one-half the European average and carries relatively little taxation. The 2011 pump price of petrol in Britain contains about 75% taxation.

5.5.4.   *US oil production and consumption*

The gap between US oil consumption and production is steadily growing (Fig. 5.8). It is seen, from Fig. 5.6, that the USA is a massive importer of oil from all of the world's major sources.

Gasoline, one of the main products refined from crude oil, accounts for just about 17% of the energy consumed in the USA. The primary use is in automobiles and light trucks, but also in boats, recreational vehicles and farm and other equipment. While gasoline is produced year-round, extra volumes are made in time for the summer driving season. Gasoline is delivered from oil refineries mainly through pipelines to a massive distribution chain serving 176,000 retail gasoline stations throughout the USA. There are three main grades of gasoline: regular, mid-grade and premium. Each grade has a different octane (fuel density) level. Price levels vary by grade, but the price differential between grades is generally constant.

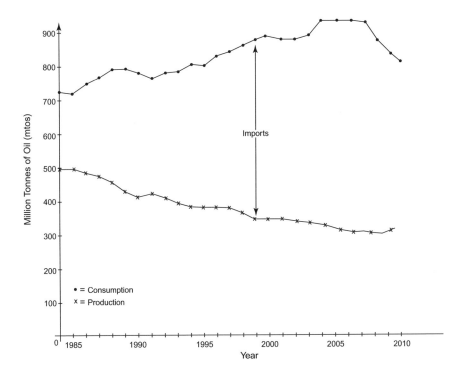

Fig. 5.8.   US oil production and consumption, 1985–2010.

The cost of producing and delivering gasoline to consumers includes the cost of crude oil to refiners, refinery processing costs, marketing and distribution costs, taxes and, finally, the retail station costs. The prices paid by consumers at the pump reflect these costs, as well as the profits (and sometimes losses) of refiners, marketers, distributors and retail station owners.

Federal, state and local taxes are a significant component of the retail price of gasoline. Taxes (not including county and local taxes) accounted for approximately 11% of the 2008 cost of a gallon of gasoline. Within this national average, federal excise taxes and state excise taxes are about equal. Also, several states levy additional state sales taxes, some of which are applied to the federal and state excise taxes. Additional local county and city taxes can have a significant impact on the price of gasoline. Refining costs and profits constitute about 15% of the retail price of gasoline. This component varies from region to region due to the different formations required in different parts of the country. Distribution, marketing and retail station costs and profits combine to make up 12% of the cost of a gallon of gasoline. Approximately one-third of the service station outlets are company stations, i.e. are owned or leased by a major oil company and operated by its employees. The remainder are operated by independent dealers free to set their own prices. The price at the pump reflects both the retailer's purchase cost for the product and the other costs of operating the service station.

It also reflects local market conditions and factors, such as the desirability of the location and the marketing strategy of the owner. When crude oil prices are stable, retail gasoline prices tend to gradually rise before and during the summer, when people drive more, and fall in the winter. Good weather and vacations cause US summer gasoline demand to average about 6% higher than during the rest of the year. If crude oil prices remain unchanged, gasoline prices typically increase by 5–6 cents per gallon during the summer [14].

The USA is the world's largest consumer of oil, both in absolute terms (Table 5.4) and (except for Singapore) in per capita terms (Table 2.4). America is the dominant world power in economic and military terms, making the President of the USA the most important and influential national leader in the world.

It was noted in Section 5.3 that the price of crude oil is usually negotiated between the major oil companies (dominated by US companies) and the OPEC consortium (dominated by Saudi Arabia). In the aftermath of the 1990s Gulf Wars, Saudi Arabia enjoys the military protection of the USA. The US-led invasion of Iraq in 2003 has not caused any great changes in oil prices. The OPEC countries outside the Middle East (principally Nigeria and Indonesia) do not appear disposed to act unilaterally to oppose or break the cartel price.

The pump price of gasoline in the USA has always been low. In 2001 the pump price fluctuated between $1.10 and $1.50 per US gallon in response to several economic factors, as well as different regional costs and levies, which was roughly one-third of the cost in Western Europe. In 2008 the cost had become $1.65/US gallon. The price increased to $2.91/gallon in May 2010 and $4/gallon in 2012. It remains, however, by far the cheapest gasoline amongst the industrial nations.

Gasoline prices in the USA are a matter for the Americans and not for foreign interference. It remains an intriguing situation, however, that the pump price of gasoline in the USA plays so prominent a part in the availability, use and price of the world's most sought-after fuel, all over the world.

The USA is a vast country, roughly the same size as Europe, and the citizens have always been mobile and had the privilege of driving private cars, often for long distances. Some annual travel information for the USA and some former EEC countries, given in Fig. 5.9, clearly shows the extensive car driving habits of Americans. Although this is slightly dated, the overall picture is still true.

The 1973 oil crisis had the ultimate major effect of reducing the sizes of cars driven by most Americans to about the same size as standard European and Japanese cars, although engine size is creeping up again. From the late 1970s there has been a significant reduction in the gasoline used per passenger-mile in the USA [15]. It is difficult to imagine that there will be any further substantial voluntary change in American driving habits. Moreover, the automobile industry in the USA is a major component of the national economy and labour force. Freight has to travel great distances to consumers because of the geography of US infrastructure, and these transport costs affect prices. In addition to energy use, oil forms

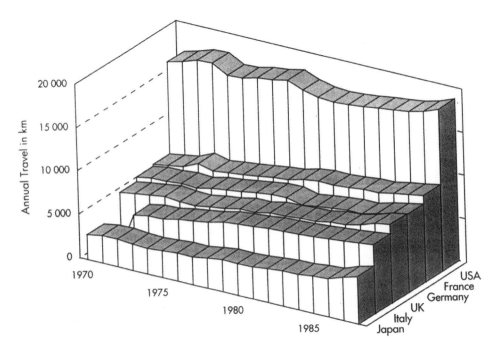

Fig. 5.9.   Annual per capita car travel, 1970–1988 [14].

a major industrial resource for the manufacture of plastics and pharmaceuticals. One might query the wisdom of the industrialized countries, including the USA, in using so much of the crude oil to create vehicle fuels.

5.6.   Unconventional Liquid Fossil Fuels

There are vast underground repositories of potential oil-bearing materials in the form of oil shale and tar sands in North America. The present (2013) very high price of conventional crude oil, sometimes exceeding $100/barrel, has caused these previously expensive sources to become financially viable.

5.6.1.   *Oil shale*

Oil shale is a fairly commonly-occurring sedimentary rock containing a solid composite of hydrocarbons called kerogen (approximate composition 80% C, 10% H_2, 6% O_2, 3% N and 1% S). Kerogen is the chemical precursor of oil and gas via a number of complex geochemical processes, mainly related to temperature and time. The shale is intractably buried into its containing rock, mud and clay. When oil shale is buried it forms wax. One of the many ancient names for oil shale is "the rock that burns". Known deposits constitute a vast reserve of potential oil-bearing

Fig. 5.10.   Oil shale deposits in the USA (estimated resources: 1.2–1.8 trillion barrels) [16].

material, but the problem of extraction poses formidable technical, environmental and economic problems. Oil from shale cannot be obtained by drilling bore-holes and collecting crude oil as it bubbles out. The necessary industrial processes are huge in scale and expense.

The former Soviet Union and China were operating major oil shale developments in the early 1980s. Up to 100,000 tonnes/year of shale oil was manufactured in Scotland in the 1950s but this was closed down in 1962. The world's largest known deposits of oil shale are in four giant basins in western USA, covering parts of the states of Wyoming, Colorado and Utah, illustrated in Fig. 5.10. [16]. At present oil shale is being mined only in Brazil, China and Estonia. Once extracted from the ground the rock can be used directly as a fuel for power plants or be processed to produce shale oil and other chemicals and materials.

Two methods have been considered for obtaining oil from shale, namely (i) mining and processing the rock and (ii) *in situ* combustion.

5.6.1.1.   *Oil shale mining*

Oil shales cannot usually be strip (i.e. open-cast) mined but require sub-surface mining, as in the UK deep coal industry. Only the richest workings justify the expense. When the extracted oil-bearing shale rock is broken up and heated to temperatures greater than 480°C the kerogen decomposes, producing an oily vapour that condenses to viscous shale oil, containing up to 80% of the original energy in the kerogen. This shale oil often contains nitrogen compounds and sulphur that need to be removed by further processing to obtain an acceptable form of crude oil [17].

Every 100 tonnes of crushed rock leaves about 90 tonnes of waste processing and, moreover, the volume of the waste is 40% greater than its original solid rock. A high energy input is required to decompose the oil shale/rock into products. Obtaining the same amount of energy from an oil shale, which typically yields a barrel of oil per tonne of rock, requires five times as much mining as coal [3]. Large quantities of shale-rich waste rock would need to be disposed of in above-ground sites, like deep coal-mining waste. This would raise the possibility of ground contamination as the rock is washed by rainwater.

A further disadvantage of mining oil shale is that pulverizing the rock creates large clouds of fine dust to rise high into the air. In the USA there would be a danger of wind-borne dust contaminating neighbouring wheat-growing states like Kansas and Nebraska. In addition, processing of the rock requires three barrels of water for each barrel of oil produced and might require drastic modification of the local water supplies. The main "oil shale deposits" occur in desert/arid areas.

5.6.1.2.   In situ *combustion*

Some of the environmental objections to the mining method described in the preceding section can be overcome by *in situ* retorting, in which the quarrying and processing is carried out underground. A cavity is hollowed out by sinking boreholes and using hydraulic power or explosives to break up the shale. Heat is then applied from above (Fig. 5.11) by injecting hot liquids or gases to combust the shale. As the fire burns down through the rubble the decomposed kerogen products flow out at the bottom and are pumped out of the mine [17, 18]. This involved industrial process is very expensive and difficult to accomplish.

The high cost of *in situ* extraction, together with the extensive environmental protection required, make oil shale mining expensive. At the current (2013) price of Middle East crude oil of about US$100/barrel, it has now become economically feasible to seriously consider this option. In the oil business there used to be a well-held theory that the major oil producers would always keep the price of conventional crude oil lower than that required to economically develop oil shale/tar sand deposits. That time appears to have passed. Oil shale mining is now a serious player in the oil business [19].

5.6.2. *Tar sands (oil sands)*

Tar sands are beds of sand impregnated with the naturally-occurring viscous petroleum material bitumen, sometimes called "sticky oil" or "heavy oil". The name "tar" is a misnomer: the sand beds do not contain tar, which is a manufactured

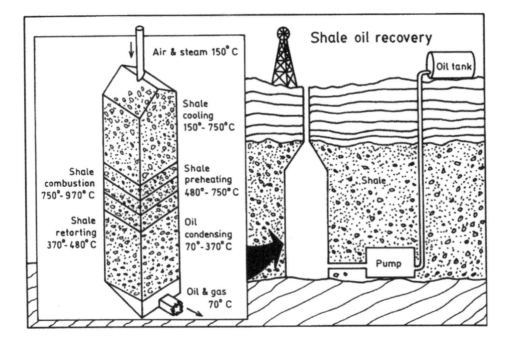

Fig. 5.11.   Shale oil recovery [18].

product. Originally the sands were "oil reservoirs", usually containing conventional oil, which had come near to the surface. This conventional oil, containing a complex mixture of hydrocarbons (including long-chain molecules), was changed by a reaction with water containing microorganisms and oxygen from the surface. In this weathering process the long-chain hydrocarbons are removed to produce tar sands or heavy oil.

Tar sands are found in more than 70 countries, but the largest known deposits are the Athabasca Tar Sands in Alberta, Canada, which contain about 75% of North America's petroleum reserves, and the Orinoco Heavy Oil Belt in Venezuela. These deposits lie underground but some of the material is sufficiently near to the surface (i.e. within 300 feet) to be accessible by open-casting mining. For the deeper deposits it would be necessary to use *in situ* methods of the type described in the preceding section.

Estimates of the world resources of oil from tar sands vary between 300–800 billion tonnes, although 90% of this would require sub-surface mining or *in situ* methods. Near-surface deposits might yield up to 30 billion tonnes [20]. This is about 3.5 times the 2009 world oil consumption, in addition to the known reserves of crude oil.

A summary of the estimated tar sands and heavy oil reserves is given as Table 5.8. It is estimated that some 66% of the world's petroleum reserves are preserved in oil sand form.

Table 5.8.   Estimated reserves of tar sands. Adapted from [21–24].

Canada (35% of reserves) [21, 22]

Athabasca-Wabiskaw oil sands — NW Alberta

The largest bitumen deposit, containing about 80% of the total, and the only one suitable for surface mining.

Cold Lake deposits — NE Alberta

Pease River deposits — NW Alberta

Reserves at least 220 billion barrels recoverable crude bitumen which amounts to three-quarters of North American petroleum reserves.

Venezuela (35% of reserves) [23]

Orinoco Oil Sands, *also known as the Orinoco Tar Sands*

Reserves estimated to be 236 billion barrels heavy crude oil which would make it one of the largest accumulations in the world.

USA [24]

Utah's Tar Sands

Eight major deposits with a combined shallow oil resource of 32.0 billion barrels of oil.

The most successful commercial production, at Athabasca, had approached 200,000 barrels per day, equivalent to 40 mtos/annum, by the early 1980s. The plant uses hot water to separate out the bitumen from the sand. This is then refined and upgraded by a fluidized-bed coking process and treated with hydrogen to remove impurities, resulting in high-grade synthetic crude oil. About two tonnes of sand are needed to produce a barrel of oil. An annual output of 40 million tonnes of oil would create almost 600 million tonnes of waste material, posing a formidable environmental waste disposal problem. Associated environmental effects include possible contamination of groundwater, air pollution and the destruction of natural drainage patterns.

In the early 1980s, estimated costs of producing synthetic crude oil from tar sands or oil shales were in the range US$40–$80 per barrel, which was barely economical even then but acceptable as prototype figures that could be expected to reduce with experience. This did, however, match the then current and anticipated prices of oil. The fall in price of Middle East crude oil to about US$20/barrel in 1984 (Fig. 5.2) made the tar sands and oil shale projects completely unviable and they were "mothballed" [25]. At the present period (2013), an international price near US$100/barrel is very profitable for synthetic oil and Canada has become a major player in the oil business [26].

5.7.   Environmental Issues

The environmental hazards of land-based oil exploration, refining and transportation include damage to wildlife preserves, interference with animal migration routes,

Fig. 5.12.   Offshore oil rig [27].

atmospheric emissions and the rehabilitation of land after mining or pipe-laying operations.

Corresponding hazards exist with regard to offshore oil platforms. A typical platform (Fig. 5.12) is an oil production factory that rests on the sea bottom on giant hydraulic legs [27]. The staff live on board for weeks at a time and essential supplies of food and mail are brought out by helicopter or boat. Sometimes crude oil from the rig sites is pumped ashore directly through pipes on the sea bed. At some sites the crude oil is pumped into oil tanker ships for distribution to refineries all over the world. About one-quarter of world oil production is from offshore fields. An offshore oilfield represents environmental risks such as contamination of the sea bed, damage to shoals of fish, interference with fish migration routes and the pollution of nearby beaches and coastlines due to leaks and spillages. A large number of rigs in a small area, such as in the North Sea or along the Gulf of Mexico and the California coastlines, can create artificial reef effects.

Table 5.9. World carbon dioxide emissions from oil use, by region, high economic growth case, 1990–2020 [8]. In million metric tons carbon equivalent.

Region/Country	History			Projections				Average annual percent change, 1999–2020
	1990	1998	1999	2005	2010	2015	2020	
Industrialized Countries	**716**	**775**	**793**	**894**	**1,008**	**1,120**	**1,235**	**2.1**
North America								
United States[a]	590	635	650	714	799	876	948	1.8
Canada	61	66	68	77	83	87	90	1.4
Mexico	65	74	76	103	127	157	197	4.7
Western Europe	**474**	**525**	**517**	**570**	**592**	**611**	**632**	**1.0**
United Kingdom	66	65	63	77	81	85	89	1.7
France	67	72	72	79	84	87	90	1.1
Germany	103	107	104	117	120	123	126	0.9
Italy	74	78	74	83	85	87	89	0.8
Netherlands	27	30	31	33	35	37	38	1.0
Other Western Europe	138	172	173	181	188	193	200	0.7
Industrialized Asia	**217**	**230**	**233**	**259**	**275**	**288**	**299**	**1.2**
Japan	179	183	185	204	215	221	225	0.9
Australasia	38	46	48	55	61	67	74	2.1
Total Industrialized	**1,407**	**1,529**	**1,543**	**1,722**	**1,875**	**2,019**	**2,166**	**1.6**
EE/FSU								
Former Soviet Union	334	148	146	207	254	323	380	4.7
Eastern Europe	66	55	55	64	75	88	98	2.8
Total EE/FSU	**400**	**202**	**201**	**271**	**330**	**411**	**479**	**4.2**
Developing Countries	**304**	**479**	**496**	**696**	**911**	**1,187**	**1,484**	**5.4**
Developing Asia								
China	94	152	160	209	280	371	474	5.3
India	45	70	73	101	136	184	226	5.5
South Korea	38	60	62	77	89	101	112	2.8
Other Asia	127	197	201	309	406	530	671	5.9
Middle East	**155**	**195**	**198**	**239**	**284**	**339**	**406**	**3.5**
Turkey	17	22	22	30	36	44	53	4.4
Other Middle East	138	173	177	210	247	295	353	3.4
Africa	**83**	**95**	**97**	**138**	**172**	**211**	**256**	**4.7**
Central and South America	**132**	**173**	**176**	**230**	**292**	**355**	**427**	**4.3**
Brazil	51	70	71	84	104	128	160	4.0
Other Central/South America	81	104	105	146	188	226	267	4.5
Total Developing	**674**	**942**	**968**	**1,303**	**1,658**	**2,091**	**2,573**	**4.8**
Total World	**2,482**	**2,673**	**2,712**	**3,297**	**3,863**	**4,520**	**5,218**	**3.2**

[a]Includes the 50 States and the District of Columbia. U.S. Territories are included in Australasia.
Notes: EE/FSU = Eastern Europe/Former Soviet Union.
Sources: **History:** Energy Information Administration (EIA), *International Energy Annual 1999*, DOE/EIA-O219(99) (Washington DC, February 2001). **Projections:** EIA, *Annual Energy Outlook 2002*, DOE/EIA-0383 (2002) (Washington, DC, December 2001) Table B19; and World Energy Projection System (2002).

An offshore oil platform, like any other component of the petroleum industry, represents an explosion risk. On the whole the safety record of the industry is very good. The worst single accident in the oil industry occurred in the North Sea field in 1988, when a catastrophic explosion and fire at the Piper-Alpha rig, operated by Occidental Petroleum, destroyed the rig and killed 167 workers, as mentioned in Section 2.6.3 of Chapter 2.

The largest-profile environmental risks of the oil industry are large-scale spillages from rigs and from oil tankers. In 2010 the world total of registered oil tankers numbered more than 3,500. The Exxon/Mobil conglomerate operated more than 500 tankers and supertankers in 1999. In Great Britain and Ireland, tanker wrecks such as the "Torrey Canyon" and "Amoco Cadiz" caused oil contamination of shorelines and widespread destruction of coastal and marine wildlife. The world's biggest oil tanker disaster was the wrecking of the "Exxon Valdez" in Prince William Sound, Alaska, in 1987. The ship ran onto rocks that punctured the double hull tanks laden with crude oil. Massive spillage occurred, necessitating a very expensive clean-up operation along the coastline.

In 2010 there was a massive spillage from British Petroleum's Deepwater Horizon Rig in the Gulf of Mexico. This caused extensive environmental damage and resulted in heavy financial penalties against the companies involved.

Sophisticated techniques have now been developed for containing, recovering and dispersing oil spills. The effects of oil contamination on the marine environment are not always irreversible. Possibly the worst feature of large-scale oil spillages is that the long-term effects are not known. Globules of oil come to rest not only on the sea bed and coastline but are swept away by ocean currents to be deposited all over the world.

Both land-based and offshore oil production contribute to environmental pollution in the form of gaseous carbon emissions. Historical and projected future carbon dioxide emissions due to oil are shown in Table 5.9 [8]. The levels of world emissions in 2010 due to oil are seen to be 3863/2814 or 37% greater than those due to coal in Table 4.8 of Chapter 4.

5.8.   Problems and Review Questions

5.1. The world price of crude oil remained stable for about 100 years, until 1973. What were the causes of the sudden change?

5.2. Use the information in Tables 5.3 and 5.4 to list, in order, the world's ten largest producers and consumers of petroleum products in 2010.

5.3. Why are the figures for all oil consumption in Asia Pacific growing so rapidly compared with other areas of the world?

5.4. In which countries of the world has yearly oil consumption increased most rapidly since 2000?

5.5. In which countries of the world has the average annual oil production increased most rapidly since 2000?

5.6. The Sultan of Brunei is one of the the world's richest people. What is the source of his wealth?

5.7. What are the world's busiest shipping routes for oil tankers and super tankers?

5.8. Most of the oil exported from the Middle East is transported in supertankers through the Straits of Hormuz [1].

    (a) How much oil was shipped through the Straits in 2000?

    (b) What effect would closure of the Straits of Hormuz have on the world economy if the closure lasted for (i) three months and (ii) three years?

5.9. If an oil supertanker can transport 200,000 tonnes of crude oil per journey, how many tanker journeys would have been needed to supply the entire USA oil import figure for the year 2010?

5.10. What were the values of UK petroleum exports in 1980, 1990, 2000 and 2009 for (a) crude oil, (b) refined petroleum products and process oils and (c) totals for petroleum?

5.11. Repeat Problem 5.10 with respect to UK petroleum imports.

5.12. Use the results of Problems 5.10 and 5.11 to calculate the net profit on the UK foreign trade in petroleum exports and imports, comparing 1980 with 2009.

5.13. What countries were the USA's biggest oil suppliers in 2010?

5.14. In the USA, oil consumption has grown steadily. In what year did the oil deficit first exceed the oil production?

5.15. Why do the North Americans have such a large per capita consumption of oil compared with the rest of the world?

5.16. Briefly speculate on the civil, political and global consequences which could conceivably arise if oil supplies to the West were (a) suddenly interrupted and (b) squeezed, forcing large price increases.

5.17. The total world consumption of oil has increased steadily since 1985. Inspite of this, the years of reserves remaining have risen significantly in the same period. Why is this so?

5.18. The "years of oil reserves remaining" at present are given in Fig. 5.3 and Table 5.2. For Western Europe the figures show only 21.7 years of reserves. What measures, if any, would you wish to see implemented in the light of this information?

5.19. In view of the impending shortfall of world oil supplies, what long-term plans should be considered?

5.20. What might be some of the effects on world oil consumption if the USA became politically isolationist and declared itself self-sufficient in energy?

5.21. Describe some of the difficulties that are experienced in extracting and refining crude oil from oil shale and tar sands.

5.22. Oil extraction and refining is responsible for almost as much carbon emission as coal and natural gas together. Yet coal is thought to be a "dirty" fuel and oil has a relatively "clean" image. Why is this so?

References

[1] Brooks, J. (1983). *Geochemical Processes*, in Petroleum Geochemistry and Fossil Fuel Exploration, Chemistry in Britain, **19**, 390–400.

[2] McVeigh, J.C. (1984). *Energy around the World*, Pergamon Press, Oxford.

[3] Foley, G. (1981). *The Energy Question*, 2nd Edition, Penguin Books Ltd, London.

[4] "BP Statistical Review of World Energy", British Petrolium plc, London, June 2011.

[5] AAPG Explorer (2001), *American Association of Petroleum Geologists*, **21**, 3.

[6] Tiratsoo, E.N. (1973). *Oilfields of the World*, Scientific Press Ltd, Beaconsfield.

[7] Sampson, A. (1975). *The Seven Sisters*, Coronet Books, Hodder and Stoughton Ltd, London.

[8] "International Energy Outlook 2001", US Dept. Of Energy Information Administration (EIA), Washington, DC, DOE/EIA-0484, March 2001.

[9] AAPG Explorer (2002).*American Association of Petroleum Geologists*, **22**,

[10] "United Kingdom, Country analysis Brief Overview", US energy Information Administration,Washington, DC, May 2013.

[11] "UK Energy Sector Indicators", Dept. of Energy and Climate Change, London, July 2010.

[12] "UK Energy in Brief", Dept. of Energy and Climate Change, London, July 2010.

[13] "International Fuel Prices 2009", Deutsche Gesellschaft für Technische Zusammenarbeit (GIZ) GmbH, 2009, http://www.giz.de/Themen/en/dokumente/gtz2009-en-ifp-full-version.pdf.

[14] "A Primer on Gasoline Prices", Energy Information Administration (EIA), US Dept. Of Energy, Washington, DC, April 2001.

[15] "Cars and Climate Change", International Energy Agency, OECD/IEA, Paris, 1993.

[16] Smith, J.W. (1980). *Oil Resources of the USA*, Mineral and Energy Resources, **23**, 6, 1–20.

[17] Harder, E.L. (1982). *Fundamentals of Energy Production*, John Wiley and Sons, New York.

[18] *The Guardian*, London, 12 May 1981.

[19] Shirley, K. (2001). *Tax Break Rekindled Interest: Shale Gas Exciting Again*, AAPG Explorer, **22**, 3, 24–25.

[20] Gray, G.R. (1981). *Encyclopedia of Energy*, McGraw-Hill Book Company, Inc., New York.

[21] Brooks, J. (2008). "Alberta Oil Sands", Government of Alberta Energy: Facts and Statistics. http://www.energy.alberta.ca/oilsands/791.asp.

[22] Brooks, J. (2008). "The Facts on Oil Sands", Canadian Association of Petroleum Producers.

[23] Brooks, J. (2009). "Estimates of Venezuela Recoverable Heavy Oil Resources of the Orinoco Belt", USGS, Fact Sheet 2009–3028, Washington, DC, USA

[24] Todd, C.M., Snarr, D.G. (2012). "Utah's Tar Sands: Developing the USA's Latest Oil Sand Resource", US Oil Sands Inc. Presentation to the Utah Governor's Energy Summit, February 2010.

[25] Stobaugh, R.B. and Yergin, D. (eds.) (1983). *Energy Future, Report of the Harvard Business School*, 3rd Edition, Vintage Books, Random House, New York.

[26] Will, G. (1996). *Energy Oil — The Jewel in Western Canada's Oil Play*, Petroleum Review, **50**, 505–507.

[27] Brookes, J. (1987). *Past, Present and Future Exploration of NW Europe*, Petroleum Review, **2**, 36–42.

CHAPTER 6

NATURAL GAS

6.1. Nature and Origin of Natural Gas

Naturally-occurring gas or natural gas is a mixture of hydrocarbon and non-hydrocarbon gases but is mostly methane (CH_4). Many natural gas deposits occur independently of oil deposits and are then sometimes called "non-associated" (with oil) deposits. Non-associated gas is derived from organic matter (often coal) by the thermal degradation and the natural degasification of coal strata deposited at greater depths and increasing geological temperature. The deposits of natural gas that occur in association with oil accumulations, either as a separate gas layer or in solution, are sometimes called "associated gas" or "condensate". In the early days of oil exploration, the fuel value of the associated gas was not always appreciated. Sometimes it was flared off by burning, a practice still widely used in the Middle East, and thereby wasted. In a modern petrochemical plant in Europe or North America a burning gas flare is likely to be a temporary safety feature.

The synthetic gas that used to be called "coal gas" or "town gas" is obtained by the industrial processing of coal and is described in Section 6.8. Coal gas and natural gas are used in the same applications but are entirely different in origin and differ in composition.

In addition to methane, which is usually between 85% and 95% of the content, natural gas also contains the hydrocarbons ethane, propane, butane and pentane (short carbon chain paraffins), plus non-hydocarbons including nitrogen, hydrogen sulphide and carbon dioxide. The precise composition depends on the particular source and nature of the gas. The "dry" natural gas sold as gaseous fossil fuel contains about 96% methane, 5.3% propane and 2.6% butane. Butane and propane may be liquefied and sold as liquid petroleum gas such as "calogas" [1]. Natural gas itself can be liquefied by refrigeration and is then known as liquid natural gas (LNG) in North America or natural gas liquid (NGL) or condensate in Europe.

Natural gas is measured by volume (not by weight) in the volumetric units of cubic metres. Slightly inconsistent conversion factors exist due to the different grades of fuel. A cubic foot is the energy equivalent of 1000 BTU or 1.05 MJ. One cubic metre has an equivalent heat energy of 1000 kcal or 3.77×10^7 J [2]. In volumetric terms it is convenient to measure industrial quantities of gas in millions (10^6), US billions (10^9) or trillions (10^{12}) of cubic feet or cubic metres. For direct comparison with the fuel energy properties of coal or oil, the large-scale use of gas can be measured in million tonnes of oil equivalent (mtoe).

6.2.   History and Development

In the 19th century the discovery and development of natural gas as a fuel first proceeded concurrently in the USA and Russia. The first natural gas pipeline, 22 km long, transported gas to Pittsburgh, USA, in 1883. During the same period the development of both oil and natural gas occurred in the Baku oilfield near the Caspian Sea in Russia [3].

The exploitation of natural gas as a fuel was dominated by the USA until the 1950s. In 1950 over 85% of the world's natural gas was produced and consumed in the USA, with Venezuela and Russia exploiting 8% and 4%, respectively. It was not until August 1959 that the first major giant Groningen gas field discovery was made in the Netherlands by a joint Esso–Shell team. The area of the North Sea between Great Britain and continental Europe has proved to contain deep buried coal layers lying underneath porous sandstone and sealed by a thick impermeable cap-rock. Natural gas was generated from these coals and has permeated into the sandstone. Countries with coastlines on the North Sea have agreed on defined areas of exploration and natural gas is being produced by Belgium, Denmark, Germany, the Netherlands, Norway and the UK. For example, the British Petroleum Company found gas 45 miles east of the Humber estuary in 1965 [1].

Since the early 1960s the countries of Eastern Europe have made massive increases in their exploitation and export of natural gas. Natural gas is expected to overtake oil as the pre-eminent source of fossil energy in the 21st century.

6.3.   Natural Gas Reserves

The known reserves of natural gas, shown in Table 6.1, are widely distributed across the world. In 2010 Russia held almost a quarter of the world reserves and 40.5% were located in the Middle East [2]. Very small proportions of natural gas reserves occur in Western Europe, Africa and East Asia.

The right-hand column of Table 6.1 gives figures for the natural gas reserve/production ratio, in years. The average value of reserves remaining for the world is 58.6 years. This compares with the value for oil of 46.2 years in Table 5.2 of Chapter 5.

Table 6.1.   World natural gas proved reserves [2].

	At end 1990	At end 2000	At end 2009	At end 2010			
	Trillion cubic metres	Trillion cubic metres	Trillion cubic metres	Trillion cubic feet	Trillion cubic metres	Share of total	R/P ratio
US	4.8	5.0	7.7	**272.5**	**7.7**	4.1%	12.6
Canada	2.7	1.7	1.7	**61.0**	**1.7**	0.9%	10.8
Mexico	2.0	0.8	0.5	**17.3**	**0.5**	0.3%	8.9
Total North America	9.5	7.5	9.9	**350.8**	**9.9**	5.3%	12.0
Argentina	0.7	0.8	0.4	**12.2**	**0.3**	0.2%	8.6
Bolivia	0.1	0.7	0.7	**9.9**	**0.3**	0.2%	19.5
Brazil	0.1	0.2	0.4	**14.7**	**0.4**	0.2%	28.9
Colombia	0.1	0.1	0.1	**4.4**	**0.1**	0.1%	11.0
Peru	0.3	0.2	0.4	**12.5**	**0.4**	0.2%	48.8
Trinidad & Tobago	0.3	0.6	0.4	**12.9**	**0.4**	0.2%	8.6
Venezuela	3.4	4.2	5.1	**192.7**	**5.5**	2.9%	*
Other S. & Cent. America	0.2	0.1	0.1	**2.3**	**0.1**	◆	22.4
Total S. & Cent. America	5.2	6.9	7.5	**261.6**	**7.4**	4.0%	45.9
Azerbaijan	n/a	1.2	1.3	**44.9**	**1.3**	0.7%	84.2
Denmark	0.1	0.1	0.1	**1.8**	**0.1**	◆	6.4
Germany	0.2	0.2	0.1	**2.4**	**0.1**	◆	6.5
Italy	0.3	0.2	0.1	**3.0**	**0.1**	◆	11.1
Kazakhstan	n/a	1.8	1.9	**65.2**	**1.8**	1.0%	54.9
Netherlands	1.8	1.5	1.2	**41.5**	**1.2**	0.6%	16.6
Norway	1.7	1.3	2.0	**72.1**	**2.0**	1.1%	19.2
Poland	0.2	0.1	0.1	**4.2**	**0.1**	0.1%	29.2
Romania	0.1	0.3	0.6	**21.0**	**0.6**	0.3%	54.4
Russian Federation	n/a	42.3	44.4	**1580.8**	**44.8**	23.9%	76.0
Turkmenistan	n/a	2.6	8.0	**283.6**	**8.0**	4.3%	*
Ukraine	n/a	1.0	1.0	**33.0**	**0.9**	0.5%	50.4
United Kingdom	0.5	1.2	0.3	**9.0**	**0.3**	0.1%	4.5
Uzbekistan	n/a	1.7	1.6	**55.1**	**1.6**	0.8%	26.4
Other Europe & Eurasia	49.7	0.5	0.4	**10.0**	**0.3**	0.2%	28.3
Total Europe & Eurasia	54.5	55.9	63.0	**2227.6**	**63.1**	33.7%	60.5
Bahrain	0.2	0.1	0.2	**7.7**	**0.2**	0.1%	16.7
Iran	17.0	26.0	29.6	**1045.7**	**29.6**	15.8%	*
Iraq	3.1	3.1	3.2	**111.9**	**3.2**	1.7%	*
Kuwait	1.5	1.6	1.8	**63.0**	**1.8**	1.0%	*
Oman	0.3	0.9	0.7	**24.4**	**0.7**	0.4%	25.5
Qatar	4.6	14.4	25.3	**894.2**	**25.3**	13.5%	*
Saudi Arabia	5.2	6.3	7.9	**283.1**	**8.0**	4.3%	95.5
Syria	0.2	0.2	0.3	**9.1**	**0.3**	0.1%	33.2

(*Continued*)

Table 6.1.   (*Continued*)

	At end 1990	At end 2000	At end 2009	At end 2010			
	Trillion cubic metres	Trillion cubic metres	Trillion cubic metres	Trillion cubic feet	Trillion cubic metres	Share of total	R/P ratio
United Arab Emirates	5.6	6.0	6.1	213.0	6.0	3.2%	*
Yemen	0.2	0.5	0.5	17.3	0.5	0.3%	78.3
Other Middle East	†	0.1	0.1	7.7	0.2	0.1%	62.1
Total Middle East	38.0	59.1	75.7	2677.0	75.8	40.5%	*
Algeria	3.3	4.5	4.5	159.1	4.5	2.4%	56.0
Egypt	0.4	1.4	2.2	78.0	2.2	1.2%	36.0
Libya	1.2	1.3	1.5	54.7	1.5	0.8%	98.0
Nigeria	2.8	4.1	5.3	186.9	5.3	2.8%	*
Other Africa	0.8	1.1	1.2	41.4	1.2	0.6%	65.7
Total Africa	8.6	12.5	14.7	520.1	14.7	7.9%	70.5
Australia	0.9	2.2	2.9	103.1	2.9	1.6%	58.0
Bangladesh	0.7	0.3	0.4	12.9	0.4	0.2%	18.3
Brunei	0.3	0.4	0.3	10.6	0.3	0.2%	24.7
China	1.0	1.4	2.8	99.2	2.8	1.5%	29.0
India	0.7	0.8	1.1	51.2	1.5	0.8%	28.5
Indonesia	2.9	2.7	3.0	108.4	3.1	1.6%	37.4
Malaysia	1.6	2.3	2.4	84.6	2.4	1.3%	36.1
Myanmar	0.3	0.3	0.3	11.8	0.3	0.2%	27.5
Pakistan	0.6	0.7	0.8	29.1	0.8	0.4%	20.9
Papua New Guinea	0.2	0.4	0.4	15.6	0.4	0.2%	*
Thailand	0.2	0.4	0.3	11.0	0.3	0.2%	8.6
Vietnam	†	0.2	0.7	21.8	0.6	0.3%	66.0
Other Asia Pacific	0.3	0.3	0.4	12.4	0.4	0.2%	20.4
Total Asia Pacific	9.9	12.3	15.8	571.8	16.2	8.7%	32.8
Total World	**125.7**	**154.3**	**186.6**	**6608.9**	**187.1**	**100.0%**	**58.6**
of which: OECD	15.7	14.7	17.0	603.8	17.1	9.1%	14.7
Non-OECD	109.9	139.6	169.6	6005.1	170.0	90.9%	83.6
European Union	3.4	3.8	2.5	86.2	2.4	1.3%	14.0
Former Soviet Union	49.3	50.8	58.4	2066.4	58.5	31.3%	77.2

*More than 100 years.
†Less than 0.05.
♦Less than 0.05%.
n/a not available.

Notes: Proved reserves of natural gas — Generally taken to be those quantities that geological and engineering information indicates with reasonable certainty can be recovered in the future from known reservoirs under existing economic and operating conditions. Reserves-to-production (R/P) ratio — If the reserves remaining at the end of any year are divided by the production in that year, the result is the length of time that those remaining reserves would last if production were to continue at that rate. Source of data — The estimates in this table have been compiled using a combination of primary official sources and third-party data from Cedigaz and the OPEC Secretariat.

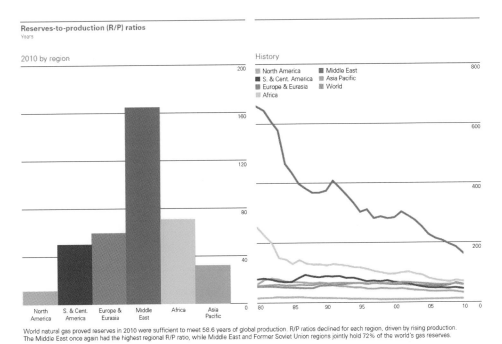

Fig. 6.1.   Natural gas reserve/production (R/P) ratios, 2010 [2].

The variation of the R/P ratio since 1985 is shown in Fig. 6.1. Global reserves have been remaining at around 60 years because new discoveries and enhanced methods of extraction have kept up with the increasing demand. This is similar to the situation with oil, shown in Fig. 5.3. As with oil, the maintenance of the R/P ratio for 25/30 years should not be interpreted to mean that the reserve will always remain at that level. The natural gas stock is depleting. For both oil and natural gas the world's biggest repository for reserves is the Middle East, where the gas accumulations are largely untouched because of the much greater profitability of oil.

The world's natural gas supply will outlast the oil by about a generation. This is particularly pertinent to the OECD countries, where the reserve figure has remained low and roughly constant since the early 1970s, and is now (2013) starting to decrease.

All of the concerns regarding world oil reserves, discussed in Chapter 5, also apply to natural gas. The world consumption of natural gas was less than that of oil in 2010 (Table 2.3) but constituted $2858.1/12002.4\,\text{mtoe} = 23.8\%$ of the world primary fuel consumption.

The depletion of the world stock of natural gas, like that of oil, is a global problem.

6.4.   Production and Consumption of Natural Gas

6.4.1.   *World natural gas production*

Natural gas is produced and consumed all over the world. Moreover, the proportion of total world primary energy consumption taken up by natural gas is projected to significantly increase with future energy demand. In the 21st century the use of natural gas is expected to overtake oil as a source of energy [4, 5].

Natural gas production is detailed in Table 6.2. Over the 11-year period 2000–2010, world gas production increased from 2178.7 to 2880.9 mtoe. This represents an increase of 32% or almost 3% per year. Production increased in many areas but most markedly in the Middle East, which more than doubled its production. China more than tripled its production during this period.

6.4.2.   *World natural gas consumption*

Figures for natural gas consumption, in mtoe, are given in Table 6.3 [2]. For the period 2000–2010 the world total increased from 2176.2 to 2858.1 mtoe, which is 13.1%, averaging 1.2% per year. In China the consumption increased more than four-fold. The trend of increased world consumption looks likely to continue, with a projected figure of 162 tcf (trillion cubic ft) in 2020, which is almost double the year 2000 total of 84 tcf. Most of the increase is expected to be in developing countries.

Although there appear to be sufficient gas resources to meet anticipated world consumption through to 2030, several regions, such as Western Europe, Mexico and the USA, are increasingly dependent on imported supplies of gas. This increase in demand for natural gas has been made possible in the last decades by the construction of long, large diameter pipelines from remote gas fields to industrial centres of gas consumption.

A comparison of the production figures of Table 6.2 with the consumption figures of Table 6.3 show that Russia, Algeria, Canada, Indonesia, the Netherlands and Norway are now net exporter countries, while France, Germany, Italy, Japan and the USA are importer countries. The trade flow of natural gas implied in the data of Tables 6.2 and 6.3 is illustrated in Fig. 6.2. In Western Europe natural gas has now overtaken coal as the second most important fuel, after oil. The price of natural gas has slowly increased over the last ten years, illustrated in Table 6.4 (page 187), and is paralleled by the price of LNG.

6.4.3.   *UK natural gas production and consumption*

The British North Sea contains an estimated 26.7 tcf of natural gas reserves. Most non-associated gas fields are located off the English coast in the Southern Gas Basin, adjacent to the Dutch North Sea sector. Key producing gas fields include

Table 6.2. World natural gas production* (mtoe) [2].

Million tonnes oil equivalent	2000	2001	2002	2003	2004	2005	2006	2007	2008	2009	2010	Change 2010 over 2009	2010 share of total
US	495.5	508.2	489.9	494.7	480.7	467.6	479.3	499.6	521.7	531.6	**556.8**	4.7%	19.3%
Canada	164.0	167.8	169.1	166.2	165.3	168.4	169.6	164.3	158.8	147.5	**143.8**	−2.5%	5.0%
Mexico	34.4	34.4	35.5	37.0	38.4	40.5	46.3	48.2	48.8	49.4	**49.8**	0.7%	1.7%
Total North America	693.9	710.4	694.4	697.9	684.4	676.5	695.3	712.0	729.3	728.5	**750.4**	3.0%	26.0%
Argentina	33.7	33.4	32.5	36.9	40.4	41.1	41.5	40.3	39.7	37.2	**36.1**	−3.0%	1.3%
Bolivia	2.9	4.2	4.4	5.7	8.8	10.7	11.6	12.4	12.9	11.1	**12.9**	16.8%	0.4%
Brazil	6.7	6.9	8.3	9.0	9.9	9.9	10.2	10.1	12.4	10.5	**13.0**	23.5%	0.5%
Colombia	5.3	5.5	5.5	5.5	5.7	6.0	6.3	6.8	8.2	9.5	**10.1**	7.2%	0.4%
Peru	0.3	0.3	0.4	0.5	0.8	1.4	1.6	2.4	3.1	3.1	**6.5**	108.4%	0.2%
Trinidad & Tobago	13.1	13.9	16.2	23.7	24.6	27.9	32.8	35.1	35.4	36.5	**38.1**	4.4%	1.3%
Venezuela	25.1	26.6	25.6	22.7	25.6	24.7	28.3	26.6	27.0	25.8	**25.7**	−0.7%	0.9%
Other S. & Cent. America	3.0	3.2	3.0	2.8	2.8	3.1	3.7	3.5	3.4	2.9	**2.6**	−9.9%	0.1%
Total S. & Cent. America	90.2	94.1	96.0	106.8	118.6	124.8	136.0	137.2	141.8	136.7	**145.1**	6.2%	5.0%
Azerbaijan	4.6	4.5	4.2	4.2	4.1	4.7	5.5	8.8	13.3	13.3	**13.6**	2.2%	0.5%
Denmark	7.3	7.5	7.5	7.2	8.5	9.4	9.4	8.3	9.1	7.6	**7.4**	−3.0%	0.3%
Germany	15.2	15.3	15.3	15.9	14.7	14.2	14.1	12.9	11.7	11.0	**9.6**	−12.7%	0.3%
Italy	13.7	12.6	12.1	11.5	10.7	10.0	9.1	8.0	7.6	6.6	**6.8**	3.6%	0.2%
Kazakhstan	9.4	9.5	9.2	11.3	18.0	20.3	21.5	24.1	26.8	29.3	**30.3**	3.3%	1.1%
Netherlands	52.3	56.2	54.3	52.2	61.6	56.3	55.4	54.5	60.0	56.4	**63.5**	12.4%	2.2%
Norway	44.8	48.5	59.0	65.8	70.6	76.5	78.9	80.7	89.4	93.4	**95.7**	2.5%	3.3%
Poland	3.3	3.5	3.6	3.6	3.9	3.9	3.9	3.9	3.7	3.7	**3.7**	0.5%	0.1%
Romania	12.4	12.2	11.9	11.7	11.5	11.2	10.7	10.4	10.3	10.1	**9.8**	−2.9%	0.3%
Russian Federation	475.7	473.6	484.9	505.4	516.0	522.1	535.6	532.8	541.5	474.9	**530.1**	11.6%	18.4%

(*Continued*)

Table 6.2.   (*Continued*)

Million tonnes oil equivalent	2000	2001	2002	2003	2004	2005	2006	2007	2008	2009	2010	Change 2010 over 2009	2010 share of total
Turkmenistan	38.3	41.8	43.6	48.1	47.5	51.3	54.3	58.9	59.5	32.7	**38.1**	16.4%	1.3%
Ukraine	14.6	14.9	15.3	15.8	16.5	16.7	16.9	16.9	17.1	17.3	**16.7**	−3.8%	0.6%
United Kingdom	97.5	95.2	93.2	92.6	86.7	79.4	72.0	64.9	62.7	53.7	**51.4**	−4.3%	1.8%
Uzbekistan	45.9	46.8	46.7	46.8	48.8	48.6	49.0	53.2	56.0	54.0	**53.2**	−1.5%	1.8%
Other Europe & Eurasia	10.0	9.8	10.1	9.6	9.9	9.8	10.3	9.7	9.2	8.8	**9.0**	3.0%	0.3%
Total Europe & Eurasia	845.0	852.0	870.8	901.7	929.0	934.2	946.5	947.9	977.8	872.8	**938.8**	7.6%	32.6%
Bahrain	7.9	8.2	8.5	8.7	8.8	9.6	10.2	10.6	11.4	11.5	**11.8**	2.4%	0.4%
Iran	54.2	59.4	67.5	73.4	76.4	93.2	97.7	100.7	104.7	118.0	**124.7**	5.6%	4.3%
Iraq	2.8	2.5	2.1	1.4	0.9	1.3	1.3	1.3	1.7	1.0	**1.1**	8.7%	◆
Kuwait	8.6	9.5	8.5	9.9	10.7	11.0	11.3	10.9	11.5	10.1	**10.4**	3.5%	0.4%
Oman	7.8	12.6	13.5	14.9	16.7	17.8	21.3	21.6	21.7	22.3	**24.4**	9.4%	0.8%
Qatar	21.3	24.3	26.6	28.3	35.3	41.2	45.6	56.9	69.3	80.4	**105.0**	30.7%	3.6%
Saudi Arabia	44.8	48.3	51.0	54.1	59.1	64.1	66.2	67.0	72.4	70.6	**75.5**	7.0%	2.6%
Syria	4.9	4.5	5.5	5.6	5.8	4.9	5.1	5.0	4.8	5.1	**7.0**	37.3%	0.2%
United Arab Emirates	34.5	40.4	39.1	40.3	41.7	43.0	44.1	45.3	45.2	43.9	**45.9**	4.5%	1.6%
Yemen	—	—	—	—	—	—	—	—	—	0.7	**5.6**	704.6%	0.2%
Other Middle East	0.2	0.2	0.2	0.2	1.4	1.7	2.4	2.7	3.3	2.8	**3.2**	15.0%	0.1%
Total Middle East	187.3	210.0	222.5	236.6	256.6	287.9	305.2	322.1	345.9	366.4	**414.6**	13.2%	14.4%
Algeria	76.0	70.4	72.3	74.5	73.8	79.4	76.0	76.3	77.2	71.6	**72.4**	1.1%	2.5%
Egypt	18.9	22.7	24.6	27.1	29.7	38.3	49.2	50.1	53.1	56.4	**55.2**	−2.2%	1.9%
Libya	5.3	5.6	5.3	5.0	7.3	10.2	11.9	13.8	14.3	14.3	**14.2**	−0.6%	0.5%
Nigeria	11.3	13.4	12.8	17.3	20.5	20.2	25.6	31.5	31.5	22.3	**30.3**	35.7%	1.1%
Other Africa	5.8	6.2	6.0	6.5	8.0	8.9	9.4	11.1	14.2	14.7	**16.1**	9.4%	0.6%
Total Africa	117.3	118.3	121.0	130.4	139.3	156.9	172.1	182.8	190.4	179.3	**188.1**	4.9%	6.5%

(*Continued*)

Table 6.2. *(Continued)*

Million tonnes oil equivalent	2000	2001	2002	2003	2004	2005	2006	2007	2008	2009	2010	Change 2010 over 2009	2010 share of total
Australia	28.0	28.9	29.0	29.4	32.2	33.5	36.2	37.7	37.4	43.1	**45.3**	5.1%	1.6%
Bangladesh	9.0	9.7	10.3	11.1	11.9	13.1	13.8	14.6	16.1	17.8	**18.0**	1.3%	0.6%
Brunei	10.2	10.3	10.3	11.1	11.0	10.8	11.3	11.0	10.9	10.3	**11.0**	6.7%	0.4%
China	24.5	27.3	29.4	31.5	37.3	44.4	52.7	62.3	72.3	76.7	**87.1**	13.5%	3.0%
India	23.7	23.8	24.8	26.6	26.3	26.7	26.4	27.1	27.5	35.3	**45.8**	29.7%	1.6%
Indonesia	58.7	57.0	62.7	65.9	63.3	64.1	63.2	60.9	62.7	64.7	**73.8**	14.0%	2.6%
Malaysia	40.7	42.2	43.5	46.6	48.5	55.0	57.0	58.1	58.2	57.7	**59.8**	3.7%	2.1%
Myanmar	3.1	6.3	7.6	8.6	9.2	11.0	11.3	12.2	11.2	10.4	**10.9**	4.9%	0.4%
Pakistan	19.4	20.4	22.1	27.4	31.0	32.0	32.5	33.1	33.8	34.6	**35.5**	2.7%	1.2%
Thailand	18.2	17.7	18.5	19.3	20.1	21.3	21.9	23.4	25.9	27.8	**32.7**	17.4%	1.1%
Vietnam	1.4	1.8	2.2	2.1	3.7	5.8	6.3	6.4	6.7	7.2	**8.4**	16.7%	0.3%
Other Asia Pacific	8.1	8.5	9.8	9.6	9.1	10.0	12.8	15.2	16.0	16.1	**15.6**	-3.4%	0.5%
Total Asia Pacific	244.9	253.8	270.2	289.4	303.7	327.5	345.4	362.0	378.7	401.7	**443.9**	10.5%	15.4%
Total World	**2178.7**	**2238.5**	**2274.9**	**2362.9**	**2431.5**	**2507.8**	**2600.4**	**2664.0**	**2763.8**	**2685.4**	**2880.9**	**7.3%**	**100.0%**
of which: OECD	973.1	995.2	985.3	991.5	989.6	976.4	991.3	1000.6	1028.8	1020.8	**1050.7**	2.9%	36.5%
Non-OECD	1205.6	1243.3	1289.7	1371.3	1441.9	1531.4	1609.1	1663.5	1735.0	1664.6	**1830.2**	9.9%	63.5%
European Union	208.7	209.6	204.9	201.2	204.6	190.8	181.2	168.7	170.5	154.4	**157.4**	2.0%	5.5%
Former Soviet Union	588.8	591.4	604.2	631.9	651.1	664.0	683.1	694.9	714.4	621.8	**682.1**	9.7%	23.7%

*Excluding gas flared or recycled.

◆Less than 0.05%.

Source: Includes data from Cedigaz.

Table 6.3.   World natural gas consumption (mtoe) [2].

Million tonnes oil equivalent	2000	2001	2002	2003	2004	2005	2006	2007	2008	2009	2010	Change 2010 over 2009	2010 share of total
US	600.4	573.9	593.7	575.3	577.3	568.5	560.0	597.3	600.6	588.3	**621.0**	5.6%	21.7%
Canada	83.4	79.4	81.2	87.9	85.6	88.0	87.3	85.7	86.0	85.0	**84.5**	−0.6%	3.0%
Mexico	36.9	37.5	40.8	45.1	48.1	48.5	54.8	56.5	59.8	59.9	**62.0**	3.4%	2.2%
Total North America	720.7	690.7	715.6	708.4	710.9	705.0	702.1	739.5	746.4	733.1	**767.4**	4.7%	26.9%
Argentina	29.9	28.0	27.2	31.1	34.1	36.4	37.6	39.5	40.0	38.8	**39.0**	0.4%	1.4%
Brazil	8.5	10.7	12.7	14.2	16.9	17.7	18.7	19.0	22.2	17.8	**23.8**	33.8%	0.8%
Chile	5.8	6.6	6.6	7.2	7.8	7.5	7.0	4.1	2.4	2.8	**4.2**	51.0%	0.1%
Colombia	5.3	5.5	5.5	5.4	5.7	6.0	6.3	6.7	6.8	7.8	**8.2**	4.3%	0.3%
Ecuador	0.3	0.3	0.2	0.2	0.3	0.3	0.7	0.5	0.4	0.4	**0.4**	−6.0%	◆
Peru	0.3	0.3	0.4	0.5	0.8	1.4	1.6	2.4	3.1	3.1	**4.9**	56.0%	0.2%
Trinidad & Tobago	9.5	10.5	11.4	13.0	12.0	13.6	18.2	18.2	19.7	18.8	**19.8**	5.5%	0.7%
Venezuela	25.1	26.6	25.6	22.7	25.6	24.7	28.3	26.6	28.3	27.5	**27.6**	0.6%	1.0%
Other S. & Cent. America	1.6	2.1	2.2	2.8	2.6	2.9	3.5	4.0	4.3	4.5	**5.0**	9.9%	0.2%
Total S. & Cent. America	86.4	90.6	91.9	97.1	105.8	110.6	121.9	121.2	127.2	121.6	**132.9**	9.3%	4.7%
Austria	7.3	7.7	7.7	8.5	8.5	9.0	8.5	8.0	8.6	8.4	**9.1**	8.6%	0.3%
Azerbaijan	4.7	6.8	6.8	7.0	7.5	7.7	8.2	7.2	8.2	7.0	**5.9**	−15.9%	0.2%
Belarus	14.2	14.1	14.5	14.3	16.1	16.5	17.1	17.0	17.3	14.5	**17.7**	22.3%	0.6%
Belgium & Luxembourg	14.0	13.8	14.1	15.0	15.2	15.4	15.4	15.3	15.5	15.7	**17.4**	10.9%	0.6%
Bulgaria	2.9	2.7	2.4	2.5	2.5	2.8	2.9	2.9	2.9	2.1	**2.3**	10.1%	0.1%
Czech Republic	7.5	8.0	7.8	7.8	8.2	8.6	8.4	7.8	7.8	7.4	**8.4**	13.7%	0.3%
Denmark	4.4	4.6	4.6	4.7	4.7	4.5	4.6	4.1	4.1	4.0	**4.5**	12.2%	0.2%
Finland	3.4	3.7	3.6	4.0	3.9	3.6	3.8	3.5	3.6	3.2	**3.5**	9.9%	0.1%
France	35.4	37.7	36.5	38.7	40.5	39.6	37.9	38.2	39.4	38.0	**42.2**	11.1%	1.5%
Germany	71.5	74.6	74.3	77.0	77.3	77.6	78.5	74.6	73.1	70.2	**73.2**	4.2%	2.6%

(*Continued*)

Table 6.3. (*Continued*)

Million tonnes oil equivalent	2000	2001	2002	2003	2004	2005	2006	2007	2008	2009	2010	Change 2010 over 2009	2010 share of total
Greece	1.8	1.8	1.9	2.2	2.4	2.4	2.8	3.4	3.6	3.0	**3.3**	8.2%	0.1%
Hungary	9.7	10.7	10.6	11.9	11.7	12.1	11.5	10.7	10.6	9.1	**9.8**	7.7%	0.3%
Republic of Ireland	3.4	3.6	3.7	3.7	3.7	3.5	4.0	4.3	4.5	4.3	**4.8**	10.8%	0.2%
Italy	58.4	58.5	58.1	64.1	66.5	71.2	69.7	70.0	70.0	64.4	**68.5**	6.4%	2.4%
Kazakhstan	8.6	9.2	13.3	15.9	22.5	24.1	25.3	23.8	24.5	22.1	**22.7**	2.9%	0.8%
Lithuania	2.5	2.5	2.6	2.8	2.8	2.9	2.9	3.3	2.9	2.5	**2.8**	14.3%	0.1%
Netherlands	35.0	36.0	35.8	36.0	36.8	35.4	34.3	33.3	34.7	35.0	**39.2**	12.1%	1.4%
Norway	3.6	3.4	3.6	3.9	4.1	4.0	4.0	3.8	3.9	3.7	**3.7**	−0.5%	0.1%
Poland	10.0	10.4	10.1	11.3	11.9	12.2	12.4	12.4	12.5	12.0	**12.9**	7.1%	0.5%
Portugal	2.1	2.3	2.8	2.7	3.4	3.8	3.7	3.9	4.2	4.2	**4.5**	6.7%	0.2%
Romania	15.4	14.9	15.5	16.5	15.7	15.8	16.3	14.5	14.3	11.9	**12.0**	0.6%	0.4%
Russian Federation	318.6	329.6	330.9	346.4	354.7	360.2	367.7	379.9	374.4	350.7	**372.7**	6.3%	13.0%
Slovakia	5.8	6.2	5.8	5.7	5.5	5.9	5.4	5.1	5.2	4.4	**5.1**	14.5%	0.2%
Spain	15.2	16.4	18.8	21.3	24.7	29.1	30.3	31.6	34.8	31.1	**31.0**	−0.3%	1.1%
Sweden	0.7	0.6	0.7	0.7	0.7	0.7	0.8	0.9	0.8	1.0	**1.4**	38.9%	0.1%
Switzerland	2.4	2.5	2.5	2.6	2.7	2.8	2.7	2.6	2.8	2.7	**3.0**	10.5%	0.1%
Turkey	13.1	14.4	15.6	18.8	19.9	24.2	27.4	32.5	33.8	32.1	**35.1**	9.2%	1.2%
Turkmenistan	11.0	11.2	11.6	12.8	13.5	14.5	16.5	19.1	18.5	17.9	**20.4**	13.5%	0.7%
Ukraine	63.9	61.9	60.9	62.1	61.7	62.1	60.3	56.9	54.0	42.3	**46.9**	11.0%	1.6%
United Kingdom	87.2	86.7	85.6	85.8	87.7	85.5	81.1	81.9	84.5	78.0	**84.5**	8.3%	3.0%
Uzbekistan	41.1	44.6	45.8	41.2	39.1	38.4	37.7	41.3	43.8	39.2	**41.0**	4.6%	1.4%
Other Europe & Eurasia	11.9	13.0	12.2	12.7	14.1	14.3	14.7	15.3	14.5	12.3	**14.1**	14.9%	0.5%
Total Europe & Eurasia	886.8	914.5	920.9	960.4	990.1	1010.5	1016.5	1029.1	1033.4	954.5	**1023.5**	7.2%	35.8%
Iran	56.6	63.1	71.3	74.6	77.9	94.5	97.8	101.7	107.4	118.2	**123.2**	4.2%	4.3%
Israel	†	†	†	†	1.1	1.5	2.1	2.5	3.7	4.1	**4.8**	17.5%	0.2%
Kuwait	8.6	9.5	8.5	9.9	10.7	11.0	11.3	10.9	11.5	10.9	**12.9**	18.8%	0.5%

(*Continued*)

Table 6.3. (*Continued*)

Million tonnes oil equivalent	2000	2001	2002	2003	2004	2005	2006	2007	2008	2009	2010	Change 2010 over 2009	2010 share of total
Qatar	8.7	9.9	10.0	11.0	13.5	16.8	17.6	17.4	17.4	18.0	**18.4**	2.0%	0.6%
Saudi Arabia	44.8	48.3	51.0	54.1	59.1	64.1	66.2	67.0	72.4	70.6	**75.5**	7.0%	2.6%
United Arab Emirates	28.3	34.1	32.8	34.1	36.2	37.8	39.0	44.3	53.5	53.2	**54.5**	2.5%	1.9%
Other Middle East	21.0	21.3	22.2	22.5	23.9	25.5	28.3	29.1	32.8	34.7	**39.6**	14.1%	1.4%
Total Middle East	168.1	186.1	195.8	206.1	222.4	251.3	262.3	272.8	298.7	309.7	**329.0**	6.2%	11.5%
Algeria	17.9	18.5	18.2	19.3	19.8	20.9	21.4	21.9	22.8	24.5	**26.0**	6.0%	0.9%
Egypt	18.0	22.1	23.9	26.7	28.5	28.4	32.9	34.5	36.8	38.3	**40.6**	6.0%	1.4%
South Africa	1.1	1.1	0.9	0.9	1.9	2.8	3.1	3.1	3.4	3.0	**3.4**	13.8%	0.1%
Other Africa	15.7	15.9	16.2	18.4	21.4	22.5	22.0	25.4	27.2	23.2	**24.4**	5.5%	0.9%
Total Africa	52.6	57.4	59.2	65.3	71.7	74.7	79.3	85.0	90.1	89.0	**94.5**	6.1%	3.3%
Australia	18.5	19.4	19.8	19.8	21.0	19.8	22.8	24.9	25.9	27.7	**27.3**	−1.2%	1.0%
Bangladesh	9.0	9.7	10.3	11.1	11.9	13.1	13.8	14.6	16.1	17.8	**18.0**	1.3%	0.6%
China	22.1	24.7	26.3	30.5	35.7	42.1	50.5	63.5	73.2	80.6	**98.1**	21.8%	3.4%
China Hong Kong SAR	2.7	2.7	2.6	1.7	2.4	2.4	2.6	2.5	2.9	2.8	**3.4**	24.3%	0.1%
India	23.7	23.8	24.8	26.6	28.7	32.1	33.5	36.1	37.2	45.9	**55.7**	21.5%	1.9%
Indonesia	26.8	27.9	29.6	31.5	29.0	29.9	29.9	28.2	30.0	33.6	**36.3**	7.8%	1.3%
Japan	65.1	66.8	65.4	71.8	69.3	70.7	75.4	81.2	84.4	78.7	**85.1**	8.1%	3.0%
Malaysia	21.7	22.7	23.6	24.6	22.2	28.3	30.4	30.1	30.4	30.3	**32.2**	6.2%	1.1%
New Zealand	5.1	5.3	5.1	3.9	3.5	3.2	3.3	3.6	3.4	3.5	**3.7**	4.2%	0.1%

(*Continued*)

Natural Gas   185

Table 6.3.   (Continued)

Million tonnes oil equivalent	2000	2001	2002	2003	2004	2005	2006	2007	2008	2009	2010	Change 2010 over 2009	2010 share of total
Pakistan	19.4	20.4	22.1	27.4	31.0	32.0	32.5	33.1	33.8	34.6	**35.5**	2.7%	1.2%
Philippines	†	0.1	1.6	2.4	2.2	3.0	2.4	2.8	3.0	3.0	**2.8**	−5.8%	0.1%
Singapore	†	0.8	3.2	3.6	4.5	6.2	6.3	7.8	7.4	7.3	**7.6**	4.2%	0.3%
South Korea	17.0	18.7	20.8	21.8	25.5	27.3	28.8	31.2	32.1	30.5	**38.6**	26.5%	1.4%
Taiwan	6.1	6.6	7.4	7.6	9.2	9.3	10.0	10.6	10.5	10.2	**12.7**	24.3%	0.4%
Thailand	19.8	22.3	24.2	25.7	26.9	29.3	30.0	31.8	33.6	35.3	**40.6**	15.0%	1.4%
Vietnam	1.4	1.8	2.2	2.1	3.7	5.8	6.3	6.4	6.7	7.2	**8.4**	16.7%	0.3%
Other Asia Pacific	3.5	3.4	3.3	3.8	4.1	4.7	4.9	5.4	5.1	4.6	**4.8**	3.6%	0.2%
Total Asia Pacific	261.7	277.2	292.2	315.8	330.9	359.0	383.4	413.7	435.6	453.5	**510.8**	12.6%	17.9%
Total World	**2176.2**	**2216.6**	**2275.6**	**2353.1**	**2431.8**	**2511.2**	**2565.6**	**2661.3**	**2731.4**	**2661.4**	**2858.1**	**7.4%**	**100.0%**
of which: OECD	1225.7	1213.2	1239.2	1260.8	1280.8	1287.8	1290.2	1336.9	1358.0	1313.9	**1397.6**	6.4%	48.9%
Non-OECD	950.5	1003.4	1036.3	1092.3	1151.0	1223.3	1275.4	1324.3	1373.4	1347.5	**1460.5**	8.4%	51.1%
European Union	396.3	406.6	406.0	425.8	437.4	444.8	438.2	433.0	440.7	412.6	**443.3**	7.4%	15.5%
Former Soviet Union	471.3	487.4	493.0	509.3	525.5	534.9	544.2	557.1	551.8	503.0	**537.1**	6.8%	18.8%

† Less than 0.05.
◆ Less than 0.05%.
Notes: The difference between these world consumption figures and the world production statistics is due to variations in stocks at storage facilities and liquefaction plants, together with unavoidable disparities in the definition, measurement or conversion of gas supply and demand data.
Source: Includes data from Cedigaz.

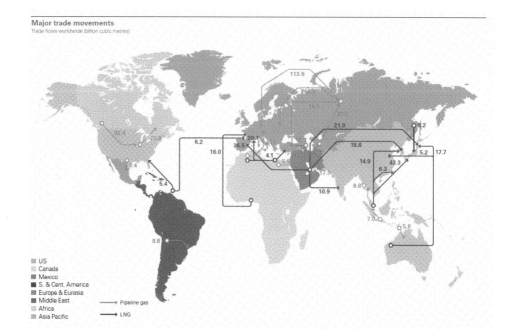

Fig. 6.2.   Major trade flows in natural gas, 2010 [2].

BP Amoco's 5.7-tcf Leman field, Chevron and Conoco's 3-tcf Brittania and Shell's 1.7-tcf Indefatigable and 0.8-tcf Clipper.

As the continental shelf reserves deplete both oil and natural gas, production has reduced since the peak of 1990, as shown in Fig. 6.3 [6]. The consequence of falling production is that the UK has now become an importer of natural gas, as shown in Fig. 6.4 [6]. The share of imports due to LNG rose from 2% in 2008 to 24% in 2009 via two new LNG terminals [6].

The UK consumption of natural gas since 1980 is illustrated in Fig. 6.5 [6]. More than one-third of the total is now used for electricity generation, which is now the largest end user.

The UK natural gas industry has completed the process of returning from public ownership to private ownership. This process began in 1986 with the privatization of the state-held gas monopoly British Gas. British Gas remained the sole natural gas supplier until 1996, when other companies entered the market. The UK's gas and electricity regulatory body, Ofgem, planned to end all price controls in British gas markets in 2001. The privatization of the UK's gas industry, leading to an increased gas supply, has helped gas to replace much of the UK's reliance on coal as a source for electricity generation. The natural gas share of fuels for electricity generation was 162.5 TWh/357.2 TWh = 45.5% in 2010 [6]. Privatization in the UK has progressed well in advance of EU requirements.

Table 6.4. Natural gas prices, 1984–2010 [2].

US dollars per million Btu	LNG Japan cif	Average German Import Price*	Natural gas UK (Heren NBP Index)[†]	US Henry Hub[‡]	Canada (Alberta)[‡]	Crude oil OECD countries cif
1984	5.10	4.00	—	—	—	5.00
1985	5.23	4.25	—	—	—	4.75
1986	4.10	3.93	—	—	—	2.57
1987	3.35	2.55	—	—	—	3.09
1988	3.34	2.22	—	—	—	2.56
1989	3.28	2.00	—	1.70	—	3.01
1990	3.64	2.78	—	1.64	1.05	3.82
1991	3.99	3.19	—	1.49	0.89	3.33
1992	3.62	2.69	—	1.77	0.98	3.19
1993	3.52	2.50	—	2.12	1.69	2.82
1994	3.18	2.35	—	1.92	1.45	2.70
1995	3.46	2.39	—	1.69	0.89	2.96
1996	3.66	2.46	1.87	2.76	1.12	3.54
1997	3.91	2.64	1.96	2.53	1.36	3.29
1998	3.05	2.32	1.86	2.08	1.42	2.16
1999	3.14	1.88	1.58	2.27	2.00	2.98
2000	4.72	2.89	2.71	4.23	3.75	4.83
2001	4.64	3.66	3.17	4.07	3.61	4.08
2002	4.27	3.23	2.37	3.33	2.57	4.17
2003	4.77	4.06	3.33	5.63	4.83	4.89
2004	5.18	4.32	4.46	5.85	5.03	6.27
2005	6.05	5.88	7.38	8.79	7.25	8.74
2006	7.14	7.85	7.87	6.76	5.83	10.66
2007	7.73	8.03	6.01	6.95	6.17	11.95
2008	12.55	11.56	10.79	8.85	7.99	16.76
2009	9.06	8.52	4.85	3.89	3.38	10.41
2010	10.91	8.01	6.56	4.39	3.69	13.47

*1984–1990 German Federal Statistical Office, 1991–2010. German Federal Office of Economics and Export Control (BAFA).
[†]Heren Energy Ltd.
[‡]Energy Intelligence Group, *Natural Gas Week*.
Note: cif = cost + insurance + freight (average prices).

In 1998, the UK–Continent Gas Interconnector pipeline was opened, with terminals at Bacton, England, and Zeebrugge, Belgium. This is the first natural gas pipeline linking the UK to the European continent. The reversible pipeline was developed by a consortium comprising British Gas (40%), BP (10%), Conoco (10%), Elf (10%), Gazprom (10%), Amerada Hess (5%), Distrigaz (5%), National Power (5%) and Ruhrgaz (5%). The pipeline was originally intended to allow continental Europe to take advantage of lower UK gas prices. At the end of 1998, British demand peaked and the pipeline is now used to send gas supplies into Britain.

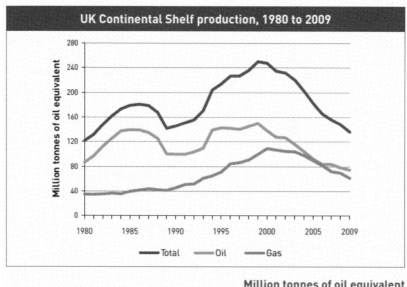

	Million tonnes of oil equivalent					
	1980	**1990**	**2000**	**2007**	**2008**	**2009**
Oil	86.9	100.1	138.3	83.9	78.6	74.8
Gas	34.8	45.5	109.3	72.1	69.7	61.5
Total	**121.7**	**145.6**	**247.6**	**156.0**	**148.3**	**136.3**

Fig. 6.3.   UK oil and natural gas production [6].

The layout of the major European pipelines is shown in Fig. 6.6 [7].

A pipeline to connect Ireland to Scottish gas sources in the Corrib field was approved in November 1999, and a plan to connect Ireland to England via Wales was announced in April 2000. A pipeline would run from Manchester, England, underground to Wales, and then under the Irish Sea to just north of Dublin. There is currently one pipeline linking Britain and Ireland, connecting Ireland to Scottish gas sources. Despite these pipeline projects, the UK remains a much smaller natural gas exporter than their North Sea neighbour Norway [5]. The country changed from being an importer to an exporter of natural gas in 1996 (Fig. 6.4) for about ten years, but is now (2013) firmly in deficit and an importer.

In 1999, 78% of home heating in the UK was fuelled by natural gas [8]. The distribution of UK natural gas consumption between different user sections is shown in Fig. 6.5. Since 1980 industrial consumption has been relatively stable, growing by 9%, while domestic consumption has grown by 49% and services consumption has more than doubled. However, since 1991 the growth in gas use has been dominated by its increasing use for electricity generation.

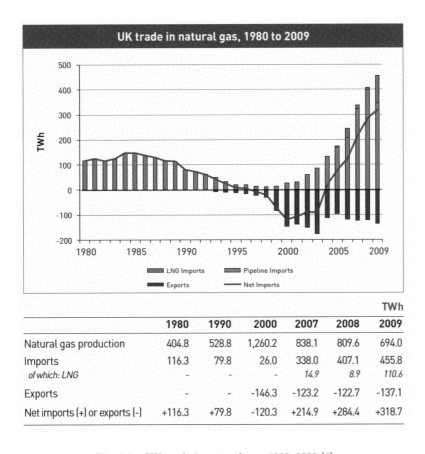

	1980	**1990**	**2000**	**2007**	**2008**	**2009**
Natural gas production	404.8	528.8	1,260.2	838.1	809.6	694.0
Imports	116.3	79.8	26.0	338.0	407.1	455.8
of which: LNG	-	-	-	*14.9*	*8.9*	*110.6*
Exports	-	-	-146.3	-123.2	-122.7	-137.1
Net imports (+) or exports (-)	+116.3	+79.8	-120.3	+214.9	+284.4	+318.7

Fig. 6.4.   UK trade in natural gas, 1980–2009 [6].

6.4.4.   *US natural gas production and consumption*

It is clear from Table 6.3 that the USA is the world's largest user of natural gas, accounting for 21.2% of world consumption in 2010. Consumption has increased by only about 2% since 2000. The main sites for the importation of natural gas are by pipelines from Canada and Mexico and the oil/gas ports of the Mexican Gulf and of New England (Fig. 6.7) [7].

In terms of per capita consumption, the North Americans (principally Canada and the USA) are also the big users, although this has declined since 1970, whereas the per capita consumption in both Western and Eastern Europe increased significantly over the same period. In 2000 the USA accounted for slightly over one-quarter of the world natural gas consumption. In 2010 this reduced to 21.7% of the world total. The world consumption of natural gas increased from 2176.2 to 2858.1 mtoe, or by 13.1%, from 2000 to 2010, whereas the US consumption increased by only 621/600.4 mtoe or 3%. The US Energy Information Administration projections for

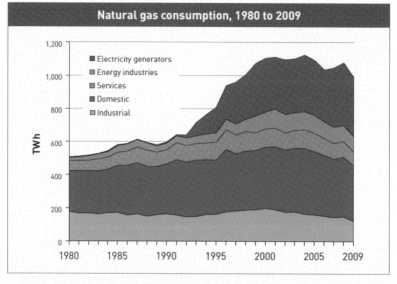

Fig. 6.5.   UK natural gas consumption, 1980–2009 [6].

	1980	1990	2000	2007	2008	2009
Electricity generators	4.0	6.5	324.6	355.9	376.8	356.2
Energy Industries	19.1	39.2	102.1	98.9	96.9	93.8
Industry	177.5	164.6	198.5	144.3	148.7	123.9
Domestic	246.8	300.4	369.9	352.9	359.6	334.8
Services	60.4	86.4	110.5	94.8	95.6	83.1
Total	**507.8**	**597.0**	**1,105.5**	**1,046.8**	**1,077.7**	**991.8**

the period 2007–2035 are given in Fig. 6.8 [8]. Most of the projected large increase is expected to be taken up by the Middle Eastern countries, who are the main sources of the gas. The projected increase by the USA is expected to be only 1–2% per year.

It is now accepted that the natural gas-fired turbine is safe, clean, more efficient than any competing technology, and uncontroversial (in comparison with nuclear power). This has prompted a big switch to gas-fired electricity-generating systems in both Europe and the USA. To build a gas-fired electricity generation plant costs about one-half the cost of a coal-fired plant of the same rating [9]. Some misgivings are starting to arise as to whether future natural gas supplies will be adequate and will last long enough to justify the present scale of financial investment in new plants [10].

The price of natural gas for the US domestic market is among the cheapest of the G7 group of nations (Table 6.4). Significantly, the USA is one of the few countries that does not tax natural gas.

Fig. 6.6.   Major European natural gas pipelines [7].

Several projects are proposed to further increase the import capacity of natural gas from Canada to the USA. Concurrently it is planned to increase exports of natural gas from the USA to Mexico. Although Mexico is rich in natural gas resources, most of this is located in south-eastern Mexico, far from the primary consuming areas of the north, and the country lacks the infrastructure to transport the gas [11].

Proven reserves of associated or condensate natural gas in 2009 were 244.7 tcf. This represents an increase of 46% over the 1999 figure, or a growth of 3.3% per year (Table 6.1).

The areas containing the largest proportions of the reserves correspond to the areas with the largest volumes of crude oil reserves [12].

6.5.   Unconventional Sources of Natural Gas

6.5.1.   *Coal-bed methane*

It is known that deposits of coal undergo natural degasification and give off (mainly) methane gas. The methane content increases with depth and with coal

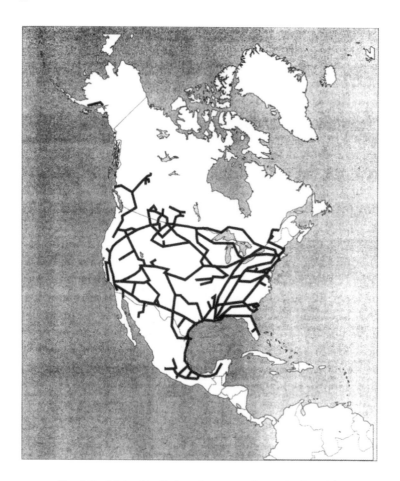

Fig. 6.7. Major North American natural gas pipelines [7].

rank, illustrated in Fig. 6.9. Coal-bed gas is stored mainly within micropores of the coal matrix in an absorbed state, and secondarily in micropores and fractures as free gas or solution of gas in water. The producibility of gas from a deposit depends on many features, including thermal maturity, gas content, coal thickness, fracture density, *in situ* stress, permeability, burial history and hydrological setting.

In deep coal mines the presence of methane can present a serious health and accident hazard, as discussed in Section 4.13 of Chapter 4. Safe underground mining procedures require that the methane in deep mine workings be flushed out with large quantities of air, at considerable expense and considerable waste of valuable natural gas. In 1980 the US Bureau of Mines reported that US mines were venting (and wasting) 256 million cubic feet per day of methane, which was equal to 0.5% of the total dry gas production.

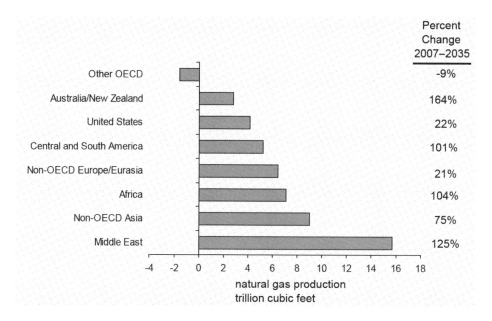

Fig. 6.8.   Projected increases in natural gas consumption, to 2035 [8].

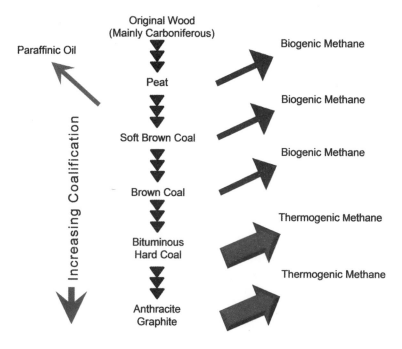

Fig. 6.9.   Development stages of coal-bed methane [13].

Table 6.5. Estimates of the major coal-bed methane gas
reserves (trillion cubic feet (tcf)) [14].

Country	Gas reserves (tcf)
Russia	600–400
China	1060–1240
USA	400–800
Canada	200–2700
Australia	300–500
Germany	100
UK	60
Kazakhstan	40
Poland	100
India	30
South Africa	30
Ukraine	60
Total	2980–9660 tcf

6.5.1.1. *World reserves of coal-bed methane*

Estimates of the world reserves of coal-bed methane vary greatly from different sources. An oil industry-based estimate is given in Table 6.5 [14]. In 2010 the world consumption of conventional natural gas was 3,169.0 billion cubic metres [2]. Since 1 cubic metre $= 35.315$ cubic feet this is equivalent to $(3169 \times 35.315)/1000 = 112$ tcf. The estimated reserves of coal-bed methane gas in Table 6.5 are therefore in tcf between $2980/112 = 26.6$ to $9660/112 = 86.3$ times the world consumption of 2010. This represents an enormous reserve of fossil energy.

The largest known resources of coal-bed methane are in Russia, China, Canada, Australia and the USA (in that order), but there are also significant resources in Germany, Poland, the UK, Ukraine, Kazakhstan, India and southern Africa. Russia also has the advantage of a well-developed infrastructure with extensive pipelines (Fig. 6.10) [7].

In conventional natural gas reservoirs, once the hole is drilled the gas production capacity is generally initially at a maximum level but declines over time. After the well starts to produce, the well pressure begins to decline and gas production decreases with time. With coal-bed methane wells, however, the well may actually increase gas production with time. This is because gas is liberated from the coal by desorption, caused by pressure release due to dewatering [15]. The total energy content of the gas obtained by coal-bed methane extraction will not normally exceed 1% of that of the host coal. Coal-bed methane extraction is a way of obtaining useful energy in a readily marketable form but is not equivalent to actually mining the coal. Nevertheless, it provides a means of extracting some energy from coal seams that cannot be mined at all.

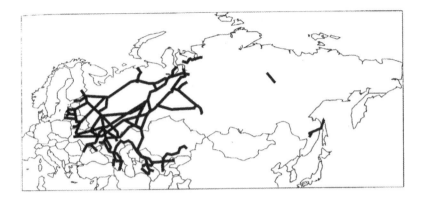

Fig. 6.10.   Major Eastern European/former Soviet Union natural gas pipelines [7].

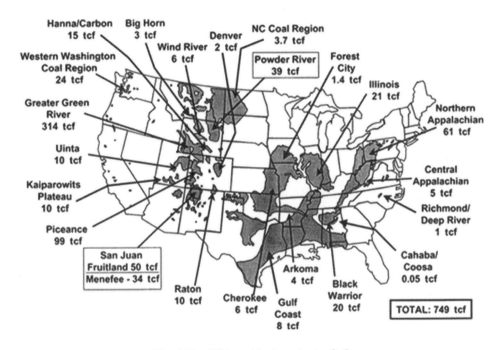

Fig. 6.11.   USA coal-bed gas basins [14].

6.5.1.2.   *US reserves of coal-bed methane*

A map of the principal US coal basins and estimated "in-place" coal-bed methane resources is given as Fig. 4.5 in Chapter 4. The coal-bed gas basins in continental USA are shown in Fig. 6.11, with numerical estimates at the particular location. Coal-bed methane production grew between the years 2000 to 2010 to become about

7% of US dry gas production. It has become an important component of the US fuel supply network that is of increasing importance.

Experience in other countries has been less encouraging, generally because their coals have low permeability compared with those in the two main coal-bed methane producing basins in the USA [15–17].

The USA and Canada are working jointly to develop the vast, deep, non-mineable coal-beds in the Canadian province of Alberta. A process of injecting carbon dioxide is used. When absorbed into the coal this displaces the trapped methane. There is an abundance of deep coal-beds in both countries, providing sites for the geographical storage of carbon dioxide from nearby coal-burning plants and thereby reducing greenhouse gas emissions [18]. Alberta is the only Canadian province to date (2013) to have commercial coal-bed methane wells.

6.5.2. *Natural gas hydrates*

A very high potential source of natural gas is from a resource known as "gas hydrates", "natural gas hydrates" or "methane gas hydrates".

These have been known about for a number of years but are now considered to be a viable source for extraction of methane gas. This source lies frozen in combustible ice crystals below the ocean floor in the Arctic regions. The crystalline solids look and behave like ice but incorporate a guest molecule (e.g. methane) as part of the structure (Fig. 6.12). During the very long geological record, catastrophic methane hydrate dissociations have been related to several abrupt climatic events.

The energy locked into natural gas hydrates has been estimated to represent double the energy potential of all the world's conventional oil, gas and coal reserves

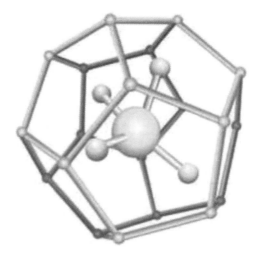

Fig. 6.12.   Gas hydrate crystal cage structure [19].

combined. No sustained exploration of this source has yet (2013) taken place. It is predicted that serious activity may begin in localized regions, such as the North Slope in Alaska, within the next ten years or so. The northern territories of Canada are thought to contain vast resources of gas hydrate.

6.5.3. *Shale gas*

Shale gas is natural gas obtained from oil shale. Ordinarily, shales have insufficient permeability to allow significant fluid flow to a well bore and are not commercial sources of natural gas. Because of the low permeability gas, production in commercial quantities requires the shale to be fractured around the well bores. Horizontal drilling is often used with shale gas wells, with lateral lengths up to 3,000 metres within the shale to create maximum bore-hole surface area in contact with the shale. This technology is necessary in order to make the shale gas production economically feasible.

Shale gas drilling and fracturing (fracking) requires large amounts of water. This may adversely affect the availability of water for other purposes. It also results in large amounts of waste water that might contain chemicals and contaminants. The mining process of hydraulic fracturing might result in accidental spillage.

The increasing demand for natural gas makes all gas sources attractive at present (2013) prices. In the year 2009 shale gas formed 14% of the US natural gas supply. This proportion is expected to rise to 45% by 2035 [11].

Several major sites are being exploited in Canada, especially in northern British Columbia [2].

6.5.4. *Tight gas*

The term "tight gas" does not denote a special type or source of natural gas but a deposit of natural gas that is tightly locked in and requires special measures to access it. Both the drilling and production operations are challenging because of the low reservoir/permeability. For example, it is necessary to drill many more wells to access a tight field because each well produces a relatively small amount of gas. Again, the wells may require having complex geometries to overcome difficulties of the geological structure [20, 21].

The various sources of natural gas for North America, including tight gas, are shown in Fig. 6.13 [22]. This correlates with the US Energy Information Administration forecast for the growth in the use of unconventional natural gas in Fig. 6.14 [8].

6.6. Environmental Aspects of Natural Gas

Natural gas is increasingly viewed as an environmentally-friendly fuel. It burns cleaner than coal or oil, causing less local air pollution. The latest figures for carbon

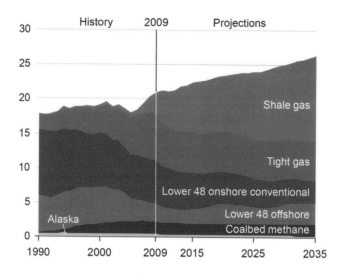

Fig. 6.13. Estimate of US natural gas resources 1990–2035 (in trillion cubic feet per year) [22].

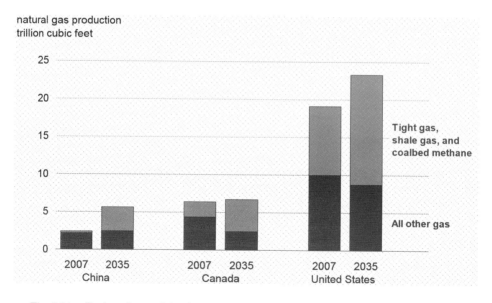

Fig. 6.14. Projected growth in the use of natural gas in China, Canada and the USA [8].

dioxide emissions from natural gas use are given in Table 6.6. A comparison with the corresponding Table 5.9 of Chapter 5 for oil shows that gas results in less than one-half of the oil emission rate at present. The future projections indicate an increasing emission rate over the next decade or so, for all areas of the world, becoming 65% of the oil emission rate by the year 2020.

Table 6.6. World carbon dioxide emissions from natural gas use, by region, reference case, 1990–2020 [11]. In million metric tons carbon equivalent.

Region/Country	History			Projections				Average annual percent change, 1999–2020
	1990	1998	1999	2005	2010	2015	2020	
Industrialized Countries								
North America	**326**	**377**	**381**	**450**	**499**	**555**	**598**	**2.2**
United States[a]	277	315	317	374	413	460	496	2.2
Canada	35	42	46	49	53	58	63	1.6
Mexico	15	20	19	26	33	37	39	3.4
Western Europe	**140**	**196**	**206**	**259**	**291**	**328**	**385**	**3.0**
United Kingdom	30	47	50	57	65	73	84	2.5
France	16	21	21	29	34	38	46	3.2
Germany	32	42	43	57	62	68	82	3.2
Italy	25	32	35	42	48	54	59	2.5
Netherlands	20	23	22	25	27	28	30	1.4
Other Western Europe	18	31	35	50	56	68	84	4.2
Industrialized Asia	**36**	**52**	**54**	**59**	**62**	**69**	**80**	**1.9**
Japan	24	38	40	43	43	48	56	1.7
Australasia	12	14	14	17	19	21	23	2.3
Total Industrialized	**503**	**626**	**641**	**768**	**852**	**953**	**1,062**	**2.4**
EE/FSU								
Former Soviet Union	369	291	294	317	351	394	440	1.9
Eastern Europe	46	36	35	46	60	78	93	4.7
Total EE/FSU	**414**	**327**	**329**	**363**	**411**	**472**	**532**	**2.3**
Developing Countries								
Developing Asia	**45**	**86**	**92**	**155**	**204**	**263**	**328**	**6.2**
China	8	13	14	31	47	76	108	10.1
India	7	13	12	19	26	36	43	6.1
South Korea	2	8	10	15	19	23	30	5.5
Other Asia	29	53	56	90	111	129	147	4.7
Middle East	**56**	**100**	**102**	**120**	**156**	**187**	**220**	**3.7**
Turkey	2	6	7	7	7	10	13	3.0
Other Middle East	54	94	96	113	148	177	207	3.7
Africa	**22**	**29**	**31**	**36**	**40**	**48**	**54**	**2.7**
Central and South America	**32**	**49**	**51**	**70**	**111**	**173**	**228**	**7.4**
Brazil	2	3	4	8	20	35	48	13.3
Other Central/ South America	30	46	47	62	92	138	180	6.6
Total Developing	**155**	**263**	**276**	**380**	**510**	**672**	**831**	**5.4**
Total World	**1,072**	**1,216**	**1,247**	**1,512**	**1,774**	**2,096**	**2,425**	**3.2**

(*Continued*)

Table 6.6.   (*Continued*)

Region/Country	History			Projections				Average annual percent change, 1999–2020
	1990	1998	1999	2005	2010	2015	2020	
Annex I								
Industrialized	488	606	622	742	819	916	1,023	2.4
EE/FSU	344	296	277	281	284	305	343	1.0
Total Annex I	**832**	**902**	**900**	**1,023**	**1,103**	**1,220**	**1,367**	**2.0**

[a]includes the 50 States and the District of Columbia. U.S. Territories are included in Australasia.
Notes: EE/FSU = Eastern Europe/Former Soviet Union.
Sources: **History**: Energy Information Administration (ElA), *International Energy Annual 1999*, DOE/EIA-O219(99) (Washington, DC, February 2001). **Projections**: EIA, *Annual Energy Outlook 2002*, DOE/EIA-0383(2002) (Washington, DC, December 2001), Table A19; and World Energy Projection System (2002).

The countries of Eastern Europe, India and China, which now rely heavily on coal for electricity generation, increasingly tend to regard natural gas as a more attractive fuel. As one might reasonably expect, there is a direct correlation between the amount of natural gas consumed and the consequent production of carbon dioxide emissions from its use.

An offshore gas platform or any facility handling natural gas is part of the petroleum industry and represents some degree of industrial hazard. The safety record of gas exploration and production, for routine industrial accidents and explosion risks, is excellent in the OECD countries. Nevertheless, gas production, like oil, is required to apply the most stringent safety practices and regulations during exploration and production to guard against leaks, fires and explosions. There has never been a major accident due to gas leakage from transmission pipes or spillage from LNG tankers. Nevertheless, any pipe leaks tend to blanket the surrounding area with gas, thus excluding oxygen, and can leave an area barren for months.

The combustion of natural gas causes the oxidation of atmospheric nitrogen. Nitrogen oxides are formed, which contribute to acid precipitation. In association with sulphur dioxide the nitrogen oxides can be toxic to plant life. The total effects of natural gas use are, however, much less severe than those of coal.

6.7.   Synthetic Gas from Coal

Until the widespread use of natural gas in the 1950s and 1960s the term "gas" meant "coal gas" or "town gas" in most countries of the world. It has long been known that coal undergoes low temperature natural degasification, described in Section 6.5 on coal-bed methane.

Coal gas was used as fuel for street lighting in several European countries before the end of the 18th century. The 19th century saw the widespread development of the coal gas industry right across the world. In the USA high temperature coal gasification was a commercial industry for 150 years, ending in the mid-1950s [23].

By the end of the 19th century the amount of coal used in coal gas production in Britain was just over 13 million tonnes, which was more than for the rest of Europe together [24].

In Britain, the old-style "gas works", with its characteristic malodorous stench, piles of coal for processing and storage gasometer tanks, had largely disappeared by the early 1960s. It is an interesting irony that world stocks of coal will long outlast the reserves of oil and natural gas. The use of coal as a base material to once again manufacture coal gas (and also possibly oil) is a possible future scenario for the early part of the 21st century.

The UK policy of drastic curtailment of the once great coal mining industry in favour of natural gas and imported coal may turn out to be short-sighted. Senior geologists in the UK have commented that it will not be possible/practicable to reopen the old deep coal mines. Things would have to start up again by drilling new shafts and boring new tunnels if the need again arises to increase the production of domestic deep coal. On the other hand, the use of imported coal preserves the indigenous coal, which can be used as a source of coal-bed methane.

All modern high temperature coal gasification processes follow similar stages to produce either low BTU, medium BTU or high BTU coal gases. Coal or coke is crushed and heated in the presence of air or oxygen. Several different industrial methods, such as the Lurgi process, have been developed in Europe. These differ only with regard to the method of bringing the reactants into contact with the coal. Gas made from passing air or an air-steam mixture over hot coal is called *low BTU* gas and has the low calorific value of $3-6 \, MJ/m^3$. This gas, which consists mainly of nitrogen plus the combustible components carbon monoxide and hydrogen, cannot be economically transported or used in equipment designed for natural gas. It can be used to power on-site industrial boilers. A useful application is for economic electric power generation using combined-cycle plants, as described in Chapter 4.

If oxygen is used, rather than air, the reaction produces a gas called *medium BTU* gas, which has a calorific value of $10-22 \, MJ/m^3$. Most of the diluting nitrogen is removed by this process, leaving mainly carbon monoxide. If methane or other hydrocarbons are added to increase the upper-end calorific properties, this fuel acquires a heating value about one-half that of natural gas. It burns rapidly and can produce a flame temperature higher than that of natural gas. The high carbon monoxide content prevents it from being distributed in the natural gas pipeline network [1].

High BTU gas is thermally equivalent to natural gas, having a calorific value of $37 \, MJ/m^3$ or $1000 \, BTU/ft^3$. It is interchangeable with natural gas and can be transmitted in the natural gas pipeline network. This synthetic natural gas (SNG) is produced by the methanation of medium BTU gas. A number of investigations are currently proceeding to develop more effective and economic methods of producing SNG. The aim is to produce high methane gas and reduce the need for oxygen in the gasifier. In the UK there is an industrial catalytic process for producing natural gas (methane) from oil.

It is possible to produce coal gas directly from a coal seam by underground gasification. A controlled fire is started underground, fed by oxygen, air or steam producing low BTU gas in the combustion area. This technique has some environmental appeal and was tried extensively in the former Soviet Union during 1930s [18]. In the UK the underground gasification process would be in direct competition with the most profitable coal mines and is not presently economical [25–27].

6.8.   Problems and Review Questions

6.1. Use the data of Tables 6.2 and 6.3 to list, in order, the world's current largest producers and consumers of natural gas.

6.2. Why did the consumption of natural gas in Eastern Europe start to fall significantly after about 1990?

6.3. In which countries of the world did the annual consumption of natural gas increase most rapidly during the 1990s?

6.4. What are the world's busiest trade routes for natural gas?

6.5. Is the UK self-sufficient in the use of natural gas? Explain using figures of UK production and consumption of natural gas from 1984 to the present time.

6.6. Is the USA self-sufficient in the use of natural gas? Explain using figures of US production and consumption from 1984 to the present.

6.7. The consumption of natural gas has been fairly constant in the UK since 2000. What has been the average annual variation?

6.8. The consumption of natural gas has very slowly risen in the USA since 2000. What has been the average annual increase?

6.9. What proportion of the world consumption of natural gas was used by the Western OECD countries in the year 2009?

6.10. What proportion of the world natural gas reserves is currently located in (a) North America, (b) Europe, (c) the former Soviet Union, (d) Middle East, (e) Africa and (f) Asia and Australia?

6.11. Why has the world stock of natural gas "proved reserves" increased since 1980 in spite of a continuous rise of consumption in the same period?

6.12. If the Middle Eastern countries become the chief exporters of natural gas by the middle of the present century, how will this gas probably be transported to Western Europe?

6.13. What is the probable chief source of synthetic natural gas?

6.14. Draw up a table of low BTU, medium BTU and high BTU gases in terms of their calorific values.

6.15. Why is natural gas projected to become increasingly popular as a prime fuel source, compared with coal, for electricity generation?

6.16. Explain the differences between natural gas, coal-bed methane and coal gas.

6.17. Why has the enormous potential of coal-bed methane, as a viable fuel source, only recently received serious attention?

References

[1] McVeigh, J.C. (1984). *Energy around the World*, Pergamon Press, Oxford.

[2] "BP Statistical Review of World Energy", British Petroleum plc, London, June 2011.

[3] Tiratson, E.N. (1972). *Natural Gas*, 2nd Edition, Scientific Press Ltd., London.

[4] "BP Review of World Gas", British Petroleum plc, London, June 1995.

[5] "United Kingdom", US Energy Information Administration, Overview/Data, Washington, DC, updated September 2011.

[6] "UK Energy in Brief 2010", Dept. of Energy and Climate Change (EEC), London, July 2010.

[7] "International Energy Outlook 1996", Energy Information Administration, US Dept. of Energy, Washington, DC, Report DOE/EIA-0484(96), May 1996, pp. 35–48.

[8] Presentation by Howard Gruenspecht, Center for Strategic and International Studies, US Energy Information Administration (EIA), Washington, DC, 25 May 2010.

[9] Fells, I. (2001). *Renewable Energy: 10% by 2010*, Engineering Science and Education Journal, **10**, 42–43.

[10] Sweet, W. (2001). *An Unnatural Rush to Natural Gas?*, IEEE Spectrum, **38**, 83.

[11] "International Energy Outlook 2013", US Dept. of Energy, Energy Information Administration (EIA), Washington, DC, DOE/EIA-0484 (2013), July 2013.

[12] "US Crude Oil, Natural Gas and Natural Gas Liquids Reserves", 2009 Annual Report, US Energy Information Administration (EIA), Washington, DC, November 2010.

[13] Brooks, J.(1983). Development Stages of Coalbed Methane, *Petroleum Geochemistry and Fossil Fuel Exploration, Chemistry in Britain*, **5**, 390–400.

[14] Ayers Jr., W.B. (2002). "Coalbed gas systems, resources and production and a review of contrasting cases from the San Juan and Powder River basins", in Law, B.E. and Curtis, J.B. (eds.). *Unconventional Petroleum Systems*, AAPG Bulletin, 86, 1853–1890.

[15] "European Energy — a focus on Coal", OCDE/AIE (1999) http://www.energy-coal-eur.com/Technology/unconventional.htm. Accessed 12 April 2001.

[16] Murchison, D. (1988). "Status of Coalbed Methane Recovery in the United States", Energy Information Administration (EIA), Natural Gas Monthly, DOE/EIA-0130 (88/09), Washington, DC, December 1988.

[17] Black, H.T. (1991). "Update on US Coalbed Methane Production", Energy Information Administration (EIA), Natural Gas Monthly, Washington, DC, 1991.

[18] "International Industry/Government Consortium Investigates Synergy Between Natural Gas Production and Reduced Greenhouse Gas Emissions", Alberta Research Council, Alberta (1998) http://www.arc.ab.ca/whatsnew/newsreleases/980129CBM.html.

[19] National Energy Technology Laboratory (2012). *Energy Resource Potential of Methane Hydrate: An introduction to the science and energy potential of a unique resource*, U.S Department of Energy, DC, http://www.netl.doe.gov/technologies/oil-gas/publications/Hydrates/2011Reports/MH_Primer2011.pdf.

[20] Howarth, R.W., Santoro, R. and Inraffea, A. (2011). *Methane and the Greenhouse Gas Footprint of Natural Gas from Shale Formations*, Springer, New York.

[21] "Unlocking Tight Gas", BP Magazine *Frontiers*, Vol. 8, August 2007.

[22] "Annual Energy Outlook, 2011", Energy Information Administration, http://www.columbia.edu/cu/alliance/documents/EDF/Wednesday/Heal_material.pdf.

[23] OGJ (1994). Worldwide Look at Reserves and Production, *Oil and Gas Journal*, **92**, 42–43.

[24] "Energy in Transition 1985–2010", Chapter 3, Report of the Committee on Nuclear and Alternative Energy Systems, National Research Council, Washington DC, 1980.

[25] Williams, T.I. (1981). *A History of the British Gas Industry*, Oxford University Press, Oxford.

[26] "Coal Bridge to the Future", Report of the World Coal Study, Ballinger Publishing Co., Cambridge, MA, 1980.

[27] Foley, G. (1976). *The Energy Question*, Penguin Books, London.

CHAPTER 7

GEOTHERMAL ENERGY

7.1. Physical Basis of Geothermal Energy

Geothermal energy is thermal energy stored in the sub-surface of the earth. It is not a renewable source because prolonged exploitation can exhaust a particular site. Nevertheless, the vast extent of energy potentially available is such that many references refer to it as if it was infinitely renewable. Energy is stored in natural underground reservoirs of steam and/or hot water, known as aquifers, and also in more solid "hot sediments" that are buried at depth or adjacent to hot spots.

Heat energy flows outwards from within the earth at the average rate of $0.063\,\mathrm{W/m^2}$. The total outward flow amounts to $32 \times 10^{12}\,\mathrm{W}$, as shown in Fig. 2.1 of Chapter 2. It is of interest that the amount of interior heat flux flowing outwards is only about one-thousandth the value of the solar energy flux falling from space onto the same area [1, 2]. The surface geothermal heat distribution is too small and too diffuse to be exploited, except in concentrated hot spots such as geysers or volcanoes.

7.2. Geological Structure of the Earth

The geological structure of the earth is illustrated in Fig. 7.1 [3]. It is believed to approximate to five concentric spheres. From the outside proceeding inwards these are the atmosphere, crust, mantle, liquid outer core (magma) and solid inner core. As one proceeds inwards the temperature and density increase. For non-volcanic (i.e. non-seismic) areas the average geothermal gradient is between $17°C$ and $30°C$ per kilometre of depth ($50°-87°F$ per mile). In volcanic areas the temperature gradient is much higher.

The earth's crust, composed of basalt, silicate rocks, is not of uniform thickness. Under the oceans the crust is about $15\,\mathrm{km}$ thick and consists of porous rock. Under the continental land masses the crust is about $35\,\mathrm{km}$ thick (Fig. 7.2) and the proportion of porous rock probably increases with depth. Between the continental

205

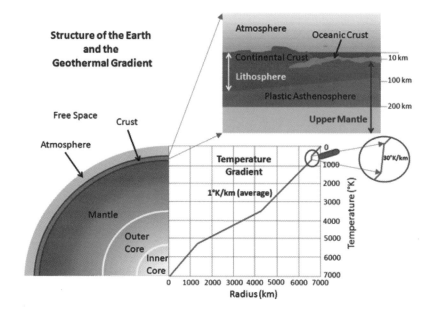

Fig. 7.1.   Structure of the earth [3].

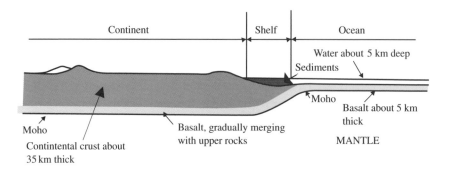

Fig. 7.2.   Schematic section through the earth's crust [4].

land mass and the ocean, the continental shelf contains a great thickness of sedimentary rocks such as sandstone or limestone [4]. Most offshore oil and natural gas exploration occurs in continental shelf areas.

The crust is not a solid annular shell but consists of several massive rigid segments, known as tectonic plates. These plates can move relative to each other (sometimes called "continental drift"). Plate tectonic theory suggests that pressure builds up at the plate boundaries and sudden movements at the plate interfaces result in earthquakes. It is difficult to control large-scale geological effects. Some serious work is under way in California, USA, using water injection to lubricate and reduce the friction on the plate edges. Usually, seismic disturbance cannot be accurately predicted.

The crust is composed of many diverse elements, and there is much variation in the chemical composition of the rocks from geographical region to region. The younger marginal platforms adjacent to a continent consist mainly of the sedimentary rocks derived from continued erosion of the continental surface. The platforms often form beds up to several miles in thickness. The oldest known rocks of the Precambrian shields generally consist of igneous rocks such as granite and highly metamorphosed gneiss (i.e. laminated) rocks.

Between the crust and the earth's mantle, about 35 km deep, is a boundary called the Mohorovicic seismic discontinuity. In Fig. 7.2 this is identified as the "Moho". This boundary layer represents some form of material and physical state different from the crust rock composition.

Below the Moho is the mantle, which forms the major part of the earth (80% by volume). It is made of highly viscous, partially molten rock with temperatures between 650°C and 1250°C and is thought to be largely composed of mixtures of magnesium and iron silicates. The upper and lower mantles (Fig. 7.1) are about 590 miles and 1,180 miles thick, respectively. Volcanic "ejecta" is composed of silicates from the mantle.

There have been many speculations about the nature of the materials which make up the inner core and outer core ("magma"). Most of the studies have pointed towards a liquid/solid metal core made up of iron or iron nickel, but these can only be inspired scientific guesses. The outer core is about 1,310 miles thick, while the inner core has a radius of about 840 miles [3]. In the core temperatures may reach 4000–7000°C (7200–12600°F).

There is presently no technology capable of directly controlling or even appropriating the enormous heat store of the core material. Cracks or fissures occur in the earth's crust, sometimes in regions where there is an underlying intrusion of magma into the crust due to tectonic plate motion. Plumes of magma ascend by buoyancy and force themselves up into the crust, bringing up vast quantities of heat [5]. In these locations volcanoes are formed. The volcanic action is impressive and formidable, sometimes awesome and beautiful, but it is not predictable or controllable. Volcanoes still remain a hazard rather than an energy opportunity. Figure 7.3 shows the earth's regions of high geothermal activity [6]. The British Isles and Ireland do not appear on this map but are known to be non-seismic.

7.3.   Origin of Geothermal Heat Flow

The classical explanation of geothermal heat flow is that heat energy is being conducted from the very hot interior regions of the earth to its surface. Through the epochs of time, the outer crust formed as a result of faster cooling rate. Masses

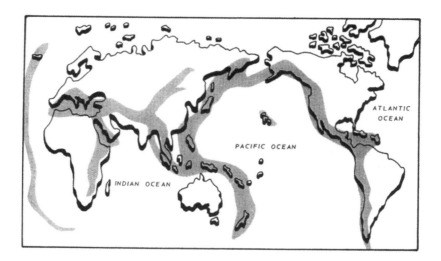

Fig. 7.3.   Regions of intense geothermal activity [5].

of molten rock circulating within the magma by gravitational action cause friction forces that generate heat [7]. A more modern theory postulates that the earth is a giant furnace. Radioactive decay of potassium, uranium and thorium in the earth's crust generates heat. Additional heat is generated by smaller concentrations of radioactive materials within the mantle below the crust [2]. The correct explanation may never be known, but might incorporate both of the present theories.

7.4.   Geothermal Energy Resources

The amount of geothermal energy within the earth is enormous. It is important to note, however, that no reliable numerical estimate of its value is available, either from individual national sources or from international sources. Many estimates have been made, with widely different results. An assessment by the World Energy Conference in 1980 for the geothermal resource base of the continental land masses to a depth of 3 km, for electricity generation purposes, at a modest extraction fraction, gave a value of 36×1020 J [8]. This was about ten times the total world primary energy consumption at that time. All numerical estimates of the geothermal energy resources should be regarded as speculative. What is undeniable, however, is that there is a vast amount of energy, most of which is presently untapped. At a depth of 6 miles, the approximate average temperature of the earth is greater than $100°$C. Interestingly, however, there are many "hot spots" on the earth which are much shallower. The total stored geothermal energy exceeds by several orders of magnitude the total fossil and nuclear resources. Only solar energy is comparable in total magnitude [6].

7.5. Geothermal Reservoirs

Water heated by the hot magma rock can rise to the surface as hot liquid springs or steam. One of the best-known of these is the Geysers of northern California, USA. Natural mineral waters have been used throughout the world for their therapeutic and medicinal values for at least 2,000 years. Mineral baths dating from the Roman era (55 BC–400 AD) may still be visited in Bath, England. Health spas using geothermally-heated water have flourished in Europe, Asia, the Middle East, North and South America, South East Asia and New Zealand. The map of Fig. 7.3 shows that geothermal energy is far more widely distributed than oil or natural gas. For instance, it might prove to be a valuable energy resource for the poor countries of East Africa.

Groundwater and snowmelt sink through the permeable rocks of the crust, such as sandstone or limestone, or penetrate through fissures into porous rock deposits. Such deposits may be several miles deep. In the porous areas the water is heated by the underlying magma and circulates by convection, sometimes circulating all the way back to the surface as water or steam (Fig. 7.4).

At the surface (atmospheric) pressure water will boil at a temperature of 100°C (212°F). Because of the high pressures within the earth at a depth of 5–6 miles, the water remains liquid and does not boil and change to steam, even at temperatures of several hundred degrees centigrade. Its density is gradually reduced due to volumetric expansion, causing it to rise to the surface by buoyancy. Sometimes the pressure conditions are such that the rising water changes to steam (with consequent reduction of pressure) as it rises and emerges from the ground through natural vents or man-made bore-holes, as steam geysers or fumaroles (Fig. 7.5) [6, 7].

Fig. 7.4.   A geyser on the mid-Atlantic ridge at Strokkur, Iceland.

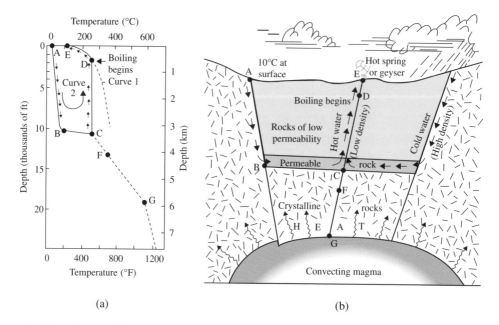

Fig. 7.5.   Geological illustration of hot-water geyser action. (a) Reference curve for the boiling point of pure water. (b) Temperature profile along a typical circulation route (from "Characteristics of Geothermal Sources", by D. E. White, in Reference [6]).

If the ascending hot water meets an impermeable rock layer, the water is trapped underground, where it fills the pores and cracks, comprising 2–5% of the volume of the surrounding rock, forming a geothermal reservoir [5]. Geothermal reservoirs can be tapped down to depths of about 8,000 feet. Even at 500°C the high pressure may prevent the reservoir water from boiling. Water or steam temperatures as high as 700°C may occur. The drilling rigs are the same land-based mobile units used in the oil and gas industries.

Large sedimentary basins may contain reservoirs at much smaller depths, where the water is thermally heated in the range 100–200°C. These may have surface outlets in the form of hot springs or boiling mud pools [8].

Hot water up to temperature of 75°C can be obtained from low-grade, semi-thermal aquifers by drilling to depths of 1.5–2 km. This corresponds to a temperature gradient of 30°C/km. Such low-grade sources can be found in non-thermal (i.e. non-seismic) areas, including southern England, but are only worth exploiting for space heating if they are near to large centres of population (Fig. 7.6) [2, 8]. In Marshwood, Dorset, England, projects have been run where cold water was injected into hot, porous sediment and recovered as hot water [9].

Some of the many industrial applications for which geothermally-heated water or steam has been used are given in Table 7.1 [4, 10]. Heat of almost any grade available from geothermal fluids (liquids or gases) can be used. It is useful

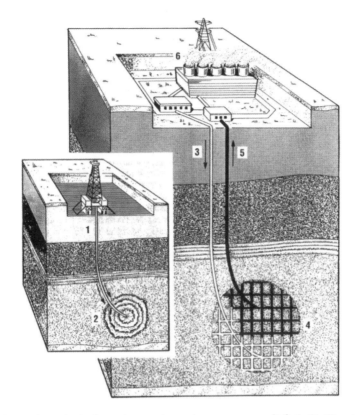

Fig. 7.6. Hot rock method of tapping geothermal energy source [14]. 1. Drilling two holes for injected water. 2. Explosive fracturing. 3–5. Injected cold water returns hot. 6. Geothermal power plant.

to remember that about 90% of the world energy consumption was used for heating purposes in 1978 and remains so now [11, 21].

The problem of geothermal application is one of availability rather than applicability. If geothermal energy was widely available it could contribute 20–50% of the energy needs of an industrialized country. But in the countries of northern Europe and the eastern parts of the Americas it is not an available source.

7.6. Locations and Types of Principal Geothermal Sources

There are four main types of geothermal sources: dry steam, wet steam, hot brine and hot rock. Hyperthermal fields, both wet and dry, are confined to the seismic areas shown in Fig. 7.3. In some locations the presence of the geothermal energy is evident through surface ports such as geysers. Good sources have also been detected in locations devoid of visible manifestations.

Table 7.1. Approximate temperature requirements (°C) of geothermal fluids for various applications [4].

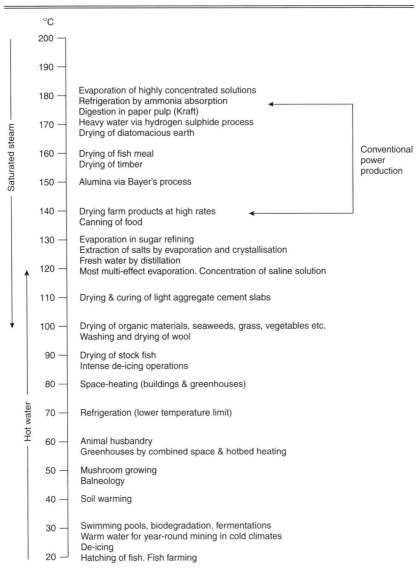

°C

Temperature	Application
200	
190	
180	Evaporation of highly concentrated solutions / Refrigeration by ammonia absorption / Digestion in paper pulp (Kraft)
170	Heavy water via hydrogen sulphide process / Drying of diatomacious earth
160	Drying of fish meal / Drying of timber
150	Alumina via Bayer's process
140	Drying farm products at high rates / Canning of food
130	Evaporation in sugar refining / Extraction of salts by evaporation and crystallisation / Fresh water by distillation
120	Most multi-effect evaporation. Concentration of saline solution
110	Drying & curing of light aggregate cement slabs
100	Drying of organic materials, seaweeds, grass, vegetables etc. / Washing and drying of wool
90	Drying of stock fish / Intense de-icing operations
80	Space-heating (buildings & greenhouses)
70	Refrigeration (lower temperature limit)
60	Animal husbandry / Greenhouses by combined space & hotbed heating
50	Mushroom growing / Balneology
40	Soil warming
30	Swimming pools, biodegradation, fermentations / Warm water for year-round mining in cold climates / De-icing
20	Hatching of fish. Fish farming

Saturated steam · Hot water

Conventional power production

7.6.1. *Dry steam sources*

Dry geothermal fields produce dry saturated steam or superheated steam at pressures of several atmospheres. These usually occur at the deepest drilling ranges. This type of deposit is the rarest but is the easiest and most economical to tap because there is little corrosion problem. The best-known example is the Geysers development near San Francisco, California, USA.

7.6.2. *Wet steam sources*

Wet geothermal fields produce pressurized water at temperatures in the range 180–370°C, which often contains impurities/contaminants such as sulphur compounds/gases. When the fluid is brought to the surface, about 10–20% of it flashes into steam due to pressure reduction. Wet steam fields appear to be about 20 times more abundant than dry steam fields. There are no known steam sources, wet or dry, in the UK and no recorded geyser activity. In Iceland, however, homes are commonly heated from geothermal sources. Also, geothermal heat is used there in greenhouses to grow tropical fruit such as bananas.

7.6.3. *Hot brine sources*

Abundant geothermal energy is known to exist worldwide within hot brine (i.e. salt-solution) deposits. A particularly large source of low temperature geothermal energy is the geopressurized brines of the sedimentary basins of the Mexican Gulf coast in the USA and Mexico. These brines may also hold very large amounts of dissolved natural gas at temperatures below 180°C. If the heat and the gas could be exploited simultaneously this would be an attractive resource [12].

The volume of brine that would need to be handled would be enormous. Where brine deposits lie under the land, the removal of massive volumes of liquid could create major ecological dislocation, such as land subsidence. Some form of liquid re-injection might be needed, as shown in Fig. 7.6. The high salinity of the brines is very corrosive to the equipment used to handle it and represents another good ecological reason for re-injection.

Some geothermal aquifers can be largely pollution-free and gaseous emissions are small. Other aquifers are offensive and environmentally unfriendly in their local impact.

7.6.4. *Dry rock sources*

Most of the potentially exploitable geothermal heat is stored in dry rocks, rather than in water or steam aquifers. The normal geothermal temperature gradient, worldwide, is 20–30°C/km, as stated in Section 7.2, above. This is sufficient to produce temperatures of 80°C for space heating at depths of 2.2 km and temperatures of up to 180°C at the accessible depth of 5.5 km. In locations of a semithermal and hyperthermal geological nature, where the geothermal gradient is higher, sometimes much higher, than 40°C/km, the subterranean hot rock is called "hot dry rock".

Various types of steam and hot-water aquifers or reservoirs are relatively easy to tap because the storage medium of the energy is also the heat transfer fluid. In order to exploit the heat energy of hot rock, however, it will be necessary to fracture the rock, to inject water as a working fluid and also to develop a network of channels through which water can flow in and out (Fig. 7.6). Fracturing of the rock can be achieved by inducing thermal stresses using cold water or by chemical or nuclear explosions. This technique is similar in principle to that proposed for shale oil extraction in Section 5.5.1 of Chapter 5.

Although a number of prototype investigations of hot rock exploitation have been reported, there are no commercial plants. Too little is yet known about the feasibility of exploiting pressurized brines or hot dry rocks to judge the ecological consequences [12].

The characteristics of the four main types of geothermal sources are summarized in Table 7.2 [11].

7.6.5.   *Molten magma*

Magma, or molten lava, is the extreme case of hot rock. It has been found, at temperatures higher than 650°C, in pools at the surface of volcanoes or in reservoirs below them. There are no practical means of extracting the heat, although some research is in progress.

Table 7.2.   Characteristics of four types of geothermal sources [11].

Type/Characteristic	Dry steam	Wet steam	Hot brine	Hot rock
Ease of utilization	Highest (greatest ease)	Somewhat difficult	Difficult	Greatest difficulty
Abundance	Lowerst	Somewhat abundant	Abundant	Great abundance
Drilling depth	8000 ft (2440 m)	3000 ft (915 m)	4500 ft (1370 m)	10000 ft (3050 m)
Pressure and temperature of steam or water	7 atmospheres, 200°C	2 atmospheres, 230°C	1 atmosphere, 240°C	400°C
Fields operating in 1978	Larderello, Italy; Geysers, USA; Matsukawa, Japan	Wairakei, New Zealand; Cerro Prieto, Mexico	None	None
Exploration sites		Kilauea Volcano, Hawaii	Imperial Valley, California	Marysville, Montana; Vallez Caldera, New Mexico

7.6.6.   *Worldwide applications*

Geothermal direct use applications provide about 10,000 thermal megawatts of power in about 35 countries. The widespread use of geothermal heat energy for direct (non-electrical) applications are illustrated in Table 7.3, representing the 1998 situation [5].

7.7.   Geothermal Prospects in the UK

7.7.1.   *Shallow drilling*

The UK does not possess geothermal sources in the form of aquifers, which contain water hot enough to provide steam for electricity generation [13]. Water with a maximum temperature of about 90°C is available and can be used for district heating or local industrial applications. With aquifer water hotter than 60°C the surface yield can be used directly in district heating schemes without the use of heat pumps.

A major geothermal project for commercial heating was undertaken in Southampton, England, in the early 1980s. At a drilling depth of 1,675 metres the water temperature was 74°C. The flow rate of test extracted water showed that the geothermal reservoir was unexpectedly small, such that the source lifetime would be 10–20 years. Because a lifetime of 25 years was the minimum needed for economic viability, the original project was scaled down in 1984 [14]. It remains in operation, however, as the only commercial use of geothermal energy in the UK. Council offices and private sector buildings are linked by a 2-km hot water main running to and from a "heat station" close to the well-head [9].

The British Geological Survey has drilled a number of 300-metre bore-holes in various parts of the UK to measure local heat flows. Potential aquifers were identified in the Wessex Basin, East Yorkshire and Lincolnshire, the Worcester Basin, West Lancashire and Cheshire, and also in Northern Ireland. Regrettably, however, the temperatures or potential yields are too low for exploitation.

The present cost of aquifer resource heat is about two-and-a-half times that from conventional commercial sources. Other difficulties are the cost of drilling "dry" wells, drilling component failures and variations of surface housing density [9].

Several areas of the UK have impermeable granite rock formations close (i.e. within a few hundred feet) to the surface, which are likely to be large heat stores. Some investigation has been made of using explosive charges to fracture granite at the bottom of 2-km bore-holes. It is estimated that it would be necessary to bore down to 6 km in Cornwall, UK, to reach rock at a temperature of 180°C, which would be suitable for electricity generation. This is not, at present, an economic proposition.

Table 7.3.   Direct (non-electrical) use of geothermal power, (MW) 1998 [5].

European Union Countries	**1,031.4**
Austria	21.1
Belgium	3.9
Denmark	0.1
France	309
Germany	307
Greece	22.6
Ireland	0.7
Italy	314
Portugal	0.8
Sweden	47
United Kingdom	2
Other European Countries	**3,614**
Bosnia and Herzegovina	33
Bulgaria	94.5
Croatia	11
Czech Republic	2
Georgia	245
Iceland	1,443
Israel	42
Hungary	750
Macedonia	75
Poland	44
Romania	137
Russia	210
Serbia	86
Slovakia	75
Slovenia	37
Switzerland	190
Turkey	160
Ukraine	12
TOTAL EUROPE	**4,645**
America	**1,908**
Canada	3
USA	1,905
Asia	**3,075**
China	1,914
Oceana	**5**
New Zealand	5
Africa	**71**
Algeria	1
Tunisia	70
TOTAL WORLD	**9,704**

An enticing longer-term prospect is the possible exploitation of deep basement rocks as hot rock sources. Where these lie close to the centres of population, with their high density heat loads, the possibility arises of both district heating and electricity generation. Little is still known about the deep geology of the UK. If the deep rocks are impermeable and if suitable drilling techniques can be developed, then hot rock geothermal energy could become available on a large scale.

The potential for geothermal energy in the UK is the equivalent of ten years of electric power. But the technological problems of extracting heat from underground aquifers and hot dry rocks are very difficult and the economics are unattractive [15].

7.7.2. *Worked example*

A geothermal aquifer supplies hot water with a well-head temperature of 75°C at the flow rate of 20 litres/s. The heat energy is used to supplement a district heating unit above a datum temperature of 40°C. If the geothermal heat is used for 170 days each year, how much oil is saved annually if the overall combustion efficiency of the oil burner is 75%? (Based on a problem in Ref. 2.)

Flow rate
$$= 20 \text{ litres/s}$$
$$= 20 \times 3600 \text{ litres/hour}$$
$$= 20 \times 3600 \times 24 \text{ litres/day}$$
$$= 20 \times 3600 \times 24 \times 170$$
$$= 294 \times 10^6 \text{ litres/working year}$$

Volume of water transferred per
  working year (1 litre $\equiv 1000\,\text{cm}^3$) $= 294 \times 106 \times 103 \times \text{cm}^3$

Mass of water transferred
  per working year (1 cm$^3 \equiv 1$ gm) $= 294 \times 10^6 \times 10^3\,\text{g}$
                                     $= 294 \times 106\,\text{kg}$

Temperature contribution above the datum $= 75 - 40 = 35°C$

Heat transferred per working year
                                       $= 294 \times 106 \times 103 \times 35$
                                       $= 10.29 \times 10^{12} \text{ cals}$

Now 1000 tonnes of oil (Ref. 3 of Chapter 5) $\equiv 10 \times 10^{12} \text{ cals}$

Oil saved per year at 100% efficiency $\equiv 10.29/10 = 1029 \text{ tonnes}$

Oil saved per year at 75% efficiency $\equiv 1029/0.75 = 1372 \text{ tonnes}$

7.8. Geothermal Uses in the USA and Elsewhere

The greatest application of US geothermal energy is in the production of electricity, which is discussed in Section 7.11. In addition there are many direct (non-electrical) uses of geothermally-heated water, described in the following sections.

7.8.1.   *Hot springs and bathing spas (balneology)* [5]

For centuries the peoples of China, Iceland, Japan, New Zealand, North America and other areas have used hot springs for cooking and bathing. The Romans used geothermal water to treat eye and skin disease and, at Pompeii, to heat buildings. Medieval wars were even fought over lands with hot springs. Today, as long ago, people still bathe in geothermal waters.

In Europe, natural hot springs have been very popular health attractions. The first known "health spa" was established in 1936 in Belgium. (One resort was named "Espa", which means "fountain". The English word "spa" came from this name.) All over Eurasia today, health spas are still very popular. Russia, for example, has 3,500 spas.

Japan is considered the world's leader in balneology. The Japanese tradition of social bathing dates back to ancient Buddhist rituals. Beppu, Japan, has 4,000 hot springs and bathing facilities that attract 12 million tourists a year. Other countries with major spas and hot springs include New Zealand, Mexico and the USA. The USA has 218 spas that use geothermally-heated water.

7.8.2.   *Agriculture*

Geothermal resources are used worldwide to boost agricultural production. Water from geothermal reservoirs is used to warm greenhouses to help grow flowers, vegetables and other crops. For hundreds of years, Tuscany in central Italy has produced vegetables in the winter from fields heated by natural steam. In Hungary, thermal waters provide 80% of the energy demand of vegetable farmers, making Hungary the world's geothermal greenhouse leader. Dozens of geothermal greenhouses can also be found in Iceland. In the western United States there are 38 geothermally-heated greenhouse complexes.

7.8.3.   *Aquaculture*

Geothermal aquaculture, the "farming" of water-dwelling creatures, uses natural warm water to speed the growth of fish, shellfish, reptiles and amphibians. This kind of direct use is increasing in popularity. In China, for example, geothermal aquaculture is growing so fast that fish farms cover almost 2 million square metres (500 acres). In Japan, aqua farmers grow eels and alligators. In the USA 28 geothermal aqua farmers in Idaho, Utah, Oregon and California grow catfish, trout, tilapia and alligators, as well as tropical fish for pet shops.

7.8.4.   *Industry*

The heat from geothermal water is used worldwide for industrial purposes. Some of these purposes include drying fish, fruits, vegetables and timber products, washing wool, dyeing cloth, manufacturing paper and pasteurizing milk. Geothermally-heated water can be piped under pavements and roads to keep them from icing over in freezing weather. Thermal waters are also used to help extract gold and silver from ore and even for refrigeration and ice-making.

7.9. Geothermal District Heating [5]

The oldest and most common use of geothermal water, apart from hot spring bathing, is to heat individual buildings, and sometimes entire commercial and residential districts. A geothermal district heating system typically supplies heat by pumping geothermal water — usually 60°C (140°F) or hotter — from one or more wells drilled into a geothermal reservoir. The geothermal water is passed through a heat exchanger, which transfers the heat to water in separate pipes that is pumped to the buildings. After passing though the heat exchanger, the geothermal water is injected back into the reservoir, where it can reheat and be used again.

In the Paris basin in France, historic records show that geothermal water from shallow wells was used to heat buildings over six centuries ago. An increasing number of residential districts there are being heated with geothermal water as the drilling of new wells progresses. The first district heating system in the USA dates back to 1893 and still serves part of Boise, Idaho. In the western United States there are over 270 communities that are close enough to geothermal reservoirs for potential implementation of geothermal district heating. Eighteen such systems are already in use in the USA, the most extensive being in Boise, Idaho and San Bernardino, California.

Because it is a clean, economical method of heating buildings, geothermal district heating is becoming more popular in many places. Besides France and the USA, modern district heating systems now warm homes in Iceland, Turkey, Poland and Hungary. The world's largest geothermal district heating system is in Reykjavik, Iceland, where almost all of the buildings use geothermal heat. The air around Reykjavik was once very polluted by emissions from reliance on fossil fuels. Since it started using geothermal energy, Reykjavik has become one of the cleanest cities in the world.

7.10. Geothermal Heat Pumps [5]

In many parts of the world the earth temperature is stable in the range 45–58°F a few feet below the surface. Animals burrow into the earth to escape the excessive winter and/or summer air temperatures. Indoor temperatures can be kept comfortable using the earth's heat by means of a geothermal heat exchange system known as a geothermal heat pump (GHP). A GHP will circulate water or other liquids through pipes buried in a continuous loop, horizontally or vertically, adjacent to a building. The system can be used for either heating or cooling.

7.10.1.   *Heating*

The difference between the earth's temperature and the colder temperature of the air is transferred through the buried pipes into the circulating liquid and hence into the building.

7.10.2.   *Cooling*

During hot weather the continuously circulating fluid in the pipes absorbs heat from the building and transfers it into the earth.

A GHP system uses only a small electric pump motor — probably much less than 1 kW rating — and is environmentally friendly.

In the USA there are estimated to be about 300,000 homes, schools and offices incorporating geothermal heat pump systems. The US Environmental Protection Agency has rated GHPs as among the most efficient and environmentally clean of heating and cooling technologies.

7.11.   Electricity Generation from Geothermal Sources

Hot water and/or steam available from geothermal sources can generate electricity via the use of steam turbines. No fossil fuel is burned. The first commercial system used steam to provide electric lighting in Larderello, Italy [16]. Until 1958 Italy was the only country where natural steam was used for electrical power generation on an industrial scale [2]. This was followed by electrical power production in New Zealand and the USA. There are major developments by Union Oil in northern California, USA, to produce geothermal energy. The largest geothermal power plant in the world is at the Geysers, near San Francisco, USA.

7.11.1.   *Worldwide geothermal electrical power production* [5, 17]

In 2010, 10709.7 MW of electrical power capacity existed from the geothermal power plants across the world. The world leader was the USA, which possessed 3086 MW (Table 7.4) or almost 29% of the world total from its 77 active plants.

Geothermal electricity is now (2013) generated in 24 countries, with the fastest growth rates in New Zealand and Iceland (Table 7.4).

Table 7.4.   Geothermal electrical power production [18].

Country	Capacity (MW) 2007	Capacity (MW) 2010
USA	2687	3086
Philippines	1969.7	1904
Indonesia	992	1197
Mexico	953	958
Italy	810.5	843
New Zealand	471.6	628
Iceland	421.2	575
Japan	535.2	536
El Salvador	204.2	204
Kenya	128.8	167
Costa Rica	162.5	166
Nicaragua	87.4	88
Russia	79	82
Turkey	38	82
Papua-New Guinea	56	56
Guatemala	53	52
Portugal	23	29
China	27.8	24
France	14.7	16
Ethiopia	7.3	7.3
Germany	8.4	6.6
Austria	1.1	1.4
Australia	0.2	1.1
Thailand	0.3	0.3
Total	**9731.9**	**10709.7**

7.11.2.   *Technologies of geothermal electrical power generation* [5]

(A) Flash steam plants. Most geothermal power plants operating today are "flash steam" power plants. Hot water from production wells is passed through one or two separators, where, released from the pressure of the deep reservoir, part of it flashes (explosively boils) to steam. The force of the steam is used to spin the turbine generator. To conserve the water and maintain reservoir pressure, the geothermal water and condensed steam are directed down an injection well back into the periphery of the reservoir, to be reheated and recycled.

(B) Dry steam plants. A few geothermal reservoirs produce mostly steam and very little water. Here, the steam shoots directly through a rock-catcher and into the turbine. The first geothermal power plant was a dry steam plant, built at Larderello in Tuscany, Italy, in 1904. The power plants at the Larderello dry steam field were destroyed during the Second World War, but have since been rebuilt and expanded. That field is still producing electricity today. The Geysers dry steam reservoir in northern California has been producing electricity since 1960. After 40 years it still produces enough electricity to supply a city the size of San Francisco.

(C) Binary power plants. In a binary power plant, the geothermal water is passed through one side of a heat exchanger, where the heat is transferred to a second (binary) liquid, called a working fluid, in an adjacent separate pipe loop. The working fluid boils to vapour, which, like steam, powers the turbine generator. It is then condensed back to a liquid and used over and over again. The geothermal water passes only through the heat exchanger and is immediately recycled back into the reservoir.

Although binary power plants are generally more expensive to build than steam-driven plants, they have several advantages: (1) the working fluid (usually isobutene or isopentane) boils and flashes to a vapour at a lower temperature than does water, so it is possible to generate electricity from reservoirs with lower temperatures; this increases the number of geothermal reservoirs in the world with electricity-generating potential; (2) the binary system uses the reservoir water more efficiently: since the hot water travels through an entirely closed system it results in less heat loss and almost no water loss; and (3) binary power plants have virtually no emissions.

(D) Hybrid power plants. In some power plants, flash and binary processes are combined. An example of such a hybrid system is in Hawaii, where a plant provides about 25% of the electricity used on the Big Island.

7.11.3.   *Locations of geothermal electricity-generating stations*

Some locations of present and intended geothermal generation sites are shown in Fig. 7.7. A comparison with Fig. 7.3 shows that these all lie in tectonic regions.

Geothermal energy remains one of the real energy prospects for developing countries. In 1976 exploration programmes had been set in motion in 35 countries [4]. By 1989 the installed capacity mix for electrical generation in Mexico included 32% from hydroelectricity and 3% from geothermal. In Mexico geothermal power has been developed to the point where the Comisión Federal de Electricidad (Mexico's state-owned electricity utility) sells 70 MWe of geothermally-generated electricity to a utility in southern California, USA [19].

In Europe the only viable sites for geothermally-generated electricity are likely to remain Iceland and Italy. There is no prospect of geothermal electricity generation in the UK.

In the USA, for example, California generates 824 MWe at the Geysers plant, 490 MWe at the Imperial Valley, 260 MWe at Coso and 59 MWe at smaller plants. Although the Geyser operates much below its capacity, it is still the world's

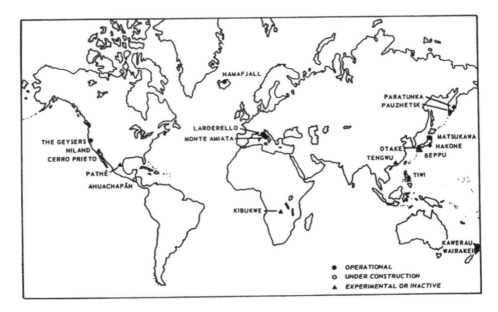

Fig. 7.7.   Geothermal electricity power stations (from "Worldwide Status of Geothermal Resources Development", by J.B. Koening, in Ref. [6] Chapter 2, pp. 15–58).

largest developed geothermal field and an outstandingly successful renewable energy project.

7.12.   Environmental Features of Geothermal Power

A number of possible environmental problems may arise in the development and use of geothermal energy sources.

7.12.1.   *Geothermal site exploration and development*

The initial exploitation of a site, especially drilling operations, is likely to cause noise, surface damage and local disruption. Access roads may need to be installed or improved to facilitate the transportation of building materials and heavy machinery. The development site will contain drilling rigs, exhaust vents, building of various kinds, machinery and pipework. Depending on location, some of this may be of objectionable visual impact.

Any geothermal development has to be associated with a heat distribution system. As with electricity, oil and gas distribution, customers want the energy but not the means of delivering it. The necessary pipelines need to pass over or under existing land and to terminate in industrial structures.

A geothermal plant sits right on top of its fuel source: no additional land is needed, such as for mining coal or for transporting oil or gas. When geothermal power plants and drill rigs are located in scenic areas, mitigation measures are implemented to reduce intrusion on the visual landscape. Some geothermal power plants use special air-cooling technology that eliminates even the plumes of water vapour from cooling towers and reduces a plant profile to as little as 24 feet in height [5].

7.12.2. *Protection of the local atmosphere*

Hydrogen sulphide gas (H_2S) sometimes occurs in geothermal reservoirs. H_2S has a distinctive rotten egg smell that can be detected by the most sensitive sensors (our noses) at very low concentrations (a few parts per billion). It is subject to regulatory controls for worker safety because it can be toxic at high concentrations. Equipment for scrubbing H_2S from geothermal steam removes 99% of this gas [5].

Early attempts at geothermal drilling in the UK were accompanied by clouds of waste steam and a strong smell of hydrogen sulphide [4]. These adverse effects can be largely contained and some geothermal aquifer sites are largely pollution-free. Gaseous emissions are then usually negligible and there are no noticeable fumes. The problem of local air pollution is always present at geothermal sites. It is claimed that air pollution standards at the Geysers field have resulted in cleaner air than before the field was exploited [2, 4].

Carbon dioxide (a major climate change gas) occurs naturally in geothermal steam but geothermal plants release amounts less than 4% of that released by fossil fuel plants. There are no emissions at all when closed-cycle (binary) technology is used [5].

7.12.3. *Protection of groundwater* [5]

Geothermal water contains a higher concentration of dissolved minerals than water from cold groundwater aquifers. In geothermal wells, pipe or casing (usually several layers) is cemented into the ground to prevent the mixing of geothermal water with other groundwater.

When highly mineralized geothermal water needs to be stored at the surface, such as during well testing, it is kept in lined, impermeable sumps. After use, the geothermal water is either evaporated or injected back in its deep reservoir, again through sealed piping.

7.12.4. *Enhancement of reservoir water*

A unique example of enhancing reservoir water is at the Geysers steam field in California, where treated wastewater from nearby communities is being piped to a steamfield and injected into the reservoir to be heated. This increases the amount of steam available to produce electricity. With this enhancement, reservoir life is increased while providing nearby cities with an environmentally-safe method of wastewater disposal.

7.12.5.   *Ecological effects of geothermal plants*

Ecological effects are specific to location. The release of some gaseous toxic emissions could destroy animal habitats and affect local flora and fauna. The US Environmental Protection Agency has conducted tests, in the vicinity of geothermal development sites, on animal tissues and animal products, as biological monitors. Concentrations of geothermal effluents in animal tissue were examined, mainly in rodents because of their importance in the food chain for snakes, hawks, owls, coyotes, foxes, etc. The samples were analysed for selected elements to confirm ongoing baseline tissue concentrations [20].

7.12.6.   *Effects on local geological structure*

Present geothermal sites use existing technologies for natural sources of steam and hot water. These do not present any geological hazard. Future geothermal exploration is likely to involve either hot dry rocks or geopressurized brines, both of which imply possible geological dislocation.

The fracturing of rocks at depth, by explosion, to provide access for large volumes of water could contaminate groundwater, cause surface subsidence or trigger seismic disturbances. If large-scale exploitation of the normal earth thermal gradient eventually becomes feasible, then either natural or induced seismic disturbances might be very serious [4, 12].

As noted in Section 7.6.3, the exploitation of geopressurized brine would involve the displacement of enormous volumes of fluid. This highly saline and polluted liquid could not be dumped into rivers or onto land surfaces after it has passed through the heat exchangers. There are few applications where the contaminated water could be passed straight into tidal estuaries. The most feasible solution would seem to be re-injection into the ground but the specific geological result of this is not known.

7.13.   Problems and Review Questions

7.1. What are the present theories to explain why temperature increases with depth of penetration into the earth?

7.2. Why does water in geothermal aquifers remain in the liquid state even though its temperature may be much higher than 100°C?

7.3. In what parts of the world would you expect to see surface evidence of geothermal energy, such as steam geysers or volcanoes?

7.4. Identify on a map the volcanoes Cotopaxi, Fujiyama, Hekla, Katmai, Kilauea, Krakatoa, Lassen Peak, Mauna Loa, Mount Etna, Mount St. Helens, Ngauruhoe, Osorno, Paricutin, Popocatépetl, Semeru and Stromboli. Comment on the overall result [14].

7.5. Why is Iceland able to grow tropical produce?

7.6. Identify the countries associated with the geothermal field sites listed in Table 7.4. Do these countries all lie in the tectonic (seismic) areas shown in Fig. 7.3?

7.7. Why is the UK unlikely to be able to use geothermal aquifer energy on any significant scale?

7.8. What are the principal forms of geothermal energy resources?

7.9. What are the disadvantages to the use of hot dry rocks as a major heat source?

7.10. Explain, using a diagram, the basis of the process of extracting heat from subterranean hot dry rocks.

7.11. What are the environmental features associated with geothermal energy exploration and development?

7.12. Why was the proposed UK geothermal scheme at Southampton curtailed from its originally estimated level of activity?

7.13. What are the main world locations for geothermally-generated electricity? Why is the UK not a viable site?

7.14. A geothermal district heating scheme issues a flow rate of 22.5 litres/s with a well-head temperature of 70°C. It supplies heat above a datum temperature of 40°C for a period of 162 days/year. If the overall combustion efficiency of the oil burner is 73%, how much oil is saved per year?

References

[1] Hubbert, M.K. (1971). "The Energy Resources of the Earth", in *Energy and Power*, Scientific American, Beaconsfield.

[2] McVeigh, J.C. (1984). *Energy around the World*, Pergamon Press, Oxford.

[3] "Battery and Energy Technologies", Electropaedia, February 2011, http://www.mpoweruk.com/geothermal_energy.htm.

[4] Armstead, H.C.H. (1983). *Geothermal Energy*, 2nd Edition, E. and F. Spon Ltd., London.

[5] "Geothermal Energy Facts (Advanced Level)", Geothermal Education Office, Washington, DC, June 2001, http://geothermal.marin.org/geoenergy.html.

[6] Kruger, P. and Otte, C. (eds.) (1973). *Geothermal Energy — Production, Resources, Stimulation*, Stanford University Press, Stanford, CA.

[7] Cheremisinoff, P.N. and Moressi, A.C. (1976). *Geothermal Energy — Technology Assessment*, Technomic Publishing Co., Inc. Westport, CT.

[8] "World Energy Resources: 1985–2020", World Energy Conference, London, 1980.

[9] "Geothermal Aquifers", Technology Status Report 016, Department of Trade and Industry, UK, April 1995.

[10] Lindal, B. (1973). "Industrial and Other Applications of Geothermal Energy", in Armstead, H.C.H. (ed.), *Geothermal Energy*, UNESCO, Paris, pp. 135–148.

[11] Dorf, R.C. (1978). "Geothermal Energy", in Dorf, R.C. (ed.), *Energy Resources and Policy*, Addison-Wesley Publishing Co., New York, 1978.

[12] "Energy in Transition 1985–2010", National Academy of Sciences, Washington, DC, 1979.

[13] Taylor, R.H. (1983). *Renewable Prospects for Britain's Utilities*, Electrical Review, **213**.

[14] "World Solar Markets", Financial Times Business Information Ltd., London, 1983 and 1984.

[15] "Renewable Energy: A Resource for Key Stages 3 and 4 of the UK National Curriculum", Energy Technology Support Unit, Harwell, Oxfordshire, 1995.

[16] Leardine, T. (1974). *Geothermal Power*, Philos. Trans. R. Soc. London, **276**, 507–526.

[17] Glassley, W.E. (2010). *Geothermal Energy: Renewable Energy and the Environment*, CRC Press, Boca Raton.

[18] Holm, A., Blodgett, L., Jennejohn, D., Gawell, K. (2010). "Geothermal Energy: International Market Update" Geothermal Energy Assn., USA, May 2010, http://www.geo-energy.org/pdf/reports/GEA_Intenationalmarket_Report.

[19] "International Energy Outlook 1999", US Energy Information Administration, DOE/EIA-0484 (99), Washington, DC, March 1999.

[20] "Geothermal Environmental Impact Assessment", EPA-600/7-78-233, US Environmental Protection Agency, Las Vegas, Nevada, December 1978.

[21] "International Energy Outlook 2013", US Energy Information Administration (EIA), Document DOE/EIA-0484(2013), Washington, DC, July 2013.

CHAPTER 8

NUCLEAR ENERGY

8.1. Basic Atomic Theory

All matter is believed to be composed of small particles, called atoms. These are the smallest particles of any chemical element possessing the chemical properties of that element. Most of the materials and substances that abound in nature or are manufactured have basic constituents, called molecules, which are compounds of different atoms.

The atom itself is not the smallest particle in nature but is believed to be comprized of various sub-atomic particles. Most of the mass of an atom is concentrated in its nucleus, which is composed of positively-charged protons and zero-charged neutrons, collectively referred to as nucleons. The nucleus is surrounded by much lighter, negatively-charged particles called electrons, which are in continuous motion in three-dimensional orbits.

In any atom there are equal numbers of protons and electrons, resulting in zero overall charge, and this number is called the "atomic number". The atomic number also represents the location of the atom in the periodic table and characterizes its chemical properties. For example, the element hydrogen is the lightest element with an atomic number of unity, while the heaviest naturally-occurring element is uranium with an atomic number of 92.

Protons and neutrons have (very nearly) equal mass but do not normally combine in equal numbers. The total number of these nucleons in an atomic nucleus is referred to as the atomic mass number in atomic mass units (amu). Each amu has the value 1.66×10^{-27} kg. It is possible for the atomic nucleus of a chemical element to have several versions, whereby the same number of protons may be combined with different numbers of neutrons. When this situation arises, the different versions of the element are called isotopes. For example, normal hydrogen has a nucleus consisting of one proton with the atomic mass number 1. However, there is also a naturally-occurring stable isotope of hydrogen called deuterium (D) that has a nucleus containing one proton plus one neutron so that its atomic mass number is 2.

Uranium has three main isotopes with the same atomic number of 92 but mass numbers of 234, 235 and 238. These isotopes are usually designated U-234, U-235 and U-238, respectively. The actual mass of a hydrogen atom, consisting of one proton plus one electron, is 1.007825 amu, while the actual mass of a neutron is 1.00867 amu.

Chemical reactions usually involve combination of the atoms of different elements to form molecules — basic building blocks — of a further different compound substance. Sometimes chemical reactions take the form of interactions between the orbiting electrons of adjacent atoms. Some examples of this are given in Chapter 12 and refer to certain semiconductor materials used in photovoltaic cells for solar energy applications. In chemical reactions the nuclei of the atoms involved remain intact.

8.2.　Basic Nuclear Theory

8.2.1.　*Nuclear fission*

Nuclear reactions are not the same as chemical reactions. They involve fragmentation by splitting the nuclei of atoms for the purpose of releasing some of the considerable binding energy. Mass is a form of energy and Einstein expressed the equivalence of mass m and energy W in the form

$$W = mc^2 \tag{8.1}$$

where m is in kilogrammes, W is in joules and c is the velocity of electromagnetic wave (light) propagation, (very nearly) 3×10^8 m/s. For example, the energy equivalent of 1 amu is

$$
\begin{aligned}
W &= 1.66 \times 10^{-27} \times (3 \times 10^8)^2 \, \text{J} \\
&= 14.94 \times 10^{-11} \, \text{J} \\
&\equiv 4.147 \times 10^{-17} \, \text{kWh} \\
&\equiv 931 \, \text{MeV}
\end{aligned}
\tag{8.2}
$$

Under appropriate physical conditions the nuclei of some heavy atoms can be fragmented if they are bombarded with neutrons; a process known as nuclear fission. Some of the consequent released energy appears in the form of heat. The best known and most used example of nuclear fission, discovered in 1938 by two German scientists, Otto Hahn and Fritz Strassman, is illustrated in Fig. 8.1 [1]. When the nucleus of a U-235 atom absorbs an extra neutron, it divides into two fragments of roughly equal mass, generating a large amount of heat and releasing either two or three

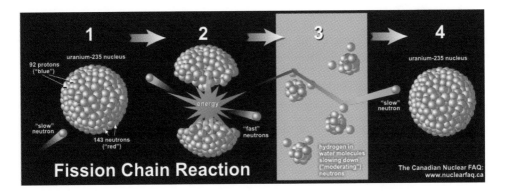

Fig. 8.1.   Uranium fission [1] (reproduced by permission of AEA technology plc).

more neutrons plus some gamma radiation. The two fission fragments are, respectively, nuclei of the elements barium and krypton. This process, shown artistically in Fig. 8.1, can be expressed in the nuclear equation

$$^{235}_{92}\text{U} + ^{1}_{0}\text{n} \rightarrow \,^{92}\text{Kr} + \,^{141}\text{Ba} + 3\text{n} + \lambda \qquad (8.3)$$

where n = neutron and λ = gamma radiation. The atomic mass numbers of the fission fragments sum to $92 + 141 = 233$, which is smaller than the mass number, 235, of the uranium. Now the amus add to 236 on both sides of Eq. (8.3). However, each fission is known to release energy equivalent to a loss of mass (called the mass defect) of 0.215 amu or $(0.215)\,(1.66 \times 10^{-27})$, which is 3.57×10^{-28} kg.

By Einstein's equation (8.1), each uranium atom fission therefore has the energy equivalent of

$$W = 3.57 \times 10^{-28} \times (3 \times 10^{8})^{2}$$
$$= 3.2 \times 10^{-11} \, \text{joules/fission}$$
$$= 200 \, \text{MeV} \qquad (8.4)$$

One kilogramme of U-235 contains 2.5×10^{24} atoms. If this is completely fissioned the energy released would be

$$W_{\text{from 1 kg of U-235}} = 3.2 \times 10^{-11} \times 2.5 \times 10^{24}$$
$$= 8 \times 10^{13} \, \text{J} \qquad (8.5)$$

In comparison, the thermal energy content of a ton of coal is about 3×10^{10} J. One kilogramme of fissioned U-235 is therefore roughly equivalent, in thermal energy terms, to $(8 \times 10^{13})/(3 \times 10^{10}) = 2700$ tons of coal. It is also roughly energy equivalent to about 2,000 tonnes of oil.

8.2.2.   *Worked examples*

Example 8.1.   One kilogramme of U-235 contains 2.5×10^{24} atoms. What mass of U-235 has the energy equivalence of 1 ton of coal?

$$\text{Thermal energy content of 1 ton of coal} \approx 3 \times 10^{10}\,\text{J}$$

$$\text{The fission of 1 atom of U-235 releases} \equiv 3.2 \times 10^{-11}\,\text{J}$$

$$\text{The fission of 1 kg of U-235} \equiv 3.2 \times 10^{-11} \times 2.5 \times 10^{24}$$

$$\equiv 8 \times 10^{13}\,\text{J}$$

Therefore, in energy terms,

$$1,\text{kg U-235} \equiv \frac{8 \times 10^{13}}{3 \times 10^{16}}$$

$$\equiv 2670\,\text{tons of coal}$$

Example 8.2.   Naturally-occurring uranium contains 0.7% of fissionable U-235 and 99.3% of largely non-fissionable U-238. Assume that, in a particular fission process, only 1% of the U-235 was fissioned. What mass of uranium ore is then the thermal energy equivalent of 1,000 tons of coal?

When pure U-235 is completely fissioned, then, from Example 8.1,

$$1000\text{ tons of coal} \equiv \frac{1000}{2670} \equiv 0.375\,\text{kg},\ \ \text{U-235}$$

If only 1% of the U-235 is fissioned, then

$$1000\text{ tons of coal} = 100 \times 0.375$$

$$\equiv 37.5\,\text{kg},\ \ \text{U-235}$$

If U-235 is contained within the uranium ore in the proportion 0.7% (without any enrichment), then

$$1000\text{ tons of coal} = 37.5 \times \frac{100}{0.7}$$

$$= 5357\,\text{kg of U-235}$$

8.3.   Radioactivity

8.3.1.   *Nature of radioactivity*

Some of the heavier chemical elements have isotopes that are intrinsically unstable. Their nuclei undergo spontaneous degeneration in order to achieve a more stable energy form. Such materials are known as radioactive and the process of degeneration is known as radioactive decay. Examples of naturally-occurring radioactive substances are radium, thorium and uranium. The decay chain of an element involves natural transmutation into a chemically different element and a succession of such changes may occur.

For a radioactive source, the following features may be of interest:

(a) energy (intensity) level;
(b) time scale of decay;
(c) nature of the radiation; and
(d) effects of the radiation.

8.3.2.   *Energy and decay rate*

The energy radiated from a radioactive source and the time rate of decay are both proportional to the number of radioactive atoms contained in it.

Let a source contain N radioactive atoms at an arbitrary instant t. The time rate of decay is then

$$\frac{dN}{dt} = -\lambda N = \text{decay rate} = \text{activity} \tag{8.6}$$

The negative sign in Eq. (8.6) indicates that the number of atoms disintegrating decreases with time in the manner shown in Fig. 8.2. The older, non-SI unit of radioactivity, the curie (Ci), was based on the activity of one gram of radium-226. The SI unit of radioactivity is the becquerel (Bq), defined as an activity of one decay per second (s^{-1}). One curie $= 3.7 \times 10^{10}$ atomic disintegrations per second. The term λ in Eq. (8.6), called the decay constant, is a characteristic of the radioactive

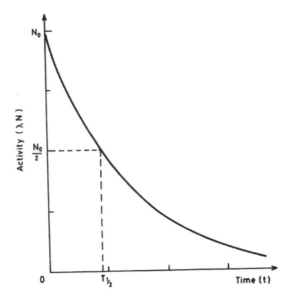

Fig. 8.2.   Nature of radioactive decay.

material and is entirely independent of the physical conditions. A mathematical expression for the activity or decay rate is obtained by integrating both sides of Eq. (8.6):

$$\int_{N_0}^{N} \frac{dN}{N} = - \int_{0}^{t} \lambda \, dt \tag{8.7}$$

If there are initially N_0 atoms present at $t = 0$, then at an arbitrary time t thereafter

$$\lambda t = -\ln \left(\frac{N}{N_0} \right) = \left(\frac{N_0}{N} \right) \tag{8.8}$$

Taking anti-logs of both sides of Eq. (8.8) gives

$$N = N_0 \, \epsilon^{-\lambda t} \tag{8.9}$$

A material with an exponential rate of decay does not reduce to zero until infinite time has passed. Equation (8.9) is zero only when $t = \infty$. In order to distinguish decay rates for different materials, it is customary to define a decay rate in terms of its "half-life", i.e. the time for one-half of the radioactivity to decay.

In Eq. (8.8), when $N = N_0/2$, the half-life $T_{1/2}$ is given by

$$\lambda t = \lambda T_{\frac{1}{2}} = \ln(2) = 0.693 \tag{8.10}$$

or

$$T_{\frac{1}{2}} = \frac{0.693}{\lambda}$$

Half-lives vary from fractions of a second to millions of years, depending on the radioactive material.

8.3.3.   *Worked examples*

Example 8.3.   How long does it take a radioactive source to decay to one-tenth of its original size? What is the value for U-235 which has a half-life of 7.1×10^8 years?

From Eq. (8.8)

$$t = \ln \frac{\left(\frac{N_0}{N} \right)}{\lambda}$$

Eliminating λ between Eqs. (8.8) and (8.10) gives

$$t = \ln \frac{\left(\frac{N_0}{N} \right)}{0.693} \cdot T_{\frac{1}{2}}$$

$$\text{When } N = \frac{N_0}{10}, \quad \ln \frac{N_0}{N} = \ln 10 = 2.3 \tag{8.11}$$

$$t = \frac{2.3}{0.693} \cdot T_{\frac{1}{2}} = 3.32 \, T_{\frac{1}{2}}$$

A radioactive source decays to one-tenth its original value in 3.32 times its half-life. For U-235, for example, $T_1 = 7.1 \times 10^8$ years. Therefore, U-235 decays to one-tenth of its size in

$$t_{\text{U-235}} = 3.32 \times 7.1 \times 10^8 \text{ years}$$
$$= 23.57 \times 10^8 \text{ years}$$
$$= 2357 \text{ million years}$$

Example 8.4. Radioactive cobalt-60 (Co-60) decays to nickel, with a half-life of 5.3 years. What is the value of the decay constant?

$$T_{\frac{1}{2}} = 5.3 \text{ years}$$

From Eq. (8.10)

$$\lambda = \frac{0.693}{T_{\frac{1}{2}}} = \frac{0.693}{5.3} = 0.131$$

The activity of this source is then

$$\lambda = 0.131 \, N \text{ curies}$$

where N is the number of atoms of the source.

Example 8.5. A radioactive source of strontium-90 will take 93.29 years to decay to one-tenth of its original activity. What is the corresponding half-life?
From Eq. (8.11), within Example 8.3

$$93.29 = \frac{\ln 10}{0.693} \cdot T_{\frac{1}{2}}$$

$$T_{\frac{1}{2}} = \frac{93.29 \times 0.693}{2.3} = 28.1 \text{ years}$$

8.4. Nuclear Radiation

8.4.1. *Forms of radiation*

There are many types of radiation. There is low-frequency, low-energy electromagnetic radiation, such as radio waves, microwaves and near visible radiation. These forms are non-ionizing and do not cause an electron to be dislodged from a target atom. There are also high-frequency forms, sources of which are all sufficiently energetic to be classified as ionizing radiation and can be harmful to humans.

In addition to electromagnetic radiation, there are various types of particle radiation, such as alpha, beta and neutron radiation.

The main forms of radiation from radioactive sources are:

(a) alpha particles (α),
(b) beta particles (β),
(c) X-radiation (X),

(d) gamma radiation (γ),

(e) cosmic radiation,

(f) neutrons (n).

(a) Alpha particles

An alpha particle (α) is a positively-charged particle consisting of two protons and two neutrons, like the nucleus of a helium-4 (^4_2He) atom. It has a mass number 4 and an atomic number 2. Alpha particles are emitted from natural heavy elements such as uranium and radium. This form of radiation has low penetrating power. For example, it will not normally penetrate the skin and is harmful only if swallowed or breathed into the body. The energy of α particles is in the range 4–6 MeV.

(b) Beta particles

A beta particle (β) is essentially a fast-moving electron ejected from the nucleus of an atom. It has no significant mass in amu but carries a negative charge. Some beta particles have greater penetrating power than alpha particles. They can penetrate the skin but can be stopped by thin layers of metal, water or glass. Like α sources, β sources are also dangerous if ingested or inhaled into the body. The relative penetrative effects of α and β radiation are depicted in Fig. 8.3 [2]. Both α and β particle radiations can be ionizing in nature: they can result in removing an electron from its associated proton in the target atom, which causes the target to change its chemical nature.

(c), (d) Gamma radiation and X-radiation

Gamma radiation (γ) and X-radiation (X) are both forms of electromagnetic radiation, like light and radio waves, but of much higher frequency (and hence shorter wavelength). They can also be thought of as beams of photon (quantum) particles, each carrying a discrete package of energy but having negligible mass. The penetrating power of γ radiation is greater than that of X-radiation (Fig. 8.3), but it can be screened by sufficient thicknesses of concrete or lead. Gamma radiation may have energy in the range 1–5 MeV. At sufficient dose rates and exposure times, both

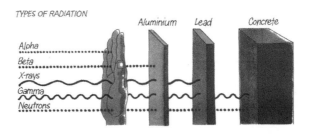

Fig. 8.3.   Types of radiation [2] (reproduced by permission of AEA technology plc).

X-rays and λ rays can produce ionizing effects and be dangerous in human tissue. Ionizing radiation impacting the human body does not cause the body to become radioactive.

(e) Cosmic radiation

Cosmic radiation consists of a variety of high-energy (i.e. low-wavelength, high-frequency) particles, including protons, which bombard the earth from outer space. They are more intense at higher altitudes than at sea level and constitute a hazard to astronauts in space and airline travelers. Cosmic radiation contributes a significant proportion of the natural background radiation (see Section 8.4.4.1).

(f) Neutron radiation

Energetic free neutrons are the most penetrative form of radiation and can be highly dangerous to the human body. In nuclear reactor systems it is necessary to shield the neutrons behind great thicknesses of concrete.

8.4.2.   Units of measurement of radiation

Of great interest, from the viewpoint of human welfare and safety, is the amount of energy deposited into a material by ionizing radiation. This is frequently called the "radiation-absorbed dose", or rad. One rad represents an energy absorption of 10^{-2} J/kg. Another unit now widely used is the gray (Gy), which represents 100 rads.

The actual quantity of X or λ radiation can be measured in roentgen (R), named after the early pioneer of medical X-rays, Wilhelm Röntgen. One roentgen is the amount of radiation required to produce a specified level of ionization of air and has a value of 2.58×10^{-4} coulombs/kg. The amount of radiation received, and the energy absorbed due to that radiation, are related by a term known as a "roentgen equivalent man", or rem. One rem is the amount of radiation, of any type, that produces the same biological effect on human tissue as one rad of gamma or X-radiation of a specific energy level. The rem was formerly the international standard unit of radiation exposure level. Since the rem represents a rather large unit of radiation, with regard to human toleration, it is common to use the term millirem (mrem), 1,000th part of a rem. For large radiation dosages it may be more convenient to use the unit sievert (Sv), which is 100 rems.

8.4.3.   Effects of nuclear radiation

For low levels of radiation exposure the biological effects are so small they cannot be detected. Radiation protection standards assume, however, that the effect is directly proportional to the dose, even at low levels. According to this "linear"

Table 8.1. Effects of nuclear radiation doses [4].

2 mSv/year	Typical background radiation experienced by everyone (av 1.5 mSv in Australia).
2.4 mSv/year	Average dose to US nuclear industry employees.
up to 2.5 mSv/year	Average dose to Australian uranium miners.
up to 5 mSv/year	Typical incremental dose for aircrew in middle latitudes.
9 mSv/year	Exposure by airline crew flying the New York–Tokyo polar route.
10 mSv/year	Maximum actual dose to Australian uranium miners.
20 mSv/year	Current limit (averaged) for nuclear industry employees and uranium miners.
50 mSv/year	Former limit for nuclear industry employees and U miners. Lowest level at which any increase in cancer is evident. It is also the dose rate that arises from natural background levels in several places. Above this, the probability of cancer occurrence (rather than the severity) increases with dose.
350 mSv in lifetime	Criterion for relocating people after the 1986 Chernobyl accident.
1000 mSv	As a dose accumulated over some time, 1000 mSv would probably cause a fatal cancer many years later in 5 of every 100 persons exposed to it (i.e. if the normal incidence of fatal cancer were 25%, this dose would increase it to 30%).
1000 mSv	As short-term dose: cause (temporary) radiation sickness such as nausea and decreased white blood cell count, but not death. Above this, severity of illness increases with dose.
5000 mSv	As short-term dose: would kill about half those receiving it within a month.
10000 mSv	As short-term dose: fatal within a few weeks.

Note: 1 mSv = 100 mrems.

theory of radiation effects, if the dose is halved the effect, or the risk of any effect, is halved [3]. Some information on nuclear radiation doses and their effects is given in Table 8.1 [4].

The International Commission on Radiological Protection (ICRP) has set a recommended maximum radiation dose, for the general public, at 0.5 rem (500 mrem or 5 mSv) per year from all sources. This is a very conservative rating, containing (necessarily) a large factor of safety [2].

Our knowledge of radiation effects is derived primarily from groups of people who have received high doses. Radiation protection standards assume that any dose of radiation, no matter how small, involves the possibility of risk to human health. However, scientific evidence does not indicate any cancer risk or immediate effects at doses below 100 mSv per year. At low levels of exposure, the body's mechanisms seem to be adequate to repair radiation damage to cells soon after it occurs [4].

8.4.4.   *Sources and amounts of nuclear radiation*

The total radiation received from radioactive sources is composed of many forms of natural radiation and man-made radiation.

8.4.4.1. *Natural radiation sources*

There are several sources of natural radiation.

(a) Radiation from space.
(b) Terrestrial radiation proceeding outwards from rocks and soil. In south-east England the terrestrial dose rate is about 40–50 mrems/year, whereas in Aberdeen, Scotland, about 500 miles north, it is 80–100 mrems/year due to granite rocks. Where the soil has a high content of uranium or thorium, as in parts of India or Brazil, the background dose rates are very much higher than in the UK.
(c) Natural radiation within the body due to small traces of radioactive potassium (K-40).
(d) Small amounts of radium in the drinking water and food chains.
(e) Small amounts of radioactive carbon in the air. The average dose from natural sources in the UK, including ingested or inhaled components, is about 200 mrems/year (2 mSv/year).

8.4.4.2. *Man-made sources*

Radioactive sources of radiation include the following items.

(f) Some building materials, such as granite or concrete.
(g) Luminous watch dials (now manufactured by other means).
(h) Television and computer screens (cathode ray screens are becoming obsolete).
(i) X-ray machines and scanners used for medical diagnosis. These contribute an average of 50 mrems/year to UK residents. For example, a chest X-ray gives about 20 mrems/year.
(j) Gamma radiation sources used in the radiotherapy treatment of cancer.
(k) Nuclear-powered space vehicles, satellites and sea-buoys.
(l) Radioactive debris in the atmosphere caused by nuclear weapons testing and past nuclear bombs. This is estimated to be less than 1 mrem/year.
(m) Nuclear power stations and radioactive waste disposal sites. These are subjected, by law, to scrupulous screening and testing by independent inspectors. During normal operation of a nuclear power plant the radiation effect on the public is negligible. In Europe and North America the activities of the nuclear power industry create an average radiation dose of 0.3 mrem/year, which is less than the radiation from a cathode ray tube device, such as a TV screen.

8.4.5. *Uses of nuclear radiation*

8.4.5.1. *Geological dating*

The radioactive decay of the natural element uranium-238 (U-238) has a half-life of 4,600 million years, which provides a basis for measuring the age of ancient rock formations. U-238 decays, via α and β particle emission, through a chain that successively forms radium, radon, polonium and finally lead-206, an isotope which is not radioactive but is known to descend from radioactive parents. The age of a rock formation can be estimated by measuring the amount of lead-206.

8.4.5.2.   *Archaeological dating*

Carbon dioxide in the air is assimilated into all living plants by photosynthesis and ingested by all plant-eating animals and other animals that feed on plant-eating animals (including humans). The atmospheric carbon dioxide contains small amounts of the radioactive isotope carbon-14 (C-14), which decays with a half-life of 5,600 years. After its death an animal or plant does not ingest any further carbon and the existing C-14 then decays without replacement. The ratio of C-14 to regular carbon then slowly decreases and provides a time-clock for estimating the age of dead organic materials, up to a few thousand years.

It should be noted that the C-14 dating method must not be regarded as infallible. C-14 is not homogeneously distributed among today's plants and animals. For example, volcanic carbon dioxide is known to suppress the C-14 level of nearby vegetation. In addition to spatial variations there are temporal variations. The injection of C-14-free carbon dioxide into the air through smoke stacks suppresses the level, while an excess of man-made C-14 is injected into the atmosphere by nuclear weapons tests.

8.4.5.3.   *Medical tracer elements*

One of the medical uses of radioactive materials is as tracer elements for diagnosis. A radioactive material such as cobalt-60 (Co-60) is introduced into the patient's suspect organ, which is then examined by X-rays. Patients with digestive tract problems may be fed a bismuth or barium meal, which enables the whole tract to be examined in order to localize the diagnosis.

8.4.5.4.   *Small nuclear power packs*

Heat from the radioactive decay of plutonium-238 ($T_{\frac{1}{2}} = 88$ years) forms the basis of long-lived nuclear batteries. Applications include heart pacemakers, navigational beacons and space satellites and vehicles. The US Apollo space missions used plutonium-powered batteries for equipment on the moon's surface. Deep space results, like the Voyager spacecraft pictures from Jupiter and Saturn, were also energized from on-board nuclear-powered batteries.

8.4.5.5.   *Biological effects on human tissue*

Radiation that causes ionization within human tissue can have very serious effects. This depends, to a certain extent, on the intensity of the radiation and the time scale over which exposure occurs.

Every cell of the body contains molecules of deoxyribonucleic acid (DNA), which store genetic information and control cell growth, function, development and reproduction. It is possible for radiation to damage the DNA so as to cause cell death or some harmful transformation (mutation). Slight injury to a chromosome may alter one or more of the inheritance characteristics passed on to the next generation. Gross chromosome injury would be likely to cause injury or death to the foetus of a pregnant woman. Hospital X-ray procedures are not normally carried out on pregnant patients for this reason.

Exposure to a large radiation dose within a short time span causes massive cell destruction. A whole body radiation dose of hundreds of rems in a few minutes causes damage to internal organs, blood cells and bone tissue such that death is likely to follow within weeks. At radiation doses of thousands of rems death is almost certain to follow within a few days. The highly destructive effects of radiation upon cell structure are used in cancer radiotherapy to try and limit or destroy cancerous tissue.

Exposure to repeated doses of low-level radiation can be insidious because the early sickness symptoms of high-level exposure do not occur. If the exposures are widespread in time, then the natural healing of any ill-effects may mitigate any cumulative effect, but some damage may still occur. It is established that overexposure to low-level radiation can cause cancer of various types in humans and shorten the lifespan of experimental animals. Delayed hereditary defects have been discovered in animals but not yet in humans. Nevertheless, with the radiation exposure of much of the world population increasing, the possibility of widespread though small genetic abnormalities poses a disturbing uncertainty.

Workers employed in hospital radiology departments, the nuclear industry and other parts of industry dealing with radioactive sources are a special case. Through proper training, protective clothing and practiced safe working procedures they may, paradoxically, be safer from radiation exposure than the general public.

8.5. Nuclear Reactors

A nuclear power station for generating electricity is similar in most respects to a coal, natural-gas or oil-fired station. Intense heat derived from the primary fuel (coal, gas, oil or uranium) is used to convert water into steam, which drives a turbine–generator system, as in Fig. 1.5 of Chapter 1. In a nuclear reactor the heat source is nuclear fission, described in Section 8.2.1 above, using uranium or plutonium as a fuel.

A great, possibly the greatest, advantage of the use of nuclear power is that no fossil fuels are burned and therefore no "greenhouse" gases such as carbon dioxide are emitted. The significance of this depends on the issue of global warming: is it a reality, a dangerous reality or a scientific misconception?

8.5.1. *Thermal (fission) reactors*

Natural uranium contains only 0.7% of fissionable U-235. The remainder of the material, 99.3% by weight, consists of the uranium isotope U-238, which absorbs neutrons without much fissioning, plus a tiny amount (0.0055%) of U-234. For use in some types of nuclear reactor the natural uranium is industrially-processed to enrich it so that it contains 2–3% of U-235.

To produce a continuous heat source, the neutrons released by fission, illustrated in Fig. 8.1, must strike other U-235 nuclei, causing further fission in a so-called chain reaction. A sustainable chain reaction requires the presence of a certain minimum or critical mass of U-235, configured in some particular design manner. In addition to a source of neutrons, a sustainable chain reaction requires that the rates of neutron production and heat generation be controlled. Also, it is necessary to slow down the velocities of the neutron products of fission by containing them in an enclosure with a moderator material, which is usually ordinary water, heavy water or graphite. Because the purpose of the reactor is heat generation, it is often called a thermal reactor.

For every unit of electrical output power MW_e required from the generator it is necessary to create about three times this value in thermal power MW_{th} in the nuclear reactor.

The principle of a boiling water nuclear reactor core is illustrated in Fig. 8.4 [5]. The nuclear fuel, usually uranium oxide (U_3O_8), is enclosed in metal cans or fuel rods and can be replaced when the active material is spent. Control rods (not the

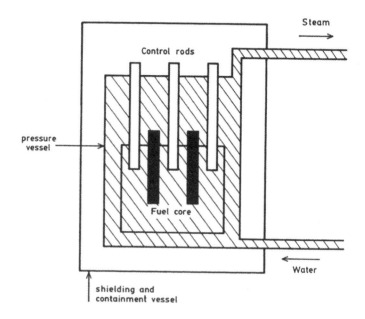

Fig. 8.4.   Basic form of boiling water reactor [5].

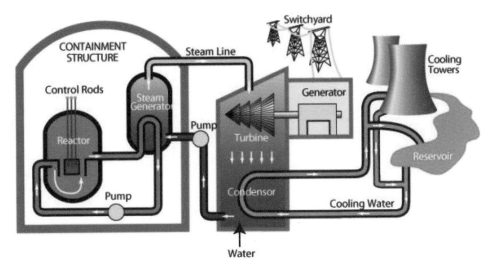

Fig. 8.5. Pressurized water reactor system [6].

same as fuel rods) of a highly neutron absorbent material such as boron or cadmium can be raised or lowered within the core and these control the rate of fission reaction and hence the heat production. When the control rods are fully inserted into the core, they absorb neutrons like a dry sponge absorbs water and can, if required, shut the reactor down.

For all types of reactors, the heat from the core has to be extracted and transferred to the steam circuit of the turbo-generator. This is done directly in the boiling water reactor (BWR), where ordinary water (H_2O) is used both as a moderator and as the coolant and heat transfer fluid (Fig. 8.4). The pressurized water reactor (PWR; Fig. 8.5) also uses ordinary water as a moderator but this is enclosed within a separate circuit from the steam circuit [6]. Both the BWR and the PWR use enriched uranium as fuel and are widely-used US designs.

In the UK the older range of nuclear reactors are called Magnox reactors, because the fuel is natural uranium clad in magnesium alloy cans with low neutron absorption. Graphite is used as a moderator within the core and heat is transferred to the external heat exchangers using high-pressure carbon dioxide (CO_2) gas. More recent nuclear stations in the UK use the advanced gas-cooled reactor (AGR) design. This is similar, in principle, to the Magnox design but uses uranium fuel enriched to 2.3% content of U-235. The fuel enrichment permits higher operational temperature and heat output than a Magnox type, requiring a smaller reactor core, and resulting in a more efficient steam cycle.

A particularly elegant and successful design is the pressurized heavy water CANDU reactor (PHWR), developed at Chalk River, Ontario, Canada, and shown in Fig. 8.6. This type of reactor uses natural uranium as fuel and high-pressure

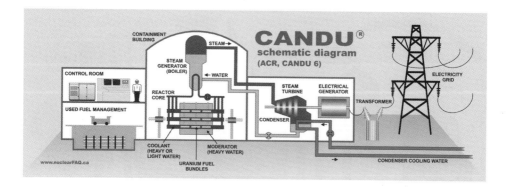

Fig. 8.6.   CANDU heavy water reactor system [1].

Table 8.2.   Performance data of thermal reactors [7].

Reactor	Magnox (UK) 600 MWe	BWR (USA) 600 MWe	PWR (USA) 700 MWe	PHWR (CANDU) (Canada) 600 MWe	AGR (UK)
Uranium enrichment % U-235	0.7 (natural)	2.6	3.2	0.7 (natural)	2.3
Coolant outlet temperature (°C)	400	286	317	305	650
Coolant pressure (1 lbf/in² absolute)	300	1050	2235	1285	600
Steam cycle efficiency %	31	32	32	30	42
Core diameter (metres)	14	3.7	3	7.1	9.1
Core height (m)	8	3.7	3.7	5.9	8.5
Moderator	graphite	water	water	heavy water	graphite
Heat transfer fluid	carbon dioxide	water	water	heavy water	carbon dioxide

heavy water (D_2O) as a moderator and heat transfer fluid. A comparison of some of the design details of various types of thermal reactors is given in Table 8.2 [7].

Historically the world's first reactor to be connected to an electricity supply network was a Magnox reactor in England in 1953. Britain and France attempted to make the Magnox gas-graphite design the world standard but this initiative ended in 1967. From the early 1970s until the mid-1980s the world market for reactors was dominated by the US light water PWR system, which was an outgrowth of the American nuclear submarine programme. Only the Canadian CANDU reactor survived as effective opposition [8].

The latest American nuclear power plant is Watts Bar 1, which opened at Spring City, Tennessee in 1996 [9, 10]. The latest British plant, Sizewell B, which is also a PWR design, rated at 1110 MW_e, went "onstream" in 1995 [11]. The prospects for nuclear power in the USA are improving. The US Nuclear Regulatory Commission

(NRC) in 2001 granted extensions to operate two reactors 20 years beyond their initial 40-year licences. Other US companies have petitioned the NRC for nuclear plant life extension [12]. Many countries are proceeding with large-scale nuclear programmes, including China, France, India, Japan, Romania, Russia, South Korea, Taiwan and Ukraine. The big future market for nuclear-generated electricity is the fast-growing Asian countries of the Pacific Rim, notably China [12, 13].

In both Britain and the USA, the nuclear industry has failed to convince a sceptical public of the safety and the economic and environmental advantages of nuclear-generated electricity. The French have continued to build nuclear stations so that now (2013) most of the electricity generated in France is from nuclear sources. Some of it is now sold to Britain using undersea cables under the English Channel.

It remains to be seen which country has chosen the wiser course. France, which has no oil, will eventually become totally nuclear. The USA and the UK are both oil-rich while the oil lasts and coal-rich until the coal runs out. What then?

8.5.2. *Uranium supplies*

Natural uranium has to be mined and there is some uncertainty about the remaining reserve stock and the economic viability of mining it. The overall uranium fuel supply situation depends on whether the uranium oxide (U_3O_8) is used on a "once through" basis only or whether the uranium fuel stock is replenished by nuclear fuel reprocessing. Another option is to implement plutonium as a fuel, obtained through the use of fast breeder reactors, described in Section 8.5.4 below. A further option is the use of thorium as a fuel.

The nuclear power industry, which largely uses nuclear fission (thermal) reactors, is in recession in the USA and the UK but uranium continues to be needed elsewhere and the demand is growing. There were 442 thermal reactors in operation around the world in 2011, including 19 in the UK and 104 in the USA (Table 8.3) [14]. A further 65 plants are under construction, including 27 in China and 11 in Russia.

A listing of the major recoverable resources of uranium oxide is given in Table 8.4 [14]. These stocks tend to deplete year by year but they are widely distributed geographically and are in accessible locations. The world demand for nuclear-generated electricity continues to increase in most areas (Table 8.5) [15].

There is no reliable data upon which one can accurately forecast a lifetime expectation of continued natural uranium supply. It is being used up and is not renewable. Even if major untapped deposits are found it is evident that the world uranium supplies will be exhausted long before the coal supplies. A pessimistic forecast would be that the uranium will not even outlast the oil. If nuclear fission (thermal) reactors are to continue to be used for electricity generation it is inevitable that this will require the widespread use of uranium fuel reprocessing and also the development of advanced forms of fast breeder reactor. This will, in turn, increase the amount of nuclear waste; an issue that is discussed in Section 8.6 below.

Table 8.3.   Nuclear power plants, operational and under construction, 2011 [14].

Country	In operation		Under construction	
	Number	Electr. net output MW	Number	Electr. net output MW
Argentina	2	935	1	692
Armenia	1	376	—	—
Belgium	7	5,863	—	—
Brazil	2	1,884	—	—
Bulgaria	2	1,906	2	1,906
Canada	18	12,573	—	—
China	11	8,438	21	20,920
Czech Republic	6	3,678	—	—
Finland	4	2,696	1	1,600
France	58	63,130	1	1,600
Germany	17	20,470	—	—
Hungary	4	1,889	—	—
India	18	3,984	5	2,709
Iran	—	—	1	915
Japan	54	46,823	1	1,325
Korea, Republic	20	17,647	6	6,520
Mexico	2	1,300	—	—
Netherlands	1	482	—	—
Pakistan	2	425	1	300
Romania	2	1,300	—	—
Russian Federation	31	21,743	9	6,894
Slovakian Republic	4	1,762	2	810
Slovenia	1	666	—	—
South Africa	2	1,800	—	—
Spain	8	7,450	—	—
Sweden	10	8,992	—	—
Switzerland	5	3,238	—	—
Taiwan	6	4,949	2	2,600
Ukraine	15	13,107	2	1,900
United Kingdom	19	10,097	—	—
USA	104	100,683	1	1,165
Total	**436**	**370,326**	**56**	**51,855**

8.5.3.   *Plutonium*

Figure 8.7 shows the continuation of the nuclear fission process illustrated in Fig. 8.1 [16]. This results in the production of radioactive plutonium-239 (Pu-239), which is an artificial chemical element isotope that does not occur in nature. Plutonium is a metal, hard and brittle like cast iron, which can be melted, moulded and machined. It can also be soft alloyed with other metals to produce wire and metal foil. Other plutonium isotopes exist, notably Pu-238, Pu-240 and Pu-241, all of which are radioactive. Plutonium is highly radiotoxic and must be used under strictly controlled and guarded conditions [16].

Table 8.4. Known Recoverable Resources of Uranium, 2011 [14][a].

Country	tonnes U	Percentage of world
Australia	1,661,000	31%
Kazakhstan	629,000	12%
Russia	487,200	9%
Canada	468,700	9%
Niger	421,000	8%
South Africa	279,100	5%
Brazil	276,700	5%
Namibia	261,000	5%
USA	207,400	4%
China	166,100	3%
Ukraine	119,600	2%
Uzbekistan	96,200	2%
Mongolia	55,700	1%
Jordan	33,800	1%
other	164,000	3%
World total	**5,327,200**	

[a] "The world's power reactors, with combined capacity of some 375 GWe, require about 68,000 tonnes of uranium from mines or elsewhere each year. While this capacity is being run more productively, with higher capacity factors and reactor power levels, the uranium fuel requirement is increasing, but not necessarily at the same rate. The factors increasing fuel demand are offset by a trend for higher burn-up of fuel and other efficiencies, so demand is stead. (Over the years 1980 to 2008 the electricity generated by nuclear power increased 3.6-fold while uranium used increased by a factor of only 2.5.)" [14].

In a thermal reactor the Pu-239 content is fissioned by neutron capture and contributes about one-third of the heat generated by the overall uranium fission process, at the same time releasing further neutrons. The fissioning of Pu-239 also creates radioactive fission products similar to those from uranium and these form part of the nuclear waste of the reactor cycle. The significance of the use of plutonium for nuclear-electric generation is largely in fast breeder reactors, described in the following section.

8.5.4. *Fast breeder reactors*

To produce a neutron chain reaction in a thermal nuclear reactor, the neutrons must be slowed down by a moderator so that they will fission the atom of U-235 in the fuel. In both natural uranium and enriched uranium fuels, the proportion of U-235 atoms is small and only a tiny part of the energy available from the uranium can

Table 8.5.   Consumption* of nuclear energy [15].

Million tonnes oil equivalent	2000	2001	2002	2003	2004	2005	2006	2007	2008	2009	**2010**	Change 2010 over 2009	2010 share of total
US	179.6	183.1	185.8	181.9	187.8	186.3	187.5	192.1	192.0	190.3	**192.2**	1.0%	30.7%
Canada	16.4	17.2	17.0	16.8	20.3	20.7	22.0	21.0	21.1	20.2	**20.3**	0.3%	3.2%
Mexico	1.9	2.0	2.2	2.4	2.1	2.4	2.5	2.4	2.2	2.4	**1.3**	−44.0%	0.2%
Total North America	197.8	202.3	205.0	201.1	210.2	209.4	212.0	215.4	215.4	212.9	**213.8**	0.4%	34.2%
Argentina	1.4	1.6	1.3	1.7	1.8	1.6	1.7	1.6	1.6	1.8	**1.6**	−11.8%	0.3%
Brazil	1.4	3.2	3.1	3.0	2.6	2.2	3.1	2.8	3.2	2.9	**3.3**	12.0%	0.5%
Chile	—	—	—	—	—	—	—	—	—	—	—	—	—
Colombia	—	—	—	—	—	—	—	—	—	—	—	—	—
Ecuador	—	—	—	—	—	—	—	—	—	—	—	—	—
Peru	—	—	—	—	—	—	—	—	—	—	—	—	—
Trinidad & Tobago	—	—	—	—	—	—	—	—	—	—	—	—	—
Venezuela	—	—	—	—	—	—	—	—	—	—	—	—	—
Other S. & Cent. America	—	—	—	—	—	—	—	—	—	—	—	—	—
Total S. & Cent. America	2.8	4.8	4.4	4.7	4.4	3.8	4.8	4.4	4.8	4.7	**4.9**	2.9%	0.8%
Austria	—	—	—	—	—	—	—	—	—	—	—	—	—
Azerbaijan	—	—	—	—	—	—	—	—	—	—	—	—	—
Belarus	—	—	—	—	—	—	—	—	—	—	—	—	—
Belgium & Luxembourg	10.9	10.5	10.7	10.7	10.7	10.8	10.6	10.9	10.3	10.7	**10.9**	2.0%	1.7%
Bulgaria	4.1	4.4	4.6	4.5	4.4	4.2	4.4	3.3	3.6	3.4	**3.5**	0.2%	0.6%
Czech Republic	3.1	3.3	4.2	5.9	6.0	5.6	5.9	5.9	6.0	6.2	**6.3**	2.9%	1.0%
Denmark	—	—	—	—	—	—	—	—	—	—	—	—	—
Finland	5.1	5.2	5.4	5.5	5.5	5.5	5.4	5.6	5.4	5.4	**5.2**	−2.8%	0.8%

(Continued)

Table 8.5. (*Continued*)

Million tonnes oil equivalent	2000	2001	2002	2003	2004	2005	2006	2007	2008	2009	2010	Change 2010 over 2009	2010 share of total
France	94.0	95.3	98.8	99.8	101.7	102.4	102.1	99.7	99.6	92.8	**96.9**	4.4%	15.5%
Germany	38.4	38.8	37.3	37.4	37.8	36.9	37.9	31.8	33.7	30.5	**31.8**	4.2%	5.1%
Greece	—	—	—	—	—	—	—	—	—	—	—	—	—
Hungary	3.2	3.2	3.2	2.5	2.7	3.1	3.0	3.3	3.4	3.5	**3.6**	2.1%	0.6%
Republic of Ireland	—	—	—	—	—	—	—	—	—	—	—	—	—
Italy	—	—	—	—	—	—	—	—	—	—	—	—	—
Kazakhstan	—	—	—	—	—	—	—	—	—	—	—	—	—
Lithuania	1.9	2.6	3.2	3.5	3.4	2.3	2.0	2.2	2.2	2.5	—	−100.0%	—
Netherlands	0.9	0.9	0.9	0.9	0.9	0.9	0.8	1.0	0.9	1.0	**0.9**	−6.1%	0.1%
Norway	—	—	—	—	—	—	—	—	—	—	—	—	—
Poland	—	—	—	—	—	—	—	—	—	—	—	—	—
Portugal	—	—	—	—	—	—	—	—	—	—	—	—	—
Romania	1.2	1.2	1.2	1.1	1.3	1.3	1.3	1.7	2.5	2.7	**2.6**	−1.1%	0.4%
Russian Federation	29.5	31.0	32.1	33.6	32.7	33.4	35.4	36.2	36.9	37.0	**38.5**	4.1%	6.2%
Slovakia	3.7	3.9	4.1	4.0	3.9	4.0	4.1	3.5	3.8	3.2	**3.3**	3.5%	0.5%
Spain	14.1	14.4	14.3	14.0	14.4	13.0	13.6	12.5	13.3	11.9	**13.9**	16.8%	2.2%
Sweden	13.0	16.3	15.4	15.3	17.3	16.4	15.2	15.2	14.5	11.9	**13.2**	11.2%	2.1%
Switzerland	6.0	6.0	6.1	6.2	6.1	5.2	6.3	6.3	6.2	6.2	**6.0**	−3.5%	1.0%
Turkey	—	—	—	—	—	—	—	—	—	—	—	—	—
Turkmenistan	—	—	—	—	—	—	—	—	—	—	—	—	—
Ukraine	17.5	17.2	17.7	18.4	19.7	20.1	20.4	20.9	20.3	18.8	**20.2**	7.5%	3.2%
United Kingdom	19.3	20.4	19.9	20.1	18.1	18.5	17.1	14.3	11.9	15.6	**14.1**	−10.1%	2.2%
Uzbekistan	—	—	—	—	—	—	—	—	—	—	—	—	—
Other Europe & Eurasia	1.5	1.6	1.8	1.6	1.8	1.9	1.9	1.9	2.0	1.9	**1.8**	−1.0%	0.3%
Total Europe & Eurasia	267.4	276.3	280.8	285.0	288.2	285.5	287.2	276.1	276.7	265.1	**272.8**	2.9%	43.6%

(*Continued*)

Table 8.5. (*Continued*)

Million tonnes oil equivalent	2000	2001	2002	2003	2004	2005	2006	2007	2008	2009	**2010**	Change 2010 over 2009	2010 share of total
Iran	—	—	—	—	—	—	—	—	—	—	—	—	—
Israel	—	—	—	—	—	—	—	—	—	—	—	—	—
Kuwait	—	—	—	—	—	—	—	—	—	—	—	—	—
Qatar	—	—	—	—	—	—	—	—	—	—	—	—	—
Saudi Arabia	—	—	—	—	—	—	—	—	—	—	—	—	—
United Arab Emirates	—	—	—	—	—	—	—	—	—	—	—	—	—
Other Middle East	—	—	—	—	—	—	—	—	—	—	—	—	—
Total Middle East	—	—	—	—	—	—	—	—	—	—	—	—	—
Algeria	—	—	—	—	—	—	—	—	—	—	—	—	—
Egypt	—	—	—	—	—	—	—	—	—	—	—	—	—
South Africa	3.1	2.6	2.9	3.0	3.4	2.9	2.7	2.8	2.7	3.1	**3.1**	−1.5%	0.5%
Other Africa	—	—	—	—	—	—	—	—	—	—	—	—	—
Total Africa	3.1	2.6	2.9	3.0	3.4	2.9	2.7	2.8	2.7	3.1	**3.1**	−1.5%	0.5%
Australia	—	—	—	—	—	—	—	—	—	—	—	—	—
Bangladesh	—	—	—	—	—	—	—	—	—	—	—	—	—
China	3.8	4.0	5.7	9.8	11.4	12.0	12.4	14.1	15.5	15.9	**16.7**	5.3%	2.7%
China Hong Kong SAR	—	—	—	—	—	—	—	—	—	—	—	—	—
India	3.6	4.3	4.4	4.1	3.8	4.0	4.0	4.0	3.4	3.8	**5.2**	37.3%	0.8%
Indonesia	—	—	—	—	—	—	—	—	—	—	—	—	—
Japan	72.3	72.7	71.3	52.1	64.7	66.3	69.0	63.1	57.0	65.0	**66.2**	1.7%	10.6%
Malaysia	—	—	—	—	—	—	—	—	—	—	—	—	—
New Zealand	—	—	—	—	—	—	—	—	—	—	—	—	—

(*Continued*)

Table 8.5.   (Continued)

Million tonnes oil equivalent	2000	2001	2002	2003	2004	2005	2006	2007	2008	2009	2010	Change 2010 over 2009	2010 share of total
Pakistan	0.2	0.5	0.4	0.4	0.5	0.6	0.6	0.6	0.4	0.6	**0.6**	−2.0%	0.1%
Philippines	—	—	—	—	—	—	—	—	—	—	—	—	—
Singapore	—	—	—	—	—	—	—	—	—	—	—	—	—
South Korea	24.7	25.4	27.0	29.3	29.6	33.2	33.7	32.3	34.2	33.4	**33.4**	◆	5.3%
Taiwan	8.7	8.0	8.9	8.8	8.9	9.0	9.0	9.2	9.2	9.4	**9.4**	0.1%	1.5%
Thailand	—	—	—	—	—	—	—	—	—	—	—	—	—
Vietnam	—	—	—	—	—	—	—	—	—	—	—	—	—
Other Asia Pacific	—	—	—	—	—	—	—	—	—	—	—	—	—
Total Asia Pacific	113.3	114.8	117.7	104.6	119.0	125.2	128.7	123.3	119.7	128.2	**131.6**	2.7%	21.0%
Total World	**584.3**	**600.7**	**610.8**	**598.5**	**625.2**	**626.8**	**635.4**	**622.1**	**619.2**	**614.0**	**626.2**	**2.0%**	**100.0%**
of which: OECD	507.4	519.7	524.8	506.0	530.7	532.5	537.7	522.0	517.0	511.5	**520.9**	1.8%	83.2%
Non-OECD	76.9	81.0	86.0	92.6	94.6	94.3	97.6	100.1	102.2	102.5	**105.3**	2.7%	16.8%
European Union	213.9	221.6	224.5	226.3	229.2	226.1	224.5	212.1	212.6	202.5	**207.5**	2.5%	33.1%
Former Soviet Union	49.4	51.2	53.4	56.0	56.4	56.4	58.4	60.0	60.0	58.8	**59.3**	0.8%	9.5%

*Based on gross generation and not accounting for cross-border electricity supply. Converted on the basis of thermal equivalence assuming 38% conversion efficiency in a modern thermal power station.

◆ Less than 0.05%.

Note: Nuclear energy data expressed in terawatt-hours is available at *www.bp.com/statisticalreview*.

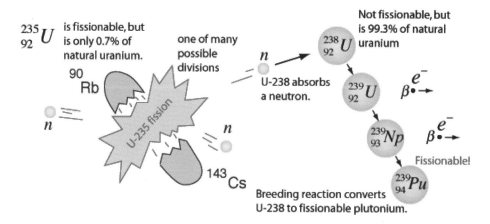

Fig. 8.7.   The formation of plutonium [17] (reproduced by permission of AEA Technology plc).

be exploited. The abundant component of natural uranium (99.3%) is U-238, which absorbs neutrons from the fission of U-235 atoms (Fig. 8.7) to produce fissionable Pu-239. Most of this plutonium remains unused in a thermal reactor cycle but can be subsequently separated out by reprocessing of the nuclear waste [17].

Great enhancement of a nuclear reactor heat generation capability is obtained by the use of fuel which contains 20–30% plutonium plus 80–70% uranium. This fuel is rich in atoms that can be fissioned by fast neutrons. A chain reaction can be sustained without the use of a moderator and the "fast reactor" core is much smaller than a thermal reactor core, being of the order of 1–4 cubic metres. As the plutonium in the core is consumed for electricity generation, the neutrons released by its fission are absorbed into a surrounding blanket of U-238, thereby creating more plutonium. The process therefore breeds plutonium and the reactor is commonly known as a "fast breeder reactor" or just "breeder reactor". By appropriate design, the amount of plutonium produced from the uranium can be made less than, equal to or greater than the original plutonium source. Utilization of the U-238 in a breeder reactor is about 60 times more efficient than in a thermal reactor [18].

The basic structure of a fast breeder reactor is shown in Fig. 8.8 [6]. Its operation is similar, in principle, to that of the thermal reactors described in Section 8.5.1. Liquid sodium, used as a coolant and heat transfer fluid, is more efficient than the water or gas of thermal reactors. The sodium becomes radioactive, with $T_{\frac{1}{2}} =$ 15 hours, so that the radiation decays to insignificance about 14 days after plant shutdown. The fuel cans and uranium blanket need to be replaced about once a year. Newly-created plutonium is separated out and can be reused.

"The construction of the fast breeder [Fig. 8.8] requires a higher enrichment of U-235 than a light-water reactor, typically 15 to 30%. The reactor fuel is surrounded by a 'blanket' of non-fissionable U-238. No moderator is used in the breeder reactor since

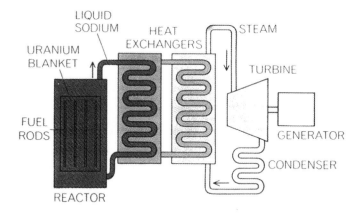

Fig. 8.8.   Configuration of a fast breeder reactor [6].

fast neutrons are more efficient in transmuting U-238 to Pu-239. At this concentration of U-235, the cross-section for fission with fast neutrons is sufficient to sustain the chain-reaction. Using water as coolant would slow down the neutrons, but the use of liquid sodium avoids that moderation and provides a very efficient heat transfer medium." [17]

The use of liquid sodium as a coolant and heat transfer medium raises questions of safety, as:

"sodium metal is an extremely reactive chemical and burns on contact with air or water (sometimes explosively on contact with water). The liquid sodium must be protected from contact with air or water at all times, kept in a sealed system. However, it has been found that the safety issues are not significantly greater than those with high-pressure water and steam in light-water reactors.
   "Sodium is a solid at room temperature but liquifies at 98°C. It has a wide working temperature since it does not boil until 892°C. That brackets the range of operating temperatures for the reactor so that it does not need to be pressurized as does a water-steam coolant system. It has a large specific heat so that it is an efficient heat-transfer fluid." [17]

Much of the pioneering work on fast breeder reactors took place in the UK. A fast reactor started up at Dounreay in Scotland in 1959 and ran until 1977. Its successors, the prototype fast reactors (PFR), rated at 250 MWe, were in operation from 1974 to 1994. The Phénix reactor in France, also rated at 250 MW$_e$, started up in 1974 and was succeeded by Superphénix, a 1200 MW$_e$ commercial size reactor. Superphénix halted production in 1996. Approximately six fast reactors operate around the world with a further one due for completion in India in 2014. Fast reactor development programmes have been considered in Germany, Japan, Russia and the USA. The cost of building a liquid-metal fast breeder reactor is higher than that of a thermal reactor of the same rating because of the need for finer engineering tolerances, adequate systems to safeguard the use of liquid metals such as radioactive sodium and also the costs of using, transporting and processing the hazardous fuel materials, notably plutonium [18].

It is the authors' view that the use of nuclear fission with fuel breeding represents one of the four great long-term energy options for the world, as discussed in Chapter 14.

8.5.5. *Reactor safety*

A nuclear reactor that is designed for electricity generation cannot explode like an atomic bomb. The fissile material in the core is insufficiently concentrated and too small in mass. Acts of deliberate sabotage might render a nuclear power station inoperable as a producer of power but could not convert it into a bomb. Also, the nuclear materials, usually uranium and/or plutonium, are not present in the form that would be directly usable in nuclear weapons.

The nuclear risks attendant in reactor operation are:

(a) failure of the steel pressure vessel or even meltdown of the core due to loss/lack of coolant,

(b) escape of radioactive material into the surrounding environment, and

(c) nuclear waste.

The engineering control systems for nuclear reactor operation are invariably duplicated and sometimes triplicated to sidetrack any equipment failures. Duplicated systems may have the additional safety feature that they overlap each other and work on alternative features. For example, the presence of reactor coolant, within a totally enclosed pressure vessel, can be sensed using both its temperature and pressure independently.

All reactor controls are designed to operate on a "fail-safe" basis. If the electricity driving the coolant pumps or heat transfer system is cut off, even from the plant standby supplies, the control rods fall under gravity into the core (Fig. 8.4) and shut the reactor down. The reactor core, contained in a steel pressure vessel, is itself located within a very thick biological shield of concrete (Fig. 8.5).

The operation of nuclear reactors is subjected to the most rigorous engineering design tolerances and to extremely stringent safety management systems. Standard safety features include the monitoring and measurement of all the physical and chemical properties of reactor materials. There is an emergency core-cooling system. Key features of a nuclear reactor system are kept in physical separation to minimize any damage due to fire or flood.

8.5.6. *Nuclear reactor accidents*

In spite of all the precautions, reactor accidents happen. For example, the consequence of significant loss of coolant is a catastrophe. In the presence of inadequate cooling, the core temperature rises rapidly. Risk exists of explosion of the pressure vessel container, which would release radioactive liquid or gas into the local environment. Slight risk also exists that the reactor core could "melt down" into a molten mass that might even burn its way through the steel container and concrete into the ground below the reactor building. It would thereby become a highly dangerous source of uncontained radioactive radiation. The local heat generated by the radiation from an uncontained core would be intense and likely to increase the spread of radiation by water, steam or gas escapes. There has never been a serious reactor coolant or containment failure, nor any reactor meltdown in the UK or France.

8.5.6.1.   *Three Mile Island*

A serious accident occurred to a 956-MWe PWR reactor at Three Mile Island, Harrisburg, Pennsylvania, USA in 1979. About an hour after a reactor was shut down out of service, a serious loss of coolant occurred due to an opened relief valve but was not noticed. Auxiliary feedwater pumps failed to circulate the necessary water due to valves that had been closed to facilitate routine repairs and maintenance. Some other serious operational errors were made. The severely damaged reactor core became exposed, with radioactive leakage into the surrounding environment [19]. There was no fatality or serious injury due to the accident at Three Mile Island, either to the workers in the industry or to the public. The damaged reactor, which suffered total meltdown, has been gradually decontaminated and is now permanently closed. The other reactor is now operating again and is licensed to generate electricity until 2034. This major accident, classified as Level 5 on the International Nuclear Event Scale (INES), was not due to design errors or equipment failures but due to human error: the safety management system was flawed. The consequent bad publicity and loss of confidence in the nuclear industry played a large part in the effective curtailment of future expansion of nuclear power generation in the USA.

8.5.6.2.   *Chernobyl*

The world's worst nuclear accident, subsequently classified as Level 7, occurred at Chernobyl in the Ukraine in 1986. At that time a major part of the electricity generated in the former Soviet Union was (and still is) nuclear-based. The standard thermal reactor systems, known as RBMK 1000, rated at 1000 MWe, are boiling-water, graphite-moderated systems using enriched uranium dioxide and ordinary (light) water cooling, similar in principle to Fig. 8.4. Each reactor system consists of two identical reactors back-to-back, rated at 500 MWe. In 1986 ten RBMK 1000 units were in service at Chernobyl, with a further eight units under construction.

On 25 April 1986 preliminary actions were begun prior to an experiment to determine if a turbine-generator would still power some reactor cooling pumps while it ran down after its steam supply was removed. The test had a legitimate purpose but was unplanned, unapproved and not properly supervised. As part of the test preliminary procedure the emergency core-cooling system was disconnected for nine hours, in contravention of written safety rules. There followed a series of major operational blunders, over several hours, in which control rod settings were inaccurate, some manual fault-trips were deliberately disengaged and some of the coolant pumps became overloaded, so that the cooling water temperature rose to nearly boiling point. A computer printout of the reactor performance that showed the need for immediate manual shutdown, because there were far too few control rods inserted into the reactor, was ignored. The automatic safety system that would have tripped (i.e. shut down) the reactor had been overridden by the operators to facilitate the proposed test. In effect, vital safety procedures were ignored because the experimentalists were under time pressure to complete their work.

On 26 April 1986, just 24 hours after the test procedures began, control rods were withdrawn and the coolant flow reduced to increase the reactor power. There was a rapid, exponential rise of reactor power and fuel temperature, generating an enormous steam pressure in the coolant circuit. An alert shift foreman ordered an emergency shutdown but was just too late. The control rods could not be fully lowered in time and the uranium fuel channel ruptured. Two steam explosions occurred, like the bursting of a steam boiler, and blew up the pressure vessel and reactor core and completely destroyed the reactor hall. Burning radioactive debris was thrown into the air and fires were started. There was a massive escape of highly radioactive gas and steam that contaminated the ground area up to about 20 miles around the site. The local city of Pripyat, which had a population of 116,000 people, was totally evacuated and remains so due to site radiation contamination.

The ancient and large city of Kiev, 80 miles away, escaped major radioactive contamination because the prevailing winds did not blow for several days [20].

The accident destroyed the Chernobyl-4 reactor and killed 31 people, including 28 from radiation exposure. A further 209 on site were treated for acute radiation poisoning, all of whom recovered. Nobody off-site suffered from acute radiation effects. However, large areas of Belarus, Ukraine, Russia and beyond were contaminated in varying degrees. The number of people since affected by the radioactive fallout is highly contentious and no figure can be regarded as definitive. The Chernobyl accident is the only case in the history of commercial nuclear power where radiation-related fatalities have occurred. There have, however, been fatalities in military and research reactor contexts [21].

Several organizations have reported on the impacts of the Chernobyl accident, but all have found difficulties in assessing the significance of what they have observed because of the paucity of reliable information on public health matters prior to 1986. In 1989 the World Health Organization (WHO) first raised concerns that local medical scientists had incorrectly attributed various biological and health effects to radiation exposure.

An International Atomic Energy Agency (IAEA) study involving more than 200 experts from 22 countries and published in 1991 was more substantial. In the absence of pre-1986 data it used a control population to compare those exposed to radiation. They found significant health disorders in both control and exposed groups, but at that stage none was radiation-related.

Studies in the Ukraine, Russia and Belarus since have been based on national registers of over one million people possibly affected by radiation. These have confirmed a rising incidence of thyroid cancer among exposed children. Late in 1995, the WHO linked nearly 700 cases of thyroid cancer among children and adolescents to the Chernobyl accident, among these some ten deaths are attributed to radiation from it [22].

So far no increase in leukaemia is discernible, but this is expected to be evident in the next few years along with a greater, though not statistically discernible, incidence of other cancers. There has been no increase attributable to Chernobyl in congenital abnormalities, adverse pregnancy outcomes or any other radiation-induced disease in the general population, either in the contaminated areas or further afield [21].

Chernobyl was a horror story come to life. It is significant to note, however, that it was not the machines that failed but the operators and designers. With better training, better plant management and strict insistence on safety procedures, it would not have happened. The nuclear industry in Britain, which is particularly well organized and well run, insist that it could not happen in the UK. Most of the electricity generated in the various countries that made up the Soviet Union is still derived from nuclear power and there are no plans to change the situation. The stricken reactor at Chernobyl is now entombed in reinforced concrete, still at its original site.

Suspicions about the safety of early Soviet reactor designs caused much misgiving amongst engineers in the West. Since the Chernobyl accident, the safety of all Russian-designed reactors has improved vastly, due largely to the development of a culture of safety encouraged by increased collaboration between East and West, and substantial investment in improving the reactors.

Since 1989 over 1,000 nuclear engineers from the former Soviet Union have visited Western nuclear power plants and there have been many reciprocal visits. Over 50 twinning arrangements between East and West nuclear plants have been put in place. Most of this has been under the auspices of the Association of Nuclear Operators, a body formed in 1989, which links 130 operators of nuclear power plants in more than 30 countries [21, 22].

8.5.6.3. *Fukushima Daiichi* [23–32]

The nuclear electricity generating station near Okuma, in Fukushima province, Japan, consisted of six boiling water reactors with a total rating of 4.7 GW.

On 11 March 2011, two of the six reactors were in cold shutdown for maintenance, but still fuelled, while another reactor had been defuelled. Three reactors were operating online.

The plant was struck by the shock from the magnitude 9.0 earthquake in the Pacific Ocean off the coast of the Tohoku region. The epicentre of the event, the most severe Japanese earthquake in recorded history, was located approximately 140 km to the north-east of Fukushima Daiichi. Seismic instrumentation automatically shut down all the plant's reactors within seconds. The quake destroyed the local power grid but power for the cooling circulation system was restored in seconds as the 12 emergency diesel generators automatically kicked in [23].

Approximately 40 minutes later, several tsunamis, one of which was estimated to be 14 metres high, overwhelmed the man-made harbour and the 5.7-metre sea wall designed specifically for tsunami defence, and flooded the power plant and entire surrounding countryside of a 670 km stretch of the Tohoku coastline, sweeping up to 10 km inland and causing damage along the entire east coast of Honshu island [24].

The flooding caused 11 of the 12 generators to fail; the survival of the 12th averted serious damage to reactors 5 and 6. Even without power, reactor 1 could have survived had there not been a bizarre misfortune: the isolation condenser (IC) system that operates only by convection and gravity had, in accordance with correct safety procedure, been turned off after the earthquake but before the tsunami hit because it was cooling the reactor too quickly (IC units in reactors 2 and 3 were operating and provided the cooling needed). Now, when the IC was needed again, there was no power to reopen its valves and, although workers eventually managed to open the valves manually, it is believed that coolant levels had fallen too far.

Restoration of power had further established contingencies in place: 11 mobile power supply trucks were immediately dispatched by the Tokyo Electric Power Company (TEPCO), but their 250-km journey was severely delayed by earthquake and tsunami damage to the roads and by traffic congestion caused by panic. The first trucks would take nearly eight hours to arrive.

Flooding had also knocked out all instrumentation, and consequently the water level in the pressure vessel could not be monitored. Workers collected car batteries and carried them across the tsunami-devastated site to the control room, connecting sufficient power to the panels, but it was now nearly six hours since power was lost. The fuel assembly was without coolant and melting, and the instruments could not accurately measure the situation, which was that the pressure vessel needed to be vented. When power trucks arrived, site damage meant 200 metres of cable had to be laid by hand, taking further hours until it was connected to an undamaged power control panel. The single fire truck not destroyed or stranded by the tsunami now pumped at first fresh water, then sea water, into the pressure vessel.

Meanwhile, another team of workers had been attempting to access the reactor building to vent the pressure vessel, first manually, then via control panels (once power was restored and the area evacuated of civilians), but they were driven back by the reading from their radiation monitors. Venting was eventually achieved using a portable air compressor and the pressure in the reactor vessel dropped.

All these heroic attempts to implement the layers of contingencies and improvise solutions on a tsunami-devastated site were thus successful, but came too late.

Hydrogen gas produced by the melted fuel assembly had leaked through the vessel's gaskets, seals and vent lines and collected at the top of the reactor building. Twenty-three hours after the earthquake struck, the hydrogen ignited. This destroyed the roof of the reactor 1 building, severed the power cable and fire hoses and covered the site with more debris, which would hamper work to stabilize the other reactors. Emergency cooling systems functioned for several days at reactors 2 and 3, but eventually failed and also suffered hydrogen explosions. Post-accident investigators believe that if reactor 1 had been saved, the others would also have survived [23, 25].

It is important to also note that other nuclear reactor plants on the Tohoku coast, including Fukushima Daini just 12 km away, were also flooded and suffered generator failures, but all managed to stabilize their reactors with emergency coolant procedures before serious damage was done. Many fossil fuel power plants, refineries and storage facilities were also damaged.

8.5.6.3.1. Consequences

The earthquake and tsunami of 11 March 2011 claimed lives to the approximate order of 20,000 (reports vary because many thousands remain "missing"). In contrast, the casualties at Fukushima Daiichi were two workers on the site killed by the earthquake and the tsunami, but none by the subsequent reactor problems.

Radiation doses for workers at the plant for March 2011 and the 14-month period thereafter, which included the investigation, salvage and clear-up of the site, have been published by TEPCO [26]. These figures show that no single worker received any dose above 200 mSv, even at the height of the crisis, and that 95% of workers reported to have received less than a 50 mSv dose for the month of March 2011. The ICRP-recommended dose limit during emergency conditions, which seeks to prevent the development of catastrophic conditions, is unrestricted if the workers are volunteers, and states that doses lower than 1000 mSv should avoid serious deterministic health effects [27].

8.5.6.3.2. Radioactive releases

Quantities of radioisotopes from fission by-products presumably escaped in particle form with the leaking hydrogen and steam, since the containment itself was not breached and yet the explosion spread isotopes into the surrounding countryside.

The release contained short-lived isotopes of iodine and tellurium, but most significantly isotopes Cs-134 and Cs-137. Fission by-product caesium-137 can reside in the top layers of soil, has a half-life of 30 years and a biological half-life (if ingested) of $(90 +/- 20)$ days, which can be reduced with treatment. It may be absorbed by plants and hence endanger the food chain, so removal is recommended. Contaminated soil is treated by removal of a few centimetres of topsoil, by deep-ploughing or adding potassium fertilizer to the soil to reduce uptake.

A range of Japanese government agencies have set up extensive and continuous monitoring procedures to test the condition of soil, air and water in the Fukushima prefecture region; their findings are freely available online [28].

Precautionary restrictions on the movement of agriculture and fishery products from the region have been taken as appropriate when monitoring reveals any contamination.

A 20-km area near the plant was evacuated of local residents, and a further area to the north-west is being monitored for an expected contribution of 1–20 mSv/year and is under requested evacuation [28]. (As a point of reference, 20 smV is the acceptable annual radiation dose for a uranium miner.) Evacuation area restrictions were gradually reduced during 2012 [26].

8.5.6.3.3.   Severity of the incident

In April 2011, the Japanese government declared that the Fukushima Daiichi crisis rated a 7 on the IAEA's INES. Level 7 is the highest rating on the scale, only previously assigned to the Chernobyl disaster.

This rating is highly questionable, both in terms of the estimated total release of radioactive material and in terms of the number of "resulting deaths from radiation" [29]. Indeed, reading the full definitions of the various levels, events at Fukushima Daiichi may be adequately described by the description of Level 4, or multiple Level 4 events [30]. The INES claims that each level is designed to represent an event "ten times more severe than the previous". On that criterion alone, when compared to the Chernobyl disaster, the radiation release from this event [25] has to be graded Level 6 or lower, and if graded according to its effects on the surrounding countryside, it must be several levels lower still. Attempts to grade this incident call into question whether the INES scale (which was created supposedly to allay public fears about genuinely minor incidents) serves any useful purpose at all.

8.5.6.3.4.   Public perception

The nuclear power industry suffers greatly from an exaggeratedly-negative public perception, fuelled primarily by anti-nuclear activist pressure groups. Most of Japan's nuclear reactors successfully survived what is now called "The Great East Japan Earthquake" with only minor damage. In the face of these natural disasters of unprecedented proportions, they survived thanks to the layers of backups and contingencies in place (many of which had been lessons learned from the Chernobyl disaster). The radiation released as a result of the one reactor that failed was brief in time-span and its effects are being continuously monitored on land, at sea and from the air.

The reactor failures themselves caused no deaths to workers on-site, and the risk of serious health problems is extremely small. The lessons learned from the failures that did occur will be invaluable to further strengthening the safety of other plants in the future.

The earthquake and tsunami of March 2011 also damaged many fossil fuel power plants and destroyed two oil refineries (at Ichihara and Sendai), whose fires burned for many days, releasing millions of tonnes of untreated hydrocarbon-burning by-products into the environment. The Fujinuma irrigation dam in Sukagawa city also failed, killing eight people [31].

The nuclear power industry remains one of the safest energy options available, both in terms of safety to the workers who provide it and in terms of safety to the general public [32].

8.6.   Nuclear Waste [33, 34]

8.6.1.   *Sources of waste*

In any section of the chemical and mining industries, the mining, processing and use of the raw material creates waste. In the nuclear industry, the collection, transportation, processing, management, storage and disposal of nuclear waste constitutes the most challenging problem of all. Nuclear waste is radioactive and is obtained by a number of different activities:

(a) uranium mining and purification,
(b) fabrication of reactor fuel,
(c) nuclear reactor operation,
(d) nuclear fuel reprocessing,
(e) nuclear reactor decommissioning.

(a) Uranium mining and purification

Large quantities of waste material are produced at uranium mines. The mined ore is brought to the surface and finely ground or milled to concentrate the uranium. This facilitates the release of radon, a radioactive gas that is part of the decay chain of uranium. The residues from the milling process, known as tailings, still contain 80% of the original radioactivity in the ore. The best treatment of the tailings would be to return them to their original underground mines, but this is not economical. In practice the wastes are covered with earth and vegetation, like coal-mining waste. Uranium-mining tailings are many times smaller than coal-mining waste for the same amount of generated electricity. There is no uranium mining in the UK.

(b) Fabrication of reactor fuel

Purification of the uranium concentrate and its enrichment to fabricate reactor fuel rods and pellets also produces waste. Enrichment plant waste usually consists of liquid effluent containing low levels of decay products.

(c) Nuclear reactor operation

The operation of a nuclear reactor results in radioactive fission products, as described in Section 8.2. These are all contained in the fuel rods, which are themselves contained within the reactor containment system during reactor operation. Some waste is produced by reactions between the neutrons and the coolant and the containment system but these remain inside the reactor until decommissioning. A small quantity of radioactive gaseous effluent is sometimes released, but this forms only about 1% of the local natural background radiation.

(d) Nuclear fuel reprocessing

The serious hazards of nuclear waste begin when spent fuel rods are removed from the reactor for disposal or, more likely, for reprocessing. Spent reactor fuel is so radioactive that it has to be handled behind thick shielding or water. Waste reprocessing involves dismantling the fuel cans and dissolving the contents in nitric acid to retrieve the reusable uranium and plutonium. There remains a liquid waste containing long-lived radioactive fission products such as strontium-90, caesium-137 and tritium, classed as high-level waste, which is concentrated and cool-stored. Various other wastes include the metal fuel cans and the cooling water or gas. These are classed as intermediate level and are also stored.

(e) Nuclear reactor decommissioning

The decommissioning of a nuclear reactor involves either "mothballing" or "entombment", both of which leave the radioactive hazard undisturbed but in need of constant monitoring. Dismantling a reactor, however, would involve handling thousands of tons of highly radioactive material and would be undertaken only after a delay of (say) 50 years to allow the bulk of the radioactivity to decay.

8.6.2.   Waste disposal

The principal aims of nuclear waste disposal are to dilute the radioactivity to a harmless level and disperse it to the environment or to isolate the radioactive materials under containment until the radioactivity has decreased to insignificance. For convenience, the waste materials are classified with regard to the level and longevity of the radioactivity as low-level, intermediate-level or high-level. Very low-level liquid and gaseous wastes are disposed of directly into the environment. Some low-level solid wastes such as soiled laboratory clothing that have only slight contamination are packaged into metal drums and buried in landfill sites. Wastes of a slightly higher radioactivity level, but still within the low-level classification, include wastes from hospitals and laboratories. These are encased in concrete and enclosed in steel drums. Such waste from the UK used to be tipped into the Atlantic Ocean, about 4 km ($2^{1}/_{2}$ miles) deep, at an internationally agreed location, about 500 miles off Lands End. The low-level group consists of 90% by volume of all the nuclear waste but at only 1% of the activity. Volume can be decreased by compacting or incineration.

Intermediate-level wastes include reactor fuel cans, fuel fragments, sludges and filters from liquid and gaseous effluent, plutonium-contaminated equipment, fuel transport flasks and reactor components. Some of this waste can be treated to extract and concentrate the long-lived radioactive constituents, discharging the rest as low-level waste. Much of the waste is encapsulated into blocks of concrete, resin or bitumen for longer-term, land-based storage and eventual disposal. Storage is usually made deep underground, at about 300 metres depth, at suitably safe geological sites. The intermediate level group consists of 7% by volume of the total waste, at an activity of 4%.

High-level nuclear waste is the most dangerous because of the high heat generation and the high level of radioactivity. Concrete-lined, stainless-steel water tanks known as cooling ponds are used to store solid waste for decades. After cooling, the fuel and weapons material can be retrieved for reprocessing. A diagram of a cooling pond is given in Fig. 8.9 [33].

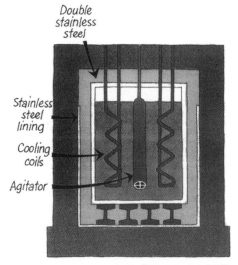

Fig. 8.9. Storage tank for high-level liquid wastes [23] (reproduced by permission of AEA Technology plc).

High-level liquid waste is cooled for several years and then solidified. In the UK and France the solidified waste is chemically incorporated into glass cylinders encased in steel cans. The radioactive solids fuse with the glass into an opaque glazed ceramic. In the USA the waste is fused with a ceramic by fluid-bed calcination. The waste containment cans are air- or water-cooled in concrete container buildings or bunkers deep underground, in salt formations or mines that prevent the intrusion of water [34]. The high-level group of waste comprises 3% by volume of the whole but at the high activity level of 95%.

The treatment, transportation and disposal of nuclear waste, especially high-level waste, is a major challenge to the nuclear industry. There have been two notable incidents involving the escape of radioactive material: at Windscale (now Sellafield), England, in 1972 and at Hanford, Washington, USA, in 1973. No long-term ill-effects seem to have accrued from either incident. The overall safety record of the UK nuclear industry is excellent.

8.6.3.   *Terrorist action*

The security of all power generation facilities is important for national security. Uranium and plutonium both require high levels of enrichment to reach weapons grade. Although no weapons-grade material is present at nuclear power facilities, precautions to protect plants are taken very seriously and are constantly reviewed by governments. Nuclear sceptics invariably claim that precautions are inadequate.

8.7.   Nuclear-Powered Electricity Generation

It takes about ten years to build a power station in the UK, from the initial design to the commissioning and running stage. Economic assessment always faces the anomaly of present and immediately ongoing building costs versus the returns on electricity sales starting ten years hence. The cost of electricity in ten years' time will depend on a host of engineering, social and political factors that are difficult to forecast accurately. Nevertheless, advance planning has to be undertaken based on assumptions and best estimates of the future, as discussed in Chapter 2. A key issue is the future demand for electricity. Obtaining unbiased economic information with regard to the nuclear power industry is particularly difficult.

A nuclear power station is relatively expensive to build compared with a fossil fuel station, because of the costs of the reactor, fuel handling and elaborate safety features. Some estimates place the building costs of a nuclear plant at twice the level of a coal-fired plant of the same rating. The financial investment is large, pointing to the need for realistic rates of interest on capital costs and on the need to avoid overrunning the construction schedules.

On the other hand, the costs of nuclear fuels are claimed to be lower than equivalent fossil fuels. With cheaper running costs, high load factors in the nuclear-generating stations and a continuing demand for electrical power, it can be seen that, overall, the cost of nuclear power could be cheaper than that of conventional fossil fuel generation.

The world consumption of nuclear fuels for the period 2000–2010 is given in Table 8.5 [15]. It can be seen that the chief users are the USA, France, Japan and Germany. World consumption increases continually due to heavy use in North America and Western Europe.

8.7.1. *Nuclear generation in the USA*

The USA expected to reduce its nuclear reliance from 20% of total electricity generation in 1999 to less than 12% in 2020. However, reductions in operating costs over the past decade have made the nuclear plants more competitive in the increasingly deregulated electricity market.

Since 1997, a wave of consolidation has occurred in the US commercial nuclear power industry through various mergers and acquisitions. The first merger occurred when the PECO Energy Company and British Energy formed a joint partnership, AmerGen, for the express purpose of buying nuclear power plants. AmerGen has purchased five nuclear power plants to date and has plans to purchase as many as 20 plants in total. AmerGen was involved in the first purchase of a US nuclear plant in its entirety in December 1999, when it bought the Clinton plant.

AmerGen and a handful of other companies are emerging as major holders of US nuclear assets. Entergy, for example, has announced plans to spend $1.7 billion over five years to build a portfolio of 12–15 nuclear power units, and Duke Energy, Constellation Energy Group and Northern States Power have also indicated interest in acquiring nuclear units. In addition, PECO and Entergy are involved in two of the largest mergers in the history of the US nuclear power industry. Unicom and PECO completed a merger in October 2000 that created the nation's largest nuclear utility. The combined company, Exelon, owns 17% of the total nuclear generation capacity, with annual revenues of $12 billion.

In another early sale, AmerGen purchased the Clinton nuclear plant for $200 million, even though it cost Illinois Power $4.25 billion to build it. In the Clinton sale, AmerGen assumed full responsibility for the decommissioning. Illinois Power ceded $98 million in decommissioning funds to AmerGen and is committed to transferring additional funds sufficient to fully fund the eventual decommissioning of the Clinton reactor.

More recently, however, prices for nuclear power assets have risen markedly. For instance, in February 2000, Entergy agreed to pay the New York Power Authority $967 million for Indian Point 3 and Fitzpatrick, a record high for nuclear sales to that date. The higher prices paid for nuclear assets in recent sales may reflect not only the quality of the assets sold but also an improved environment for nuclear power in the USA [12].

In 2011 the net electrical output from the 104 nuclear plants was 100740 MW, which was about 20% of the total generation.

8.7.2.   *Nuclear generation in the UK* [35]

In 1995, the UK government announced that it would privatize its more modern nuclear stations while retaining the ownership of older stations. In 1996, the more modern stations were privatized and British Energy became the holding company of Nuclear Electric and Scottish Nuclear, which merged in 1998 to form British Energy Generation, the nation's largest private nuclear generator and the world's first wholly privatized nuclear utility. They rebranded in July 2011 to become EDF Energy Nuclear Generation Ltd and operate eight power stations in the UK (as well as several in the USA through its AmerGen subsidiary, which is jointly owned with PECO). Each station consists of two advanced gas-cooled reactors, except Sizewell B, which is a modern pressurized water reactor. Nuclear power stations were not privatized simultaneously with non-nuclear stations. In October 2013, a new plant, to be named Hinkley Point C, was given approval. Because of limited domestic coal and gas reserves, new construction is under discussion, at least to maintain nuclear's market share, as older nuclear plants are retired. Of the UK's 33 reactors, 26 are of the old Magnox design. Six of the Magnox reactors are being decommissioned, as well as the Dounreay prototype fast reactor. The remaining Magnox plants are run by the state-owned British Nuclear Fuels. British Nuclear Fuels operates the Sellafield reprocessing plant and is one of only two companies in the world that provide reprocessing and recycling technologies. Decommissioning of the older Magnox buildings and plant began in 2003.

In 2011, the total electrical energy supplied in the UK was 325.86 TWh, of which 62.7 TWh, 19.24 %, was from nuclear stations [36].

8.8.   Nuclear Fusion

8.8.1.   *Basic theory* [37]

Nuclear fusion is the term describing the union or fusion of two or more light atomic nuclei. If the positive electrostatic repulsive forces of two nuclei can be overcome to permit fusion, large amounts of energy are released.

Many different nuclear fusion reactions occur in the sun and other stars. The principal process taking place in low-mass main sequence stars such as the sun is a series of fusion events called the "proton–proton" cycle (or chain), in which hydrogen nuclei (protons) fuse to produce helium-4 via deuterium and helium-3, or via other less prevalent reactions. A temperature of 10^7 K is required for proton fusion: all other fusion processes require higher temperatures. The combined equation for the cycle is shown in Eq. (8.12).

$$4\,^1\mathrm{H} \rightarrow \,^4\mathrm{He} + 2\mathrm{e}^+ + 2\nu + 27.6\,\mathrm{MeV} \tag{8.12}$$

It would be ideal for energy production if such a process could be used on a small scale on earth, but the proton–proton cycle relies on a weak interaction taking

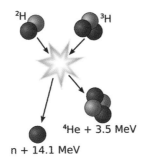

Fig. 8.10.   Illustration of the deuterium–tritium fusion.

place, in which a proton decays to a neutron to allow deuterium to be produced for the next step in the chain. The half-life of such a decay is 10^9 years, which is not a problem in the sun's core where protons are so abundant — indeed, the sun would have exhausted its hydrogen long ago were this delay not at work — but in an earth-based reactor only tiny quantities of matter are used and the delay is prohibitive.

The first choice for a fusion reaction to release energy for large-scale electricity generation should be the one with the lowest temperature threshold, but since proton–proton is not an option the best candidate reactions involve the light isotopes of hydrogen, being deuterium and tritium.

The deuterium–tritium (D–T) reaction (Fig. 8.10) produces helium-4 plus energy and a neutron.

$$^2D + {}^3T \rightarrow {}^4He + {}^1n + 17.6\,\text{MeV} \tag{8.13}$$

The helium-4 particles produced in the D–T reaction are ionized and possess momentum, which means that they are alpha particles. Indeed, they contribute to the self-heating of the D–T plasma. Once they lose their energy they pick up free electrons and revert to neutral helium-4.

The energy carried by the neutrons is converted to heat by slowing down the neutrons in a lithium blanket surrounding the plasma (Fig. 8.11), which is then transferred to a steam boiler [37]. The neutrons react with the lithium, producing ^{4}He and tritium, which is recovered and fed back into the reactor as fuel.

$$^1n + {}^6Li \rightarrow {}^4He + {}^3T + 4.8\,\text{MeV} \tag{8.14}$$

In effect the tritium produced is recycled back to be used in a subsequent reaction. The net equation of the D–T fusion is obtained by combining Eqs. (8.13) and (8.14) to obtain

$$^2D + {}^6Li \rightarrow 2\,{}^4He + 22.4\,\text{MeV} \tag{8.15}$$

Equation (8.15) shows that deuterium and lithium are the basic fuels for D–T fusion. They are both naturally occurring, abundant and stable (i.e. non-radioactive).

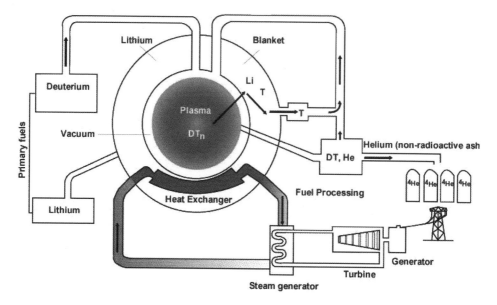

Fig. 8.11. Nuclear fusion heat production (reproduced by permission of the JET http://www.
jet.efda.org/documents/presentations/borba.pdf). Fuel cycle integrated in plant: D and Li as basic
fuel, T bred inplant by neutron Li reactions.

8.8.2. *Nuclear fusion materials*

8.8.2.1. *Supply of deuterium*

Deuterium ($^2_1$H or D) is a stable isotope of hydrogen that occurs naturally in sea
water and has atomic mass number 2, as discussed in Section 8.1. Deuterium can
fuse with deuterium, but the energy yield is better from a D–T reaction, illustrated
in Fig. 8.10.

Deuterium constitutes $\frac{1}{7000}$ of the hydrogen in seawater. Complete fusion of the
deuterium in 1 cubic metre of seawater would yield 12×10^{12} joules of energy, which
is equivalent to about 2,000 barrels of oil. Rather more startling is the fact that
each cubic kilometre of ocean contains enough deuterium that, if fully converted to
heat by nuclear fusion, would be roughly energy equivalent to the earth's known
oil reserves. Also, compared with the cost of coal, natural gas or oil, deuterium is
virtually free.

8.8.2.2. *Supply of lithium*

Lithium is a light metal, also present in seawater. Its main source, however, is
from rocks that are available by mining from secure, land-based sources. Identified
resources total 4 million tons in the USA and approximately 29 million tons in other
countries. In the year 2010 the biggest producer countries were Chile and Australia
(Table 8.6) [38].

Table 8.6. World mine production and reserves of lithium [38].

	Mine production (tonnes)		Reserves (tonnes)[a]
	2009	2010	
USA	W	W	38,000
Argentina	2,220	2,900	850,000
Australia	6,280	8,500	580,000
Brazil	160	180	64,000
Canada	310	—	—
Chile	5,620	8,800	7,500,000
China	3,760	4,500	3,500,000
Portugal	—	—	10,000
Zimbabwe	400	470	23,000
World Total	18,800[b]	25,300[b]	13,000,000

a — imports — exports + adjustments; b: excludes US production. W — information withheld (company proprietary data).

The USA remains (2013) the leading consumer of lithium minerals and compounds and the leading producer of value-added lithium materials. Because only one company produces lithium compounds from domestic resources, reported production and value of production data cannot be published. An estimation of value for the lithium mineral compounds produced in the USA is extremely difficult because of the many compounds used in a wide variety of end uses and the great variability of the prices for the different compounds. Some details of the production and reserves of lithium are given in Table 8.6 [38]. The present consumption is used mainly for industrial lubricants and ceramics. Major automobile companies are developing lithium batteries for hybrid electric vehicles.

8.8.2.3. *Supply of tritium*

Tritium ($^3_1$H or T) is a radioactive isotope of hydrogen with the mass number 3 and half-life $T_{1/2} = 12.3$ years. It does not occur naturally except when produced by a high-energy interaction, such as a cosmic ray collision with atmospheric nitrogen. Tritium may be produced by bombarding lithium with neutrons or it may be collected from heavy water reactors, where small amounts are produced when deuterium captures neutrons.

Tritium is a weak beta-emitter that is much used in nuclear weapons. Because it has a short half-life of 12.3 years, the weapons in store in nuclear arsenals require to be periodically replenished.

It is a vital component of the fuel system of a nuclear fusion reactor. A D–T facility burns tritium at a rate of 55.8 kg per 1000 MW of fusion power per year [39, 40]. The only credible supply of tritium for D–T fusion development is from the Canadian CANDU fission reactors, which produce a few kg of tritium/year and at great expense. This amount is far below the level necessary to support a programme of nuclear fusion electricity generation.

Any type of fusion reactor will require several months' supply of tritium just to bring the breeding process on stream [41]. The production of tritium from lithium, shown in Eq. (8.14), in fission reactors involves the use of highly energetic neutrons that can render other materials radioactive. Currently (2013) much research is being conducted to determine how various materials cope with long-term neutron bombardment to determine which materials would be best for fusion power plant reaction chambers.

The development of D–T fusion must be carefully planned worldwide, taking into account the available tritium. Experiments without breeding should be of low power. Techniques and programmes that breed significant quantities of tritium are needed [41]. The ITER programme, described in Section 8.8.4.2 below, must incorporate tritium breeding.

8.8.3.   *Nuclear fusion reactors*

In order to generate electricity from nuclear fusion, three scientific and engineering challenges have to be overcome [37]:

(a) to create and heat a plasma requires temperatures of the order 10^8 K,
(b) to hold enough plasma away from the container walls for long enough to permit abundant reactions to occur, and
(c) to design a practical, safe and economic fusion reactor.

8.8.3.1.   *Nuclear plasma properties*

The nuclear fusion of hydrogen into helium takes place naturally in the sun and other stars, at core temperatures of the order 10^7 K. To fuse deuterium and tritium into helium, displayed in Eq. (8.12), requires temperatures of the order 10^8 K, which is ten times the sun's core temperature. This has to be maintained for a minimum time period t_m of the order of 1–2 sec.

In the most likely scenario for a fusion power plant, a D–T mixture is admitted to the evacuated reactor chamber and there ionized and heated to thermonuclear temperatures. The fuel is held away from the chamber walls by magnetic forces long enough for a useful number of reactions to take place. The charged helium nuclei that are formed give up energy of motion by colliding with newly-injected cold fuel atoms, which are then ionized and heated, thus sustaining the fusion reaction. The neutrons, having no charge, move in straight lines through the thin walls of the vacuum chamber with little loss of energy.

The neutrons and their 14 MeV of energy are absorbed in a "blanket" containing lithium that surrounds the fusion chamber. The neutrons' energy of motion is given up through many collisions with lithium nuclei, thus creating heat that is removed by a heat exchanger, which conveys it to a conventional steam electric plant [33].

For a continuous steady-state operation the central plasma ion density in nuclei/m^3 must be such as to satisfy a relationship

$$[t_m][\text{nuclei/m}^3] > 2\text{--}3 \times 10^{20} \qquad (8.16)$$

If Eq. (8.16) is expressed in terms of atomic nuclei per cubic centimetre,

$$[t_m][\text{nuclei/cm}^3] > 2\text{--}3 \times 10^{14} \qquad (8.17)$$

For a realistic minimum plasma pulse time of 1 sec, the necessary density of the D–T gaseous mixture must be of the order 10^{14} atoms/cm^2, which is about one-millionth the density of a solid material.

8.8.3.2. *Terrestrial fusion*

In the core of the sun and other stars, intense pressure keeps the plasma density high enough to sustain fusion. For earthbound fusion, the strategy is to introduce a small number of ions into a restricted volume and excite them to extreme temperatures. Several strategies have been researched to achieve ignition, but by far the most successful to date has been the tokamak.

Originating in the former Soviet Union, the tokamak is a toroidal (ring-shaped) vacuum chamber (Fig. 8.12) [37]. A toroidal magnetic field is created around the major axis by current coils wound around the sectional circumference. The magnetic field confines the plasma, both constricting it and holding it away from the container walls. A second, poloidal magnetic field is produced by the moving plasma of charged particles itself, with the result that the plasma ions are "pinched" away from the walls and follow a helical path around the torus (Fig. 8.13) [37].

Stellarator fusion projects also use magnetic confinement, but do not require a toroidal current.

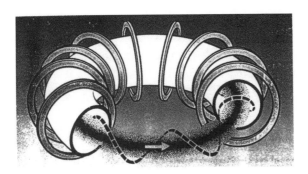

Fig. 8.12.   Detail of the tokamak helical magnetic field [26] (reproduced by permission of UK AEA).

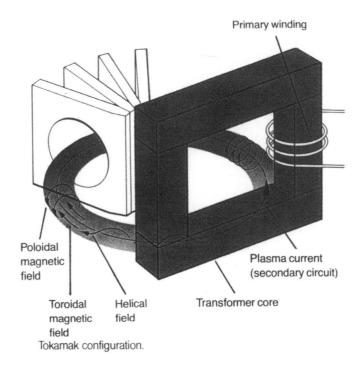

Poloidal magnetic field

Plasma current (secondary circuit)

Toroidal magnetic field

Helical field

Transformer core

Tokamak configuration.

Fig. 8.13.   Creation of a toroidal magnetic field [26] (reproduced by permission of UK AEA).

Another strategy, inertial confinement fusion, uses small pellets, about 1 mm in diameter, containing about 10 mg of D–T fuel. These are irradiated by high-energy lasers, ion beams or laser-produced X-rays to rapidly heat the fuel, causing implosion and core compression to 10^8 K. The intention is to create a reaction time so fast, of the order of one nanosecond, that the inertia of the fuel itself provides the necessary confinement.

8.8.4.   *Fusion reactor research* [42]

It is seen in Eq. (8.15) that each fusion reaction produces 22.4 MeV of heat energy. There must be sufficient fusion events per second to sustain the plasma and an excess of reactions satisfying Eq. (8.17) if there is to be net heat energy increase.

8.8.4.1.   *JET*

In Europe the research is based on the Joint European Torus (JET) project, formerly located in England.

JET successfully achieved a 10^8 K plasma in 1986. In 1997, the project set a world record by generating 16 MW of D–T fusion power at a fusion energy gain (Q) factor of 0.7, and has sustained steady fusion power of 4 MW for 4 seconds.

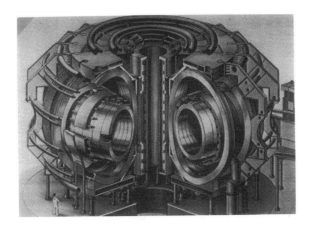

Fig. 8.14. Scale model of the vacuum chamber of a JET fusion experiment [26] (reproduced by permission of UK AEA).

JET has reached the end of its primary mission and is regarded as an unqualified success, which has paved the way for the next, far larger experimental tokamak called ITER (see Section 8.8.4.2), but JET will continue to operate as a fusion research facility preparing for ITER operation by collecting data, experimenting with materials, training scientists and gaining experience handling fuel. A scale model of the vacuum chamber of a JET fusion experiment is shown in Fig. 8.14 [37].

The plasma chamber is evacuated to the order of 5×10^{-6} Pa to prevent atmospheric gases interfering with the desired reactions.

The wall of the plasma chamber is covered in carbon fibre composite materials that can withstand high temperatures (up to 1800 K where the plasma touches the wall) and neutron bombardment. Strong magnetic fields of the order of 3.5 Tesla are produced along the toroidal radius by the rapid transfer of energy stored in large flywheels spun up to 700 rpm.

A second, more compact tokamak (MAST) on the same site as JET is used for diagnostics and for testing materials exposed to plasma.

8.8.4.2. *ITER*

ITER, a collaboration between the European Union, China, India, Russia, Korea, Japan and the USA, is under construction at Cadarache in the south of France.

Its plasma volume will be 830 m³, its 13 tesla toroidal field will be produced by superconducting magnets and its divertor component is being designed to tolerate 3300 K.

The vacuum region will be surrounded by a 0.5–1.0 metre thick blanket of lithium (Fig. 8.11) to capture the neutrons and exchange the heat so produced with cooling water, which could then be used to help to produce steam to drive turbines in the same manner as conventional power stations. Tritium purged from the lithium blanket will be recovered for re-use. ITER's "first plasma" is projected for 2019 and first D–T operation for 2027. Ultimately its goal is to sustain reactions for up to one minute and reach a fusion energy gain factor of Q = 10.

Concurrent with ITER, the International Fusion Material Irradiation Facility (IFMIF) is planned to test the resilience of materials when subjected to sustained neutron bombardment. Its first tests are due to begin in 2017.

ITER and IFMIF will pave the way for the first full-scale fusion power station (DEMO) and the first commercial fusion power station (PROTO). Current projections for DEMO are that construction will begin in 2024 and contributions to the power grid will start by 2040, whilst projections place PROTO post-2050.

8.8.5.   *Advantages of nuclear fusion as an energy option*

As an energy option, nuclear fusion has a compelling set of advantages over conventional fuels, renewable options and over nuclear fission.

8.8.5.1.   *Fuel supply advantages*

Since only tiny amounts of fuel are needed, expensive and carbon-producing transport of mass qualities of raw materials is unnecessary. For comparison, a 1 GW coal-fired plant requires 2.7 million tonnes of fuel per year but a comparable fusion plant will require only 250 kg of fuel per year.

Of the fuel materials themselves, the supply of deuterium is effectively infinite since it may be extracted from sea water, as discussed in Section 8.8.2.1.

The supply of lithium is discussed in Section 8.8.2.2, above. Opinion is divided as to whether lithium supplies could cope with an explosion in demand for battery-powered vehicles, but the amounts of lithium that will be needed per person per unit of fusion energy generated is negligible.

A fully-operating fusion power plant will of course breed its own tritium via the lithium blanket, but some amount will still be required to start the reactions that begin the breeding process.

8.8.5.2.   *Zero emissions/green energy*

A certain "carbon cost" is incurred in the building and decommissioning of any power plant due to building materials, transportation etc., but the energy-producing process itself produces no carbon dioxide, carbon monoxide or any other greenhouse gases. The only by-product of the process is a tiny amount of the inert gas helium. Low-carbon energy is extremely attractive to policy-makers deeply concerned about greenhouse gas production levels in an increasingly industrializing world.

8.8.5.3. *Zero high-level nuclear waste*

The relatively low volumes of high-level radioactive waste produced by fission power stations are discussed in Section 8.6.2 above. Nevertheless, the public perception of nuclear energy is that such waste is a consequence so highly undesirable as to disqualify it, regardless of the fact that the alternatives are far more polluting and far less "green" in terms of carbon emissions. Fusion carries no such burden because the fusion process produces no high-level nuclear waste, nor any high-activity isotopes. Reactor components when decommissioned will be treated as medium and low-level waste and stored on-site for the few years it will take any short half-life isotopes present to decay to natural background levels. The total amount of waste expected to be produced during the 20-year operational lifetime of ITER could be contained in a single 10-metre cube.

8.8.5.4. *Zero risk of nuclear catastrophe*

Public perception also associates nuclear energy with an unacceptable risk of catastrophic failure. A fusion power plant carries none of the risk associated with nuclear fission because if for any reason the plasma chamber was breached, the chamber would be flooded with air and reactions would not be able to take place; all energy-producing events would cease instantly. If power loss caused the magnetic fields to fail, ions would no longer be accelerated so would not be able to react. If the plasma were to come in contact with the chamber wall, the mass of heated plasma is so low that it does not have sufficient heat capacity to damage the chamber (indeed during normal operation, some plasma does touch the wall near the helium and unused fuel extraction "divertor" with no adverse consequences). This would still be the case if fusion continued but the water supply to the heat exchanger failed. If tritium were somehow to escape into the environment, the amounts used are too small and its half-life too low to contribute more than a tiny addition to natural background radiation.

8.8.5.5. *Modest land use and plentiful site options*

A fusion power plant will occupy approximately the same amount of land space as a conventional power station. It will not require large areas of land use like hydro-electric schemes nor a network of towers like wind energy projects. Locations near a water supply are required, as is true for conventional and nuclear fission plants, but the absence of toxic emissions and zero risk of radioactive leakage mean locations may be nearer to population centres rather than requiring remote locations. Fusion plants will not be constrained by a need to be located near fuel sources such as coal fields or sustainable forests for transport economy. Overall, site options for fusion plants should therefore be plentiful.

8.8.5.6.   *No intermittency of supply*

The supply of energy from a fusion power station will have no dependency on weather conditions, unlike wind, solar and wave energy options.

8.9.   Problems and Review Questions

8.1. What is the thermal energy equivalent, in tons of coal, to the complete fissioning of 1 kg of uranium-235?

8.2. The fissioning of one atom of U-235 has an energy 200 MeV. What is the energy per fission in (i) joules and (ii) kWh?

8.3. Enriched uranium contains 2.3% of U-235 in a mixture with U-238. Assume that, in a particular fission process, 1% of the U-235 is fissioned. What mass of uranium ore has the thermal energy equivalent of 1,000 tons of coal? How does this compare with the corresponding calculation for natural uranium?

8.4. How long does it take a radioactive source of half-life $T_{\frac{1}{2}}$ to decay to (i) one tenth and (ii) one hundredth of its original value N_o?

8.5. List the features of a radioactive source that you would consider to be of interest in studying its properties.

8.6. Uranium-235 has a half-life of 710 million years. How long does it take to decay to one-tenth of its original level of activity?

8.7. The radioactive element caesium has a half-life $T_{\frac{1}{2}}$ of 30 years. What is the value of its decay constant?

8.8. For cobalt 60 the radioactive decay constant is found to have the value of 0.131. Calculate the half-life.

8.9. How long does it take a radioactive source of half-life $T_{\frac{1}{2}}$ to decay to one-thousandth of its original value?

8.10. Define numerical relationships between the following radiation units: rad, rem, millirem, gray, sievert.

8.11. How much energy is absorbed by a human body in a radiation dose of 100 rads?

8.12. List various forms of natural radiation. What is the average annual dose from natural sources in the UK?

8.13. List various forms of man-made radiation. What is the average annual dose-rate in the UK caused by the nuclear industry? What is the ICRP-recommended maximum radiation dose, from all sources?

8.14. Describe, using a diagram, the principle of operation of a nuclear reactor for use in electricity generation. In particular, describe the functions of the core moderator, control gear and coolant.

8.15. In Great Britain and in the USA the use of nuclear energy for electricity generation stalled for several years in the latter part of the 20th century. Why was this so?

8.16. The continued long-term use of nuclear fission energy for electricity generation will require the use of fast breeder reactors or plutonium reactors. Why is this so?

8.17. In the UK the fossil fuel part (i.e. about 90%) of the electricity generation industry has smoothly passed from public to private (public limited company) ownership. There still remains difficulty about privatization of the nuclear generation industry. Why is this so?

8.18. Describe the three main classes of nuclear waste and how these are treated or disposed of.

8.19. Discuss the difference between nuclear fission and nuclear fusion.

8.20. Explain why the deuterium–lithium nuclear fusion process takes place via an intermediate reaction involving tritium.

8.21. Each nuclear fusion of deuterium and lithium releases 22.4 MeV of thermal energy. What is the equivalent energy in (i) joules and (ii) kilowatt hours?

8.22. Explain why, from the point of view of fuel supply, nuclear fusion represents such a highly appealing option.

8.23. The fission of one atom of uranium is equivalent to an energy of 200 MeV, whereas the fusion of deuterium and lithium releases 22.4 MeV. Does this mean that nuclear fission is $200/22.4 \approx 9$ times better than fusion?

8.24. List the advantages and disadvantages involving nuclear fusion as a means of generating electricity.

8.25. What are the major technical features involved in the development of nuclear fusion reactors for electricity generation?

References

[1] "CANDU Nuclear Power Technology", Atomic Energy of Canada Ltd., Chalk River, Ontario, Canada, 1981.

[2] "The Effects and Control of Radiation", United Kingdom Atomic Energy Authority (UKAEA), March 1983.

[3] "Radiation and Life", Uranium Information Centre Ltd., Melbourne, Australia, April 2000, http://www.uic.com.au/ral.htm.

[4] "Radiation and the Nuclear Fuel Cycle", Nuclear Issues Briefing Paper 17, Uranium Information Centre Ltd., Melbourne, Australia, March 2001, http://www/uic.com.au/nip17.htm.

[5] Dorf, R.C. (1978). *Energy, Resources and Policy*, Addison Wesley Publishing Co., Reading, MA.

[6] "Energy Pressurized-Water Reactor and Reactor Vessel", US Energy Information Administration, Washington, DC, August 2012, http://www.eia.gov/cheaf/nuclear/page/nuc_reactors/pwr.html.

[7] "Nuclear Power Reactors", UKAEA and Nuclear Power Company, Ltd., London, 1977.

[8]   McVeigh, J.C. (1984). *Energy around the World*, Pergamon Press, Oxford.

[9]   "Nuclear Reactors Built, Being Built or Planned 1993", Report DOE/OSTI-8200-R57, US Department of Energy, Washington, DC, June 1994.

[10] "World Nuclear Outlook", Report DOE/EIA-0436(94), Energy Information Administration, US Department of Energy, Washington, DC, December 1994.

[11] "Nuclear Power in the United Kingdom", Department of Trade and Industry, London, September 1995.

[12] "International Energy Outlook 2001", Energy Information Administration, DOE/EIA-0484, Washington, DC, March 2001.

[13] "Itam 5-Uranium", Uranium Information Centre Ltd., Melbourne, Australia, 2001.

[14] "Supply of Uranium", World Nuclear Association, London, February 2011, http://www.world-nuclear.org/info/Nuclear-Fuel-Cycle/Uranium-Resources/Supply-of-Uranium/.

[15] "BP Statistical Review of World Energy 2011", British Petroleum Company plc, London, June 2011.

[16] "Plutonium", United Kingdom Atomic Authority (UKAEA), London, 1981.

[17] "Fast Breeder Reactors", Nave, R., Georgia, February 2011. http://hyperphysics.phy-astr.gsu.edu/hbase/nucene/fasbre.html.

[18] "Fast Reactors: Potential for Power", United Kingdom Atomic Energy Authority, (UKAEA), London, 1983.

[19] Moss, T.H. and Sills, D.L. (eds.) (1981). *The Three Mile Island Nuclear Accident*, New York Academy of Sciences, New York.

[20] Mould, R.F. (1988) *Chernobyl — The Real Story*, Pergamon Press, Oxford.

[21] Chernobyl Forum Expert Group (2006). "Environmental consequences of the Chernobyl Accident and their Remediation: Twenty Years of Experience", International Atomic energy Agency, Vienna.

[22] Sweet, W. (1996). "Chernobyl's Stressful After-Effects", *IEEE Spectrum*, **33**, 27–34.

[23] Stickland, E. (2011). *24 hours at Fukushima*, IEEE Spectrum, New York, http://spectrum.ieee.org/energy/nuclear/24-hours-at-fukushima/0.

[24] "Magnitude 9.0-near the east coast of Honshu, Japan", US Geological Survey, Memphis, TN, March 2011, http://earthquake.usgs.gov/earthquakes/recenteqsww/Quakes/usc0001xgp.php#summary.

[25] "IAEA international fact finding expert mission of the Fukushima Dai-ichi NPP accident following the Great East Japan earthquake and tsunami", Report to IAEA member states, Document cn200, Vienna, June 2011, http://www-pub.iaea.org/MTCD/Meetings/PDFplus/2011/cn200/documentation/cn200_Final-Fukushima-Mission_Report.pdf.

[26] "Fukushima Daiichi Status Report", IAEA, Vienna, July 2012, http://www.iaea.org/newscenter/focus/fukushima/statusreport270712.pdf.

[27] "Protecting People Against Radiation Exposure in the Event of a Radiological Attack", ICRP Publication 96, ann. ICRP, **35**, 2005, http://www.remm.nlm.gov/ICRP_guidelines.htm.

[28] "Final report on the international mission on remediation of large contaminated areas off-site the Fukushima Dai-ichi NPP": IAEA, October 2011, http://www.iaea.org/newscenter/focus/fukushima/final_report151111.pdf.

[29] "Fukushima accident upgraded to severity level 7", Strickland, E., IEEE Spectrum, New York, April 2012. http://spectrum.ieee.org/tech-talk/energy/nuclear/fukushima-accident-upgraded-to-severity-level-7.

[30] "INES: The International Nuclear and Radiological Event Scale", IAEA, 2008 edition, Vienna, 2012. http://www-pub.iaea.org/MTCD/publications/PDF/INES-2009_web.pdf.

[31] Harder, L.F, Kelson, K.I., Kishida, T. and Kayen, R. (2011). *Preliminary observations of the Fujinuma dam failure following the March 11 2011 Tohoku offshore earthquake, Japan*, Geotechnical Extreme Events Reconnaissance, http://www.geerassocia-tion.org/GEER_Post%20EQ%20Reports/Tohoku_Japan_2011/QR5_Preliminary%20 Observations%20of%20Fujinuma%20Dam%20Failure_(06-06-11).pdf.

[32] "Safety of Nuclear Power Reactors", World Nuclear Association, London, May 2012. http://www.world-nuclear.org/info/inf06.html.

[33] "The Management of Radioactive Wastes", 2nd Edition, United Kingdom Atomic Energy Authority (UKAEA), London, March 1983.

[34] "Energy in Transition 1985–2010", National Academy of Sciences, Washington, DC, 1979.

[35] "United Kingdom", US Energy Information Administration, Washington, DC, September 2011, http://www.eia.gov/cabs/uk.html.

[36] "UK Energy in Brief 2012", Dept. of Energy and Climate Change, London, July 2012.

[37] Shultz, J.K., Faw, R.E. (2008). *Fundamentals of Nuclear Science and Engineering*, 2nd Edition, CRC Press, New York, USA.

[38] "Mineral Commodity Summaries 2011", Lithium, US Geological Survey, Washington, DC, January 2011, http://minerals.usgs.gov/minerals/pubs/commodity/lithium/mcs-2011-lithi.pdf.

[39] "The Trouble with Tritium", Focus Fusion Association, Middlesex, NJ, July 2006, http://focusfusion.org/index.php/site/article/the_trouble_with_tritium.

[40] Willms, S. (2004). "Tritium Supply Consideration", Los Alamos National laboratory, ITER Test Blanket Module Meeting, February 23–25.

[41] Evans, J. H. (2009). "Where Will We Find the Tritium to Fuel Hybrid Reactors?" *Nature*, **460**, 571.

[42] "World Energy Outlook 2012 Factsheet", International Energy Agency, Paris, France, 2012.

CHAPTER 9

WATER ENERGY

The use of a moving stream of water as a source of energy has been known for several thousand years. Early water turbines were used for irrigation purposes. In Europe and in North America water mills have been used for hundreds of years as sources of motive power for the grinding or milling of grain. The principle, illustrated in Fig. 9.1, is that a stream of moving water impinges on the blades or vanes of a horizontal axis water-wheel. Energy is transferred from the moving water to the wheel, which can be used as a source of slow speed mechanical energy. Many different designs and topologies have been investigated to determine the optimum form of water-wheel configuration. Many different techniques have been used to control the amount and velocity of the moving water so as to maximize the energy collection process. Some of these are discussed below.

9.1. Hydroelectric Power Generation

The principal modern use of water power internationally is as a source of prime power for the generation of electricity. Figure 9.2 illustrates the principle of water falling from a high reservoir onto the blades of a water turbine that directly rotates the shaft of an electricity generator. The earliest example of hydroelectric generation in England was in 1880. This was a private plant rated at 5 kW that lighted a picture gallery 1.5 km away. In 1881 electricity was generated from the River Wey in Surrey, England, and used for street lighting in the town of Godalming. Concurrently, in 1882, a hydroelectric station rated at 25 kW began operation on the Fox River in Appleton, Wisconsin, USA, feeding two paper mills plus residential lighting.

The world's first large-scale hydroelectric power station was built at Niagara Falls, New York, USA, in 1895, with a total capacity of 8200 kW. Canada had a 9.3-MW plant in operation at Niagara Falls in 1903 [1]. The vast scale of large modern plants is illustrated in Fig. 9.3, which shows the Shasta Dam in California, USA. This has a spillway three times higher than Niagara Falls and holds 4.5 million acre-feet of water in the reservoir [2].

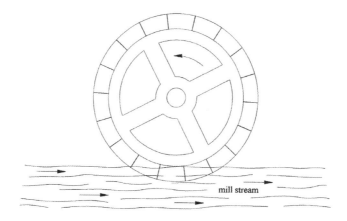

Fig. 9.1.   Principle of the water mill.

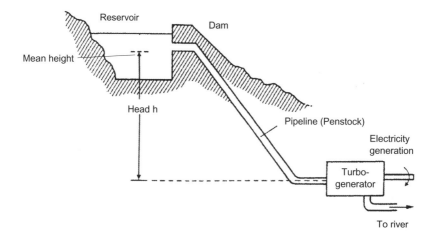

Fig. 9.2.   Principle of hydroelectric power generation [1].

The largest producers of hydroelectric generation are listed in Table 9.1. China now generates 22.25% of its electricity from hydro sources and has a further 50 MW of plant under construction [5]. Norway produces almost all its electricity from hydro sources and Brazil now produces 85.56%.

A more complete listing of hydroelectricity consumption is given in Table 9.2 [6]. World consumption has increased steadily since 1999 and, with an increase of 1.5%, in 2009 was the world's fastest growing major fuel for a second consecutive year [6]. This was in spite of a reduction of 1.1% in total global primary energy consumption due to the economic recession.

The Scandinavian countries and Canada generate much of their electricity from hydropower. In Canada the various provincial electricity supply organizations

Fig. 9.3.   The Shasta Dam, California, USA [2].

Table 9.1.   Large hydroelectricity producers 2009 [3–5].

Country	Annual hydro electricity generation (TWh)	Installed capacity (GW)	% of total capacity
China	652.05	196.79	22.25
Canada	369.5	88.974	61.12
Brazil	363.8	69.080	85.56
USA	250.6	79.511	5.74
Russia	167	45.000	17.64
Norway	140.5	27.528	98.25
India	115.6	33.600	15.80
Venezuela	85.96	14.622	69.20
Japan	69.2	27.229	7.21
Sweden	65.5	16.209	44.34

are called hydroelectric power companies. For example, the Canadian province of Quebec is served by the giant electricity company known colloquially as "Quebec Hydro". The USA obtains about 6% of its electricity from hydroelectric stations. There is a small amount of natural hydropower in Scotland but virtually none in England. The largest hydroelectric power plant in the world is located at Churchill Falls, Labrador, Canada, and produces more than $5000\,\mathrm{MW_e}$ of electric power. Hydropower is doubling in capacity throughout the world about every 20 years.

Table 9.2.   Hydroelectricity consumption* [6].

Million tonnes oil equivalent	2000	2001	2002	2003	2004	2005	2006	2007	2008	2009	2010	Change 2010 over 2009	2010 share of total
US	63.0	49.6	60.4	63.0	61.3	61.8	66.1	56.6	58.2	62.5	**58.8**	−6.0%	7.6%
Canada	80.8	75.0	79.1	76.1	76.6	82.1	80.2	83.6	85.3	83.6	**82.9**	−0.8%	10.7%
Mexico	7.5	6.4	5.6	4.5	5.7	6.2	6.9	6.1	8.8	6.0	**8.3**	38.9%	1.1%
Total North America	151.2	131.1	145.1	143.6	143.6	150.1	153.2	146.3	152.3	152.1	**149.9**	−1.4%	19.3%
Argentina	6.5	8.4	8.1	7.7	6.9	7.9	9.8	8.5	8.4	9.2	**9.2**	−0.2%	1.2%
Brazil	68.9	60.6	64.7	69.2	72.6	76.4	78.9	84.6	83.6	88.5	**89.6**	1.3%	11.6%
Chile	4.3	4.9	5.2	5.2	4.9	6.0	6.6	5.2	5.4	5.6	**4.9**	−13.0%	0.6%
Colombia	6.9	7.1	7.6	8.1	9.0	9.0	9.7	9.5	10.4	9.3	**9.1**	−1.6%	1.2%
Ecuador	1.7	1.6	1.7	1.6	1.7	1.6	1.6	2.0	2.6	2.1	**2.0**	−6.4%	0.3%
Peru	3.7	4.0	4.1	4.2	4.0	4.1	4.4	4.4	4.3	4.5	**4.4**	−1.9%	0.6%
Trinidad & Tobago	—	—	—	—	—	—	—	—	—	—	—		—
Venezuela	14.2	13.7	13.5	13.7	15.9	17.5	18.5	18.8	19.6	19.5	**17.4**	−10.7%	2.2%
Other S. & Cent. America	18.3	16.9	17.8	18.1	17.8	18.3	18.5	19.3	19.4	19.2	**20.7**	7.3%	2.7%
Total S. & Cent. America	124.6	117.2	122.8	127.8	132.7	140.6	148.0	152.4	153.9	157.9	**157.2**	−0.4%	20.3%
Austria	9.5	9.5	9.0	8.1	7.7	7.7	7.2	7.7	7.9	8.2	**7.8**	−3.8%	1.0%
Azerbaijan	0.3	0.3	0.5	0.6	0.6	0.7	0.6	0.5	0.5	0.5	**0.8**	49.3%	0.1%
Belarus	†	†	†	†	†	†	†	†	†	†	†	—	◆
Belgium & Luxembourg	0.1	0.1	0.1	0.1	0.1	0.1	0.1	0.1	0.1	0.1	**0.1**	21.0%	◆
Bulgaria	0.5	0.2	0.5	0.7	0.7	1.0	0.9	0.6	0.6	0.8	**1.3**	59.4%	0.2%
Czech Republic	0.5	0.6	0.6	0.4	0.6	0.7	0.7	0.6	0.5	0.7	**0.8**	14.0%	0.1%
Denmark	†	†	†	†	†	†	†	†	†	†	†	10.5%	◆

(*Continued*)

Table 9.2. (*Continued*)

Million tonnes oil equivalent	2000	2001	2002	2003	2004	2005	2006	2007	2008	2009	2010	Change 2010 over 2009	2010 share of total
Finland	3.3	3.1	2.4	2.1	3.4	3.1	2.6	3.2	3.9	2.9	**3.2**	10.6%	0.4%
France	15.3	17.0	13.9	13.5	13.5	11.8	12.7	13.2	13.7	13.1	**14.3**	9.7%	1.8%
Germany	4.9	5.3	5.4	4.3	4.7	4.6	4.4	4.6	4.5	4.2	**4.3**	2.8%	0.6%
Greece	0.9	0.6	0.8	1.2	1.2	1.3	1.5	0.8	0.9	1.3	**1.7**	29.0%	0.2%
Hungary	†	†	†	†	†	†	†	†	†	0.1	†	−34.7%	◆
Republic of Ireland	0.2	0.1	0.2	0.1	0.1	0.1	0.2	0.2	0.2	0.2	**0.1**	−39.1%	◆
Italy	10.0	10.6	8.9	8.3	9.6	8.2	8.4	7.4	9.4	11.1	**11.2**	0.7%	1.4%
Kazakhstan	1.7	1.8	2.0	2.0	1.8	1.8	1.8	1.8	1.7	1.6	**1.5**	−5.0%	0.2%
Lithuania	0.1	0.2	0.2	0.2	0.2	0.2	0.2	0.2	0.2	0.3	**0.3**	12.9%	◆
Netherlands	†	†	†	†	†	†	†	†	†	†	†	7.1%	◆
Norway	32.2	27.4	29.4	24.0	24.7	30.9	27.1	30.6	31.8	28.8	**26.7**	−7.2%	3.4%
Poland	0.9	1.0	0.9	0.7	0.8	0.9	0.7	0.7	0.6	0.7	**0.8**	17.1%	0.1%
Portugal	2.7	3.3	1.9	3.6	2.3	1.2	2.6	2.3	1.7	2.0	**3.8**	88.2%	0.5%
Romania	3.3	3.4	3.6	3.0	3.7	4.6	4.2	3.6	3.9	3.6	**4.6**	27.4%	0.6%
Russian Federation	37.4	39.8	37.1	35.7	40.2	39.5	39.6	40.5	37.7	39.9	**38.1**	−4.4%	4.9%
Slovakia	1.1	1.2	1.2	0.8	1.0	1.1	1.0	1.0	1.0	1.0	**1.3**	22.3%	0.2%
Spain	7.7	9.3	5.2	9.3	7.2	4.0	5.8	6.2	5.3	6.0	**9.6**	60.9%	1.2%
Sweden	17.8	17.9	15.0	12.1	13.7	16.5	14.0	15.0	15.7	14.9	**15.1**	1.5%	2.0%
Switzerland	8.3	9.4	8.0	7.9	7.6	7.1	7.0	8.0	8.2	8.1	**8.2**	0.9%	1.1%
Turkey	7.0	5.4	7.6	8.0	10.4	9.0	10.0	8.1	7.5	8.1	**11.7**	44.3%	1.5%
Turkmenistan	—	—	—	—	—	—	—	—	—	—	—		
Ukraine	2.6	2.8	2.2	2.1	2.7	2.8	2.9	2.3	2.6	2.7	**2.9**	10.2%	0.4%
United Kingdom	1.2	0.9	1.1	0.7	1.1	1.1	1.0	1.2	1.2	1.2	**0.8**	−32.4%	0.1%
Uzbekistan	1.3	1.2	1.6	1.7	1.6	1.4	1.4	1.4	2.6	2.6	**2.5**	−1.5%	0.3%
Other Europe & Eurasia	17.4	16.9	16.7	17.7	18.8	19.0	18.3	17.3	18.1	19.6	**22.3**	13.5%	2.9%
Total Europe & Eurasia	188.6	189.2	176.2	169.0	180.3	180.1	176.9	179.3	182.1	184.0	**195.9**	6.4%	25.3%

(*Continued*)

Table 9.2.   (*Continued*)

Million tonnes oil equivalent	2000	2001	2002	2003	2004	2005	2006	2007	2008	2009	2010	Change 2010 over 2009	2010 share of total
Iran	0.9	0.9	1.8	2.2	2.7	3.0	4.2	4.1	1.7	1.5	**2.2**	47.2%	0.3%
Israel	—	—	—	—	—	—	—	—	—	—	—	—	—
Kuwait	—	—	—	—	—	—	—	—	—	—	—	—	—
Qatar	—	—	—	—	—	—	—	—	—	—	—	—	—
Saudi Arabia	—	—	—	—	—	—	—	—	—	—	—	—	—
United Arab Emirates	—	—	—	—	—	—	—	—	—	—	—	—	—
Other Middle East	1.0	1.0	1.1	1.0	1.3	1.1	1.2	1.1	0.9	0.7	**0.9**	26.2%	0.1%
Total Middle East	1.8	1.9	2.9	3.2	4.0	4.1	5.4	5.1	2.6	2.1	**3.0**	40.5%	0.4%
Algeria	†	†	†	0.1	0.1	0.1	†	0.1	0.1	0.1	†	−49.4%	◆
Egypt	3.2	3.3	3.2	2.9	2.9	2.9	2.9	3.5	3.3	2.9	**3.2**	10.0%	0.4%
South Africa	0.3	0.5	0.5	0.2	0.2	0.2	0.3	0.6	0.2	0.2	**0.3**	17.7%	◆
Other Africa	12.9	13.5	14.6	14.8	16.0	16.6	16.9	17.1	17.5	18.9	**19.6**	3.7%	2.5%
Total Africa	16.4	17.3	18.3	18.0	19.1	19.8	20.1	21.2	21.1	22.2	**23.2**	4.5%	3.0%
Australia	3.6	3.6	3.6	3.7	3.6	3.6	3.6	3.3	2.7	2.6	**3.4**	29.8%	0.4%
Bangladesh	0.2	0.2	0.2	0.3	0.3	0.3	0.3	0.3	0.3	0.3	**0.3**	0.7%	◆
China	50.3	62.8	65.2	64.2	80.0	89.8	98.6	109.8	132.4	139.3	**163.1**	17.1%	21.0%
China Hong Kong SAR	—	—	—	—	—	—	—	—	—	—	—	—	—
India	17.4	16.3	15.5	15.7	19.0	22.0	25.4	27.7	26.0	24.0	**25.2**	4.9%	3.2%
Indonesia	2.3	2.6	2.3	2.1	2.2	2.4	2.2	2.6	2.6	2.6	**2.6**	0.5%	0.3%
Japan	18.5	18.6	18.9	21.1	21.1	17.9	20.4	17.5	17.5	16.5	**19.3**	17.0%	2.5%
Malaysia	1.7	1.5	1.2	1.3	1.3	1.2	1.6	1.5	2.0	2.0	**2.1**	8.0%	0.3%

(*Continued*)

Table 9.2. (*Continued*)

Million tonnes oil equivalent	2000	2001	2002	2003	2004	2005	2006	2007	2008	2009	2010	Change 2010 over 2009	2010 share of total
New Zealand	5.5	4.9	5.6	5.3	6.1	5.3	5.3	5.3	5.1	5.5	**5.5**	1.1%	0.7%
Pakistan	4.0	4.1	4.6	5.8	5.5	6.9	6.8	7.1	6.1	6.4	**6.4**	0.3%	0.8%
Philippines	1.8	1.6	1.6	1.8	1.9	1.9	2.2	1.9	2.2	2.2	**1.8**	−20.3%	0.2%
Singapore	—	—	—	—	—	—	—	—	—	—	—	—	—
South Korea	0.9	0.5	0.7	1.1	1.0	0.8	0.8	0.8	0.7	0.6	**0.8**	32.9%	0.1%
Taiwan	1.0	1.1	0.6	0.7	0.7	0.9	0.9	1.0	0.9	0.8	**0.9**	11.9%	0.1%
Thailand	1.4	1.4	1.7	1.7	1.4	1.3	1.8	1.8	1.6	1.6	**1.2**	−24.8%	0.2%
Vietnam	3.3	4.2	4.1	4.3	4.3	3.7	4.5	5.1	5.9	6.8	**6.3**	−7.5%	0.8%
Other Asia Pacific	4.9	5.0	4.9	5.4	5.5	5.9	6.3	6.4	6.8	6.8	**7.4**	8.3%	0.9%
Total Asia Pacific	116.7	128.4	130.8	134.3	153.7	163.9	180.8	192.2	212.9	218.0	**246.4**	13.0%	31.8%
Total World	**599.4**	**585.1**	**596.2**	**595.9**	**633.3**	**658.6**	**684.4**	**696.5**	**724.7**	**736.3**	**775.6**	**5.3%**	**100.0%**
of which: OECD	310.1	288.5	293.3	287.8	292.7	295.2	299.4	292.0	301.6	299.4	**309.5**	3.4%	39.9%
Non-OECD	289.2	296.6	302.8	308.1	340.7	363.4	385.0	404.5	423.1	436.9	**466.1**	6.7%	60.1%
European Union	81.8	85.7	72.4	70.7	73.4	69.5	69.6	70.0	73.1	74.1	**83.0**	12.1%	10.7%
Former Soviet Union	52.1	54.2	52.0	51.3	56.3	55.9	55.6	56.4	54.1	56.2	**55.9**	−0.5%	7.2%

*Based on gross primary hydroelectric generation and not accounting for cross-border electricity supply. Converted on the basis of thermal equivalence assuming 38% conversion efficiency in a modern thermal power station.

†Less than 0.05.

◆Less than 0.05%.

Notes: Hydroelectricity data expressed in terawatt-hours is available at *www.bp.com/statisticalreview*.

The potential for further increased development is enormous, especially in the developing countries.

Hydroelectric projects can be enormous in their scale and size. For example, the Hoover Dam on the Colorado River, USA, is 726 feet high, 660 feet thick at the bottom and 45 feet thick at the top. Lake Mead, contained by the dam, is 110 miles long and holds 26×10^6 acre-feet (315×10^{12} tons) of water. A total of 17 generators have a plant capacity of $1345\,\mathrm{MW_e}$.

To a visitor accustomed to thermal or to fossil-fuel-burning power stations, a hydroelectric generation power plant looks strange: it has no factory chimneys and no strangely-shaped cooling towers billowing steam into the atmosphere.

The construction costs of building a hydroelectric plant are larger than the costs of a thermal plant of the same rating. This is because of civil engineering costs such as clearing the reservoir site, building the dam, relocating any buildings or farms that lie within the water basin, lengths of the access roads on the site, etc.

On the other hand, the running costs of a hydroelectric plant are the cheapest option available because the fuel is free. Since the tendency is for fossil fuels such as coal, oil and natural gas to increase in price, the likelihood is that hydroelectric schemes will gain an increasing cost advantage.

The potential hydro capacity in various countries is given in Fig. 9.4 [7].

9.1.1. *Principles of hydroelectric plant operation*

The natural hydrological weather cycle consists of evaporation, precipitation and surface run-off. During evaporation, water vapour is convected to high altitudes and gains heat energy and gravitational potential energy due to its mass and height. On precipitation this potential energy and heat energy are released. Large masses of moving water are sources of kinetic energy. Large masses of stationary water are sources of stored potential energy.

In order to ensure an adequate supply of water, most hydroelectric schemes use dams for retaining a reservoir of water. This reservoir is replenished by rainfall from the surrounding countryside, so the depth of water varies with the local climate. Most hydropower water reservoirs are pleasant and attractive places that fulfil recreational and environmental functions, in addition to their primary purpose.

Let the average head (e.g. height) of water in the reservoir be h metres, (Fig. 9.2). For a mass of water m kilogrammes, the potential energy in watt-seconds or joules is

$$W_P = \text{Gravitational potential energy} = mgh \qquad (9.1)$$

where g is the gravitational acceleration in m/s^2. The energy W_P can also be expressed in terms of the volumetric flow rate Q and the density ρ of the water

$$W_P = Q\rho gh \qquad (9.2)$$

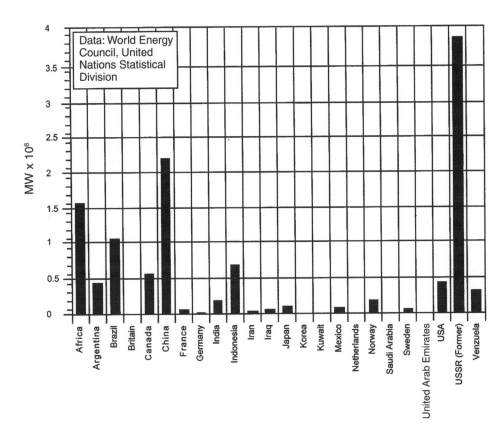

Fig. 9.4. Hydroelectricity potential capacity [7].

where Q is in m^3/s.

It is clear from Eq. (9.2) that the potential energy of stored water can be released either by using a large volumetric flow rate and a small head height or a small volumetric flow rate falling through a large height.

Now the potential energy lost in falling is equal to the kinetic energy KE gained by the water plus the energy losses incurred.

$$\text{KE} = \text{Kinetic energy} = \frac{1}{2}mV^2 \qquad (9.3)$$

where V is the velocity of the water impinging on the blades of the turbine. Neglecting the losses and equating Eqs. (9.1) and (9.3) gives a maximum possible value for the velocity

$$V_{\max} = \sqrt{2gh} \qquad (9.4)$$

Some of the energy in the reservoir, or head water, is lost due to friction and turbulence at the inlet and outlet pipes and also in the penstock or downpipe. In many systems the penstock follows the contours of the ground surface, which is of much greater length than the vertical head height. Most of the kinetic energy at the turbine blades is imparted to its shaft, with an efficiency η_t greater than 90%. Energy from the water is thereby converted directly into rotational mechanical energy, without any intervening energy conversion process. Large electricity generators of the size used in hydroelectric systems are also highly efficient, with efficiency values η_g of the order 95% or greater. The combination efficiency $\eta_t\eta_g$ between the water power at the turbine blades and the electrical output power is of the order 85%. Because of the various water flow losses, however, the ratio of the electrical output power to the reservoir water power is of the order 75–80%. This overall efficiency η_0 figure is much greater than can be obtained by a wind turbine system or by a heat–work cycle involving a boiler plus a steam engine, a gas or diesel engine, or a nuclear reactor system.

9.1.2.   *Types of hydraulic turbine*

9.1.2.1.   *Impulse turbines*

In the most common form of water turbine, known as the impulse turbine, a jet or jets of pressurized water are directed against vanes or cups placed on the perimeter of a wheel (Fig. 9.5). The consequent force on the rotor is intermittent and impulsive. Many different configurations of turbine blade have been used, such as those of Fig. 9.6, in which the jet of water is applied tangentially to the turbine rotor.

The most successful of the impulse-type water turbines uses the Pelton wheel (Fig. 9.7) [8]. Water jets are directed against a structure of double hemispherical cups cast on to the turbine rotor. The high-velocity water jets are designed to strike the cups perpendicularly at the speed that will result in the maximum transfer of momentum. Pelton wheel turbines are suitable for systems with large heads of water, 1,200 feet and higher. For example, there is a 6-jet, 429-rpm system, rated at 29 MW (39000 HP), operating with a head of 1,233 feet, and developing an efficiency

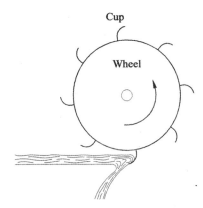

Fig. 9.5.   Principle of the impulse turbine.

SHAFT

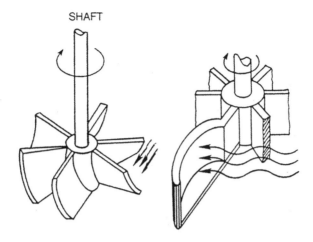

Fig. 9.6. Configurations of impulse turbine blades [5].

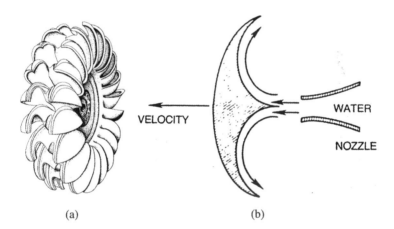

VELOCITY

WATER

NOZZLE

(a) (b)

Fig. 9.7. Pelton wheel [8]. (a) Turbine runner construction and (b) water deflection action.

of 60% at 1/10 rated output. Large commercial installations typically have full-load efficiencies of the order 90%, whereas small installations are of the order 50% [8].

9.1.2.2. *Reaction turbines*

When the head of water available is in the range $5\,\text{m} \leq \text{h} \leq 400\,\text{m}$, a Pelton wheel system would require too many jets. From Eq. (9.2) it is seen that the volumetric flow rate of the water at the turbine blades must then be increased considerably. This can be achieved using, for example, the structure illustrated in Fig. 9.8 [8]. The periphery of the turbine rotor is encased in a housing that permits continuous fluid pressure against the blades. The water enters radially but leaves axially, parallel

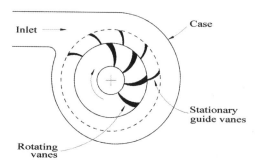

Fig. 9.8.   Structure of the reaction (radial flow) Francis turbine [8].

to the shaft. Because the water exerts continuous pressure on the blades, these react continuously, giving rise to the generic name "reaction turbine". In addition to the kinetic energy extracted from the water flow, a pressure head exists across the turbine, also giving a component of potential energy.

A typical installation of this so-called mixed-flow radial turbine, or Francis reaction turbine, has a diameter of 19 feet and delivers 104 MW (140000 HP) at 100 rpm, using a water head of 162 feet.

9.1.2.3.   *Axial flow turbines*

When only a relatively small head of water is available, of the order $3\,\mathrm{m} \le h \le 30\,\mathrm{m}$, a large volumetric flow rate is needed in order to develop significant levels of power. This can be realized by the use of an axial flow type of turbine, utilizing a propeller [8].

The principle of the axial flow propeller type of turbine is illustrated in Fig. 9.9 [8]. The inlet water flow is restricted by a nozzle that increases the water pressure by the Venturi effect. Water flows in the axial direction of the propeller. Guide vanes (not shown in Fig. 9.9) impart a whirl or rotary component of force to the blades, resulting in rotation. If the propeller blades are of fixed pitch they have a high conversion efficiency at full load but a poorer performance, typically 50% efficiency, at part load. The use of variable pitch blades, as in the Kaplan turbine, gives high efficiency, typically 90%, at all levels of output but is much more expensive.

9.1.3.   *Pumped storage systems*

The load on electricity generation and distribution systems is spread unevenly throughout a 24-hour day. Usually the peak load demand occurs during the daytime, and the night-time load is much smaller. A feature of this load spread is that some very expensive generation equipment is only lightly loaded or is totally unused during the hours of darkness, which is uneconomical. The economics of electricity generation requires a maximum utilization of the plant, preferably at its rated

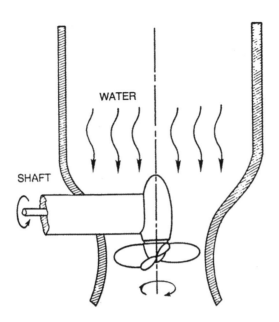

Fig. 9.9. Axial flow (propeller) turbine [5].

operating levels. Because of the underuse of generation equipment at night-time, it is possible in the UK to buy "off-peak" electricity at reduced tariffs.

A principal use of the night-time spare generation capacity is in pumped storage schemes. Generators that would otherwise be unused are operated to pump water up from turbine level to high storage reservoirs. Electric motors are used to drive the water pumping units, which act as water turbines in reverse. Stored water is then available as a top-up for driving the water turbines during the peak demand hours of daylight, which represents a saving of plant costs. In order to be economical, the energy saving due to pumped storage generation by day has to exceed the cost of pumping the water into the reservoir by night, including the amortized costs of the pumped storage installation.

The pump efficiency η_p used to pump water into the storage reservoirs is about 90%. Combining this efficiency factor with the overall efficiency η_0 from Section 9.1.1 gives a total efficiency η_{ps} in the presence of pumped storage.

$$\eta_{ps} = \eta_0 \eta_p \qquad (9.5)$$

If the overall efficiency η_0 without pumped storage is 80%, then, with pumped storage,

$$\eta_{ps} = 0.8 \times 0.9 = 0.72 \text{ per unit} \qquad (9.6)$$

In spite of the significant reduction of plant efficiency with pumped storage schemes, they still represent a great cost saving because they eliminate the need for additional conventional generation plants.

The world's largest pumped storage scheme is at Ludington, Michigan, USA, which takes its water from Lake Michigan and delivers a peak capacity of $1900\,\text{MW}_\text{e}$ of electrical power from its 2 mile × 1 mile reservoir. In Europe, the largest pumped storage scheme is at Dinorwig in North Wales, where a power station is built inside an excavated mountain. The upper reservoir is at a height of 568 metres above the turbines. Six turbo-generator units are each rated at $300\,\text{MW}_\text{e}$. Full output can be delivered for five hours and the power station can come on-stream within ten seconds of demand.

9.1.4.   *Worked examples on hydroelectric power generation*

Example 9.1.   A water reservoir contains a head of water 400 metres above the turbine level. The overall efficiency of the penstock, turbine and generator is 80%. It is required to generate 300 MW of electrical power. What is the necessary mass flow rate of water through the turbines?

The necessary input power is

$$P_\text{in} = \frac{P_\text{out}}{\eta} = \frac{300\,\text{MW}}{0.8} = 375\,\text{MW} = 375 \times 10^6\,\text{J/s}$$

The stored water in the reservoir must contain the potential energy (PE) to deliver this power.

Now, from Eq. (9.1)

$$\text{PE} = mgh$$

The power associated with this $\text{PE} = m \times 9.81 \times 400\,\text{J/s}$ when m is the mass flow rate in kg/s.

Therefore

$$m \times 9.81 \times 400 = 375 \times 10^6$$

or

$$m = 95.6 \times 10^3\,\text{kg/s}$$

Since a cubic metre of water weighs 1000 kg, the volumetric flow rate is

$$Q = 95.6\,\text{m}^3/\text{s}$$

Example 9.2.   In the pumped storage scheme at Dinorwig, North Wales, the upper reservoir is at a height of 568 metres above the turbine house. If losses are neglected, what is the velocity of the water arriving at the turbine blades?

From Eqs. (9.1) and (9.2)

$$mgh = \frac{1}{2}mV^2$$

Velocity V then has its maximum theoretical value $V_{\max}$

$$V_{\max} = \sqrt{2gh} = \sqrt{2 \times 9.81 \times 568}$$
$$V_{\max} = \sqrt{11144} = 105.6 \, \text{m/s}$$

Example 9.3. In a pumped storage scheme of height 400 metres, the combined efficiency of the turbine, electric generators and storage pump motors is 70%. If a water volumetric flow of $50 \, \text{m}^3/\text{s}$ strikes each turbine blade, calculate the electrical output power and the water velocity of impact.

A cubic metre of water weighs 1000 kg, so that

$$Q = 50 \, \text{m}^3/\text{s} \equiv 50000 \, \text{kg/s mass flow rate}$$

From Eq. (9.1) the gravitational potential energy is

$$\text{PE} = mgh$$

When the mass flow rate is given in terms of mass per unit time, the power potential is

$$\text{Potential power of the stored water} = \frac{\text{PE}}{\text{time}} = 50 \times 10^3 \times 9.81 \times 400$$
$$= 196.2 \times 10^6 \, \text{J/s}$$
$$= 196 \, \text{MW}$$

At an efficiency η of 70% the output power from each generator is

$$P_{\text{out}} = \eta \times \text{power potential} = 0.7 \times 196 = 137.2 \, \text{MW}$$

From Eq. (9.4), the final velocity of the water, neglecting losses, is

$$V = \sqrt{2gh} = \sqrt{2 \times 9.81 \times 400} = \sqrt{7848} = 88.6 \, \text{m/s}$$

Example 9.4. The mean height of the feeder reservoir in the Dinorwig pumped storage hydroelectric scheme is 568 metres. At rated load the overall efficiency is 86%. If the plant operates for five hours, delivering 1750 MW, what mass of water has passed through the turbines? What has been the flow rate?

Solution

$$P_{\text{out}} = 1750 \, \text{MW}$$
$$P_{\text{in}} = \frac{1750 \times 10^6}{0.86} = W = 2035 \times 10^6 \, \text{J/s}$$

The stored water in the reservoir must contain the potential energy to deliver this power. But

$$\text{PE} = mgh$$

The power associated with this potential energy is $PE = m \times 9.81 \times 568$ when m is the mass flow rate in kg/s

$$m = \frac{1750 \times 10^6}{0.86 \times 9.81 \times 568}$$

$$= 0.365 \times 10^6$$

$$= 365{,}000 \, \text{kg/s}$$

$$\text{But } 1 \, \text{m}^3 \equiv 1000 \, \text{kg/s}$$

$$\therefore \text{Flow rate} = 365 \, \text{m}^3/\text{s}$$

In five hours,

$$m = 365 \times 5 \times 3600$$

$$= 6.574 \times 10^6 \, \text{m}^3$$

$$= 6.574 \times 10^9 \, \text{kg}$$

From the mass conversion factor in Table 1.4,

$$1 \, \text{kg} = 2.20462 \, \text{lb}$$

$$= \frac{2.20462}{2240} \, \text{tons}$$

In five hours, the mass of water through the turbine is

$$m = 6.574 \times 10^9 \times \frac{2.20462}{2240} = 6.47 \times 10^6 \, \text{tons}$$

9.2.   Tidal Power Schemes

Tidal energy, sometimes called lunar energy, is the most predictable form of the various renewable energy sources. Not only is the timing predictable but the scale of the event and its potential energy content can be accurately calculated.

Tides are caused by gravitational and kinematic forces due to motions of the earth, the moon and the sun. At any point on the earth's surface, on land or at sea, there is a gravitational effect, depending on the positions and distances of the sun and moon. The moon orbits the earth with a period of approximately 655.7 hours. But the earth itself rotates about its tilted axis every 23 hours, 56 minutes, 4 seconds (to the nearest second). The combined effect of these two motions creates a tidal period of approximately 24.8 hours, which contains two complete tidal cycles.

In the North Atlantic Ocean there are two crests and two troughs of moving water circling the ocean in an anticlockwise direction once a day. Consecutive crests and troughs are separated by (almost) six and one-quarter hours.

When the earth, sun and moon are almost in line (Fig. 9.10), the tidal effect on earth is a maximum and results in what are called "spring tides" [9]. This term is

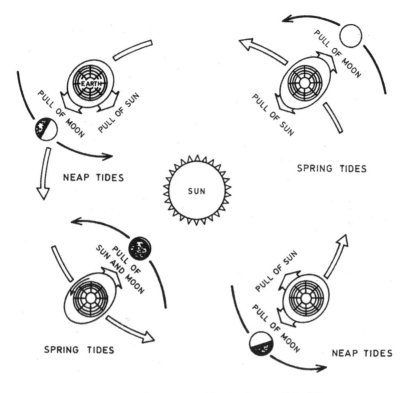

Fig. 9.10. Gravitational basis of ocean tides [9].

not associated with the calendar season of spring time. Spring tides occur regularly every two weeks throughout the year.

When the sun and moon are perpendicular to each other, relative to earth (Fig. 9.10), the tidal effect on earth is a minimum and results in so-called "neap tides". Spring tides have roughly twice the amplitude of neap tides.

The difference in water level between high tide and low tide is known as the "tidal range". In the mid-oceans this is less than 1 metre. There are some large areas of water such as the Great Lakes of North America and the Mediterranean Sea where the tidal effect is negligible.

Closer to land, on the continental shelves, the tidal range becomes about 2 metres. A typical range in many European estuaries is 4 metres. Great opportunities for using tidal energy arise in certain river estuaries, bays and basins because the tidal range can be increased to about 16 metres by shelving, funnelling and other geographical effects. An incoming tidal wave from the sea may be reinforced by tidal waves reflected from the banks. It is the slow-moving wave nature of tidal water that makes the tides so predictable.

9.3.   Methods of Utilizing Tidal Power

In tidal power schemes the flow of water caused by the tides is harnessed to pass through hydroelectric turbines that drive electric generators.

Three different schemes can provide engineering methods of utilizing the power in the tides [10]:

(a) tidal lagoons,
(b) tidal stream farms, and
(c) tidal barrages.

Each of these different options is considered in the following sections.

9.3.1.   *Tidal lagoons* [10]

A tidal lagoon can be created by building dam walls in the sea to establish an area like an artificial estuary. Necessary conditions for viability are that the water must be shallow and the tidal range must be large.

Two main locations for establishing large tidal lagoons in Britain, each of area about $400 \, km^2$, are the Wash, on the east coast, and the waters of Morecombe Bay, off Blackpool on the west coast. Smaller facilities could be built in north Wales, Lincolnshire, south-west Wales, and east Sussex.

At present (2013) there are no working tide pools (tidal lagoons) in the UK.

9.3.2.   *Tidal stream farms* [10–12]

It is possible to build arrays of water turbines, underwater in tidal streams, in a similar fashion to wind farms on land surfaces. A dam is then unnecessary.

The first underwater turbine or tidal stream generator was a 300-kW machine installed in 2003 near the Norwegian city of Hammerfest. Detailed power production figures have not been published [10]. A 35-kW turbine was installed in the Mississipi River, in Hastings, Minnesota, USA in 2008.

Although the seas and oceans are in constant motion, the velocities of the motion in most locations are far too slow to offer useful levels of power. Intense and suitable power densities can be found where currents are accelerated between islands or around headlands. Seawater is 832 times the density of air. The kinetic energy available from a 5-knot ocean current is equivalent to the very large wind velocity of 270 km/h (168 mph). A water current of 4.5 knots (2.25 mph) has a power density of about 6 kW/m, while a wind-stream of over 40 knots (20 m/s) would be needed to realize a similar power density. A typical marine current turbine can have a rotor-swept area of diameter only 40% the value of a wind turbine of the same rating. Although a rotor for extracting energy from flowing water can be readily designed and built, the difficulty of installing it securely is great. Much research

Fig. 9.11. Artistic impression of underwater turbines [11].

and feasibility exploration is going on all over the world. One such topic, from the UK, is illustrated in Fig. 9.11 [11].

9.3.3. *Tidal barrages*

Large water basins may be created by the damming or partial damming of rivers or estuaries. The dams are usually referred to as water barrages or, more simply, as barrages. The erection of a water barrage and its attendant turbine houses, sluice gates, navigable channels, etc., is obviously a major civil engineering task. It is possible to use both the inflow (flood) tides and the outflow (ebb) tides for electricity generation (Fig. 9.12) [13]. Two-directional turbines are more expensive than single-flow (usually ebb flow) directional systems.

Hundreds of potentially useful sites for tidal power exploitation using tidal barrages exist in different locations across the world. The basic essential prerequisite is an adequate tidal range. Table 9.3 shows some characteristics of the best-known tidal power opportunities [1, 14, 15]. The Rance River project in France, operating successfully since 1966, uses a dam wall 725 metres long and delivers an average electrical power of 160 MW$_e$. The only modern tidal plant in North America is the Annapolis Royal plant in Nova Scotia, Canada. This 20 MW facility uses the outflow tide, which has a head of 7 metres, to produce an annual energy generation of 50 GWh.

The nearby Bay of Fundy, containing the Cumberland Basin and the Minas Basin, could deliver 5000 MW and involve a barrage (dam wall) of length 8 km

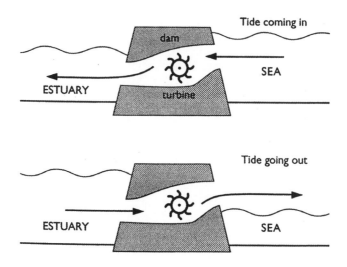

Fig. 9.12. Tidal power generation [12].

outfitted with 97 sluice gates (Table 9.3). The working systems in Table 9.3 mainly use single-basin, two-way flow schemes.

In the UK interest has been devoted to the River Severn estuary, lying between south Wales and the south-west area of England. A sorry tale of missed opportunity has existed since the first favourable UK government report was published in 1933. Further studies and feasibility reports have proposed different schemes and detailed sites. A joint government industry proposal in 1983 was for a barrage of length 13 km. Some detail of one possible UK option is shown in Fig. 9.13. In 1984 a private consortium suggested an alternative site using a 6.7 km barrage. The enormous cost of a tidal barrage scheme is such that only government sponsorship could support it. The most recent proposal, in 2010, was rejected because of the expense.

9.3.3.1. *Power available from a tidal barrage*

Water to the high tide level is permitted to flow upstream through the turbines and sluices. This is retained in the basin until the ebbing tide has created a sufficiently large tidal range, and then released, through the turbines, downstream according to electricity demand. For water control purposes the upstream water behind the barrage can also be released through sluices without energizing the turbines. Let the tidal range of water available be designated R, as in Fig. 9.14. If the surface area of retained water in the basin is A square metres, then

$$\text{Volume of water in the basin} = AR \tag{9.7}$$

If the density of the water is ρ, then

$$\text{Mass of retained water} = \text{volume} \times \text{density} = AR\rho \tag{9.8}$$

Table 9.3. Major world tidal power sites [1, 14, 15].

					Actual and Potential		
	Mean range (m)	Basin area A (km^2)	Barrage length L (m)	L/A (m/km^2)	Theoretical mean power (MW)	Actual mean power (MW)	Annual production (1000 MWh)
North America	5.5	262	4270	16.3	1800	378	15800
Passamaquoddy							
Cobscook	5.5	106			722		6330
Annapolis	6.4	83			765		6710
Minas-Cobequid	10.7	777	8000	10.3	19900	4743	175000
Amherst point	10.7	10			256		2250
Shepody	9.8	117			520		22100
Cumberland	10.1	73			1680		14700
Petitcodiac	10.7	31			794		6960
Memramcook	10.7	23			590		5170
South America							
San Jose,	5.9	750			5870		51500
Argentina							
England							
Severn	9.8	70	3500	50	1680	370	14700
France							
Aber-Benoit	5.2	2.9			18		158
Brest		92	3640	40		211	
Arguenon	8.4	12			446		3910
Frenaye	7.4	22			148		1300
La Rance	8.4	1.1	725	33	349	160	3060
Rotheneuf	8.0	610			16		140
Mont St Michel	8.4	49	23500	39	9700	5252	85100
(Chausey)							
Somme	6.5				466		4090
Ireland							
Strandford	3.6	125			350		3070
Lough							
FSU							
Kislaya	2.4	2	30.5	15.25	2	1.8	22
Lumbouskii Bay	4.2	70			277		2430
White Sea	5.65	2000			14400		126000
Mezen Estuary	6.6	140			13700		12000
Australia							
Kimberley	6.4	600			630		5600
						$\sim$62000	$\sim$560000

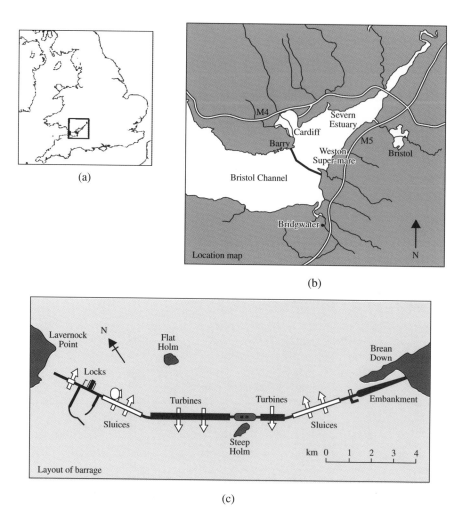

Fig. 9.13.   Proposed barrage scheme for the Severn Estuary, England. (a) National area [16], (b) barrage location [17], and (c) barrage layout [17].

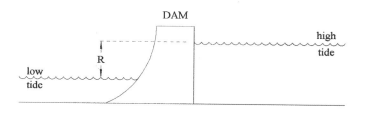

Fig. 9.14.   Definition of "tidal range".

Now the force available due to controlled motion of the water is

$$\text{Force} = \text{mass} \times \text{acceleration}$$

$$F = AR\rho g \tag{9.9}$$

where g is gravitational acceleration constant. If losses are neglected, the potential energy of the retained water is equal to the work done by the water in flooding into or ebbing out of the dam basin

$$W = \text{force} \times \text{distance}$$

$$\therefore\ W = AR\rho g \cdot R/2 \tag{9.10}$$

since R/2 is the mean vertical distance travelled by the water. It is convenient to rewrite Eq. (9.10) in a more compact form. For each tide

$$W = \frac{1}{2} \cdot \rho g A R^2 \tag{9.11}$$

It is clear from Eq. (9.11) that the tidal range R is the dominant design variable. Now the tide floods and ebbs twice during each tidal day so that there are four tides every 24.813 hours. The theoretical maximum power is therefore given by

$$P_{\text{max}} = \frac{4 \times \frac{1}{2}\rho g A R^2}{24.8 \times 3600} \tag{9.12}$$

If $\rho = 1000\,\text{kg/m}^3$ and $g = 9.81\,\text{m/s}^2$, then

$$P_{\text{max}} = 0.22\,AR^2\ \text{MW} \tag{9.13}$$

where A is km^2 and R is in metres.

Because of various losses, the power actually available is about 25% of the value implied in Eqs. (9.12) and (9.13)

$$P_{\text{actual}} = \frac{P_{\text{max}}}{4} = 0.056\,AR^2 \tag{9.14}$$

Some numerical calculations involving Eq. (9.14) are given in Section 9.3.3.5.

9.3.3.2. *Costs of tidal barrage schemes*

The main cost of building a tidal barrage scheme is the cost of the dam or barrage across the bay. Roughly speaking, the cost of construction is proportional to the length of the dam wall, L. Since the extractable power is proportional to the basin surface area A, shown in Eq. (9.14), good design requires the combination of low L and high A. A useful "figure of merit" in design is to minimize the value of the ratio L/A. Values of this ratio are given in Table 9.3 for various active and proposed systems. If length L is in metres and area A is in square kilometres, the ratio L/A must be less than the value 80 for financial viability.

$$\frac{L}{A} < 80 \qquad (9.15)$$

It is seen in Table 9.3 that the proposed Severn Barrage in the UK has a relatively high value $L/A = 50$. This reflects the high construction costs per unit energy output for a relatively small system.

The ratio of barrage length L in metres to the actual power delivered P_{actual} in megawatts is L/P_{actual} and should also be as small as possible.

9.3.3.3.   *Combination of a pumped storage facility with a tidal barrage scheme*

A pumped storage system can be incorporated with a tidal barrage scheme by the use of a double water basin. Water can be released from the high-level basin to the low-level basin and out to the sea through two sets of generators. The system is illustrated diagrammatically in Fig. 9.15, where a head of water h exists over and above the value R due to the tidal range, in a basin of the same area.

The potential energy available during emptying of the tidal range R plus the pumped storage h is, by implication from Eq. (9.11),

$$\text{PE}_{\text{total}} = \frac{1}{2}\rho g A (R + h)^2 \qquad (9.16)$$

The additional component of energy input that is needed to pump the water above the high tide level, to a level h metres, is

$$\text{PE}_{\text{pump}} = \frac{1}{2}\rho g A h^2 \qquad (9.17)$$

Now the net gain of energy due to the use of pumped storage can be expressed as an energy balance equation

> Net gain of energy due to pumped storage
>
>   = Total energy available in the presence of pumped storage

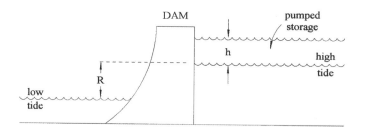

Fig. 9.15.   Pumped storage contribution to tidal range.

- Energy due to tidal basin
- Energy input to raise the water level to h (9.18)

Substituting Eqs. (9.11), (9.16) and (9.17) into (9.18) gives

$$W_{\text{net gain}} = \frac{1}{2}\rho g A[(R+h)^2 - R^2 - h^2] = \frac{1}{2}\rho g A[2Rh] \tag{9.19}$$

It is expressive to rewrite Eq. (9.19) in terms of the basic tidal energy equation (9.11)

$$W_{\text{net gain}} = \frac{1}{2}\rho g A R^2 \left[\frac{2h}{R}\right] \tag{9.20}$$

In order to gain the maximum benefit from the pumped storage addition, the ratio h/R must be made as large as possible.

If the pump-generator system has an efficiency of k per unit, the energy needed to pump the water into the storage area is increased. Equation (9.17) then becomes

$$PE_{\text{pump }k} = \frac{\frac{1}{2}\rho g A h^2}{k} \tag{9.21}$$

The net gain of energy is then

$$W_{\text{net gain }k} = \rho g A \left[(R+h)^2 - R^2 - \frac{h^2}{k}\right] = \frac{1}{2}\rho g A \left[2Rh + h^2 \left(\frac{k-1}{k}\right)\right] \tag{9.22}$$

The $(k-1)/k$ term in Eq. (9.22) is negative for all $k < 1$. In terms of the basic tidal energy equation, Eq. (9.22) can be rewritten as

$$W_{\text{net gain }k} = \frac{1}{2}\rho g A R^2 \left[\frac{2h}{R} - \frac{h^2}{R^2}\left(\frac{1-k}{k}\right)\right] \tag{9.23}$$

When the efficiency is 100%, $k = 1$ and Eq. (9.23) reduces to Eq. (9.20). Numerical examples of pumped storage tidal schemes are given in Section 9.3.3.5.

9.3.3.4. *Features of tidal barrage schemes*

A large tidal barrage scheme is likely to have a major impact in its local area, in addition to providing a diversified supply of renewable energy for electricity generation. McVeigh [1] lists various relevant factors, which are included in the following list.

(a) The construction industry would benefit from major contracts, involving much additional labour.
(b) Manufacturing industry in the vicinity of the water reservoir and its fill streams may need to adopt higher standards of cleanliness and effluent control.

(c) There would be vastly enhanced recreational opportunities (i.e. sailing, fishing, camping, etc.) in the region of the reservoir and barrage. For example, the reservoir basin of a Severn Barrage scheme in the UK could be developed into the largest water sports centre in Europe.

(d) A barrage across an estuary could carry a public road for vehicles and pedestrians.

(e) The water levels, both upstream and downstream of the barrage, could be affected.

(f) Water flow rates upstream of the barrage could be reduced, which might affect drainage and irrigation of the adjacent land.

(g) Changed water flow rates downstream of the barrage would affect the whole estuary. If effluent is now deposited directly or after treatment into the estuary, then its clearance to the sea might be reduced.

(h) The entire aquatic ecosystem will be affected by any changes of salinity or turbidity of the water.

(i) Sedimentation may occur. If it occurs in the basin there may be a reduction of basin volume. On the seaward side, sedimentation from upstream may stay deposited rather than be swept out to sea. Serious sedimentation could affect the navigability of any part of the waterway.

(j) The local sea defences would be enhanced. A barrage system would result in greater flood control and reduce the risk of storm damage along the estuary coastline.

(k) The navigation of ships would be affected. For example, the presence of locks could slow down the journey passage time. On the other hand, the deep water navigation channels might be more predictable.

(l) A total barrage across a river or estuary would affect the migration routes for fish. Swimming routes for fish upstream would likely be feasible through the sluices in ebb generation schemes. Swimming routes downstream would need the provision of channels to bypass the turbine ducts.

(m) The permanent employment possibilities, once the barrage is in operation, are limited. There would be a large work force during construction and this would move on after completion of the building work.

(n) In an environmentally-aware world the political kudos for building a major clean, non-polluting, infinitely renewable source of electricity is considerable.

(o) Electricity is likely to remain the major preferred power source for both industrial and domestic markets. It is in the national interest to supplement the major (fossil fuel) generation with different sources of supply.

9.3.3.5.   *Worked examples on tidal barrage schemes*

Example 9.5.   A small tidal energy scheme has a basin measuring $3\,km$ times $20\,km$. If the average tidal range is 1.5 metres, calculate the maximum electrical power available.

From Eq. (9.13) the maximum theoretical power is

$$P_{\text{max}} = 0.22 \times 3 \times 20 \times (1.5)^2 = 29.7\,\text{MW}$$

The actual power available, from Eq. (9.14), is one-quarter of the theoretical value

$$P_{\text{actual}} = 0.056 \times 3 \times 20 \times (1.5)^2 = 7.42\,\text{MW}$$

Example 9.6. A tidal-power barrage scheme has a basin area of $50\,\text{km}^2$. The tidal range R, representing the difference between the high and low tide levels, is 11.5 metres. Both the ebb and flood tides can be harnessed for power raising and there are two tides in each tidal day of 24.814 hours. The density of sea water is $1000\,\text{kg/m}^3$ and the acceleration of mass due to gravity is $9.81\,\text{m/s}^2$.

(a) Calculate the total theoretical energy for each ebb or flood tide.
(b) If 25% of the theoretical energy is extractable, calculate the power generation capacity per day.
(c) What is the maximum length of dam wall that would be considered financially viable for this barrage area?
(a) The energy/tide is given by Eq. (9.11).

$$W = 2 \times 1000 \times 9.81 \times 50 \times 10^6 \times (11.5)^2 = 32.45 \times 10^{12}\,\text{joules}$$

(b) Assuming that there are two tides/day and that both the ebb and flood tides are usable, then

$$P_{\text{max}} = \frac{4 \times \frac{1}{2}\rho g A R^2}{24.8 \times 3600} = 0.22 A R^2\,\text{MW}$$

If 25% of the power is extractable, then

$$P_{\text{actual}} = \frac{0.22}{4} \times 50 \times (11.5)^2 = 363.7\,\text{MW}$$

(c) For financial viability the inequality Eq. (9.15) must be applied

$$\frac{\text{length of wall (m)}}{\text{area of basin (km}^2)} < 80$$

In this case, $A = 50\,\text{km}^2$ so that

$$\text{length of wall} < 80 \times 50 < 4000\,\text{m (2.49 miles)}$$

Example 9.7. A tidal barrage scheme has a tidal range of R metres. The scheme is to be enhanced by the addition of a pumped storage facility that effectively increases the basin height by h metres above the high tide level. If $h = R$ calculate the extent of the additional energy capacity of the enhanced scheme due to the pumped storage contribution.

The net gain due to pumped storage is given in Eq. (9.20). If $h = R$ it is seen that

$$W_{\text{net gain}} = \frac{1}{2}\rho g A R^2 \left[\frac{2h}{R}\right] = \frac{1}{2}\rho g A R^2 [2]$$

Due to the pumped storage component the net gain of energy is *twice* the original basin energy. In other words, the new basin energy is three times the original value.

Example 9.8. Recalculate the additional energy capacity of the pumped storage scheme of Example 9.7 if the pump motor operates at 50% efficiency.

In Eq. (9.23), k = 0.5. The net energy gain is therefore

$$W = \left[\frac{2h}{R} - \frac{h^2}{R^2} - 1\right] = \frac{1}{2}\rho g A R^2$$

$$= -\frac{1}{2}\rho g A R^2 \left(\frac{h}{R} - 1\right)^2$$

$$= 0, \quad \text{when } h = R$$

The result is equal to the basic energy gain per tide in Eq. (9.11). In other words, with a pump of such low efficiency as 50% the advantage of the pumped storage energy is neutralized by the pump losses, when $h = R$.

9.4. Wave Power

Ocean waves are generated by the wind. Since the wind energy is produced by solar energy, it follows that wave energy is an indirect form of solar energy. Local variations in wind velocity are compensated by short-term energy storage in the high inertia of the water. In effect, the ocean mass acts to significantly smooth out the effects of wind variations.

An estimate of the global annual average wave power in kW/m of wave-front is given in Fig. 9.16. The western coastline of the British Isles is the most favourable location in the world, considering both the wave power potential and the long coast-line. More detail of this is given in Fig. 9.17, which shows the location of the UK ocean weather ship (OWS) India, south of Iceland [9]. Wave power calculations in the UK are usually based on data collected by this ocean weather station.

Measurements in the early 1980s showed that at UK inshore sites feasible for the deployment of wave-energy devices, power levels of the order 40–50 kW/m were available in 50 metre depth of water. Lower levels of 25 kW/m are available off the north-east coast of England and the south-west coast of Wales (Fig. 9.17). Power capabilities of this order are of considerable interest for electrical power generation.

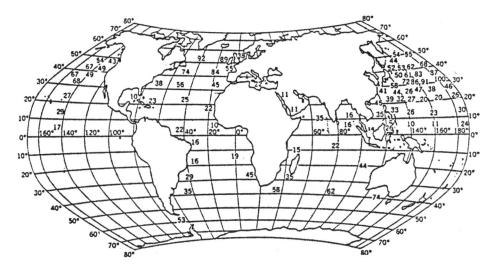

Fig. 9.16.   Estimate of global annual average wave power (kW per metre of wavefront) [17].

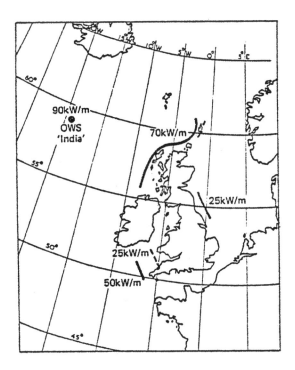

Fig. 9.17.   Position of ocean weather ship (OWS) India and likely wave power sites [9].

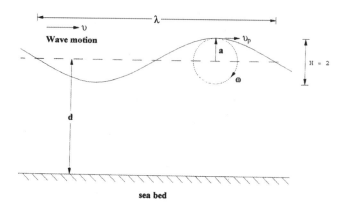

Fig. 9.18.   Sinusoidal surface water wave.

9.4.1.   *Basic properties of ideal deep-water waves* [8]

Proposed wave power systems are usually designed to operate in deep water, where the mean depth of the sea bed is greater than half a wavelength of the water wave. The surface waves are sinusoidal in nature. Individual particles of water describe circular paths (Fig. 9.18) of amplitude that are independent of the wavelength λ. Usually a $< \lambda/10$. Although a wave progresses across the water surface, with horizontal velocity v, the individual particles describe their circular paths in a fixed location and do not progress.

Consider the ideal case of a sinusoidal wave of peak-to-peak amplitude H and wavelength λ travelling across the ocean surface at a horizontal velocity v (Fig. 9.18). The singular frequency f and periodic time T of this monochromatic wave are related to λ and v by

$$v = \lambda f = \frac{\lambda}{T} \tag{9.24}$$

If v is in m/s and λ in metres the frequency f has the dimension of reciprocal seconds (s^{-1}) or hertz. Let the angular frequency ω of the wave be given, as usual with wave motions, by

$$\omega = 2\pi f = \frac{2\pi}{T} \tag{9.25}$$

Note that the angular frequency of the surface travelling wave ω is the same as the angular frequency of rotation of a particle path. The instantaneous velocity of a particle $v_p = a\omega$ at the top of its circular path, however, is not the same as the wave velocity v. The wavelength of a surface travelling wave can be shown from wave theory to be, if $H \ll d$,

$$\lambda = \frac{2\pi}{g} \tag{9.26}$$

Combining Eqs. (9.25) and (9.26) gives an expression for the periodic time T in terms of wavelength.

$$T = \sqrt{\frac{2\pi\lambda}{g}} \tag{9.27}$$

In the North Atlantic, for example, the waves have periodic times of typical value 10 seconds. In Eq. (9.27) this corresponds to a wavelength of 156 metres.

The horizontal component of the wave surface velocity can be obtained by combining Eqs. (9.24), (9.25) and (9.26)

$$v = \frac{\omega\lambda}{2\pi} = \frac{\omega}{N} = \frac{gT}{2\pi} = g\sqrt{\frac{\lambda}{2\pi g}} \tag{9.28}$$

It is of interest to note that the velocity v is independent of the wave amplitude H.

A wave of fixed-value, single frequency, such as that of Fig. 9.18, is known as a monochromatic wave and is characterized by a property called the wave number N.

$$N = \frac{2\pi}{\lambda} = \frac{\omega}{v} \tag{9.29}$$

The total energy content W due to the equal components of potential energy plus kinetic energy, in each wavelength, per unit width of the wave-crest (i.e. the width of the wave perpendicular to the page in Fig. 9.18) of an individual wave is found to be

$$W = \frac{\rho g H^2}{8} \text{ J/m/}\lambda \tag{9.30}$$

For a wavelength λ the theoretical total energy content per metre of wave-crest width may be written as

$$W = \frac{\rho g H^2 \lambda}{8} \text{ J/m} \tag{9.31}$$

Combining Eqs. (9.25), (9.26) and (9.31) gives other expressions for the total energy content per-unit wave-crest width.

$$W = \frac{\pi \rho g^2 H^2}{4\omega^2} \text{ J/m} = \frac{1}{16\pi}\rho g^2 H^2 T^2 \text{ J/m} \tag{9.32}$$

The theoretical maximum power P_{ideal} corresponding to the total energy content, per-unit crest width, under ideal conditions, is obtained by dividing Eq. (9.32) by time T.

$$P_{\text{ideal}} = \frac{\rho g^2 H^2 T}{16\pi} \text{ W/m} \tag{9.33}$$

Using the values $\rho = 1000\,\text{kg/m}^3$ and $g = 9.81\,\text{m/s}^2$, (9.33) gives

$$P_{\text{ideal}} = 1915 H^2 T \text{ W/m}$$

A more useful form is

$$P_{\text{ideal}} = 1.915 H^2 T \text{ kW/m} \tag{9.34}$$

9.4.2.   *Power extractable from practical deep-water waves*

In any given location the resultant surface of the sea is called a dispersive wave and is due to a complicated combination of waves having different wavelengths, directions and time-phase displacements. The power extractable in the direction of the overall average wave motion, for random waves, can be shown mathematically to be one-half of the power of an individual ideal wave. Although the waves are travelling across the water surface at an individual velocity v, defined in Eq. (9.28), the energy content of the group of waves is transmitted at only one-half of this velocity. This "energy" velocity $v/2$ is sometimes called the group velocity.

From Eq. (9.31), the practical power extractable per metre of wave-front can be deduced to be

$$P_{\text{pract}} = \frac{1}{8}\rho g H^2 \cdot \frac{v}{2} \tag{9.35}$$

Because the average horizontal velocity v of ideal energy transmission is halved, the corresponding periodic time is doubled to $2T$.

If the ideal energy Eq. (9.32) is divided by $2T$, the theoretical average power content of the group of practical deep-sea waves is found to be

$$P_{\text{pract}} = \frac{W}{2T} = \frac{1}{32\pi}\rho g^2 H^2 T \text{ W/m} \tag{9.36}$$

Using the values $\rho = 1000\,\text{kg/m}^3$ and $g = 9.81\,\text{m/s}^2$ in Eq. (9.36) gives a value for the theoretical group average power

$$P_{\text{pract}} = 0.96 H^2 T \text{ kW/m} \tag{9.37}$$

But, in reality, even this reduced value is not available. In mid-ocean the resulting wave effect at a location is due to a combination of so many influences that it cannot be accurately calculated mathematically. With a complex, multi-variable situation like this the best way to determine the overall effect is to measure it repeatedly and use statistical data based on measured previous performance.

A set of data measured on-site by the UK OWS India is reproduced in Fig. 9.19 [8]. Each co-ordinate space represents a data reading and the number represents the average number of occurrences per 1,000 samplings in the one-year experiment period. For example, at the intersection $H = 4\,\text{m}$ and $T = 9.25\,\text{s}$ is the number 19, which means that waves with these parameters were measured 19 times in the experimental period time. At the location of this weather station the measured power per metre of wave front is found to be

$$P_{\text{meas}} = 0.55 H^2 T \tag{9.38}$$

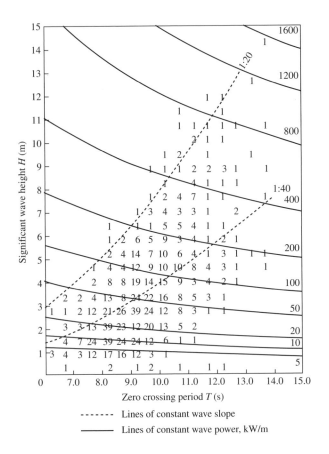

Fig. 9.19.   Scatter diagram of wave height and frequency at OWS India [8].

This measured value of potential wave power is seen to be about 30% of the maximum value predicted by ideal wave theory in Eq. (9.34) and 57% of the more practical value given by Eq. (9.37). The data of Fig. 9.19 suggests that, in that particular location, waves of common incidence have values $T = 8 - 10\,\mathrm{s}$ and $H = 2 - 3\,\mathrm{m}$, having an associated power of about $50\,\mathrm{kW/m}$ of wave-crest length. But the actual efficiency of extraction, incorporating outage time due to breakdown and maintenance, is likely to be about 30%. A reasonable estimate of the power actually extractable is therefore

$$P_{\mathrm{extract}} = 0.3 P_{\mathrm{meas}}$$

$$= 0.3 \times 0.55 H^2 T$$

$$= (0.16 - 0.2) H^2 T \ \mathrm{kW/m} \tag{9.39}$$

The power actually extractable is seen to be roughly 10% of the power content of an individual ideal wave, given in Eq. (9.34). Roughly speaking, waves from the Atlantic striking the western shorelines of Great Britain and Ireland have crest–trough values

H of 2–4 m and a periodic time $T \approx 10$ s. The actual power extractable is obtained by substituting typical values into Eq. (9.39), giving

$$P_{\text{extract UK}} \approx 15 - 20 \, \text{kW/m} \tag{9.40}$$

9.4.3. *Worked examples on wave energy*

Example 9.9. The periodic time of a typical North Atlantic wave is 10 s. Calculate the associated frequency, wavelength and velocity.

With $T = 10$ s, from Eq. (9.25)

$$\omega = \frac{2\pi}{T} = \frac{2\pi}{10} = 0.63 \, \text{radians/s}$$

$$f = \frac{\omega}{2\pi} = \frac{0.63}{2\pi} = \frac{1}{10} \, \text{Hz}$$

from Eq. (9.27)

$$\lambda = \frac{gT^2}{2\pi} = \frac{9.81 \times 10^2}{2\pi} = 156.1$$

from Eq. (9.24)

$$v = \frac{\lambda}{T} = \frac{156.1}{10} = 15.61 \, \text{m}$$

Example 9.10. The waves striking the shores of the western coast of Scotland contain an estimated theoretical maximum power of 70 kW/m of crest length. If a typical periodic time is 11 s, what is the corresponding wave height?

Using Eq. (9.34),

$$P_{\text{max}} = 1.915 H^2 T \, \text{kW/m}$$

The peak-to-peak wave height is therefore given by

$$H^2 = \frac{70}{1.915 \times 11} = 3.32 \, \text{m}^2$$

$$\therefore \; H = \sqrt{3.32} = 1.82 \, \text{m} (5.98 \, \text{ft})$$

It should be noted, however, that only about 10% of the theoretical maximum power is likely to be extractable by wave-energy systems.

Example 9.11. What value is the extractable power from a deep-sea wave of wavelength 150 metres and height 1.5 metres if $g = 9.81 \, \text{m/s}^2$?

From Eq. (9.27),

$$T = \sqrt{\frac{2\pi\lambda}{g}} = \sqrt{\frac{2\pi 150}{9.81}} = \sqrt{96} = 9.8 \, \text{s}$$

Using Eq. (9.39), if the trough–crest height is $1.5 \times 2 = 3\,\mathrm{m}$,

$$P_{\text{extract}} = 0.2H^2T \text{ kW/m} = 0.2(3)^2 9.8 = 17.64\,\mathrm{kW/m}$$

This compares with the value, from Eq. (9.37),

$$P_{\text{pract}} = 0.96H^2T = 84.7\,\mathrm{kW/m}$$

and the value, from Eq. (9.34),

$$P_{\text{ideal}} = 1.915H^2T = 169\,\mathrm{kW/m}$$

The extractable power is seen to be $17.64/169 = 10.4\%$ of the total power capacity of an ideal water wave.

9.4.4.   *Types of wave power converters*

Most of the wave-energy conversion devices and systems that have been actively considered consist of structural elements that move due to the water wave motion. Energy is extracted from the waves and is converted to electrical power by the relative motions and forces between different elements of the structures. These various forms of wave-energy converter are aligned in wide rows perpendicular to the incident wave direction. The size and geometry of a design must be appropriate for the range of water waves experienced at the particular location.

The best-known wave-energy converter is the Salter duck [1, 18, 19]. Duck sections (Fig. 9.20) are mounted along a fixed central spine. Each section has an asymmetric cam which extracts energy by semi-rotary motion, about the fixed spindle, in response to the incident water waves.

The cam is profiled so that its front surface moves with the water of the oncoming wave, while the back surface does not disturb the water behind. It therefore absorbs energy from the approaching waves but does not transmit a wave behind it. In

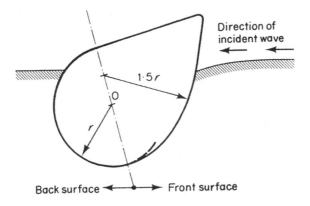

Fig. 9.20.   Principle of the Salter duck [19].

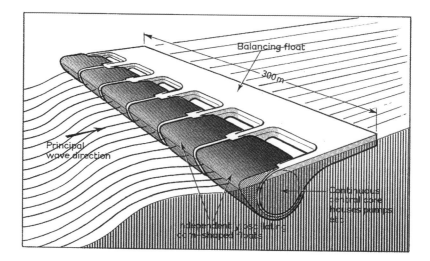

Fig. 9.21.   Salter duck assembly for large-scale power generation [20].

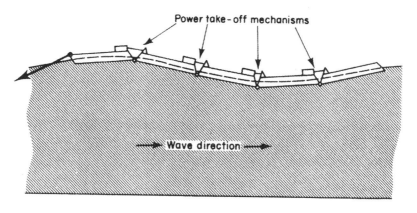

Fig. 9.22.   Cockerell wave contouring raft [1].

motion, the bobbing up and down of the cam tails is very similar to a flock of ducks on the water. To be practical, under sea conditions, a Salter duck would require a wave frequency $\omega = 0.8$ radians/s. From Eq. (9.25) this corresponds to a wave periodic time $T = 2\pi/\omega = 8$ s. The duck structure would then need to be the large estimated value of 15 metres in diameter and would be very expensive to engineer (Fig. 9.21) [18]. This scheme is probably best thought of as future technology rather than present technology.

Another UK invention is the Cockerell raft, which consists of a structure of floating rafts or pontoons. Adjacent rafts are hinged and relative motion between the rafts is used as a source of mechanical power (Fig. 9.22). To be used in realistic sea conditions, the two hinged rafts would need to be between 80 and 120 metres long. The displacement per metre length of structure is then likely to be 100–200 tonnes of steel and concrete. This obviously represents a major engineering project, involving a great use of raw materials, high energy use in fabricating the structure and skilled engineering in constructing the system and retrieving its collected energy. The Cockerell raft is probably too expensive to have any serious prospect of large-scale implementation.

Scale model tests have been performed on both the Salter and Cockerell systems but no high power implementation has yet taken place, nor is likely to do so.

A further range of proposed designs involves the oscillations of columns of water. The vertical motion of the contained water (Fig. 9.23) due to wave action causes compression of the air above it, as in a diving bell, and the compressed air can be used to power a turbine [21]. An experimental prototype has been in operation on the shoreline of Islay (pronounced 'eye-la'), Scotland, since 1991. The wave-energy converter is built into a natural rock gully. Water in the concrete housing (Fig. 9.23) rises and falls like a piston, forcing air in and out of the capture chamber through a Wells turbine. Electrical output powers of 60 kW have been obtained and delivered to the national grid in calm sea conditions, and up to 600 kW short-term output during severe storms.

Fig. 9.23. Islay shoreline wave device [13, 21]. (1) Piston action of the wave motion. (2) Alternative compression and expulsion of air. (3) Air turbine rotation. (4) Generated electricity piped to the grid. (5) Human operator.

Fig. 9.24.   Artistic impression of SEA Clam assembly [20].

A promising offshore wave-energy project appears to be the SEA Clam, developed at the University of Coventry, UK [22]. A flotilla of flexible air bags are attached to the face of a moored floating spine (Fig. 9.24), which is the stabilizing component. Compressed air created by the wave action is forced from the bags through a turbine housing into and out of the hollow spine, allowing an interchange of air between the clam bags. Each clam element has only two moving parts: the air bags, which are proposed to be manufactured using a type-cord reinforced fabric, and the turbo-generator rotor. The basic structure would be unmanned and simple but robust. Since the working fluid is compressed air in a closed circuit, there is no risk of equipment damage due to contamination by sea water.

A specification for a clam unit to deliver up to 10 MW of power would require the unit to be 275 in (902 ft) long, with an external section about 15 metres by 13 metres. An artist's drawing of a clam section (Fig. 9.25) clearly shows the turbine channel and illustrates the necessary vast scale of the structure. Large-scale application of the clam would involve the use of many of such 10-MW devices, each of which has a displacement of 44,000 tonnes.

A variation of the SEA Clam idea is the Pelamis Wave Energy Converter (Fig. 9.26) [23]. Build in Scotland, the prototype device is installed three miles off the coast of Aguçadoura, Portugal, and consists of a snake of four articulated units, rated at 750 kW (peak). The units are 3.5 metres diameter and 140 metres long, weighing 700 tons, of which 350 tons is ballast. Since each unit uses 350 tons of carbon steel it gives a weight-to-power ratio of 500 kg per kW (peak) [10].

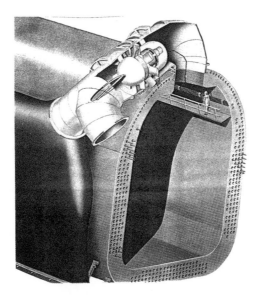

Fig. 9.25.   Artistic impression of SEA clam section [22].

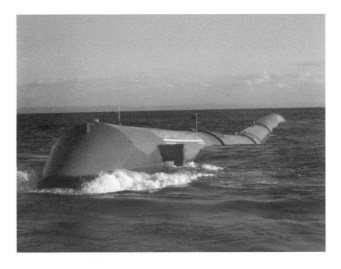

Fig. 9.26.   Pelamis wave energy converter [21].

Many other forms of wave-energy machines have been proposed. Some of these are fixed and massive immovable structures, mounted on the sea bed, designed to capture large volumes of water for use in water turbines. As with the relative motion systems of Salter and Cockerell and compressed air methods such as the SEA Clam and the oscillating water column, however, they largely remain expensive, small-scale experimental prototypes or ingenious ideas that have never been tested at all.

9.4.5. *Worked examples on wave-energy converter devices*

Example 9.12.   It is estimated that a realistic figure for the power extractable from water waves along the western coastlines of Scotland and Ireland is about 20 kW/m length of the wave power capture system. What length of ideal wave power extractor, aligned perpendicularly to the incoming waves, would be required to realize a power of 3 MW, equivalent to that of the Orkney wind turbine system?

The Orkney wind power system is designed to deliver a rated load of 3 MW of electrical power. In order to match this from a wave power system the necessary energy-capturing structure would need to have

$$\text{Length} = \frac{3 \times 10^6}{20 \times 10^3} = 150 \, \text{m} = 496 \, \text{ft}$$

For example, the Salter converter assembly of Fig. 9.21 is shown having a capture length of 300 metres. This would extract up to 6 MW of wave power, but be subject to various losses.

Example 9.13.   The Cockerell raft and the SEA Clam systems for wave-energy capture and conversion involve constructions of displacement about 150 tonnes per metre of capture length. A realistic value for the extractable power from the waves, in a favourable location, is 20 kW/m. What length of ideal extractor structure would be needed to realize 1 MW of power and what is the structure displacement?

$$\text{Required length} = \frac{10^6}{20 \times 10^3} = 50 \, \text{m}$$

At a rate of 150 tonnes/m the corresponding displacement of structural materials is

$$\text{Displacement} = 50 \times 150 = 7500 \, \text{tonnes}$$

Example 9.14.   A proposed SEA Clam wave-energy converter system consists of separate units, each 275 metres long, which can realize up to 10 MW of power. How many clam units and what approximate length of coastline would be needed to generate 1000 MW of power?

To generate 1000 MW, if all the units were in operation and working to capacity, would require

$$\text{Number of clam units} = \frac{1000 \times 10^6}{10 \times 10^6} = 100 \, \text{units}$$

If each unit was sited end-to-end this would require a sea length of

$$\text{Length of clam units} = 100 \times 275 = 27500 \, \text{m} = 27.5 \, \text{km}$$

It is highly probable, however, that there would be generous spacing between the units to permit the passage of boats and to prevent shading effects. Since the units are all arranged to be at the same small angle to the sea-wave fronts (illustrated in Fig. 9.24), the overall length of the barrage of converters might be about 50 km. Therefore

$$\text{Length of coastline} \approx 50 \, \text{km}$$
$$\approx 50 \times 0.62 = 31 \text{ land miles}$$
$$\approx 50 \times 0.54 = 27 \text{ nautical miles}$$

In a practical engineering situation it is likely that (say) one-third of the clam units might be inoperative because of repair or maintenance. This implies that there would need to be one-third more units occupying one-third more space. A more realistic estimate of the requirements for 1000 MW generation is therefore

$$\text{Number of units} \approx 130 - 140$$
$$\text{Length of coastline} \approx 65 \, \text{km}$$
$$\approx 40 \, \text{miles}$$

9.4.6. *Features of wave power systems: Summary*

No large-scale, floating, wave power installations have yet (2013) been built. Although there can be no dispute about the amount of potential energy available in the waves, there is much misgiving about the viability of extracting the energy. In particular, the large size of the necessary structures and the financial costs of construction and implementation are formidable. Some of the implications of the use of large-scale wave-energy schemes are listed below [1]. The features are not listed in order of significance but in an arbitrary order.

(a) The seas, around Great Britain in particular, frequently represent a hostile environment. It would be necessary for the wave-energy structures to be extremely strong, rugged, heavy and expensive.
(b) Fouling of the structures due to salt water pollution could be a serious problem, requiring ongoing maintenance.
(c) The turbine and generator systems would require ongoing maintenance. A wave power structure would not provide suitable living accommodation. Servicing of the structures would have to be carried out by shore-based personnel. Difficulties of access to the rigs would be encountered due to bad weather.
(d) Ecological effects would be experienced at the shoreline due to the changed energy of the waves reaching the shore.
(e) A rate of extractable power of the order 20 kW/m, Eq. (9.40), ideally corresponds to about 1 MW for 50 metres of wave-front. In order to generate power of a few MW (i.e. the order of power available from a large wind turbine system), the structure would require to be 100–200 m in length. This would be a massive construction with a displacement weight between 10,000 and 20,000 tonnes, or about the displacement of a small cargo ship or a small oil tanker.

(f) The necessary large structures may require vast amounts of steel and concrete in their construction. These raw materials are energy-intensive in their industrial preparation. In other words, there would be a large energy investment to be repaid before the rig operated at an energy profit.

(g) A structure of useful size for generating electricity would have a large visual impact if it was close to shore. Example 9.14 shows that a 1000 MW system would occupy roughly 40 miles of coastline.

(h) A structure of useful size for generating electricity would be a navigational hazard to shipping. Sea lanes would need to be planned, marked by buoys and continuously monitored.

(i) A wave-energy converter structure could limit the access of fishing boats within its vicinity and/or interfere with the trawling process.

(j) The deposition of particles along the local leeward shore or on the sea bed might be affected.

(k) The overall conclusion is that large-scale wave-energy extraction is likely to be formidable in its difficulties. At the present period (2013) it does not appear, to the authors, to be an attractive option.

9.5.   Problems and Review Questions

Problems on hydroelectricity

9.1. From the information in Table 9.1, list the half-dozen countries that consumed the most hydroelectric power in 2009.

9.2. In which countries has the consumption of hydroelectric power (a) increased, (b) decreased, most significantly in the period 1999–2009?

9.3. What was the proportion of the world total primary energy consumption taken by hydroelectricity in the year 2009?

9.4. A hydroelectricity supply system has an overall efficiency of 82%. If the effective head of water is 500 metres, calculate the volumetric flow rate needed to generate 300 MW of electric power.

9.5. Show that, if losses are neglected, the velocity of the water striking the blades of a hydroelectric turbine is proportional to the square root of the head height, independently of the flow rate.

9.6. The volumetric flow rate of water in a hydroelectric scheme is $50 \, \text{m}^3/\text{s}$. The overall efficiency of the turbine, generator and pumped storage motor is 72%. If the electric power output is to be 150 MW, what head of water is required? ($\rho = 1000 \, \text{kg/m}^3$, g $= 9.81 \, \text{m/s}^2$.)

9.7. What types of water turbine would be likely to be used in locations where the available head of water is: (a) 10 metres, (b) 100 metres, (c) 1,000 metres?

9.8. The mean height of the feeder reservoir in the Dinorwig pumped storage hydroelectric scheme is 568 metres. At rated load the overall efficiency is 86%.

If the plant operates for 5 hours, delivering 1750 MW, what mass of water has passed through the turbines? What has been the flow rate?

9.9. (Adapted from Worldwatch Institute [5].) Estimate the hydroelectric potential of an area or location, chosen from an atlas giving geophysical data. Use the following technique and choose a location X.

  (i) What is the lowest altitude of X?
  (ii) What area of X lies more than 300 metres above the lowest level?
  (iii) What is the annual rainfall on the high parts of X?
  (iv) If all of the rainfall ran to the lowest level, what amount of potential energy per year in MW would be given up by the moving water?
  (v) What factors would prevent all of the rainfall being converted to electricity?
  (vi) Estimate the fraction of the rainfall potential energy that might be convertible to electricity.
  (vii) If your selection location X already contains a hydroelectric power station, compare your estimate of its potential power capacity with the station rating. Comment on any large differences.

Problems on tidal power schemes and pumped storage schemes

9.10. List the advantages and disadvantages of a tidal barrage scheme as a source of electrical power.

9.11. In a pumped storage scheme combined with a tidal power project, water is pumped from the high tide level to an additional height of 1 metre. The tidal range R is 6 metres. Calculate the proportion of extra energy gained: (a) neglecting losses, (b) if the pump motor is 80% efficient.

9.12. If the efficiency of the pump-motor in a pumped storage tidal system is fixed in value at any level k below 100%, show that the net energy gain will be a maximum when $h = kR/(1 - k)$.

9.13. The mean tidal range of the Rance scheme is 8.45 metres and its basin area is $22\,\mathrm{km}^2$, as given in Table 9.2. If the mean output is 75 MW, what proportion of the theoretical power capacity does this represent?

9.14. Calculate the electrical power output from the proposed Severn estuary tidal power scheme in Britain if the mean tidal range is 8.8 metres and the basin area is $50\,\mathrm{km}^2$.

9.15. A certain tidal power scheme has a water basin with a V-shaped vertical cross-section. Show that the theoretical maximum power capability, for ebb-tide operation, is proportional to the cube of the maximum depth of water.

9.16. In North America the Bay of Fundy-Passamaquoddy area near the border of New Brunswick and Nova Scotia in Canada, with Maine in the USA, is considered to be a prime tidal power site. It has a basin area of 700 km^2 with an average tidal range of 10.8 metres. Calculate the theoretical maximum power capability and the estimated realistic power available.

Problems on wave energy

9.17. For a monochromatic (single frequency) water wave of period 8.5 seconds, calculate the corresponding frequency, wavelength and velocity of the wave motion.

9.18. If the water wave of Problem 9.14 has a crest–trough height of 4 metres, calculate the theoretical power and the realistic extractable power from a random group of such waves.

9.19. What is the power extractable from a deep-sea wave system of wavelength 140 metres and height 3 metres?

9.20. The western coast of Scotland is incident by Atlantic waves of theoretical maximum power 70 kW per metre of wave width. If a typical wave height is 2 metres, what are the corresponding frequency and periodic time? Estimate the realistic power available onshore.

9.21. List and briefly discuss the main features of proposed wave power systems.

9.22. Estimate the necessary length of a proposed wave power receiver station to collect 10 MW of usable power if the maximum theoretical power is 70 kW/m.

References

[1] McVeigh, J.C. (1984). *Energy around the World*, Pergamon Press, Oxford.

[2] "BP Statistical Review of World Energy 1997", British Petroleum Company plc, London, June 1997.

[3] "Norway and the Environment: Binge and Purge", The Economist, January 22, 2009.

[4] China Electricity Council (2011). "The Overview of China Electricity Industry, Jan–Nov 2010", Market News, 13 January 2011, http://www.cec.org.cn.

[5] "Use and Capacity of Global Hydropower Increases", Worldwatch Institute, Washington, DC, January 2012.

[6] "BP Statistical Review of World Energy", British Petroleum plc, London, June 2010.

[7] The World Resources Institute (1994). *World Resources 1994–95: A Guide to the Global Environment*, Oxford University Press, New York.

[8] Twidell, J. and Weir, T. (1986). *Renewable Energy Resources*, E. And F. Spon, London.

[9] Swinden, D.J. (1980). *Energy*, Lecture Support Notes, England.

[10] Mackay, D. (2009). *Sustainable Energy — Without the Hot Air*, Cambridge University Press, Cambridge.

[11] Fraenkel, P. (2001). "Electricity from Marine Currents", *Renewable Energy World*, **4**, 144–147.

[12] Perlman, S. (2000). "Treading Water", Electrical Generating Systems Assn., http://www.egsa.org/powerline/current/feature.htm.

[13] "Renewable Energy — A Resource for Key Stages 3 and 4 of the UK National Curriculum", Renewable Energy Enquiries Bureau, ETSU, Harwell, 1995.

[14] Dorf, R.C. (1977). *Energy Resources and Policy*, Addison Wesley Publishing Co., Reading, MA.

[15] King Hubbert, M. (1969). *Resources and Man — Energy Resources*, National Academy of Sciences, W.H, Freeman and Co., San Francisco.

[16] "UK Severn Barrage Could be Cheaper than Offshore Wind", Offshore Wind, 2013, http://www.offshorewind.biz/2013/02/04/uk-severn-barrage-could-be-cheaper-than-offshore-wind/.

[17] Shepherd, W. and Shepherd, D.W. (2003). *Energy Studies*, 2nd Edition, Imperial College Press, London.

[18] Salter, S.H. (1974). "Wave Power", *Nature*, **249**, 720–724.

[19] Harker, J.H. and Blackhurst, J.R. (1981). *Fuel and Energy*, Academic Press, New York.

[20] CEGB Research, No. 2 London, England, May 1975, p. 33.

[21] "Renewable Energy Review", Department of Energy, UK Issue 17, 1991.

[22] "Sea Clam–Wave Energy Converter", Sea Energy Associates Ltd., University of Coventry, England, 1981.

[23] "Wave Power — an Introduction", Alternative Energy and Fuels, September 2008, http://www.alternative-energy-fuels.com/water/wave-power/wave-power-an-intro.

CHAPTER 10

WIND ENERGY

10.1.   Nature and Origin of the Wind

10.1.1.   *Atmospheric pressure* [1]

The wind is the motion of a mass of air. For the purpose of using wind energy it is normally the horizontal component of the wind that is of interest. There is also a vertical component of the wind that is very small compared with the horizontal component, except in local disturbances such as thunderstorm updrafts.

At the earth's surface the atmospheric pressure is measured in the unit called the pascal (Pa) and has an average value 101.325 kilopascal (kPa), which is sometimes called "one atmosphere". Another unit of pressure used for meteorological calculations is the millibar (mbar). There are exactly 100 Pa per millibar so that one atmosphere is about 1000 mbars. On a map regions of equal atmospheric pressure are identified by isobar lines such as those illustrated in Fig. 10.1. A close concentration of isobar lines indicates a high pressure gradient or region of rapid pressure change. Wind speed is directly proportional to the pressure gradient.

The atmospheric pressure varies from place to place and from day to day, caused by the combined effects of solar heating and the rotation of the earth. As the earth spins, anticlockwise, the atmospheric air surrounding it is dragged round with it at different levels depending on altitude. The mix of air forms turbulence, causing wind at the earth surface.

An additional feature is the inertial force known as the Coriolis force, which occurs in rotational systems. When air moves over the surface of the earth as it rotates, instead of travelling in a straight line the path of the moving air veers to the right. The effect is that air moving from an area of higher pressure to an area of lower pressure moves almost parallel to the isobars. In the northern hemisphere the wind circles in a clockwise direction towards the area of low pressure but in the southern hemisphere the wind circles in an anticlockwise direction.

Fig. 10.1.   Atmospheric pressure isobars for North America, April 2008 [1].

10.1.2.   *Atmospheric density* [2]

From the viewpoint of energy conversion the most important properties of the wind at a particular location are the velocity of the air stream and the air density. The air density varies with altitude and with atmospheric conditions such as temperature, pressure and humidity. At sea level, and at standard atmospheric temperature and pressure, the density (ρ) is

$$\rho = 1.201 \, \text{kg/m}^3 \text{ at 1000 millibars (29.53 inches of mercury)}   \text{ or}$$
$$101.3 \, \text{kPa pressure and temperature 293 K.}$$

In the UK a useful figure for the atmospheric air density is

$$\rho = 1.29 \, \text{kg/m}^3 \; (0.08 \, \text{lb/ft}^3) \tag{10.1}$$

In the USA a commonly quoted figure for sea level under dry conditions at a temperature of $0°C$ (273 K) is

$$\rho = 1.275 \, \text{kg/m}^3 \tag{10.2}$$

There is a lot of local variation of values of air density in different areas of the world. It is found that the temperature, pressure and density of the air decrease with altitude. For wind turbine applications the range of interest is mostly within a couple of hundred feet of ground level. Within this range it is adequate to use the density values in Eqs. (10.1) and (10.2) or their local alternatives at other locations.

10.2. The Availability of Wind Supply

In most geographical locations there is a characteristic pattern of wind velocity over the year. There are typical patterns of wind flow on a seasonal basis. At any given site there may be significant variations of wind on a daily (sometimes hourly) basis, both with regard to the magnitude and direction of the wind, but the annual features are fairly consistent.

10.2.1. *Global survey*

A very broad representation of the average overall global wind is shown in Fig. 10.2. In the northern hemisphere the wind power density is of the order 2–3 times greater in the winter than in the summer. In Europe the prevailing wind (i.e. the direction of the wind for most of the time) is westerly. It flows from west to east. The wind-map shows that the western coastlines of North America, South America, northern Europe and Africa have great potential for the exploitation of wind energy. The annual average wind at any site depends on:

(a) geographical position;
(b) the detailed location (in particular, the altitude and the distance from the sea);

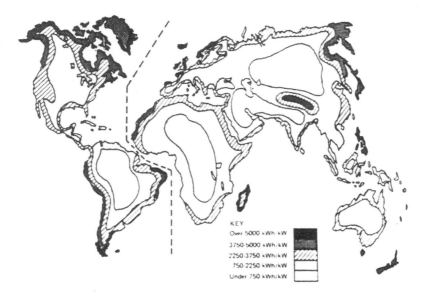

Fig. 10.2. Annual availability of wind energy in different parts of the world [1].

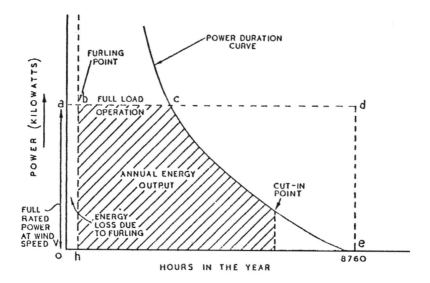

Fig. 10.3.   Estimation of annual wind energy output [1, 3].

(c) the exposure and possible screening due to any surrounding hills, vegetation, buildings; and

(d) the shape of the land surface in the immediate area.

In order to gain altitude and thereby increased wind speed, turbine sites are often chosen on hill tops.

10.2.2.   *Energy content of the wind*

A rough estimation of the wind energy capability of any site is provided by the wind speed-duration data. The average wind speed in mph is plotted against the duration for the $(365 \times 24) = 8{,}760$ hours of the year in Fig. 10.3. It should be noted that a characteristic of power versus time means that the area under the curve has the dimension of power multiplied by time, which is energy. For wind energy use, the power versus time availability characteristic should contain the largest possible area.

The cut-in speed eliminates the hours representing too low a wind speed. Correspondingly, there are periods where the wind speed is too high and the turbine operation has to be discounted. The hours representing this condition are shown to the left of the "furling point" in Fig. 10.3.

A measure of the perception of wind of different speeds is provided by the Beaufort scale, shown in Table 10.1. For example, a wind speed of 25 mph is described as a "strong breeze", in which it would be difficult to manipulate an umbrella.

For most locations, even in the southern hemisphere, the wind speed in the months October–March is significantly greater than in the periods April–September.

Table 10.1. The Beaufort wind force scale [1].

Beaufort Scale	Description	When you see this	Wind speed mph	Wind speed kmh
0	Calm	Smoke goes straight up. No wind present.	Less than 1	Less than 2
1	Light Air	Direction of the wind is shown by smoke drift, but not by a wind vane.	1–3	2–5
2	Light Breeze	Wind is felt on your face. Leaves rustle. Wind vane moves.	4–7	2–5
3	Gentle Breeze	Small twigs and leaves move steadily. Wind extends small flag straight out.	8–12	12–19
4	Moderate Breeze	The wind raises loose paper and dust. Small tree branches move.	13–18	20–29
5	Fresh Breeze	Waves form on lakes and ponds. Small trees sway.	19–24	30–39
6	Strong Breeze	Large tree branches move. Umbrellas become hard to use. Wires whistle.	25–31	40–50
7	Moderate Gale	Hard to walk against the wind. Whole trees are in motion.	32–38	51–61
8	Fresh Gale	Very difficult to walk against the wind. Twigs break from trees.	39–46	62–74
9	Strong Gale	Roof shingles are torn away. Small damage to buildings occur.	47–54	75–87
10	Whole Gale	Trees are uprooted.	55–63	88–101
11	Violent Storm	Widespread damage from the wind.	64–72	102–116
12	Hurricane	Widespread destruction from the wind.	73+	117+

Since the energy potential and the power content of the wind are proportional to the cube of the wind speed there will obviously be great differences in the energy viability of the different locations.

10.2.3. *Wind energy supply in Europe* [3]

The wind energy resources over Europe (Fig. 10.4) are influenced by three major factors:

1. large temperature differences between the polar air in the north and sub-tropical air in the south;
2. the distribution of land and sea with the Atlantic Ocean to the west, Asia to the east, and the Mediterranean Sea and Africa to the south; and
3. major geographical features such as the Alps, the Pyrenees and the mountains of Scandinavia.

The British Isles, consisting of the United Kingdom (UK) of Great Britain (England, Scotland, Northern Ireland and Wales) and the Republic of Ireland, form one of the

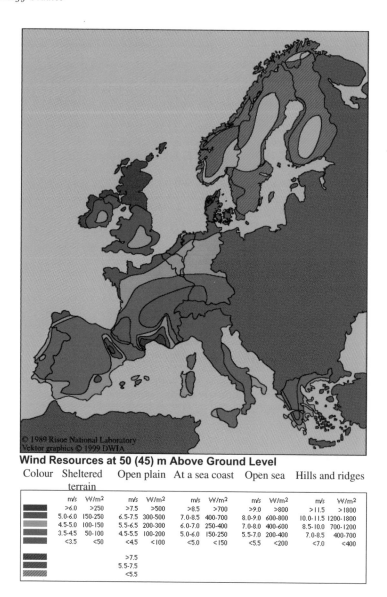

Wind Resources at 50 (45) m Above Ground Level

Colour	Sheltered terrain		Open plain		At a sea coast		Open sea		Hills and ridges	
	m/s	W/m²	m/s	W/m²	m/s	W/m²	m/s	W/m²	m/s	W/m²
	>6.0	>250	>7.5	>500	>8.5	>700	>9.0	>800	>11.5	>1800
	5.0-6.0	150-250	6.5-7.5	300-500	7.0-8.5	400-700	8.0-9.0	600-800	10.0-11.5	1200-1800
	4.5-5.0	100-150	5.5-6.5	200-300	6.0-7.0	250-400	7.0-8.0	400-600	8.5-10.0	700-1200
	3.5-4.5	50-100	4.5-5.5	100-200	5.0-6.0	150-250	5.5-7.0	200-400	7.0-8.5	400-700
	<3.5	<50	<4.5	<100	<5.0	<150	<5.5	<200	<7.0	<400
			>7.5							
			5.5-7.5							
			<5.5							

Fig. 10.4.   European wind resource map [4].

windiest regions on earth. Contours of the mean annual wind speed, measured at the agreed standard height of 33 feet (10 metres) above the ground, are given in Fig. 10.5 [4]. The wind speed increases greatly at higher levels of elevation.

In general the coastal areas are windier than inland areas. The prevailing wind is westerly, from the Atlantic Ocean, creating the high average value of 17.5 mph along the western coastlines of Scotland and Ireland. Incidentally, the same regions offer great potential for water wave energy.

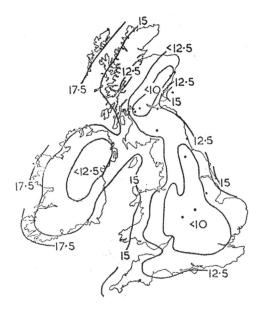

Fig. 10.5. Isovent contours of UK mean wind speed (mph) [3] (reproduced by permission of the UK Meteorological Office).

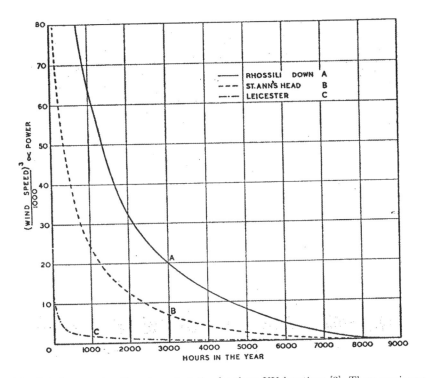

Fig. 10.6. Wind power-duration characteristics for three UK locations [3]. There are important seasonal variations in the UK availability of wind energy.

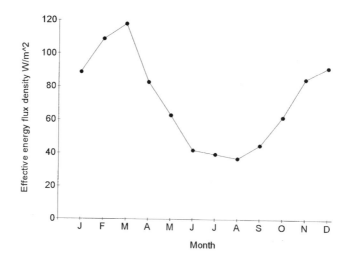

Fig. 10.7.   Wind availability in the UK (mean wind velocity $= 6\,\text{m/s} = 13.4\,\text{mph}$) [3].

Tremendous local variations of wind energy occur even in a relatively small country like Great Britain. In Fig. 10.6, units of wind power are plotted against hours of wind availability per year for three different UK locations. Obviously the preferred location is Rhossili Down, which is on the west coast of Scotland, whereas a location in the English midlands, such as Leicester, is completely unsuitable.

In northern temperate latitudes the highest daily average wind speeds occur in winter, with maximum values usually being in January, and the lowest in July/August. The seasonal variation between winter and summer is of the order 2:1. The average wind energy flux density on a typical month by month basis for the UK is given in Fig. 10.7 [3]. It is of immense significance in the UK that the availability of wind energy closely matches the national demand for electricity. It is also relevant that the profile of UK wind energy input is in anti-phase with the profile of solar radiation input.

10.2.4.   *Wind energy supply in the USA*

The mean annual wind speeds across the USA are shown in Fig. 10.8. The predominant feature is the southerly winds blowing north through Texas to North Dakota along the entire north–south dimension (about 1,500 miles). This is reflected in the wind power density data, which shows that the centre and western states of the USA have much greater wind energy potential than the eastern states. Many of the eastern states have no significant wind energy potential at all. The figures show that the central states of (from south to north) Texas, Oklahoma, Kansas, Nebraska and the Dakotas have the potential to supply almost all of the USA's demand of electricity by wind energy alone. The potential is obviously such that the US is wise to consider wind energy as a major renewable resource [5].

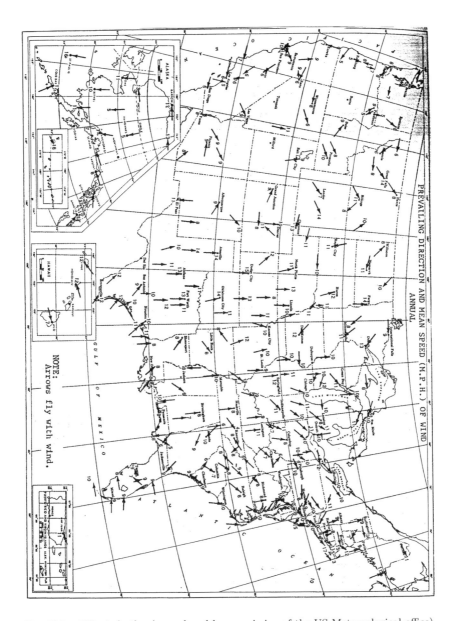

Fig. 10.8.   US wind atlas (reproduced by permission of the US Meteorological office).

10.3.   Power Available in the Wind [2, 4]

10.3.1.   *Theoretical power available*

Consider a smooth and laminar flow of wind passing perpendicularly (normally) through an element of area A of any shape, having thickness x, shown in Fig. 10.9.

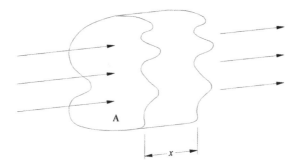

Fig. 10.9.   Element of space through which the air flow passes [2].

If the air mass is m and it moves smoothly with an average velocity V, the motion of the air mass has a kinetic energy (KE).

$$KE = \frac{1}{2}mV^2 \qquad (10.3)$$

The mass m of air contained in an element of volume Ax is given, in terms of density ρ, by

$$m = \rho Ax \qquad (10.4)$$

Combining Eqs. (10.3) and (10.4) gives, for the kinetic energy associated with this mass and volume of air,

$$KE = \frac{1}{2}\rho AxV^2 \qquad (10.5)$$

It is seen that the right-hand side of Eq. (10.5) represents a force $\frac{1}{2}\rho AV^2$ multiplied by a distance x. Now, the KE passing through the element per unit time is equal to the power rating:

$$P_{\text{element}} = \frac{d(KE)}{dt} = \frac{1}{2}\rho AxV^2\frac{dx}{dt} \qquad (10.6)$$

But the average time rate of change of the displacement, dx/dt, is the average wind velocity V

$$\frac{dx}{dt} = V \qquad (10.7)$$

The average power in the wind P_{w} is obtained by combining Eqs. (10.6) and (10.7) to give

$$P_{\text{w}} = \frac{1}{2}\rho AV^3 \qquad (10.8)$$

Equation (10.8) is the basis of all wind power and energy calculations. The most obvious feature is that the wind power is proportional to the cube of the average

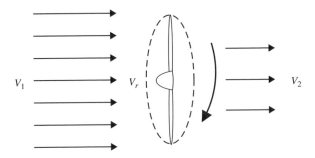

Fig. 10.10. Rotor of a wind converter.

wind speed. It is clear that the average wind speed is, by far, the dominant consideration in wind turbine location.

10.3.2. *Maximum theoretical power extractable from the wind*

Let a flow of smooth, steady air with an upstream average velocity V_1 impinge upon the rotor of a wind machine, as illustrated in Fig. 10.10. Some of the energy from the wind is transferred to the wind machine rotor so that the smooth, steady air far downstream flows at a smaller average velocity V_2. The kinetic energy reduction of the air flow, of mass m, per unit time is

$$\text{Kinetic energy} = \text{KE} = \frac{1}{2}mV_1^2 - \frac{1}{2}mV_2^2$$
$$= \frac{1}{2}m(V_1^2 - V_2^2) \tag{10.9}$$

In the process of extracting energy from the wind, the wind velocity V_r that actuates the rotor is less than the upstream "free wind" velocity V_1. With an ideal, lossless system all the energy reduction in the air stream is transferred to the rotor of the wind machine. The downstream average velocity V_2 is then smaller than the actuating speed V_r at the rotor.

Combining Eqs. (10.4) and (10.7) for the air stream at the rotor blades gives an expression for the time rate of air mass transferred.

$$\frac{dm}{dt} = \rho A \frac{dx}{dt} = \rho A V_r \tag{10.10}$$

The power at the rotor is the time rate of kinetic energy transferred,

$$P_r = \frac{d(KE)}{dt} \tag{10.11}$$

Substituting Eqs. (10.9) and (10.10) separately into Eq. (10.11) gives

$$P_r = \frac{1}{2}\frac{dm}{dt}(V_1^2 - V_2^2) = \frac{1}{2}\rho A V_r(V_1^2 - V_2^2) \tag{10.12}$$

Now the air mass passing through the rotor undergoes not only an energy reduction but a reduction of linear momentum

$$\text{reduction of linear momentum} = m(V_1 - V_2) \tag{10.13}$$

The time rate of change of momentum reduction is a force, of value

$$\frac{d}{dt}(m(V_1 - V_2)) = \frac{dm}{dt}(V_1 - V_2) = \rho A V_r (V_1 - V_2) \tag{10.14}$$

By equating the rate of change of kinetic energy transfer in Eq. (10.12) with the power associated with the rate of change of momentum from Eq. (10.14), it is found that the wind velocity V_r at the rotor may be expressed as

$$V_r = \frac{V_1 + V_2}{2} \tag{10.15}$$

In this idealized model of air flow the wind velocity at the rotor is therefore the average of the upstream and downstream steady wind velocities. Substituting for V_r from Eq. (10.15) into Eq. (10.12) gives an expression for the power extractable from the wind by the rotor. Using the symbol P_{ex} to denote extracted power,

$$P_{ex} = \frac{1}{4}\rho A(V_1 + V_2)(V_1^2 - V_2^2)$$

$$= \frac{1}{4}\rho A V_1^3 \left[1 + \frac{V_2}{V_1} - \left(\frac{V_2}{V_1}\right)^2 - \left(\frac{V_2}{V_1}\right)^3\right]$$

$$= \frac{1}{4}\rho A V_1^3 \left[1 + \frac{V_2}{V_1}\right]\left[1 - \left(\frac{V_2}{V_1}\right)^2\right] \tag{10.16}$$

The value of the wind velocity ratio V_2/V_1 that results in maximum power transfer is calculated by differentiating Eq. (10.16) with respect to (V_2/V_1) and equating to zero. This results in a quadratic equation showing that to maximize P_{ex} the ratio V_2/V_1 must have the values either $\frac{V_2}{V_1} = \frac{1}{3}$ or $\frac{V_2}{V_1} = -1$.

The negative option is meaningless, so that the correct solution is

$$\frac{V_2}{V_1} = \frac{1}{3} \tag{10.17}$$

Substituting Eq. (10.17) into Eq. (10.16) gives an expression for the maximum possible power extraction, under ideal conditions:

$$P_{ex}(\text{max}) = \frac{8}{27}\rho A V_1^3 = \left(\frac{16}{27}\right)\frac{1}{2}\rho A V_1^3$$

$$= (0.593)\frac{1}{2}\rho A V_1^3 \tag{10.18}$$

Table 10.2.   Maximum theoretical power extractable by ideal wind machine [2] ($\frac{0.593}{2}\rho AV^3$).

Wind speed (mph)	Power (kW) from circular area of different diameter (ft)				
	12.5	25	50	100	200
10	0.38	1.5	6	24	96
20	3.08	12.3	49.2	196	784
30	10.4	41.6	166.4	666	2664
40	24.6	98.4	393.6	1574	6296
50	48.2	192.8	771.2	3085	12,340
60	83.2	332.8	1331	5326	21,300

The very important result in Eq. (10.18) is sometimes referred to as Betz's law, being named after the German astroscientist Albert Betz of Göttingen. This states that "even with ideal wind energy conversion the maximum power transferrable is only 0.593 or 16/27 of the total power in the wind". In reality only a fraction of this theoretical maximum power is realized, even by the best-designed turbines.

Table 10.2 shows the values of the maximum theoretical power obtainable from a range of wind-turbine sizes at typical wind speeds [3]. It is notable how large a circular area must be used to generate any useful amount of power. For example, in a 10-mph wind, which is a light breeze, a swept area of 25-foot diameter would realize only a maximum theoretical value of 1.5 kW (and a practical value of roughly one-half of that). This immediately points to the difficulty of using wind energy for domestic use in urban areas: the swept area required is too large to be practicable. A further issue is that the lifetime energy gathered by a small wind turbine may well be smaller than the energy used in its manufacture.

The authors do not regard small wind turbines as a viable contribution to energy saving or as a useful method of large-scale electricity generation. Large wind turbines, on the other hand, are a very viable option.

Wind speed is usually expressed in miles per hour (mph) but also often in metric units of metre/sec (m/s). It is useful to use the conversion factor

$$1\,\text{mph} = 0.447\,\text{m/s} \tag{10.19}$$

10.3.3.   Practical power extractable from the wind

The power actually extractable by a wind turbine is much less than the maximum theoretical value defined in Eq. (10.18). A practical wind machine experiences air drag and air friction on the rotor blades, causing heat losses. In addition, the rotation of the blades causes swirling of the air and eddies, which reduce the torque imparted to the blades. The net effect of the various losses is incorporated into a parameter called the power coefficient C_p. With an upstream velocity V_1, the extractable power P_{ex} can be written as

$$P_{ex} = C_p \frac{1}{2} \rho A V_1^3 \tag{10.20}$$

Coefficient C_p is seen to be the ratio of the power in the wind P_w, in Eq. (10.8), and the power extracted P_{ex}, in Eq. (10.20). It therefore represents the efficiency of the turbine rotor:

$$C_p = \frac{P_{ex}}{P_W} = \text{turbine rotor efficiency} \tag{10.21}$$

Parameter C_p is a dimensionless variable. By comparison of Eqs. (10.20) and (10.16), it is seen that C_p can be expressed in terms of the upstream and downstream average wind speeds.

$$C_p = \frac{1}{2} \left[1 + \frac{V_2}{V_1} \right] \left[1 - \left(\frac{V_2}{V_1} \right)^2 \right] \tag{10.22}$$

For the ideal theoretical case, when $V_2/V_1 = 1/3$, coefficient C_p has a maximum or Betz's law value of 0.593. But for practical wind turbines, the value is usually in the range $0 \le C_p \le 0.4$. With a value $C_p = 0.4$, for example, the power available from the wind is 0.4/0.593 or about 67% of the ideal theoretical value and is 40% of the total power in the wind. Power coefficient C_p has a value that depends on the wind average velocity, the turbine rotational velocity and also on turbine blade design parameters such as the pitch angle.

The basic operating characteristic for any rotational mechanical machine is the shaft torque T versus the shaft rotational speed ω. A separate T–ω characteristic is obtained for each different value of wind speed at the turbine rotor. The value of the torque at zero speed is the starting or stall torque, caused by friction, which has to be overcome before rotation will commence. Stall torque increases, for any wind turbine, as the wind velocity V increases.

The shaft power is the product of shaft torque T and the shaft speed ω. In SI units there is no constant of proportionality, so that

$$P = T\omega \tag{10.23}$$

where P is in watts when T is in newton-metres and the rotational speed ω is in radians per second.

10.3.4.   *Tip-speed ratio*

In order to express the power coefficient C_p in terms of both the upstream wind velocity V and the blade rotational velocity ω, a parameter called the tip-speed ratio (TSR) is defined. Figure 10.11 illustrates the main physical parameters, including

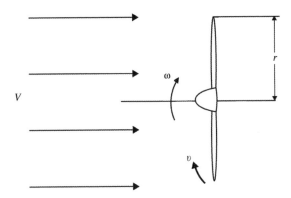

Fig. 10.11.   Motion of a two-blade propeller [2].

the blade radius r. The instantaneous velocity v of the blade tip is related to the angular velocity of rotation ω by the relationship

$$v = r\omega \tag{10.24}$$

For a blade of radius r, the TSR is defined as

$$TSR = \frac{v}{V} = \frac{r\omega}{V} \tag{10.25}$$

The blade rotational velocity n in rpm is related to the angular velocity ω in radians per second by the relationship

$$\omega = \frac{2\pi n}{60} \tag{10.26}$$

If C_p is plotted against V, there will be a different characteristic for every value of ω. Similarly, if C_p is plotted against ω, there will be a different characteristic for every value of V. The characteristic of C_p versus TSR is a "universal" curve that subsumes values of both ω and V. Good rotor design requires that the maximum value of the power coefficient C_{pm} occurs near to the design-rated value of the rotational speed. Typical characteristics for various different types of wind turbine are shown in Fig. 10.12. The maximum ideal efficiency characteristic for propeller machines is asymptotic to the Betz's law value of 0.593. It can be seen that the most efficient forms of wind converter are the propeller type, for which $0.4 \leq C_{pm} \leq 0.5$. In addition, the maximum value of the power coefficient is designed to occur in the range of TSR, namely $4 < TSR < 7$. A more detailed performance characteristic for propeller and Darrieus machines is shown in Fig. 10.13 [6]. The peak value of C_p is seen to be approaching 0.4, which is typical of small wind converters.

The power extractable from a freely flowing stream of wind, with a power coefficient $C_p = 0.4$, is shown in Fig. 10.14, using logarithmic scales on both axes.

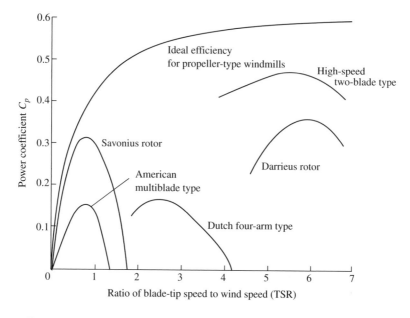

Fig. 10.12.   Power coefficient versus tip-speed ratio for various converters.

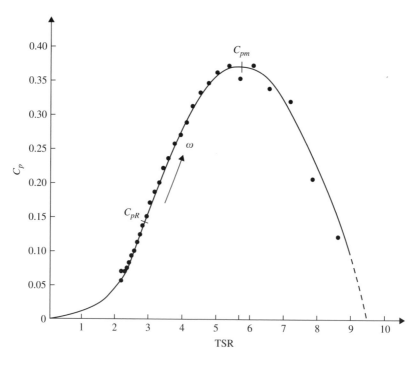

Fig. 10.13.   Power coefficient versus tip-speed ratio for Darrieus and propeller machines [6].

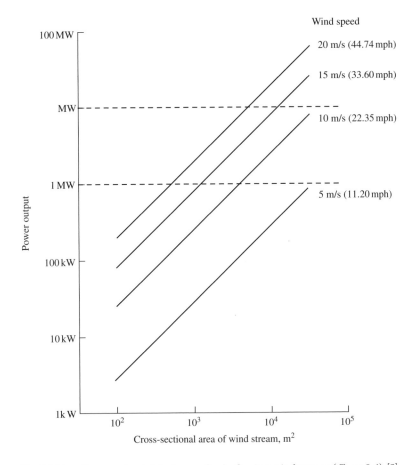

Fig. 10.14. Power extractable from a freely flowing windstream ($C_\text{p} = 0.4$) [2].

10.4. Mechanical Features of Wind Machines

10.4.1. *Axial thrust (pressure)*

The action of the wind-stream onto the rotating propeller, as in Fig. 10.3, is to create a pressure force acting along the horizontal shaft, called the thrust, *Th*. A detailed aerodynamic analysis (not given here) shows that the thrust may be expressed as

$$\text{Axial thrust} = Th = \frac{1}{2}\rho A V_1^2 \left[1 - \left(\frac{V_2}{V_1}\right)^2\right] \tag{10.27}$$

The thrust may be expressed in terms of the extracted power P_ex by comparing Eqs. (10.27) and (10.16):

$$Th = \frac{P_\text{ex}}{V_1 \left[\frac{V_1+V_2}{2}\right]} = \frac{P_\text{ex}}{V_1 V_r} \tag{10.28}$$

Like the extracted power, the thrust per unit area in the wind-stream is determined entirely by the wind velocities. The thrust has to be counteracted by the end bearings on the propeller shaft.

10.4.2.   The "yaw" effect

The wind at a given site is subject to rapid and frequent changes of direction. But to maintain efficient operation, the turbine propeller plane must remain perpendicular to the wind direction. This requires that the turbine assembly be free to rotate about a vertical axis; a phenomenon that aeronautical engineers call the "yaw" effect. With good bearings, a machine can be pivoted to swivel under the influence of a vane or a rudder wheel mounted downwind, as illustrated in Figs. 10.15 and 10.16.

In large modern wind turbines, a weather vane monitors the wind direction and an electric yaw drive is used to swivel the propeller plane broadside onto the wind.

10.4.3.   Gyroscopic forces and vibrations

Yawing rotation about the vertical axis while the rotor is turning about its horizontal axis encounters strong gyroscopic forces. These forces have to be transmitted through the bearings and propeller shaft, causing high stresses and vibrations. For this reason, the propeller blades of large machines are made of a lightweight material such as a composite plastic like fibreglass rather than metal.

The action of rotation of the blades results in periodic vibrations. With a downwind-designed machine, which is characteristic of many large systems, each rotating blade passes through the wind shadow of the tower once per rotation. This results in a sudden reduction of air pressure on each blade followed by a sudden increase of air pressure, as it emerges from the shadow of the tower. The result is to apply a bending moment on each blade at its root or hub joint in alternate directions. Continual flexing of the propeller blades at every rotation produces fatigue stresses in the materials. With two-blade propellers, sometimes the whole rotor is mounted on a single-shaft hinge, allowing fore-aft rotation or "teetering" to reduce out-of-plane bending moment fluctuations. In order to minimize the vibration problem, some wind machine designers prefer to use three-blade propellers rather than two-blade propellers, even with the additional cost of the extra blade.

10.4.4.   Centrifugal forces

The rotation of the blades of a wind turbine causes outward acting centrifugal forces. This phenomenon can be experienced by tying a weight at the end of a string and swinging it around. The outward acting force depends directly on the mass or weight and on the speed of rotation. Calculation of the centrifugal forces on

Fig. 10.15.   American farm windmill used for pumping water [7].

a wind turbine tending to pull the rotating propeller blades out of their sockets is complicated because the weight is distributed non-uniformly along the length of the blade. A simple calculation that assumed all the weight to be concentrated at a fixed radius of rotation would give inaccurate results. In large modern wind turbines, the blades are large and heavy. Moreover, the cost of the blades and propeller unit is a significant portion of the total system cost.

The amount of power taken from the wind at a fixed wind velocity can be adjusted by varying the pitch angle of the propeller blades. This is realized by rotating part of the propeller arms in their sockets, such as adjusting a screw or bolt. In effect, this changes the force and torque exerted on the rotating propeller.

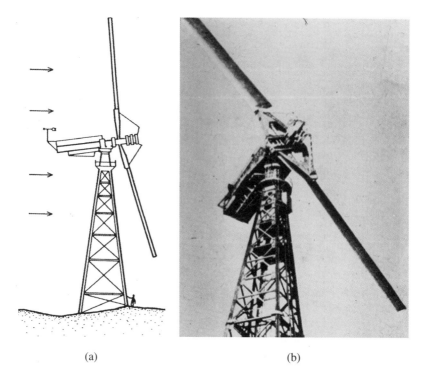

(a)   (b)

Fig. 10.16.   Smith–Putnam wind machine [3], (a) diagrammatic (b) on-site photograph.

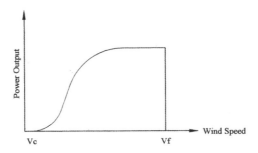

Fig. 10.17.   Effect of feathering the propeller [2]. V_c = cut-in speed; V_f = furling speed.

The same principle is used in landing a propeller-driven aeroplane to change the thrust on the blades and thereby reduce the speed. The use of the same technique enables the power extracted from a propeller to be kept constant over a range of wind speeds, illustrated in Fig. 10.17. When the wind reaches a maximum acceptable level known as the furling velocity V_f, the pitch angle of the blades can be adjusted so that zero power is extracted. In severe wind conditions, some form of mechanical brake is also applied.

10.4.5.   *Solidity factor*

The solidity factor is defined as the total blade area of the rotor divided by the area swept normal to the wind. In a horizontal axis, a propeller machine for example, an efficiently-designed aerofoil intercepts a large area of wind with a small area of blade. It therefore has a low-solidity factor, which is highly desirable in high-rotational speed systems.

Turbines with high-solidity factor usually suffer from a high degree of aerodynamic interference between the blades, which results in low values of TSR and power coefficient C_p. Examples are the Savonius rotor and the American farm multi-blade type, with the typical performance characteristic given in Fig. 10.12. Wind turbines with high solidity usually operate at low-rotational speeds but have high-starting torques. They are used for direct mechanical applications such as water pumping but are not usually suitable for driving electric generators. For the purpose of electricity generation, it is usual to use low-solidity machines, such as the two-blade propeller, in order to utilize high-operational speeds and high values of power coefficient.

10.4.6.   *Two rotor blades or three rotor blades?*

Most large modern wind turbines are horizontal axis propeller machines, having either two blades or three blades on the rotor. There are long-standing and ongoing differences of view amongst wind and aeronautical engineers as to the merits of the two designs. The maximum achievable values of power coefficient C_p over a range of values of TSR have been calculated under the idealized condition of no aerodynamic drag for the rotors with several blade numbers.

Within the normal working range of TSR, a three-blade propeller has a slightly larger value (e.g. 5%) of power coefficient. But most two-bladed wind turbines use a higher value of TSR than most three-bladed machines. There is little practical difference in the maximum achievable C_p between two- and three-bladed designs, assuming no drag.

Three-blade machines have the advantage that the polar moment of inertia with respect to yawing is constant, which contributes to smooth operation. A two-bladed rotor has a lower moment of inertia when the blades are vertical than when they are horizontal, creating rotation imbalance. An important consideration in selecting the number of blades is that the blade root stress increases with blade number for a turbine of given solidity. In general, increasing the design TSR entails decreasing the number of blades [5].

10.4.7.   *Shaft torque and power*

Most wind energy systems are used to generate electricity. The wind turbine is usually coupled to a generator directly, as in Fig. 10.18(a), or via a gearbox to step up the generator shaft speed, as in Fig. 10.18(b). For this reason, the generator

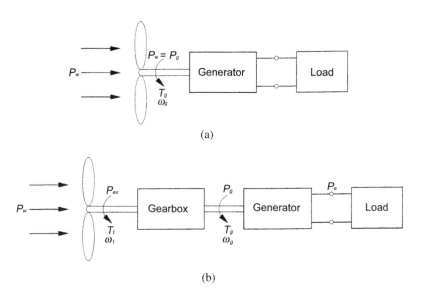

Fig. 10.18.   Power train for wind-powered electricity generation [1]. (a) Direct-on load. (b) Gear box system.

is usually mounted at the top of the supporting tower along with the gearbox. Electric cables run down the tower to connect the generator to its electrical load on the ground below. The torque, speed and power of a rotating shaft are linked by the relationship, given in Eq. (10.23).

In Fig. 10.18(b), the shaft torque into the gearbox T_t from the turbine shaft is given by

$$T_{\text{t}} = \frac{P_{\text{ex}}}{\omega_{\text{t}}}$$

(10.29)

Similarly, the torque on the gearbox output shaft into the generator T_g is given by

$$T_{\text{g}} = \frac{P_{\text{g}}}{\omega_{\text{g}}}$$

(10.30)

Considerable torsional shear stress is imposed on a shaft due to rotational forces. For a solid cylindrical shaft subjected to a torque T, the torsional stress f_s at any arbitrary radius r_s (Fig. 10.19) is given by

$$f_{\text{s}} = \frac{T \cdot r_{\text{s}}}{J} \ N/m^2$$

(10.31)

where J is the polar (area) moment of inertia having the dimension $(\text{mass})^4$ or m^4. For a solid cylindrical shaft of outer radius r_o, the polar moment of inertia can be shown to be

$$J = \frac{\pi r_o^4}{2} \ m^4$$

(10.32)

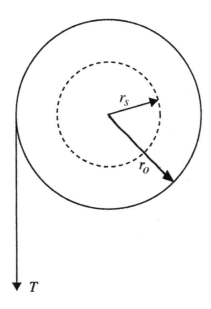

Fig. 10.19.   Wind machine shaft.

Combining Eqs. (10.31) and (10.32) gives an expression for the shear stress at the surface of a solid cylindrical shaft of radius r_o:

$$f_s = \frac{2T}{\pi r_o^3} \; N/m^2 \tag{10.33}$$

10.5.   Fixed Rotational Speed or Variable Rotational Speed?

The ambient wind in any location is variable for both wind speed and direction. In addition, a turbine is subject to turbulent wind gusts of a transient and unpredictable nature. A design choice has to be made between operating a turbine at variable rotational speed following the wind and regulating the speed of rotation to create a fixed speed or a choice of two (usually) different fixed speeds of rotation. For either option, the turbine must be capable of being completely stalled into total immobility at some predetermined safe maximum operating speed.

Any wind energy system design must aim to optimize the annual energy capture at its given site. In order to operate at its highest efficiency (i.e. with maximum power coefficient C_p), a turbine must operate at its optimum value of TSR as the wind speed varies, as illustrated in Fig. 10.13. The best condition to be aimed for in design is for the turbine to operate, at all wind speeds, at a value of TSR at or close to the value that results in maximum power coefficient. But since the wind speed varies, the design issue is therefore either: (a) to operate the turbine rotor at a fixed speed by (say) adjusting the pitch angle of the turbine propellers as the wind speed changes; or (b) to permit the rotational speed to change, following the variable wind speed.

10.5.1.   *Constant speed operation*

Operation at a constant speed may be realized by one of the two basic control methods.

1. By varying the pitch angle of the propeller blades as the wind speed varies. This can be achieved by rotating either the whole propeller blades or the tips of the propeller blades in their sockets. This form of control is usually called "pitch angle control".
2. By the use of a propeller of fixed pitch angle but where the propeller surfaces are designed to introduce stall over a range of wind speeds. This form of design is usually referred to as "stall regulation" or "stall control".

The two design methods lead to very similar turbine power characteristics, as shown in Fig. 10.20. At low wind speeds (1–3 m/s), the turbines are shut down. Start-up begins at a cut-in speed between 2.5 and 5 m/s. Rated wind speed, at which the nominal output is reached, is in the range of 12–15 m/s. Below the nominal wind speed, the aim is to maximize the turbine rotor efficiency [8].

When the turbine rotational speed is constant, a coupled AC generator will operate at a fixed frequency, which can be synchronized to the frequency of the electrical system to which it is connected. It is not then necessary to use any form of electronic frequency changer as a decoupling device, as is required in variable speed systems. This results in a simpler and cheaper electrical system.

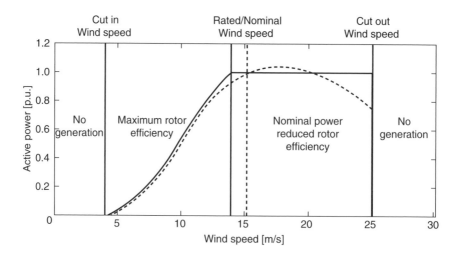

Fig. 10.20.   Active power versus wind speed [6], — pitch controlled and ---- stall controlled.

10.5.2.   *Variable speed operation*

From Eq. (10.25) it is seen that to keep the TSR constant, at its desired value, it is necessary to control the turbine rotational speed ω in proportion to the wind speed V. Alternatively, a turbine can be operated at one of two fixed rotational speeds, which gives a TSR closer to the optimal than a single fixed speed, but is slightly less efficient.

Variable speed control is more expensive than fixed speed control but is found to yield 20%–30% more energy. The additional expense of variable speed control arises from the need for a controlled and variable-pitch propeller and also from the need for a power electronic frequency controller between the variable frequency generator output and the fixed frequency bus bars of the load system.

10.6.   Efficiency Considerations in Wind-Powered Turbine Systems [2]

A power flow diagram for a basic wind converter system is given in Fig. 10.18. The combination of turbine, gearbox and generator is sometimes called a power train. The stage efficiencies of the various stages are given by

$$\text{turbine efficiency} = \frac{P_{ex}}{P_w} = C_p \tag{10.34}$$

$$\text{gearbox efficiency} = \frac{P_g}{P_{ex}} = \eta_{gb} \tag{10.35}$$

$$\text{generator efficiency} = \frac{P_e}{P_g} = \eta_g \tag{10.36}$$

The overall efficiency η of the three-stage system of Fig. 10.18, from input to output, is

$$\eta = \frac{\text{electrical output power}}{\text{power available in the wind}}$$

$$= \frac{P_e}{P_w} = \frac{P_{ex}}{P_w} \cdot \frac{P_g}{P_{ex}} \cdot \frac{P_e}{P_g} \tag{10.37}$$

In terms of the individual stage efficiencies, from Eqs. (10.34)–(10.36), the overall efficiency η can be written as

$$\eta = \frac{P_e}{P_w} = C_p \cdot \eta_{gb} \cdot \eta_g \tag{10.38}$$

For small wind energy installation, up to a few kW output rating, the overall efficiency η is of the order 20%–25%.

The electrical output power may be written in terms of the total power in the wind P_w and the wind velocity V, utilizing Eq. (10.38) and substituting from the basic Equation (10.8).

$$P_e = C_p \cdot \eta_{gb} \cdot \eta_g \cdot P_W = C_p \cdot \eta_{gb} \cdot \eta_g \cdot \frac{1}{2}\rho A V^3 \tag{10.39}$$

Values of the power coefficient C_p for the two-blade propeller type of the wind turbine vary with size and rating. They are found to be in the ranges shown below.

$$\begin{aligned}
C_p &= 0.4\text{--}0.5 \text{ for large machines } (100\,\text{kW--}3\,\text{MW}) \\
&= 0.2\text{--}0.4 \text{ for small machines } (1\,\text{kW--}100\,\text{kW}) \\
&= 0.35 \text{ typically for small machines} \tag{10.40}
\end{aligned}$$

Modern types of mechanical gearbox have efficiencies that depend both on the size (rating) and on the speed of rotation. At the rated speeds, it is found that the gearbox efficiencies lie in the following ranges:

$$\begin{aligned}
\eta_{gb} &= 80\%\text{--}95\% \text{ for large machines} \\
&= 70\%\text{--}80\% \text{ for small machines} \tag{10.41}
\end{aligned}$$

Electrical generators have efficiencies that increase with rated size. The process of converting rotational mechanical energy to electrical energy is inherently more efficient than wind-turbine conversion or than any process involving heat to work (i.e. thermodynamic) conversion. For operation at rated output:

$$\begin{aligned}
\eta_g &= 80\%\text{--}95\% \text{ for large machines} \\
&= 60\%\text{--}80\% \text{ for small machines} \tag{10.42}
\end{aligned}$$

It is notable that large systems are more efficient than small systems. In general, an increase of scale is accompanied by an increase of efficiency. Typical values of overall efficiency η can be obtained by substituting typical values of stage efficiencies into Eq. (10.36).

For large machines,

$$\eta = 0.45 \times 0.9 \times 0.9 = 0.36$$

For small machines,

$$\eta = 0.35 \times 0.75 \times 0.7 = 0.18 \tag{10.43}$$

10.7.   Wind Energy Turbines

10.7.1.   *19th-century windmills*

The use of wind power, mainly for the pumping of water, was very extensive in Europe and North America during the 19th century. It is estimated that, since the mid-19th century, more than six million machines of the American farm type, shown in Fig. 10.15, have been used in the USA [9]. The metal-blade rotor is roughly 12–16 feet in diameter, with a tail-vane stabilizer to keep the multi-sails upstream and is ideal for remote farm locations. Each individual turbine is rated at less than one horsepower (HP). Typically for this type of machine the solidity factor is high, being of the order 0.7.

By 1850, the use of windmills in the USA provided about 109 kWh of energy per year [10]. This type of small system continued into the 1930s in the USA, both for pumping water and to power small electric generators, rated up to 2 kW, used for charging lead-acid storage batteries. These, in turn, provided power for the lighting of barns and animal sheds.

In Europe, corresponding developments took place in several countries, notably Denmark, Germany, the Netherlands, Russia and Great Britain. For example in Denmark, by the end of the 19th century about 2,500 industrial windmills were in operation, having a total rating of about 40000 HP (30 MW). Most of these were in rural areas. Also, there were approximately 4,600 windmills on Danish farms, used for threshing and milling grain and for the pumping of water. By the 1930s, industrial applications were much reduced but the number of small farm units had increased to about 16,000 [9]. Denmark has always been a world leader in the design and use of wind-energy systems and remains so to the present day.

10.7.2.   *Early 20th-century wind energy turbines*

Towards the end of the 19th century and into the 20th century, the application of wind energy changed from milling grain and water pumping to the generation of electricity. This required the development of turbines of low solidity factor and much higher speeds than were realizable by multi-blade farm windmills.

The first use of a large windmill to generate electricity in the USA was a system built in Cleveland, Ohio, in 1888 by Charles F. Brush. Brush's machine had a multi-bladed rotor, 17 metres in diameter, with a large fan tail. It was the first wind machine to incorporate a step-up gearbox (gear ratio 50:1) and drove a direct current (DC) generator at 500 rpm. The generator was rated at 12 kW and operated for 20 years [4]. By 1920, the two dominant rotor configurations, fan-type and sail-type, had both been found to be inadequate for generating appreciable amounts of electricity. The further development of wind-powered electrical systems in the USA was inspired by the design of aeroplane propellers and aerofoil wing sections [11].

By the mid-1920s, small (1–3 kW) wind-powered DC generators were in widespread use in rural areas, driven by low-maintenance three-blade propeller turbines mounted on tall (e.g. 70 foot) towers. The predominant companies in the USA were Parris-Dunn and Jacobs Wind-Electric. These were gradually forced out of business by the customer demand for larger amounts of grid (utility) supplied electricity. The escalating price of oil in the latter years of the 20th century has caused a comeback and modern successors of the early machines can now be seen all over the country.

In 1891, Professor Poul La Cour in Denmark developed the world first wind-powered electricity-generating system to employ aerodynamic design principles. The machines incorporated low-solidity, four-blade rotors 75 feet in diameter that were designed with primitive aerofoil sections [11]. By about 1920, the use of 25-kW, high-speed wind generators was common in Denmark. As in America, however, cheaper fossil-fuel steam plants put the wind-powered electricity generation industry out of business for many years.

Large-scale wind-energy conversion systems based on aerodynamic designs were first undertaken in Russia in 1931. The 100-kW wind turbine had a generator and controls mounted at the top of a 100-foot tower. The heel of the inclined strut was mounted on a carriage that ran on a circular track to keep the rotor facing into the wind [8]. This turbine was connected by a 6300-V transmission line to a 20-MW steam-powered station 20 miles away. Subsequent experimental large-scale systems in Denmark, France, Germany, Great Britain and the USA all failed to result in practical economic designs.

The first successful large wind turbine was the Smith–Putnam machine built at Grandpa's Knob, Vermont, USA in 1941 [12]. This privately-funded venture proved to be the prototype and inspiration for what has now become a vast industry. As with all horizontal axis propeller machines, the generator and gearbox were mounted on the turbine shaft in a housing at the top of the tower (Fig. 10.16). The Smith–Putnam machine was a two-blade propeller downstream system, rated at 1.25 MW, with a blade diameter of 53.34 metres (175 feet). It operated at a constant rotational speed of 28 rpm. For 35 years, it held the record as the world's largest wind machine system. The electric generator was a synchronous machine that fed electrical power directly into the Central Vermont Public Service Corporation electricity grid.

The Smith–Putnam machine suffered two mechanical failures. After a main bearing replacement there was a spar failure, causing one of the propeller blades to fly off. The operating company decided that a repair would be uneconomical and the venture was closed down in 1945. Although the Smith–Putnam machine operated for only 18 months in all, it was a proving ground for high-power wind generation systems. The mechanical failures were due to the limitations of the knowledge of the materials available at the time and not to the basic system design. Better materials are now available. Also, the engineering knowledge about the bearing design and about the fatigue failure of metals makes it unlikely that the Smith–Putnam type of failure would occur in modern wind turbines. Although the Smith–Putnam project was superficially a failure, it worked well for long enough to prove that large-scale, wind-powered electricity was feasible. All of the principal engineering challenges were overcome, opening the way to future developments.

Considerable work was done on wind-powered electricity generation in Great Britain in the late 1940s and the 1950s [3]. For example, in 1950, the North Scotland Hydroelectric Board commissioned an experimental 100-kW, three-blade machine. This operated for short periods in 1955 coupled to a diesel-powered electricity network but had to be shut down because of operational problems [10].

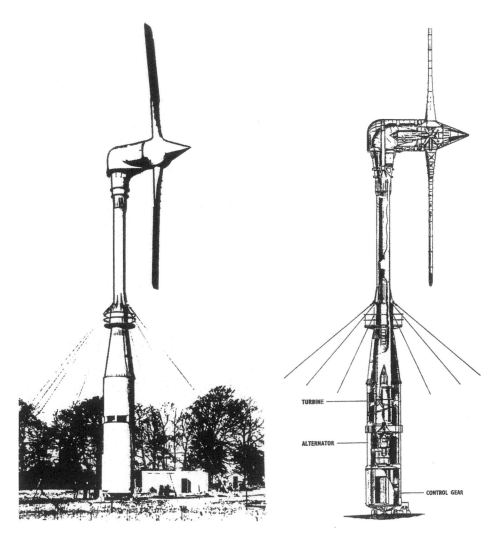

Fig. 10.21.   Enfield–Andreau wind turbine [3] 100 kW (33 mph rate wind speed).

In the 1950s, the Enfield Electric Cable Company built a unique 100-kW wind-powered generator (Fig. 10.21), at St. Albans, England. The rather dramatic aeronautical-looking design was due to a Frenchman named Andreau. This consisted of a hollow tower 85 feet high and a hollow rotor with openings at the blade tips. A pressure differential drove air from the openings near the base of the tower, up the tower, through an air turbine and out of the rotor blade tips. The design proved to be of low efficiency and was moved to the windier coast of North Wales. Because of the local environmental objections, it was then sold on to Algeria, where it operated successfully for a number of years.

Fig. 10.22.   Restored Gedser wind turbine, 200 kW at 33.6 mph [13].

10.7.3.   *Later 20th-century wind energy turbines*

After the Second World War, through the 1950s, a series of experimental wind-energy turbines were developed in Denmark. The latest in the series was the 200-kW Gedser mill (Fig. 10.22), which was a horizontal-axis three-blade propeller machine.

This operated until 1968 but was then closed down because it was uneconomic compared with fossil-fuel-powered steam plants. The energy crisis precipitated by the sudden increases in the price of oil in 1973 caused a change of policy so that the Gedser mill was refurbished and reinstated in operation in 1977 [10].

Several three-blade propeller designs of wind-operated electric generators were developed in France during the period 1958–1966. One unit, near Paris, was rated at 800 kW, with a rotor of 100-foot diameter mounted on a 100-foot tower. This operated at the constant speed of 47 rpm, with its synchronous alternator, operating at 1000 rpm and 3000 V, connected to a 50-Hz and 60-kV electricity grid via a step-up transformer and a 15-km transmission line. A further unit, in southern France, operated at 56 rpm with a 70-foot diameter rotor. This used an asynchronous generator with a nominal speed of 1530 rpm [10].

In Germany, Professor Ulrich Hütter developed a series of advanced horizontal-axis designs that utilized aerofoil-type fibreglass and plastic blades with variable pitch, using diameters as large as 110 feet. The two-blade propeller design sought to reduce bearing and structural failures by techniques of load shedding. One of the most innovative of such schemes was the use of a hinge bearing at the rotor hub that allowed the rotor to "teeter" in response to wind gusts and vertical wind shear. The German design features were later used by US designers [11].

10.8. Modern Large Wind Power Installations

10.8.1. *Review of some installations*

The design construction and operation of large (i.e. a few megawatts (MW) down to hundreds of kilowatts (kW)) wind turbine systems for electricity generation is now a well-established technology. Most large-scale systems use horizontal axis and two-blade propeller designs, but three-blade systems are also widely used. Although the basic scientific and engineering problems of turbine design and location have been solved, there still remain major problems arising from political, environmental and economic considerations.

A crisis in the price and world supply of oil from the Middle Eastern oil-producing countries in the early 1970s was a great incentive to the industrialized countries of Western Europe and North America to investigate alternative supplies of energy. Since the early 1970s, the USA, in particular, has made a massive investment of effort in the development of wind turbines and wind-powered electricity generation schemes. This has been organized via the US Electrical Development Association (ERDA).

In 1974, the ERDA Model Zero (MOD-0) 100-kW machine was reported [14]. This was a two-blade variable-pitch propeller downstream turbine with a diameter of 38.1 metres (125 feet). The AC synchronous generator was driven via a step-up

Fig. 10.23.   American NASA MOD-2 machine, 2.4-MW, two-blade propeller.

gearbox and fed electrical power directly into the local grid. Subsequent MOD-0 machines were upgraded to 200-kW rating.

The MOD-1 wind turbine system, commissioned in 1979, became, at the time, the world's largest machine. This was a two-blade propeller downwind machine, rated at 2 MW, with a blade diameter of 60.96 metres (200 feet). The MOD-1 design was followed in 1980 by the first MOD-2 machine, rated at 2.5 MW, with a blade diameter of 91 metres (298.6 feet) and upstream orientation (Fig. 10.23). A group of three MOD-2 machines on a fixed site, with a hub height of 60 metres and providing a test bed for examining the effects of machine clustering, was operating in 1984. A MOD-5 programme involving two-blade upwind machines rated at 7.3 MW, with a blade diameter of 122 metres (400 feet), was planned for the late 1980s. This plan was abandoned, not for technical reasons but because the reduced price of oil and the reduced demand for electricity made it then uneconomical [15].

Large wind machines have a cut-in speed V_c (Fig. 10.17) of about 10 mph and rotate at speeds up to about 100 rpm. For economic reasons, large systems need to operate in locations where the average wind speed is greater than 15 mph.

Wind turbine design and development in Europe are dominated by Denmark and Germany. During the 1980s, work concentrated on turbines with ratings of several hundreds of kW. These are now being up-scaled to 800–900 kW. The first generation of 1.5-MW machines in 1995–1996 was successful, and by the year 2002 there were 1,100 machines of this class in operation. A total of 1,496 turbines of average rating of 1150 kW were installed in Germany during the year 2000. The largest wind farm in Europe is an array of offshore 2 MW three-blade propeller turbines at Middelgrunden, Denmark [16]. In 2010, Denmark obtained 20% of its electricity from the wind and in 2004 was the world's leading manufacturer and exporter of wind turbines [17].

One of the largest individual wind turbines now operating, very successfully, in the British Isles is situated in Orkney. Located off the north coast of Scotland, the Orkney Islands are in one of the windiest locations in the world, with average wind speeds of the order 17 m/s (17/0.447 = 38 mph). The high wind speed permits the Enercon 70 2.3-MW upstream machine to use the relatively small blade diameter of about 60 metres (197 feet) at a hub height of about 46 metres (151 feet). Electrical power generated by the Orkney machine is used in the islands and replaces the expensive diesel–electric generation previously used [14]. An impression of the size and internal complexity of a large turbine can be gauged from Fig. 10.24 [14]. Wind speeds up to 60 mph can be utilized. The physical scale of modern turbines is illustrated in Fig. 10.25 [18].

An assessment of the necessary scale of the wind generation can be made by comparing the output power, 3 MW, of modern high-efficiency turbines with a conventional large power station that produces 1000–2000 MW. To replace the electrical output power of a 1000-MW fossil fuel or nuclear station would require 435 Orkney-size machines or 400 ERDA MOD-2 machines. To take account of the necessary spacing between large wind turbines in the same cluster (i.e. approximately 1,500 feet for 2-MW machines), 1000 MW of generation could require as much as 500 square miles of ground site. This could create local ecological disruption and aesthetic complications. In the USA, the necessary tracts of land are available in the area of the windy western Great Plains. In the much smaller countries of western Europe, land is scarcer, dearer, and in demand for other uses, such as farming. The land area required for wind-generated electricity can also be used if it is suitable for the farming of crops or grazing of animals.

In April 2001, the UK government accepted bids for sites designed to produce 1500 MW of wind-powered generating capacity. Further, in December 2003, bids were received for 15 additional offshore sites with a generating capacity of 7000 MW. The total project is intended to satisfy the residential electrical power requirements of 10 million people [17].

Most of the 28400 MW of generation capacity on land now has been developed in Europe and much exploration is taking place in various offshore sites. For example, the Nysted wind farm in Denmark consists of 72 Bonus turbines, each of 2.3-MW capacity [19]. Also, the Arklow Bank wind farm, off the Irish coast, uses turbines of greater capacity than 3 MW [20]. As the capacity of the turbines continues to increase, the challenges of interconnection and energy storage will become more demanding.

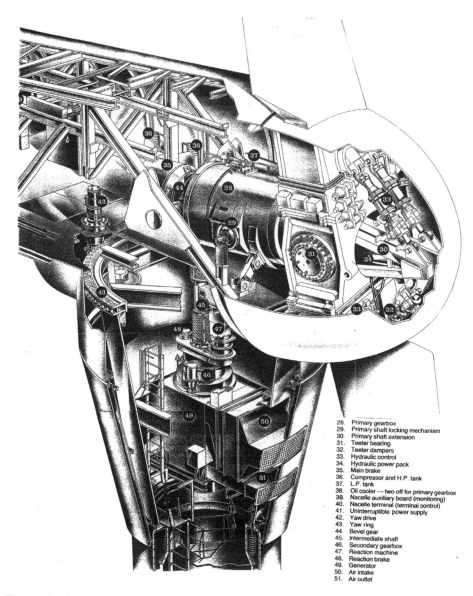

28. Primary gearbox
29. Primary shaft locking mechanism
30. Primary shaft extension
31. Teeter bearing
32. Teeter dampers
33. Hydraulic control
34. Hydraulic power pack
35. Main brake
36. Compressor and H.P. tank
37. L.P. tank
38. Oil cooler — two off for primary gearbox
39. Nacelle auxiliary board (monitoring)
40. Nacelle terminal (terminal control)
41. Uninterruptible power supply
42. Yaw drive
43. Yaw ring
44. Bevel gear
45. Intermediate shaft
46. Secondary gearbox
47. Reaction machine
48. Reaction brake
49. Generator
50. Air intake
51. Air outlet

Fig. 10.24.   Artistic impression of the internal construction of a 3-MW wind machine (reproduced by permission of the Renewable Energy Enquiries Bureau, Energy Technology Support Unit, Harwell, UK).

The modern wind-generation industry, with its aeronautical rotor designs, was born in California, the USA, during the early 1980s. When the US Department of Energy (DOE) released its first wind-energy resource inventory in 1991, it pointed out that three wind-rich states — North Dakota, Kansas, and Texas — had enough harnessable wind energy to satisfy the nation's electricity needs.

Fig. 10.25.   Bovin wind farm, France [15] N80/2500 kW turbines.

Advances in wind turbine design since then enable turbines to operate at lower wind speeds, which increases the efficiency of conversion and also enlarges the regime of wind capture. In 2004, new wind turbine towers were 100-metres tall, with much longer blades than the designs of 10 years earlier. This results in approximately tripling the amount of harvestable wind. Since the early 1980s wind-generated electricity in the USA reduced in cost from \$0.38/kWh to \$0.04/kWh in the prime wind states [17].

The USA is the world's greatest economic and military power. Its economy is greatly tied to a carbon-dominated energy market and particularly to the price of oil, of which it is the world's largest per-capita consumer [18]. The USA in 2009 had the greatest installed capacity of wind generation, as shown in Table 10.3. However, China by 2012 had a greater wind utilization growth rate than the USA. Regardless, the very large land area of the USA includes many sites suitable for large wind farms, such as the inaptly named Carbon County, Wyoming. It remains

Table 10.3.   Installed Windpower Capacity (MW) 2002–2009 [21].

Rank	Nation	2002	2003	2004	2005	2006	2007	2008	2009	1 Yr % growth	5 Yr avg. % growth
—	**World**	**31,180**	**39,295**	**47,693**	**59,024.10**	**74,150.80**	**93,926.80**	**121,187.90**	**157,899**	**30.3**	**25.3**
—	European Union	—	—	—	40,722	48,122	56,614	65,255	74,767	12.1	—
1	United States	4,685	6,370	6,725	9,149	11,603	16,818.80	25,170.00	35,159	39.7	31.6
2	Germany	12,001	14,609.10	16,628.80	18,427.50	20,622	22,247.40	23,902.80	25,777	7.8	10.3
3	China	468	567	764	1,266	2,599	5,912	12,210.00	25,104	105.6	84.8
4	Spain	4,830	6,202	8,263	10,027.90	11,630	15,145.10	16,740.30	19,149	14.4	22
5	India	1,702	2,110	3,000	4,430	6,270	7,850	9,587.00	10,925	14	35.4
6	Italy	785	904	1,265	1,718.30	2,123.40	2,726.10	3,736.00	4,850	29.8	32.8
7	France	148	248	386	757.2	1,567	2,455	3,404.00	4,410	29.6	68.9
8	United Kingdom	552	648	888	1,353	1,962.90	2,389	3,287.90	4,070	23.8	38.4
9	Portugal	194	299	522	1,022	1,716	2,130	2,862.00	3,535	23.5	57.1
10	Denmark	2,880	3,110	3,124	3,128	3,136	3,125	3,160.00	3,465	9.7	0.3
11	Canada	236	322	444	683	1,460	1,846	2,369.00	3,319	40.1	49.1
12	Netherlands	682	908	1,078	1,224	1,559	1,747	2,225.00	2,229	0	19.6
13	Japan	334	506	896.2	1,040	1,309	1,528	1,880.00	2,056	9.4	30
14	Australia	103	197.2	379	579	817.3	817.3	1,494.00	1,712	14.6	49.9
15	Sweden	345	404	452	509.1	571.2	831	1066.9	1,560	28.4	21.4
16	Ireland	137	186	338.9	495.2	746	805	1,244.70	1,260	54.6	46.3
17	Greece	276	365	473	573.3	757.6	873.3	989.7	1,087	13.3	22.1
18	Austria	139.3	415	606	819	964.5	981.5	994.9	995	1.4	19.1
19	Turkey	19.4	20.6	20.6	20.1	64.6	206.8	333.4	801	61.2	74.5
20	Poland	28.2	58.2	58.2	73	153	276	472	725	71	52
21	Brazil	22	23.8	23.8	28.6	236.9	247.1	338.5	606	37	70.1
22	Belgium	44	68	95	167.4	194.3	286.9	383.6	563	33.7	41.3
23	New Zealand	35	36.3	168.1	168.2	171	321.8	325.3	497	1.1	55.1
24	Taiwan	0	0	13	103.7	187.7	279.9	358.2	436	28	129.0
25	Norway	97.3	100	270	268	325	333	428	431	28.5	33.7

(Continued)

Table 10.3. (*Continued*)

Rank	Nation	2002	2003	2004	2005	2006	2007	2008	2009	1 Yr % growth	5 Yr avg. % growth
26	Egypt	69	180	145	145	230	310	390	430	25.8	16.7
27	Mexico	0	0	2.2	2.2	84	85	85	402	0	
28	South Korea	16	18.7	22.5	119.1	176.3	192.1	278	348	44.7	71.6
29	Morocco	53.9	53.9	53.9	64	64	125.2	125.2	253	0	18.4
30	Hungary	1.2	2	3.5	17.5	60.9	65	127	201	95.4	129.4
31	Czech Republic	3	10	16.5	29.5	56.5	116	150	192	29.3	71.9
32	Bulgaria	0	10	10	14	36	56.9	157.5	177	176.8	73.6
33	Chile	1.3	2	2	2	2	20.1	20.1	168	0	58.6
34	Finland	41	51	82	82	86	110	140	147	27.3	22.4
35	Estonia	2	3.7	22.1	33	33	58.6	78.3	142	33.6	84.1
36	Costa Rica	69.9	69.9	69.9	71	74	74	74	123	0	1.1
37	Ukraine	46	56.4	68.8	77.3	85.6	89	90	94	1.1	9.8
38	Iran	11	11	25	31.6	47.4	66.5	82	91	23.3	49.4
39	Lithuania	0.2	2	7	7	55	52.3	54.4	91	4	93.6
40	Romania	1	1	0.6	0.9	2.8	7.8	76	—	874	138
41	Luxembourg	16.1	21.5	35.3	35.3	35.3	35.3	35.3	—	0	10.4
42	Latvia	24	24	26.7	27.4	27.4	27.4	30	—	9.5	4.6
43	Argentina	25.7	25.7	25.6	26.8	27.8	29.8	29.8	—	0	3
44	Philippines	0	0	25.2	25.2	25.2	25.2	25.2	—	0	—
45	South Africa	13	16.4	16.55	16.6	16.6	16.6	21.8	—	31.3	5.9
46	Jamaica	0	0	20.7	20.7	20.7	20.7	20.7	—	0	—
47	Guadeloupe	0	0	20.5	20.5	20.5	20.5	20.5	—	0	
48	Uruguay	0.2	0.2	0.15	0.2	0.2	0.6	20.5	—	3320	152.4
49	Tunisia	19	10	20	20	20	20	20	—	0	14.9
50	Colombia	0	19.5	19.5	19.5	19.5	19.5	19.5	—	0	0
51	Croatia	0	0	6	6	17.2	17.2	18.2	—	5.8	—
52	Russia	10.7	10.8	10.8	14	15.5	16.5	16.5	—	0	8.8

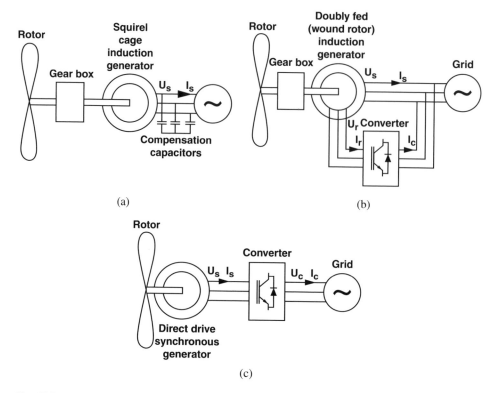

Fig. 10.26.   Main types of wind generator system [1]. (a) Fixed speed with directly grid-coupled induction generator. (b) Variable speed with doubly fed induction generator. (c) Variable speed based on a direct drive system and synchronous generator.

an interesting question as to how the use of the renewable forms of energy, notably wind energy, will advance in competition with the large and traditional carbon-based industries of coal, oil and natural gas.

10.8.2.   *Types of wind generator systems* [1]

The voltage and frequency of the electrical grid into which the incoming wind generator is to be connected are rigidly fixed. To accommodate variations of voltage and frequency in the incoming generators it is necessary to use various interconnection techniques. The main types of wind generator systems used in Europe are shown in Fig. 10.26 [8].

10.8.2.1.   *Fixed-speed and directly coupled cage induction generator*

In this scheme, the turbine rotor is coupled to an induction machine cage rotor, via a gearbox (Fig. 10.26(a)). The induction generator stator winding is directly connected to the electrical grid. An induction generator slip varies with the generated power so that the speed is not strictly constant but undergoes variation of about 1–2%. Because of the leading vars delivered by the generator, interpreted in power system terms as a sink of lagging vars, the generator is almost always power factor (PF) compensated by terminal capacitors.

A sophisticated variation of this system uses fast power electronics to create instantaneous variations of the rotor resistance. The commercial system uses the trade name OptiSlip and has been installed in turbines in the range 600 kW–2.75 MW. The use of the variable rotor (secondary) resistance permits transient speed increases up to 10% of the nominal value.

Fixed-speed wind turbine systems are relatively simple in design and are slightly cheaper than variable speed systems. Because the rotor speed is fixed, while the wind speed varies, the fluctuations in wind speed create torque fluctuations in the power train. The resulting power fluctuation can cause variations of grid voltage resulting in lamp "flicker".

10.8.2.2. *Variable-speed, doubly fed induction generator*

An induction machine in which both the stator (primary) and the rotor (secondary) have three-phase windings can be operated as a doubly fed induction generator (DFIG). The wind-turbine rotor drives the induction machine rotor via a gearbox (Fig. 10.26(b)). The induction machine stator is connected directly to the electrical grid. Decoupling is achieved between the fixed frequency of the grid and the variable (slip) frequency of the induction generator rotor by means of a power electronic switching converter, making a variable speed of generator rotation possible. It is common to use an inverter switching logic that produces output voltage waveforms known as pulse-width modulation (PWM).

The doubly fed connection has been extensively used as an energy-saving device in induction motor drives, where it is usually described as "slip-energy recovery". The power electronic converter in the rotor circuit usually needs to be rated lower than the generator rating, which is a cost saving on the expensive solid-state switches used. Depending on the speed range used, the power converter might be rated at about only one-third the rating of the generator. For example, if the drive speed range is 0.75–1.25 times synchronous speed, then the converter needs to be rated at only 0.25 times the machine rating. The system is very versatile and permits independent control of both the active and the reactive powers.

Because of the effective decoupling, the rotor can act as an energy storage system, absorbing the torque pulsations created by wind gusts. There is a requirement for periodic slip-ring maintenance in the rotor circuit.

10.8.2.3. *Variable-speed, direct drive synchronous generator*

In the scheme shown in Fig. 10.26(c), the synchronous generator rotor is mechanically coupled directly to the wind turbine. The direct drive design requires a relatively heavy generator and a larger rated power electronic converter, through which 100% of the generated power must pass, compared with the DFIG described in the previous subsection.

In an induction generator, all of the excitation is provided from the stator winding. The necessary level of air-gap flux density can be provided, from the stator, by the use of a relatively small air gap. By comparison, synchronous machines have their excitation systems on the rotor and operate with larger air gaps.

10.9. Vertical Axis Wind Machines

Most of the earliest historical accounts of wind machines refer to vertical axis structures used for pumping water. Small vertical axis machines with rotating vanes or rotating cups are now very widely used for instrumentation purposes, such as wind measurement. Currently, research is being devoted to vertical axis wind turbine systems for electricity generation, especially in the low- and medium-power ranges (a few tens of watts up to tens of kilowatts). The effective rotor surfaces move in the wind direction, rather than perpendicular to it. It is a feature of vertical axis machines that they accept the wind forces equally well from any direction. The issues of upstream, downstream, tower shadow and yaw that occur in horizontal axis propellers do not arise. The orientation of the blades into the wind is not required. Turbine power coefficients tend to be low: usually less than one-third, that is the theoretical maximum. With a vertical axis machine the electric generator can be mounted on the ground at the bottom of the shaft.

10.9.1. *The Savonius design*

The most basic of the modern types of vertical axis wind turbines is the Savonius rotor (Fig. 10.27). This consists of a hollow cylinder that is cut along its long axis to form two semi-cylinders. The two halves are mounted into a rough "S" shape so

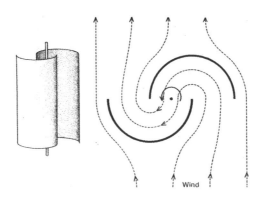

Fig. 10.27.   Savonius rotor and its air-stream flow [24].

that the wind flows through the cavity, being directed from the back of the concave side onto the inside of the convex side, resulting in rotation [22–24].

Savonius wind machines have a low cut-in speed and can operate in winds as low as 5 mph. This makes the machine suitable for electricity generation in low-power applications such as individual domestic installations. The machine is particularly suited to locations of variable wind direction. A Swiss company markets a 6-kW version of the Savonius machine. The peak efficiency of this form of turbine is about 30% (see Fig. 10.12) and the tip-speed ratio is low.

A disadvantage of the Savonius design is its high-solidity factor. Also, the machine is heavy if metal vanes are used. Because of the nature of the construction, the vane or sail area cannot be modified, so the machine may need to be tied stationary in high winds.

10.9.2. *The Darrieus design*

Much attention and research effort has been devoted to the Darrieus wind machine. This looks like an egg-beater or food-mixer and consists of two or three vertically mounted vanes with aerofoil cross-sections. The shape of the vanes (Fig. 10.28) is the natural shape that a flexible cable, such as a skipping rope, would adopt if it was swung horizontally about the bottom and top pivots.

Unlike the Savonius rotor, the Darrieus machine is not self-starting. The fixed-pitch rotor must be in rotation before the wind exerts a driving force on it. In practical designs, a Savonius rotor is often incorporated onto the Darrieus shaft to provide a starting torque. At high speeds, a Darrieus machine produces far more power than a Savonius machine and has a much higher tip-speed ratio (Fig. 10.12).

Darrieus machines have been studied for single-dwelling domestic housing applications. A blade diameter of about 15 feet is needed to produce 1 kW of output

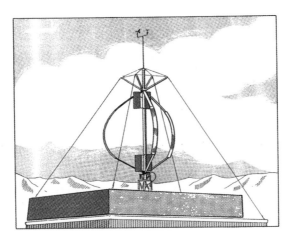

Fig. 10.28.   Darrieus rotor.

power. The Darrieus design has a peak efficiency of about 35%. A government-sponsored and large-scale Darrieus turbine was developed at the Sandia National Laboratories, Albuquerque, New Mexico, the USA, in the early 1980s. This used a blade diameter of 55 feet to develop a power of 80 kW. Another three-blade Darrieus machine in the USA had developed 500 kW. The world's largest Darrieus project, in Quebec, Canada, has developed a 4-MW machine. This is comparable in rating with the largest horizontal axis propeller systems.

As an alternative to Savonius assisted start-up, a Darrieus system can be started by using its coupled generator as a starter motor, taking electrical power from the battery or electricity supply.

Darrieus machines have a low-solidity factor, comparable with that of a horizontal axis propeller system, which makes for an economical use of aerofoil materials. The centrifugal forces of rotation exert tensile stresses on the blades, which may be metallic or made of fibreglass. The forces on the blades are similar in some respects to the aerodynamic forces on an aeroplane wing section or aerofoil.

It should be noted that the detailed operation of both the Savonius and Darrieus designs are very complex. Analyses cannot be undertaken using the comparatively simple equations of Section 10.3 above.

10.9.3. *Other forms of vertical axis machine*

There are many different forms of modern vertical axis machine. The most promising of these are derivations of the Darrieus principle. For example, Musgrove has designed machines with rotors of a two-blade or three-blade "H" configuration. Two vertical blades can either use a fixed tilt angle with rigid fixing to the hub or be hinged for variable tilt angles. At high speeds, the variable tilt blades move outwards due to centrifugal action and act to govern the speed of rotation, eliminating the danger of over-speeding [25].

10.10. Worked Numerical Examples on Wind Turbine Operations

Example 10.1. Use the data in Fig. 10.14 to predict the approximate diameter, in feed, of a wind-stream of circular cross-section that contains an extractable power 1 MW in a wind speed of 38 mph, $C_p = 0.4$.

The 1 MW coordinate in Fig. 10.14 intersects the line representing 15 m/s (33.6 mph) at a cross-sectional area of approximately 1100 m². Interpolating between coordinates on a non-graduated logarithmic scale requires careful attention. The value 38 is almost midway between the calibrated characteristic parameters of 44.74 and 33.6 in Fig. 10.14 but the 38 mph characteristic does not lie midway between the two calibrated characteristics.

By estimation, 1 MW of extractable power, at 38 mph, corresponds to a wind-stream area of $1000\,\text{m}^2$.

$$A = 1000\,\text{m}^2$$
$$= 1000 \times \left(\frac{39.37}{12}\right)^2 = 10764\,\text{ft}^2$$

But

$$A = \frac{\pi D^2}{4}$$
$$D^2 = \frac{4 \times 10764}{\pi} = 12732\,\text{ft}^2$$
$$D = 112.84\,\text{ft}$$

Example 10.2.   The largest wind turbine in the British Isles to date (2013) is the 3-MW generator system in the Orkney Isles, north of Scotland. Use the data of Fig. 10.14 to roughly verify its swept diameter for operation in a 38 mph wind.

The Orkney wind-generator system is rated at 3 MW of electrical output power. To realize this, the extractable power from the wind P_{ex} would need to be (say) 20% or 0.75 MW bigger. Interpolating between the 1 MW and 10 MW gradations in Fig. 10.14 for 3.75 MW is about one-half of the way and corresponds to a horizontal intercept of roughly A = $3000\,\text{m}^2$. The actual design diameter of the Orkney machine is 60 metres, giving a swept area:

$$A = \frac{\pi D^2}{4} = \frac{\pi \times 3600}{4} = 2827.4\,\text{m}^2$$

Despite the limitations of reading from a small-scale logarithmic data sheet, it can be seen that there is good correlation.

Example 10.3.   A wind turbine of the two-blade propeller type is designed to have its maximum power coefficient value at a TSR = 6 when the wind velocity is 25 mph. If the blade diameter is 100 feet, what is the recommended speed of rotation?

$$V = 25\,\text{mph} = 25 \times 0.447 = 11.18\,\text{m/s}$$

From Eq. (10.24)
    With TSR = 6

$$V = r\omega = V \times \text{TSR} = 6 \times 11.18 = 67.1\,\text{m/s}$$
$$D = 100\,\text{ft} = 100 \times \frac{12}{39.37} = 30.5\,\text{m}$$
$$R = \frac{D}{2} = 15.25\,\text{m}$$

Speed of rotation:

$$\omega = \frac{V}{r} = \frac{67.1}{15.25} = 4.4\,\text{rad/s}$$

From Eq. (10.26)

$$N = \frac{60\omega}{2\pi} = 42\,\text{rpm}$$

Example 10.4.   Evaluate typical values of overall working efficiency for (a) large wind systems and (b) small wind machine systems, assuming realistic values of the various stage efficiencies.

(a) For large wind-turbine systems, typical stage efficiencies are:

from Eq. (10.40) $C_p = 0.42$
from Eq. (10.41) $\eta_{gb} = 0.85$
from Eq. (10.42) $\eta_g = 0.92$.

Then, from Eq. (10.38), $\eta = 0.42 \times 0.85 \times 0.92 = 33\%$.
(b) For small wind-turbine systems, typical stage efficiencies are:

from Eq. (10.40) $C_p = 0.3$
from Eq. (10.41) $\eta_{gb} = 0.75$
from Eq. (10.42) $\eta_g = 0.7$
from Eq. (10.38) $\eta = 0.3 \times 0.75 \times 0.7 = 16\%$.

There is seen to be an overall efficiency advantage of the order 2:1 in using large-scale wind generation.

Example 10.5.   Wind-turbine units are rated at 2 MW in a rated wind speed of 13 m/s. The stage efficiencies are $C_p = 0.32$, $\eta_{gb} = 0.94$, $\eta_g = 0.96$. What is the necessary swept area? If the rotor is a two-blade propeller (horizontal axis), what is the diameter? ($\rho = 1.29\,\text{kg/m}^3$.)

$$\eta = 0.32 \cdot 0.94 \cdot 0.96 = 0.29$$
$$P_e = 2\,\text{MW}$$
$$\therefore\ P_W = \frac{P_e}{\eta} = \frac{2 \times 10^6}{0.29} = 6.9 \times 10^6\,\text{W}$$

Now:

$$P_W = \frac{1}{2}\rho A V^3$$

$$6.9 \times 10^6 = \frac{1}{2} \times 129 \times A \times (13)^3$$

$$A = \frac{2 \times 6.9 \times 10^6}{1.29 \times 13^3} = 4870\,\text{m}^2$$

For a circular area:

$$A = \frac{\pi D^2}{2}, \quad D = 78.8\,\text{m} \; (258\,\text{ft})$$

The comparatively large diameter is because of the low value of the turbine power coefficient.

Example 10.6. A generator driven by a wind turbine is required to deliver 1 MW of power at the generator terminals. The turbine is a two-blade propeller rotating about a horizontal axis and the maximum permitted shear stress of the turbine shaft is $55 \times 10^6\,\text{N/m}^2$. The rotor is designed to operate at a rotational speed of 22 rpm.

(a) If the turbine delivers its rated power at a wind average speed of 25 mph, calculate the corresponding diameter of the propeller and its TSR, assuming a typical value for the overall efficiency. The air density may be assumed to have a value $1.29\,\text{kg/m}^3$.

(b) Calculate the torque on the turbine shaft and the necessary shaft diameter.

(a)
$$P_{\text{elect}} = 1.0\,\text{MW}$$

$$\text{Let } \eta = 0.35\,\text{overall}$$

$$\therefore P_{\text{wind}} = \frac{P_{\text{elect}}}{\eta} = \frac{1.0 \times 10^6}{0.35} = 2.86 \times 10^6\,\text{W}$$

$$25\,\text{mph} \equiv 25 \times 0.447 \equiv 11.175\,\text{m/s}$$

Now wind power $P_{\text{w}} = \frac{1}{2}\rho A V^3$

$$\therefore A = \frac{2 \times 2.86 \times 10^6}{1.29 \times (11.175)^3} = 3171.8\,\text{m}^2$$

$$\text{But } A = \frac{\pi D^2}{4}$$

$$D = \sqrt{\frac{4 \times 3172}{\pi}} = \sqrt{4038.7}$$

$$= 63.55\,\text{m} \; (208.51\,\text{ft})$$

$$\text{TSR} = \frac{r\omega}{V}$$

$$r = \frac{D}{2} = 31.38\,\text{m}$$

$$\omega = 22 \times \frac{2\pi}{60} = 2.3\,\text{rad/s}$$

$$\therefore \text{TSR} = \frac{31.78 \times 2.3}{11.175} = 6.54$$

(b)
$$T = \frac{P}{\omega} = \frac{2.86 \times 10^6}{2.3} = 1.24 \times 10^6 \, \text{N/m}$$

$$f_s = \frac{Tr_s}{J} = \frac{Tr_s}{\frac{\pi r_s^4}{2}} = \frac{2T}{\pi r_s^3}, \quad \text{where } r_s = \text{shaft radius}$$

Now stress

$$\therefore \ r_s^3 = \frac{2T}{\pi f_s} = \frac{2 \times 1.24 \times 10^6}{\pi \times 55 \times 10^6} = 0.01 \, \text{m}^3$$

$$\therefore \ r_s^3 = \frac{10}{1000}$$

$$r_s = \frac{\sqrt[3]{10}}{10} = 0.215 \, \text{m} \ (8.46 \, \text{inches})$$

Example 10.7.   The Orkney wind machine in Great Britain is rated at 3 MW at the rated wind speed of 17 m/s, with a blade diameter of 60 metres. What is the power coefficient of the turbine? ($\rho = 1.29 \, \text{kg/m}^2$.)

$$P_{\text{W}} = \frac{1}{2}\rho A V^3$$

$$= \frac{1}{2} \times 1.29 \times \frac{\pi 60^2}{4} \times 17^3 = 8959810 \, \text{W}$$

$$P_{\text{e}} = 3 \, \text{MW}$$

Overall efficiency is then, from Eq. (10.38)

$$\eta = \frac{3 \times 10^6}{8959810} = 0.335$$

$$\text{But } \eta = C_{\text{p}} \cdot \eta_{\text{gb}} \cdot \eta_{\text{g}}$$

If $\eta_{\text{gb}} = 0.90$ and $\eta_g = 0.90$ the turbine power coefficient is

$$C_{\text{p}} = \frac{0.335}{0.9 \times 0.9} = 0.414$$

Example 10.8.   A report by the Electrical Research Association of England suggested that there are about 1,500 UK land-based sites, having wind speeds of 20 mph, suitable for wind turbine-generator systems. What portion of the UK demand for electricity could be supplied [26]?

Assume that one could mount one 3-MW turbine or three 1-MW turbines in each location. If all the turbines in all the locations were working to capacity at the same time — an unlikely eventuality — the total power available would be

$$P_{\text{total}} = 3 \times 1500 = 4500 \, \text{MW}$$

This is equivalent to the power output of four or five medium-large fossil power stations. There are 8,760 hours/year. If all of the systems operated to capacity for 24 hrs/day and 365 days/year, the energy produced would be

$$W_{total} = 365 \times 24 \times 4500 = 39.42 \times 10^6 \, \text{MWh/yr}$$

If one includes consideration of the down time for repair and maintenance and also of periods of slack wind, then about one-third of this might be available,

$$W_{avail} = 13 \times 10^6 \, \text{MWh/yr} = 13 \, \text{TWh/yr}$$

In 2009 the total UK domestic electricity consumption was 122.5 TWh, which was 37.1% of the total electricity consumption [27]. The wind supply that was reported above would contribute 13/122.5 or 10.6% of the domestic demand and 13/330.5 or 3.93% of the total electricity demand.

10.11. Problems and Review Questions

10.1. Use the information in Eq. (10.16) to show that maximum power is extracted from a wind-stream when the upstream velocity V_1 is three times the downstream velocity V_2.

10.2. If the wind speed in a certain location is doubled, how does this affect the power output of a wind generator?

10.3. Use the characteristics of Fig. 10.14 to estimate the necessary swept diameter of a large wind turbine to generate 10 MW of electrical power in a 25-mph wind.

10.4. Explain, using a diagram, the term "tip-speed ratio" (TSR). If the optimum TSR = 6.0 for a two-blade propeller of diameter of 180 feet, what speed of rotation in rpm must be used in a 20-mph wind?

10.5. A large propeller-type wind turbine has a diameter of 200 feet. If the speed of rotation at full load is regulated to 32 rpm when the wind speed is 30 mph, what is the value of the TSR?

10.6. What are the extreme limits of overall efficiencies for (a) large wind-turbine systems and (b) small wind-turbine systems, indicated by the stage efficiency values of Section 10.6?

10.7. It is required to generate 1200 W of electrical power at the terminals of a generator driven through a gearbox by a wind machine. The location is such that the wind supply is of smooth laminar flow with an average speed of 17.5 mph. Assign typical efficiency values to the components of the system and calculate the blade diameter required for a good quality two-blade propeller type wind machine (air density $= 1.29 \, \text{kg/m}^3$).

10.8. A two-blade propeller wind turbine is coupled to a 10-kW electric generator. It is desired to generate 10 kW at the load terminals. The average wind speed is 8 m/s and can be considered as ideally smooth. Assign typical values of efficiency to the turbine and the generator and calculate the necessary diameter of the area swept out by the rotating blades (air density $= 1.29\,\text{kg/m}^3$).

10.9. A two-blade propeller is used as a wind turbine directly on the shaft of a small electric generator. Assign typical efficiencies to the wind turbine and the generator and calculate the blade diameter required to generate 500 W at the load terminals in a wind of average speed 15 mph ($p = 1.29\,\text{kg/m}^2$).

10.10. A two-blade propeller wind machine has a blade diameter 3.5 metres and a power coefficient $C_p = 0.36$. What average wind speed in mph would result in 1 kW of power generation if the generator has 70% efficiency?

10.11. (a) A two-blade propeller wind turbine has a blade diameter of 4 metres. What average wind speed would result in a power output of 500 W at the generator terminals, assuming typical efficiencies for the generator, the gearbox, and the turbine?

(b) At your calculated wind speed, what is the TSR if the turbine shaft speed is 40 rpm?

(c) Calculate the diameter of the turbine shaft if the maximum permitted shear stress is $55 \times 10^6\,\text{N/m}^2$.

10.12. A wind turbine system consists of a two-blade propeller, rotating about a horizontal axis, driving an electrical generator via a gearbox. The generator is required to deliver its rated capacity of 2 MW with the wind turbine rotating at 18 rpm. The maximum permitted shear stress of the solid steel turbine shaft is $55 \times 106\,\text{N/m}^2$. The design estimate for the overall efficiency is 30%.

(a) Calculate the torque on the turbine shaft at rated load and the necessary shaft diameter.

(b) Define the term "tip-speed ratio" (TSR) and explain its use in wind energy calculations.

(c) Calculate values for the TSR and the required diameter of the propeller if the rated turbine power is developed in an average wind speed of 27.5 mph (air density $= 1.29\,\text{kg/m}^3$).

(d) Estimate realistic and consistent figures for the turbine power coefficient, the gearbox efficiency, and the generator efficiency.

10.13. A solid-steel cylindrical turbine shaft has a diameter of 12 inches. If the shear stress coefficient $f_s = 55 \times 10^6\,\text{N/m}^2$, what is the maximum permitted shaft torque?

10.14. The maximum safe rotational speed for a certain wind turbine is 35 rpm. How is this maximum speed retained in high winds? What problems would arise if the speed of rotation became excessive? What mechanism is used on the cupola of a Dutch-type windmill to keep the rotating blades facing upstream?

10.15. The jet stream over the North Atlantic Ocean travels from west to east at 100 mph. How does this affect (i) the air speed and (ii) the ground speed of 500-mph high-flying jetliners crossing the ocean between Europe and the USA?

10.16. On a certain day, the North Atlantic jet stream travels from west to east at 95 mph. If the average air speed of a jetliner is 445 mph, how long will the flight take to travel from Manchester, UK, to Atlanta, Georgia, the USA, a distance of 3,400 miles?

10.17. A two-blade propeller, horizontal axis wind turbine system has a diameter of 150 ft. It is designed to deliver its rated power at a wind speed of 23.5 mph using a tip-speed ratio, TSR = 5.5. The power train (turbine + gearbox + generator) has an overall efficiency of 23.5%.

(a) What is the rotational velocity in rpm for rated power operation?
(b) Explain how you would reduce the rotational velocity, if required.
(c) Calculate the electrical power output for rated operation.
(d) If the wind speed increased by 5%, what effect will this have on the output power delivered?

$$\rho = 1.29 \, \text{kg/m}^3, \quad 1 \, \text{mph} \equiv 0.447 \, \text{m/s}$$

10.18. A two-blade propeller, horizontal axis wind turbine system is rated at 250 kW and has a blade diameter of 100 feet. It is designed to have its maximum efficiency at a tip-speed ratio (TSR) = 6.

(a) Explain, using a diagram, the meaning of the term "tip-speed ratio".
(b) What is the recommended speed of turbine rotation, in rpm, when the wind velocity is 25 mph (11.2 m/s)?
(c) The overall efficiency of the power train (turbine + gearbox + generator) is designed to be 30% for operation at and near its rated output. What value of wind speed in mph would be required to deliver 250 kW of electrical power at the generator terminals? [$\rho = 1.29 \, \text{kg/m}^3$, $1 \, \text{mph} \equiv 0.447 \, \text{m/s}$.]

10.19. Table 10.2 incorporates the maximum power in kW realizable by an ideal wind turbine, two-blade propeller machine of a swept diameter of 200 feet, rotating at a 30-mph wind. The actual power extractable under full-load conditions is 75% of the ideal value in the table. The shear stress coefficient of the turbine shaft is $f_s = 55 \times 10^6 \, \text{N/m}^2$. [$1 \, \text{mph} \equiv 0.447 \, \text{m/s}$.]

(a) How much power is delivered by the turbine?
(b) If the design value of the TSR = 6, what must the turbine rotational velocity in rpm be at full load?
(c) What is the maximum permitted torque on the turbine shaft? If the shaft is a solid-steel cylinder, what must its minimum diameter be?
(d) If the turbine is connected to a gearbox and a generator with a combined efficiency of 81%, what will be the electrical power output?

References

[1] Shepherd, W. and Zhang, Li. (2010). *Electricity Generation Using Wind Power*, Imperial College Press, London.

[2] Shepherd, W. and Shepherd, D.W. (2003). *Energy Studies*, 2nd Edition, Imperial College Press, London.

[3] Golding, E.W. (1955). *The Generation of Electricity by Wind Power*, E. and F.N. Spon Ltd., London.

[4] "European Wind Energy Atlas", Risø DTU National Laboratory, Danish Wind Industry Association, 2003, http://www.wind-energy-the-facts.org/en/appendix/appendix-a.html.

[5] Elliott, D.L. and Swartz, M.N. (1993). *Wind Energy Potential in the United States*, National Wind Technology Center, Washington, DC.

[6] Johnson, G.L. (1985). *Wind Energy Systems*, Prentice Hall Inc., Englewood Cliffs, NJ.

[7] "12' Aermotor Water Pumping Windmill", American Windmills, 2012, http://www.windmills.net/images/water_pumping/reno1.jpg.

[8] Slootweg, H.E. de V. (2003). *Inside Wind Turbines*, Renewable Energy World, **6**, 31–40.

[9] McVeigh, J.C. (1984). *Energy around the World*, Pergamon Press, Oxford.

[10] Eldridge, F.R. (1980). *Wind Machines*, 2nd Edition, Van Nostrand Reinhold Co., New York.

[11] Doge, D.M. (2002). "Illustrated History of Wind Power Development, Part 2: 20th Century Developments", http://telosnet.com/wind/20th.html.

[12] Putnam, P.C. (1948). *Power from the Wind*, Van Nostrand, New York.

[13] "The Wind Energy Pioneers: The Gedser Wind Turbine", Danish Wind Industry Association, 2000, http://www.windpower.org/en/pictures/juul.htm.

[14] Reed, J.W., Maydew, R.C. and Blackwell, B.F. (1974). *Wind Energy Potential in New Mexico*, SAND-74-0077, Sandia National Labs, Energy Report, New Mexico, USA.

[15] "World Financial Markets", Financial Times Business Information Ltd, London, 1983, 1984.

[16] Renewable Energy World (2001). Vol. 4, No. 3.

[17] Brown, L.R. (2004). *Europe Leading World into Age of Wind Energy*, Earth Policy Institute, USA, http://www.earth-policy.org/plan_ b_ updates/2004/update37.

[18] Renewable Energy World (2004). Vol. 7, No. 2, p. 23.

[19] Renewable Energy World (2004). Vol. 7, No. 1, p. 105.

[20] Renewable Energy World (2004). Vol. 7, No. 3.

[21] "Wind Power by Country", World Wind energy Association, 2010, www.wwindea.org.

[22] Burton, T., Sharpe, D., Jenkins, N. and Bossanyi, E. (2001). *Wind Energy Handbook*, John Wiley and Sons, Chichester.

[23] Freris, L.L. (1990). *Wind Energy Conversion Systems*, Prentice Hall Inc., Englewood Cliffs, NJ, USA.

[24] Inglis, D.R. (1978). *Windpower and Other Energy Options*, University of Michigan Press, Ann Arbor, MI.

[25] Musgrove, P.J. and Mays, I.D. (1978). "The Variable Geometry Vertical Axis Windmill", *Proc 2nd Int Symp Wind Energy Systems*, Cranfield.

[26] "Feasibility of Large Wind Turbines in the UK", Electricity Research Association (ERA), Leatherhead, 1974.

[27] "UK Energy in Brief, 2008", Dept. of Trade and Industry, London, 2009.

CHAPTER 11

SOLAR HEATING OF WATER OR AIR

11.1. Radiation from the Sun

Our sun is a fairly typical star that astronomers call a G-type "main sequence star". It exists in what we believe to be an average galaxy called the Milky Way, which is rich in gas and stellar dust and has a diameter of about 100,000 light years.

The sun is a large mass of burning gases, mainly hydrogen and helium, with a diameter of about 1.4 million km (860,000 miles). At its interior core, occupying about one-quarter of the cross-section diameter, the intense heat is estimated to be at a temperature of 20 million kelvins (2×10^7 K). In this so-called plasma region, atomic hydrogen fuses to form stable helium, with the generation of high-frequency gamma radiation and the release of large amounts of energy that fuel the ongoing chemical reaction.

Energy generated within the solar core is radiated and convected to the solar surface, which, in the present context, may be regarded as a uniform spherical black-body radiator at a temperature of about 5800 K.

Solar radiation spans a large portion of the whole electromagnetic spectrum. For example, the sun is a very strong radio source. However, radio waves have negligible energy. The vast majority of the sun's radiated energy is concentrated between wavelengths 300 nm ($0.3\,\mu$m) and 3000 nm ($3\,\mu$m), with the spectral distribution shown in Fig. 11.1. This range of wavelengths extends both into the near ultraviolet and into the near infrared regions. Interestingly, integration between the limits of optically visible light (400–750 nm) reveals that it contributes about 45% of the total solar radiated energy per unit area at the earth's mean distance of 150 million km (93 million miles) from the sun.

All of the radiation travels at the velocity of electromagnetic wave propagation c to satisfy the relationship

$$c = \lambda f \tag{11.1}$$

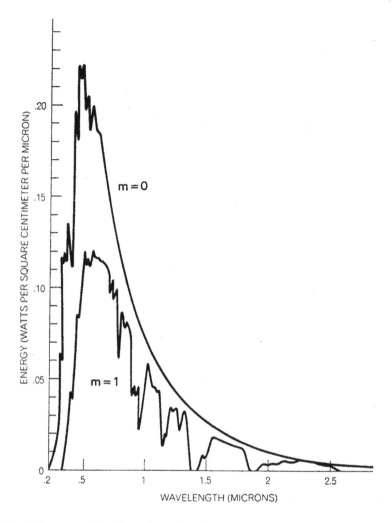

Fig. 11.1.   Solar spectral irradiance (reproduced by permission of Scientific American, USA).

where f is the frequency in hertz and λ is the wavelength in metres. Velocity c has the value 2.998×10^8 m/s. The region of visible light lies in the frequency range $4.0 - 7.5 \times 10^{14}$ Hz.

The larger spectral contour of Fig. 11.1 represents the radiated solar energy falling normally (i.e. perpendicularly) onto a receiver just outside the earth's atmosphere. It is sometimes called the $m = 0$ characteristic because, in that location, there is zero mass of air between the transmitter (the sun) and the spectral receiver (the earth). Outside the earth's atmosphere the sun power or solar insolation flux has a value of 1.36 kW/m^2 or 0.136 W/cm^2, which is referred to as the solar constant. At sea level with zero zenith angle, the column of air through which the radiation passes reduces the average received power to about 1 kW/m^2. This is shown as the

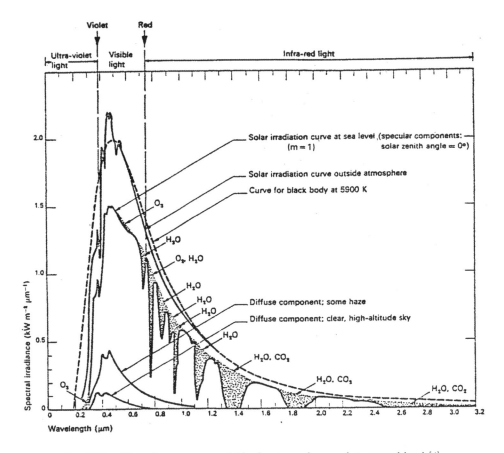

Fig. 11.2.   The solar spectrum outside the atmosphere and at ground level [4].

$m = 1$ characteristic in Figs. 11.1 and 11.2. At various different wavelengths, the received ground-level radiation has been differently inhibited due to absorption by water vapour or atmospheric gases such as oxygen (O_2) or carbon dioxide (CO_2).

The intensity of the radiation or solar insolation on a surface depends on:

(i) obliquity, the angle between the surface and the beam of radiation, and
(ii) the length of air mass through which the radiation has to pass.

Both of the factors depend on the altitude of the sun above the ground. In northern Europe the maximum solar altitude is 15° in midwinter, which corresponds to an air mass $m = 4$.

In the UK the average annual solar insolation at ground level is about $100\,\mathrm{W/m^2}$.

In the summer months the peak insolation rises to about $1\,\mathrm{kW/m^2}$ but this exists only for about an hour at midday.

The passage of the solar radiation through the earth's atmosphere is illustrated in Fig. 11.3. Some radiation is received at ground level by direct and uninterrupted transmission and this is referred to as the direct component of radiation. Some

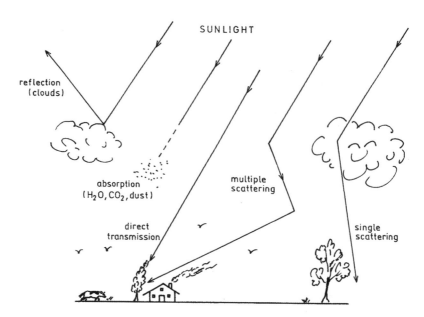

Fig. 11.3.　Interaction of sunlight with atmosphere (reproduced by permission of Dr M. Munroe).

component of the radiation is reflected off the clouds and back into space, while another component is absorbed by atmospheric dust or gas clouds. The presence of gas molecules or dust particles in the atmosphere cause some rays of sunlight to be deflected by scattering. Each collision causes the incoming radiation to lose some of its energy, which has the consequent effect of increasing the wavelength. The sum of the various effects of reflection, scattering and absorption is known as the diffuse component of radiation. The relative amounts of direct and diffuse radiation vary greatly in different parts of the world and for different times of the year. In the UK the direct and diffuse components of energy are roughly equal in magnitude, as illustrated by the month-by-month data of Fig. 11.4 [1].

Additional information about the extent of solar energy falling onto a horizontal surface, at ground level, is given in Table 11.1 [2].

The diffuse radiation is assumed to be uniform in all directions and is defined as D. If the direct-incident radiation is I, then the total global radiation on a G is given by [3]:

$$G = D + I \sin \gamma \qquad (11.2)$$

The unit of energy in Table 11.1 is the kilowatt hour (kWh), which is the energy unit used in UK domestic electricity meters. In electrical energy measurements the kilowatt hour is often called a unit of electricity. By contrast, the energy unit in Fig. 11.4 is the megajoule (MJ) ($1\,\mathrm{kWh} = 3.6\,\mathrm{MJ}$). Table 11.1 shows that in many parts of the northern hemisphere, the solar energy total in a year at ground level is about $900\,\mathrm{kWh/m^2}$ [2, 3]. This is a tremendous amount of energy, with enormous potential.

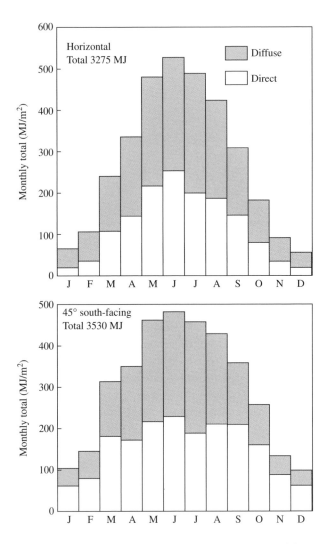

Fig. 11.4. Solar radiation data for London, England [1].

The total amount of solar energy falling on the earth, as discussed in Section 2.1.1 of Chapter 2, is staggeringly large: more than enough for present human needs and amply sufficient to meet any anticipated future demand. For the USA or any of the European countries the average annual solar insolation is several hundred times the total energy needs. But serious difficulties exist with regard to the efficient and economic energy extraction, conversion and storage, which inhibit the attractiveness of solar energy. The main problems are the low intensity of the radiation at ground level and its intermittent nature. Nonetheless, the sun is a source of free, non-polluting and infinitely renewable energy.

Table 11.1. Solar energy intensity values (on horizontal surface) [2].

Location	Latitude	kWh/m²					
		Max. bright day		Min. dull day		Yearly total	
		(1)	(2)	(1)	(2)	(1)	(2)
Equator	0°	6.5	7.5	5.8	6.8	2200	2300
Tropics	23.5°	7.1	8.3	3.4	4.2	1900	2300
Mid-earth	45°	7.2	8.5	1.2	1.7	1500	1900
Central UK	52°	7	8.4	0.5	0.8	1400	1700
						(900 due to atmospheric conditions)	
Polar circle	66.5	6.5	7.9	0	0	1200	1400

(1) Direct sunlight
(2) Direct + diffuse sunlight

11.2. Seasonal Variation of Solar Radiation

As the earth circuits the sun every $365\frac{1}{4}$ days, it makes one complete rotation every day about its axis. The axis of rotation is tilted at 23.44° from the normal to the ecliptic plane (Fig. 11.5) [4]. This is why the latitude lines marking the Tropic of Cancer and the Tropic of Capricorn are each displaced by 23.44° from the earth's equator. The solar radiation strikes the equatorial regions of the earth much more squarely (i.e. it has a low zenith angle) than it does the polar regions. For this reason the solar flux at ground level is roughly dependent on latitude. In Fig. 11.6 the mean annual radiation is divided into zones of ground-level intensity [3,5]. It can be seen that the regions of greatest mean annual solar intensity coincide largely with desert regions, where the radiation falls on land masses.

The range $100 \, \text{W/m}^2$ in the projection of Fig. 11.6 includes northern Europe and the British Isles. But it also includes the permafrost regions of the northern territories of Canada and the Siberian territories of Russia. The winter air temperatures in Great Britain, for example, are very much milder than northern Canada and Russia because of the moderating effects of ocean currents, notably the Gulf Stream. For Great Britain the mean annual solar power received is about $100 \, \text{W/m}^2$ on average, with a monthly distribution that peaks in June. By comparison, Egypt and Sudan (for example), in northern Africa, receive about $300 \, \text{W/m}^2$ and the month-by-month distribution is much flatter.

A listing of the solar power in W/m^2 falling on many of the principal cities of the world is given in Table 11.2 [5]. The solar power figures, city-by-city, correlate directly with the map of Fig. 11.6. The listing does not include the many great cities of Asia, Australia and New Zealand.

Contours of total solar radiation for the British Isles are given in Fig. 11.7 for astronomical high summer (the summer solstice is between June 21 and 22) and

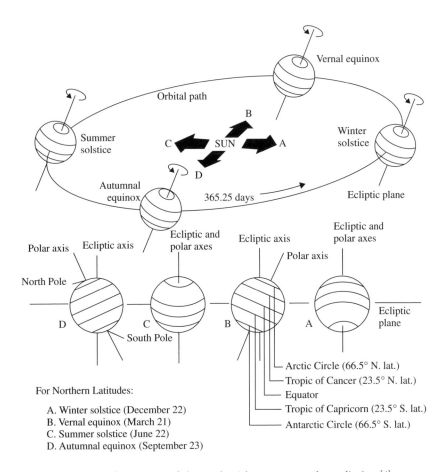

For Northern Latitudes:

  A. Winter solstice (December 22)
  B. Vernal equinox (March 21)
  C. Summer solstice (June 22)
  D. Autumnal equinox (September 23)

Fig. 11.5. Orientation of the earth with respect to solar radiation [4].

astronomical midwinter (the winter solstice is between December 21 and 22) [1]. The ground-level solar radiation is higher in the west than the east in England and Wales because of cloud cover and the prevailing westerly winds. The most striking feature of the data of Fig. 11.7 is that there is a magnitude factor of 10 to 1 between the radiation figures for midsummer and midwinter.

In Britain the solar radiation is strongest in the summer at the time of the year when the energy demand is least, in both the domestic and industrial sectors. One might describe the situation by saying that, in Britain, the incoming solar radiation is in time anti-phase with the demand for electricity (shown in Fig. 3.8 of Chapter 3). It should be noted that there are other countries where the levels of incoming solar radiation match the energy demand. In the southern areas of the USA, for example, the demand for electricity is much greater in the summer than in the winter because of the widespread use of air-conditioning (cooling) and refrigeration equipment.

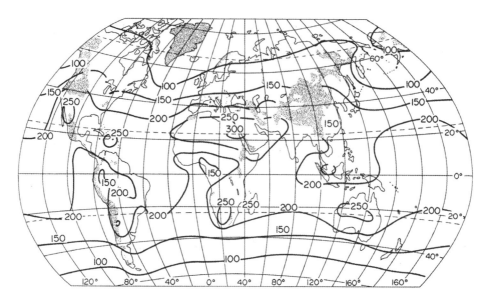

Fig. 11.6.   Annual mean global solar radiation on a horizontal plane at the earth's surface, W/m^2 [6].

In the plot of solar power input versus time, the area under the curve has the dimension of power times time, which is energy. The striking contrast of the energy content of the incoming UK solar radiation between summer and winter is illustrated by the relative areas under the two curves of Fig. 11.8.

A feature of the British climate is the large variations of solar radiation that exist from day to day. Figure 11.9 shows actual recordings of the solar input at a particular location in northern England from sunrise to sunset. The frequent dips in the traces (Fig. 11.9(b)) are due to cloud motion across the sky. The energy input on a horizontal surface, obtained by calculating the areas under the curves of Fig. 11.9, demonstrates typical climatic behaviour. This is given in Table 11.3. At the same period of the year there can be a 3:1 difference in the solar energy input, dependent on local weather conditions.

11.3.   Classification of the Collection of Solar Energy

Solar energy collection systems fall into three broad categories, as illustrated in Fig. 11.10. These categories are:

(i) the direct conversion of sunlight into electricity by the use of photovoltaic cell;
(ii) the direct heating of air or water to give a thermal output below 150°C (this area of thermionic technology is oriented towards domestic applications in the heating of buildings and/or hot water supply); and

Table 11.2.   Average solar power density for cities of the world [5].

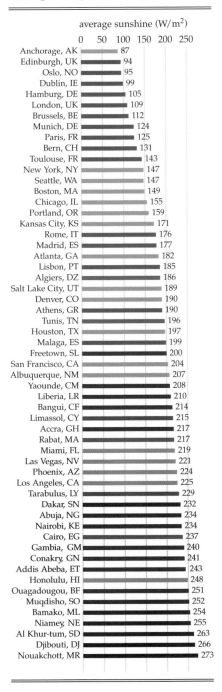

average sunshine (W/m^2)

Anchorage, AK	87
Edinburgh, UK	94
Oslo, NO	95
Dublin, IE	99
Hamburg, DE	105
London, UK	109
Brussels, BE	112
Munich, DE	124
Paris, FR	125
Bern, CH	131
Toulouse, FR	143
New York, NY	147
Seattle, WA	147
Boston, MA	149
Chicago, IL	155
Portland, OR	159
Kansas City, KS	171
Rome, IT	176
Madrid, ES	177
Atlanta, GA	182
Lisbon, PT	185
Algiers, DZ	186
Salt Lake City, UT	189
Denver, CO	190
Athens, GR	190
Tunis, TN	196
Houston, TX	197
Malaga, ES	199
Freetown, SL	200
San Francisco, CA	204
Albuquerque, NM	207
Yaounde, CM	208
Liberia, LR	210
Bangui, CF	214
Limassol, CY	215
Accra, GH	217
Rabat, MA	217
Miami, FL	219
Las Vegas, NV	221
Phoenix, AZ	224
Los Angeles, CA	225
Tarabulus, LY	229
Dakar, SN	232
Abuja, NG	234
Nairobi, KE	234
Cairo, EG	237
Gambia, GM	240
Conakry, GN	241
Addis Abeba, ET	243
Honolulu, HI	248
Ouagadougou, BF	251
Muqdisho, SO	252
Bamako, ML	254
Niamey, NE	255
Al Khur-tum, SD	263
Djibouti, DJ	266
Nouakchott, MR	273

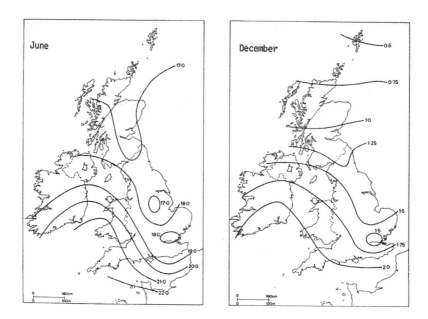

Fig. 11.7.   Average daily solar radiation in the UK, MJ/m^2 [1].

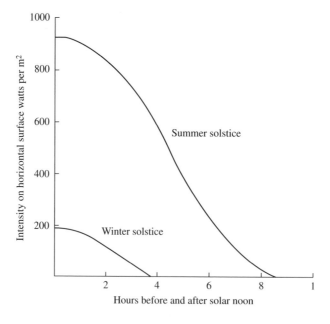

Fig. 11.8.   Solar power density (W/m^2) versus time for the UK.

(a)

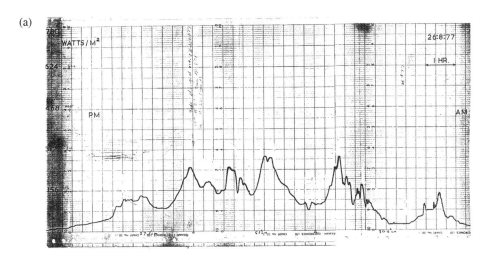

(b)

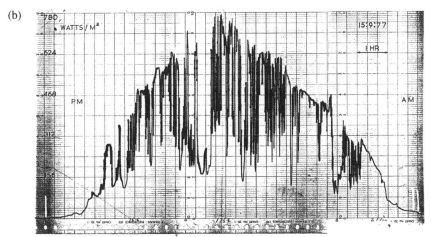

Fig. 11.9. Solar radiation recordings in central England, W/m². (a) Overcast. (b) Bright and sunny. (Ref: M. Munroe, PhD thesis, Bradford, UK, 1977.)

(iii) thermionic concentrator systems for producing high temperature (i.e. above 150°C) heat energy for, for example, thermodynamic (i.e. heat–mechanical work) systems.

Various applications of solar thermionic systems are discussed in Sections 2.4 and 2.5 of Chapter 2. Photovoltaic conversion is discussed in Chapter 12.

11.4.   Solar Water Heating (Domestic)

There are many Roman remains in various parts of Europe and the Mediterranean countries, some now 2,000 years old, that demonstrate that the Romans used solar

Table 11.3. Measured solar data at the University of Bradford, England, 1977.

Date	Climate condition	Hours of daylight	Peak insolation W/m^2	Energy content kWh/m^2
26/08/1977	overcast and rain	15	202	1.3
07/09/1977	intermittent sunshine	14	450	2.6
15/09/1977	bright and sunny	13	780	4.5

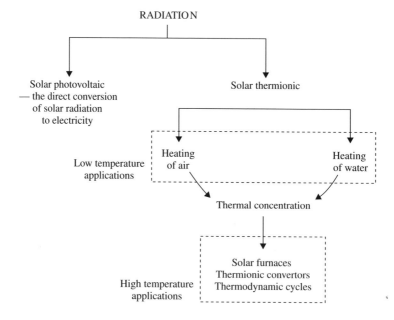

Fig. 11.10. Categories of solar energy collection.

water heating. Open water channels feeding the bath houses were lined with grooved black slate to increase the heat absorption.

About one-quarter of all the energy consumed in the USA and European Community countries at present is used for low-grade heating or cooling purposes. Much of this is used to heat the air in buildings — usually called space heating — or to heat water for domestic purposes. The most common form of domestic solar water heater employs the "greenhouse" effect, described below.

11.4.1. *Operation of a garden greenhouse*

Glass has a high coefficient of transmittance for radiation within the spectral range $0.3 \leq \lambda \leq 3$ microns (μm). Within this waveband about 90% of the incident radiation is transmitted, although the transmittance is not equal at all wavelengths. For radiation of wavelength $\lambda > 3\,\mu$m the transmittance falls below 25% and in the

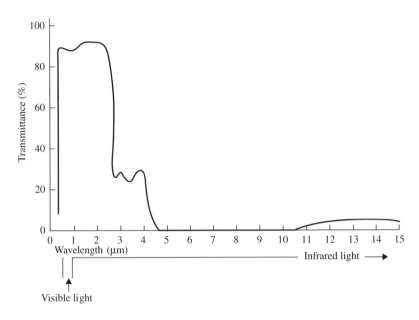

Fig. 11.11.   Transmittance of window glass as a function of wavelength [4].

region $4 \le \lambda \le 15\,\mu\mathrm{m}$ the transmittance is virtually zero (Fig. 11.11). In other words, the transmittance properties of glass are such that glass is a low-pass filter in terms of radiation wavelength. The pass band of the radiation transmission includes the range of human visibility and much of the useful infrared range.

Much of the incident radiation in the infrared region, with $\lambda > 4\,\mu\mathrm{m}$, is reflected back from the greenhouse surface (Fig. 11.12), but gives up some of its energy to the glass.

When solar radiation passes through the glass, collisions between some photons of the electromagnetic radiation and some molecules of the glass cause a certain amount of the incident energy to be dissipated in heating the glass. The loss of energy suffered by the incoming radiation is accompanied by an increase in wavelength. After the increased wavelength radiation enters the greenhouse enclosure (Fig. 11.12), some of it undergoes reflections or refractions. Each of these changes of direction is accompanied by a proportion of energy loss and wavelength increase. The net effect is that the increased wavelength prevents much of the internal radiation from being re-transmitted outwards. It undergoes multiple reflections, scattering and absorption until its incident energy is all dissipated. The overall effect is a temperature rise until the incoming energy balances the heat losses due to conduction, convection and re-radiation. At night, when the incoming radiation ceases, a lot of the greenhouse heat energy is lost by re-radiation outwards and the temperature falls.

In Great Britain one of the common hobbies is gardening. The ground temperature and air temperature during the winter can sometimes fall below the freezing

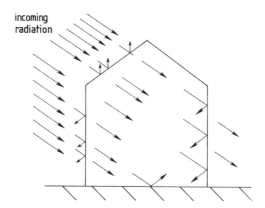

Fig. 11.12.   Radiation pattern illustrating the greenhouse effect.

point of water (0°C or 32°F). When this happens many forms of plants would suffer severe damage, due to ice and snow or simply due to freezing cold air. Experienced gardeners know which plants to leave outdoors over the winter period and which must be taken indoors.

Many gardeners use a glass-enclosed shed, called a "greenhouse" because the contents are the green foliage of living and growing plants. Some greenhouse structures are built as conservatories and share a common wall with the main house. Other greenhouses are small glass enclosures that are physically independent of other buildings. It is an interesting experience to walk into a UK greenhouse in midwinter when the outside temperature is below the freezing point. Even without any form of local heating the air temperature inside the greenhouse remains well above freezing, due to solar radiation, and is perceptibly warmer than outside. In the summertime, greenhouses experience a large solar gain of energy so that they must be shaded and ventilated if excessive temperatures are to be avoided.

11.4.2.   *Solar flat-plate collectors*

Most domestic solar water heater collectors in the UK are referred to as flat-plate collectors. The term "flat-plate collector" is rather misleading and is used to define a wide variety of collectors having different geometries (some of them being far from flat), different heater fluids and different methods of heat transfer.

A typical collector panel is illustrated in Fig. 11.13. The collector is a metal plate that is ribbed or corrugated so that a fluid (usually water) passes through it or over it. A common form of structure is similar to that of a central heating radiator in a pumped water system. The surface of the solar collector is exposed to the incoming radiation and is coated black for maximum radiation absorption. Usually the collector is contained within some enclosure that has a glass or transparent plastic cover to retain its collected heat. As described in the previous section, the

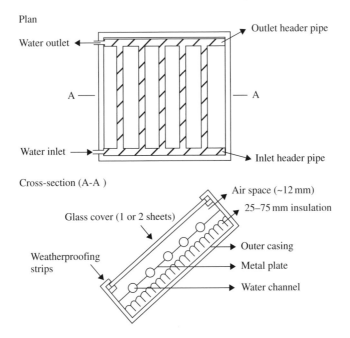

Fig. 11.13. Structure of flat-plate solar collector.

greenhouse effect can cause high temperatures to exist within the collector enclosure. Heat is transferred from the metal of the collector to an outside energy reservoir by pumping water through the collector. Even so, on a warm summer day a collector becomes much too hot to be touched by hand. Collectors are often roof-mounted and inclined at roughly the latitude angle minus ten degrees to be perpendicular to the incident radiation.

In flat-plate collectors the incoming radiation is not concentrated optically as in the higher temperature collector systems described in Section 11.5 below. The collector area is equal to the absorber area because both functions are incorporated into the same collector device (i.e. the flat plate). Using the terminology of Section 11.5, one could say that a flat-plate collector has a concentration ratio of unity.

About one-half of the incident energy falling onto the collector is subsequently re-emitted. The absorbance (or absorptance) of solar energy causes the blackened collector temperature to rise and its re-emitted radiation then has a slightly shorter wavelength. It is found that some of the energy re-radiated from the black collector has wavelengths in the range $4 \leq \lambda \leq 15\,\mu$m and cannot exit through the glass cover of the container. Part of this entrapped radiation is reflected from the glass cover sheet back onto the collector plate.

The radiation emittance property of an ideal black-body radiation varies with the frequency (or wavelength) of the radiation and also depends on the black-body

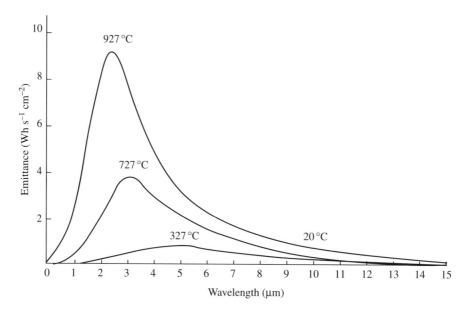

Fig. 11.14.   Emittance of a black body versus wavelength [4].

temperature. For example, the radiation surface of the sun behaves like a black body with a temperature of 5800 K and its emittance versus wavelength characteristic is given in Fig. 11.2. Similarly, emittance characteristics for several other temperatures are given in Fig. 11.14 [4]. The feature whereby the emittance can be related to the temperature of the black radiating body is a useful measure of the effectiveness of a solar collector.

Selective coatings are available for the solar collector, to increase its absorbance and reduce its re-emittance. Even without a glass cover the deposition of thin films of a metal such as beryllium can greatly improve the collector performance. The use of the cover, however, contributes to the vital greenhouse effect and gives thermal insulation against loss of the heat to the ambient air by conduction and convection.

A listing of some relative selective industrial coatings is given in Table 11.4, in which the solar emittance is also defined in terms of black-body temperature. The aim is to achieve maximum absorbance α and minimum re-emittance ε so that the ratio α/ε is a maximum.

Selective coatings are also available in the form of paints. These tend to be very expensive compared with standard decorating paints. The authors have found that for domestic use and for student projects, any standard form of matt black paint gives good results and represents good value for money.

The efficiencies of solar water collectors with different enclosure coverings are shown in Fig. 11.15, in which the independent (x-axis) variable is the difference $(T_c - T_a)$ between the collector fluid temperature T_c and the ambient air temperature

Table 11.4. Solar absorptance (α) and infrared emittance ($^\varepsilon$IR) for various selective coatings.

Material	Solar absorptance (α)	IR emittance ($^\varepsilon$IR)	Performance factor α/ε
Copper oxide on copper	0.90	0.12	7.5
Black nickel on copper	0.90	0.08 (573 K)	11
Black chrome on copper	0.95	0.12	7.92
Silicon on silver	0.76	0.06 (773 K)	12
Nonmetallic black surfaces: asphalt slate, carbon	0.92	0.94	0.98
Flat black paint	0.97	0.86	1.13
3M Velvet-black paint	0.98	0.90	1.09
Grey paint	0.75	0.95	0.79
Red brick	0.55	0.92	0.6
Concrete	0.60	0.88	0.68
Galvanized steel	0.65	0.13	5
Aluminium foil	0.15	0.05	3
ZrNy on Ag	0.85	0.03 (600 K)	23

This table is slightly modified from an original given in: *Engineers' Guide to Solar Energy*, by Y. Howeel, J. A. Bereny, Solar Energy Information Services, PO Box 204, San Mateo, CA 94401, USA, 1979.

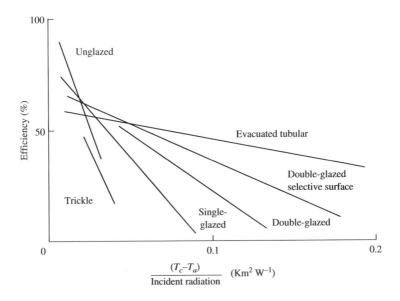

Fig. 11.15. Efficiency of solar water collectors [3].

T_a divided by the incident radiation in W/m^2 [3]. The glazing should possess the properties of resistance to ultraviolet ray deterioration, good thermal stability, durability in harsh weather and mechanical strength and rigidity, combined with low cost.

When only a small temperature rise is required, as in swimming pool heating, high efficiency can be obtained by using solar water collector panels that are completely unglazed. In the UK this form of swimming pool heating is widely used during the summer months. In the USA the largest end use for solar collectors shipped during 1999 (95% of total shipments) was for heating swimming pools. Shipments of solar thermal collections in the USA increased steadily over the period 1974–2009 [7].

In domestic hot water systems the maximum required working temperature is usually not more than 6°C. Temperature rises of 20–4°C may be desired. For such applications the solar collector enclosure should be thermally insulated and some form of glazed cover is required. Such collectors usually operate with an efficiency of the order 40–50%.

Under steady-state conditions the heat energy delivered by a flat-plate solar collector is the heat energy absorbed by the plate minus the heat losses from the plate to the surroundings. This can be expressed in mathematical form by the empirical equation below [3]:

$$Q = F[G\tau\alpha - U(T_c - T_a)] \tag{11.3}$$

where

F = design factor of the collector construction (collector heat removal efficiency factor)

G = total normal incident radiation, W/m^2

Q = quantity of heat collected per unit area of collector

T_a = ambient temperature outside the collector enclosure, K

T_c = collector fluid temperature, K

U = heat loss coefficient per degree of temperature difference T_c–T_a

α = absorbance of collector plate (the fraction of radiation reaching the absorber plate that is actually absorbed). Absorbance properties of various materials are listed in Table 11.4

τ = transmittance of enclosure cover (fraction of G transmitted through the cover plates). Transmittance properties of various materials are listed in Table 11.5

$\tau\alpha$ = design factor, which takes into account multiple reflections and absorption in the cover glazing

The thermal performance is defined by the three design factors: F, $\tau\alpha$ and U. Values of these are given in Table 11.6, for a range of commercial flat-plate collector systems.

The thermal efficiency of the collector is the proportion of the total incident power G that is converted into usable heat. If the collector thermal efficiency is denoted by η_c, then

$$\eta_c = \frac{Q}{G} = F\left[\tau\alpha - \frac{U}{G}(T_c - T_a)\right] \tag{11.4}$$

Table 11.5. Solar transmittance (τ) for various cover-plate materials: the exact value depends on the cover thickness and the direction of the solar beam.

Material	Transmittance (τ)
Crystal glass	0.91
Window glass	0.85
Polymethyl methacrylate (acrylic)	0.89
Acrylite	
Lucite	
Plexiglass	
Polycarbonate	0.84
Lexan	
Merlon	
Polyethylene terephthalate (polyester)	0.84
Mylar	
Polyvinyl fluoride	0.93
Tedlar	
Polymide	0.8
Kapton	
Polyethylene	0.86
Fluorinated ethylene propylene (fluorocarbon)	
FEP Teflon	0.96
Fibreglass-reinforced polyester	0.87
Kalwall	
Fibreglass-reinforced, acrylic-fortified	0.86
polyester with polyvinyl fluoride weather surface	
Tedlar-clad Filon	

Ref: *Engineers' Guide to Solar Energy*, Howell, Y., Bereny, J. A. Solar Energy Information Services, PO Box 204, San Mateo, CA 94401, USA, 1979.

Equations (11.3) and (11.4) are valid for any self-consistent set of units. It is seen from Eq. (11.4) that the efficiency varies inversely with the temperature difference $T_c - T_a$, as indicated by the characteristics of Fig. 11.15.

Another empirical form for the efficiency relationship of a glazed collector when only the input radiation and temperature values are known is

$$\eta_c = 0.78 - \frac{7.7(T_c - T_a)}{G} \tag{11.5}$$

where T_c, T_a are in degrees centigrade or degrees Kelvin and G is in W/m^2 [3]. Note that Eq. (11.5) cannot be made the subject of detailed conceptual analysis. For example, a direct examination of Eq. (11.5) suggests that the maximum collector efficiency is 78%. This would be true mathematically if $T_c = T_a$. But if $T_c = T_a$ there is no thermal gain and the collector is not giving any output, so the system efficiency is zero. The use of Eq. (11.5) is restricted to practical situations and practical ranges of collector operation.

Table 11.6.   Performance parameters of certain commercial solar flat-plate collectors.

Collector source	Absorber material	Absorber surface coating	Transparent covers	F_R	U_L $\frac{Btu}{hft^2F}$	U_L $\frac{W}{m^2C}$	τ_α	α	ε	τ
NASA/Honeywell	Aluminium	Black nickel	2 Glass	0.94	0.56	3.2	0.74	0.95	0.07	0.78
Marshall Space Flight C.	Aluminium	Black nickel	2 Tedlar	0.95	0.69	3.9	0.56	0.73	0.1	0.77
NASA/Honeywell	Aluminium	Black paint	1 Glass	0.9	1.3	7.4	0.89	0.97	0.97	0.92
NASA/Honeywell (mylar honeycomb)	Aluminium	Black paint	2 Glass	0.96	0.57	3.2	0.77	0.97		0.79
NASA/Honeywell	Aluminium	Black paint	2 Glass	0.93	0.8	4.6	0.76	0.97	0.97	0.78
PPG Industries	Aluminium	Black paint	2 Glass	0.85	1.1	6.2	0.73	0.95	0.95	0.77
Owens-Illinois (evacuated tube)	Glass	Selective surface	1Glass	0.75	0.2	1.1	0.72	0.8	0.07	0.9
Solaron (data furnished by manufacturer). Heat transfer fluid is air.	Steel	Black paint	2 Glass	0.67	0.77	4.4	0.73	—	—	—

Ref.: *Solar Heating and Cooling of Residential Buildings: Design of Systems*, Government Printing Office, Washington DC, 1977. (Report on work at Colorado State University.)

If $G\tau\alpha \leq U(T_c - T_a)$ in Eq. (11.4) the efficiency is mathematically negative. This means that the collector losses are greater than the incoming solar energy. The physical conditions under which this is likely to occur are when the collector mass is warm from previous operation, the ambient air temperature is low and the incoming radiation falls to a low value.

A detailed design of a solar collector is outside the scope of the present book. Such design is partly empirical and requires knowledge of the scope and range of the design factors U and F, which are usually determined by experience. The various terms of Eq. (11.4) do, however, suggest several strategies that might be employed in improving the efficiency of the collector. It is seen that various terms in Eq. (11.4) must increase or decrease, as follows, if an increase of η_c is to be realized:

design factor F, ↑

effective incident radiation G, ↑

transmittance τ, ↑

absorptance (absorbance) α, ↑

heat loss coefficient U, ↓

temperature difference $(T_c - T_a)$, ↓

collector fluid temperature T_c, ↓

Suggested methods of change and the likely effects on collector performance are summarized in Table 11.7. Any change that will reduce the heat losses due to conduction, convection or re-radiation is likely to increase the collector working efficiency. If any proposed change involves increased expenditure it will be necessary to undertake an economic analysis to assess if such a change represents an overall cost saving. For example, is the increased installation cost justified by reduced running costs? Over what period of time would the investment be recovered?

Table 11.7.   Improving the efficiency of flat-plate solar collectors.

Desired parameter change	Method of change	Effect on the collector
F ↑	improved geometry of collector	better heat transfer from radiation to collector
G ↑	concentration of incident radiation — south facing and tracking	higher thermal input
τ ↑	selective windows	reduced radiation losses
α ↑	selective absorber on collector	increased absorbance and reduced emittance
U ↓	e.g. double glazing, better insulation	reduced conduction and convection losses
$(T_c - T_a)$ ↓ T_c ↓	reduced working fluid temp	reduced conduction and radiation losses

11.4.3.   *A typical domestic solar water heating system*

The solar radiation in any particular location undergoes daily and seasonal variations. Sometimes the solar input is not adequate for the intended use and a back-up supply of energy is required. On other occasions the solar input energy may exceed the immediate demand and the excess energy must be wasted or put into storage. The successful long-term (i.e. weeks and months) storage of heat energy is a scientific, engineering and economic problem that has not yet been solved. For domestic hot water systems it is usually required that the heat energy be stored for only a few hours.

A typical form of solar-assisted domestic hot water system is shown in Fig. 11.16 [1]. The collector should be mounted facing due south, in the northern hemisphere, and inclined at roughly the latitude angle minus 10°. For example, central UK is of latitude 54°N, so a collector should be inclined at about 44° to the horizontal. A relevant numerical example is given as Example 11.4. Where a precise spatial orientation is not possible it is a reasonable compromise to mount the collector on the sloping roof of a house, provided that the roof is at least partly south-facing

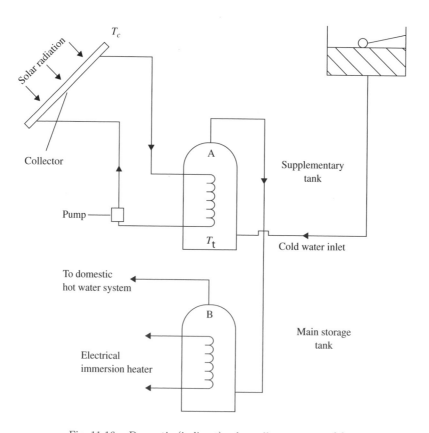

Fig. 11.16.   Domestic (indirect) solar collector system [1].

and is not screened by other buildings or trees. The collector fluid is restricted inside a self-contained pipe system and heat is transferred using a copper coil pipe immersed inside a supplementary domestic storage tank. The water, or other fluid, in the solar collector circuit can be circulated using a small electric pump, of the size used in domestic central heating systems (rated at tens of watts). This should be temperature-actuated to switch on when the collector fluid temperature T_c exceeds the supplementary tank temperature T_t. Suitable pumps, with temperature-sensitive controllers, can be purchased commercially.

With the "indirect" solar collector system of Fig. 11.16, the fluid in the solar collector circuit is completely separate from the water of the domestic hot water system. Where a solar collector system is added (retrofitted) to an existing house or other building, it is good practice to install a supplementary storage tank, as in Fig. 11.16.

In the system of Fig. 11.16, water from the cold water supply is fed into the bottom of supplementary tank A. This water is heated due to heat transfer from the solar heated water in the solar coil. The solar pre-heated water is then piped to the main storage tank B, from which it can be drawn off for domestic use. It is customary in the UK to use an electrical "immersion" heater as a back-up facility. Alternatively, the main storage tank water may be heated via a gas-heated or oil-heated furnace. The pre-heating created by the solar collector system means that less fuel is used in raising the water temperature in the main storage tank.

It is possible to implement systems using only one storage tank. The solar collector circuit and the building heating system both heat the water in the same storage tank, which acts as a heating source for the hot water supply [8].

Some solar collector systems do not use a pump but rely on the natural convection of the heated water to circulate itself. This is sometimes called a thermosyphon effect. In Japan, for example, there are several million small solar water heating installations, with roof-mounted storage tanks and inclined flat-plate collectors, which rely upon thermosyphon circulation.

A typical installation on UK domestic premises requires 1 square metre of collector area plus about 10 gallons (45.4 litres) of storage for each person in the household. The supplementary storage tank would normally be of capacity 30–50 imperial gallons, which is the same size as the standard storage cylinder tank of a UK domestic hot water central heating system. The two storage tanks of an indirect system do not need to be located in adjacent spaces. The main storage tank is likely to be in a clothes-airing cupboard. In many cases it is found convenient to locate the supplementary storage tank out of sight, in (say) a building roof space. The supplementary tank and its pipework must be well cladded (i.e. lagged) to prevent heat loss [8].

Commercial solar water heating systems are expensive to install in Western countries. An installation for a typical dwelling house in the UK or USA would cost several thousand pounds sterling or several thousand US dollars. A closely considered cost–benefit analysis should first be undertaken to ensure that the "payback period" is acceptable. It is likely that the entire installation cost would need to be paid "up front". The customer has to choose either the loss of capital and interest from savings or, more expensively, borrowing money at commercial rates [9, 10]. Customers like short payback periods of less than five years.

In order to be economically viable, a domestic installation in the UK would need to be in the southern part of the country and in a household that uses a lot of hot water. One might raise the issue as to whether an environmentally-friendly government should offer tax incentives to citizens to use renewable energy. Or is it more important to protect the industries involved in fossil fuel extraction and use?

11.4.4. *Worked examples involving solar flat-plate collectors*

Example 11.1.   A solar collector measures $2\,\mathrm{m} \times 1\,\mathrm{m}$ and has a capacity of 5 imperial gallons. After exposure to sunlight for 1 hour the mean temperature rise of the static water was $11.5°C$. During this exposure of the solar collector a solarimeter was found to record the mean voltage $V = 4.5\,\mathrm{mV}$.

(a) Calculate the heat energy absorbed by the water.
(b) What is the rating in watts of an electric lamp, equivalent to the energy absorbed?
(c) Calculate the heat energy falling on the collector if the solarimeter calibration in $\mathrm{W/m^2}$ is 78 times the solarimeter voltage in mV.
(d) What is the efficiency of the collector under these conditions?

(a) An imperial gallon of water weighs $10\,\mathrm{lb}$.

$$\text{Quantity of heat } Q = \text{Mass of water} \times \text{temp rise} \times \text{specific heat}$$
$$\text{Mass} = 5 \times 10 \times 454\,\mathrm{g}$$
$$Q = 5 \times 10 \times 454 \times 11.5 \times 1$$
$$= 261058 \text{ calories}$$

By Joule's law

$$\text{Heat energy } W = 4.2\,Q$$
$$= 4.2 \times 261058 \text{ joules}$$
$$= 1096444 \text{ joules}$$

(b)

$$\text{Power} = \frac{\text{energy}}{\text{time}} = \frac{1096444}{1 \times 60 \times 60} = 304.6\,\mathrm{W}$$

(c)

$$\text{Input power} = 78 \times 4.5 \, \text{W/m}^2$$
$$= 78 \times 4.5 \times 2 = 702 \, \text{W}$$
$$\text{Input energy} = 702 \times 3600 = 2527200 \text{ joules in 1 hour}$$

(d)

$$\text{Efficiency} = \frac{304.6}{702} = \frac{1096444}{2527200} = 43.3\%$$

which is a typical value.

Example 11.2. A flat-plate solar collector feeds a storage tank of capacity 50 imperial gallons (227 litres). The flow rate of the pumped water is 20×1^{-6} cubic metres per second. The ambient air temperature is 15°C. For 6 hours of operation the temperature rise of the collector water averages 22°C. If the temperature rise in the storage tank is 16.5°C, what is the system thermal efficiency? What is the power rating of the collector?

$$\text{Volume of water transferred} = \frac{20}{10^6} \times 6 \times 3600 \, \text{m}^3$$

$$= \frac{20}{10^6} \times 6 \times 3600 \times 10^6 \, \text{m}^3$$

$$\text{Mass of water transferred} \, (1 \, \text{g} \equiv 1 \, \text{cm}^3) = \frac{20}{10^6} \times 6 \times 3600 \times 10^6 \, \text{g}$$

$$\text{Quantity of heat} = \text{mass} \times \text{temp rise} \times \text{specific heat}$$
$$= \frac{20}{10^6} \times 6 \times 3600 \times 10^6 \times 22 \times 1$$
$$= 9.504 \times 10^6 \, \text{cal}$$

$$\text{Heat energy collected} = 4.2 \, Q = 4.2 \times 9.504 \times 10^6$$
$$= 39.91 \times 10^6 \text{ joules}$$

Energy transferred to storage tank, at a system efficiency $X\%$

$$= X/100 \times 39.91 \times 10^6 \text{ joules}$$

$$\text{Capacity of storage tank} = 227 \text{ litres} = 227 \, \text{kg}$$
$$= 227000 \, \text{cm}^3$$

Amount of heat needed to raise tank temperature by 16.5°C

$$= 227,000 \times 16.5$$
$$= 3.745 \times 10^6 \, \text{cal}$$

$$\text{Efficiency of collector system} = \frac{\text{heat transferred to tank}}{\text{heat collected}}$$

$$= \frac{3.745 \times 10^6}{9.504 \times 10^6} = 39.4\%$$

$$\text{Power rating of collector} = \frac{\text{energy collected}}{\text{time of collection}}$$

$$= \frac{39.91 \times 10^6}{6 \times 3600} = 1848\,\text{watts}$$

Example 11.3.

(a) A glass-covered, thermally insulated, flat-plate solar collector is used to heat the water indirectly in a domestic storage tank of 50 imperial gallons (227 litres) capacity. Water is pumped through the collector at $20 \times 1^{-6}\,\text{m}^3/\text{s}$ on a warm, sunny day when the mean temperature difference between the inflow and out-flow is 17°C. If the effective area of the collector is $3\,\text{m}^2$ and it is 50% efficient, what is the temperature rise of the water in the storage tank after 4 hours? What is the power rating per unit area of the solar collector?

(b) If the efficiency of the collector system in part (a) above is reduced to 42%, what water flow rate would be needed to cause the same temperature rise in the storage tank?

(a) Mass of water through the collector in 4 hours

$$= \frac{20}{10^6} \times 10^6 \times 4 \times 3600\,\text{g}$$

$$\text{Heat transferred to collector water} = \frac{20}{10^6} \cdot 10^6 \cdot 4 \cdot 3600 \cdot 17$$

$$= 4896100\,\text{cal}$$

Energy collected in 4 hours $= 4896000 \times 4.2 = 20563000$ joules

Energy transferred to storage tank $=$ 50% of energy collected

$$= 10281000\,\text{joules}$$

Mass of water in storage tank $= 50\,\text{gal} = 227\,\text{litres} = 227\,\text{kg}$

$$\text{Temperature rise of storage tank} = \frac{\text{heat transferred from collector to storage}}{\text{mass of water}}$$

$$= \frac{\frac{1}{2} \cdot 4896000}{227000} = 10.8°\text{C}$$

$$\text{Power rating of collector} = \frac{\text{energy collected}}{\text{time of collection}} = \frac{20563000}{4 \times 3600}$$

$$1428\,\text{W} = \frac{1428}{3} = 476\,\text{W/m}^2$$

(b) If the temperature rise of the storage tank is fixed, so is the ratio heat transferred/mass of water. If the heat transferred is reduced by a factor 42/50,

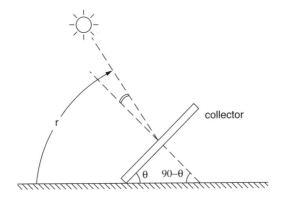

Fig. 11.17.   Inclination of a flat-plate solar collector [11].

the mass of water (i.e. pumping rate) must be increased by the same factor.

$$\text{New flow rate} = \frac{50}{42} \times 20 \times 10^{-6}\,\mathrm{m^3/s}$$

$$= 23.81 \times 10^{-6}\,\mathrm{m^3/s}$$

Example 11.4.   A solar collector is mounted at an inclination of 45° to the horizontal. If the sun rises to an inclination of 54° above the horizon, what proportion of the radiation is then falling normally onto the collector [11]?

The angle of solar inclination $= \gamma$ above the horizontal. The angle of inclination of the solar collector $= \theta$. Radiation falls normally (perpendicularly) onto the collector, illustrated in Fig. 11.17, at an angle.

$$\text{Angle of normal incidence} = \gamma - (90° - \theta) = \gamma + \theta - 90°$$

In this case $\gamma = 54°$, $\theta = 45°$ so that

$$\text{Angle of normal incidence} = 54 + 45 - 90 = 9°$$

Radiation onto the collector $= \cos 9° = 0.988\,\mathrm{p.u.}$

11.5.   Solar Water Heating (Industrial)

Many industrial processes involve the conversion of water into steam and thence into mechanical work. This requires operating temperatures much higher than those achievable by flat-plate collectors. For instance, electrical power generation requires fluids of temperatures 120–320°C (250–600°F). To realize such high temperatures by solar heating it is necessary to use various forms of concentrator systems. Curved reflectors or lenses can be used to concentrate the solar radiation from a large collector area onto a smaller absorber area. Flat-plate solar collectors are inadequate by themselves but sometimes are incorporated to provide a level of pre-heating [12].

A maximum amount of energy can be collected if some form of solar tracking system is used. Various collector dishes or reflectors are driven by small electric motors so that they track the sun across the sky during daylight hours. This involves an elaborate computer-controlled tracking system that is expensive. Such systems are not commercially viable at present but many exist as experimental prototypes in different countries.

In order to provide some measure of effectiveness of solar concentrator systems, the term "concentration ratio" (CR) is used.

$$\text{Concentration ratio} = \frac{\text{collector area}}{\text{absorber area}} \qquad (11.6)$$

In existing systems, ratios in the range of 10:1 to 10000:1 are generally realized. As discussed in Section 11.4.2, a flat-plate solar collector does not normally incorporate any form of radiation concentration so that its concentration ratio is unity.

11.5.1.   *Solar tracking systems*

11.5.1.1.   *Parabolic dish collector*

A highly effective solar collector has the arrangement of a paraboloid-shaped dish, with the principle illustrated two-dimensionally in Fig. 11.18 [13]. The dish has to be mounted on a "ball and socket" type of joint because it must be able to swivel to any position and at any angle in order to track the sun. This involves moving the reflector through two degrees of freedom, or two axes of rotation, representing azimuth and elevation control. The absorber is fixed in position at the focus of the collector (reflector). With a large diameter reflector concentration, ratios higher than 10000:1 can be realized. The parabolic dish collector uses largely direct radiation, rather than diffuse radiation. If the solar mirror is mounted on an equatorial (like an astronomical telescope), the drive motor needs only one degree of freedom.

11.5.1.2.   *Solar power tower*

An alternative to the single collector system of Fig. 11.18 is to use a system of tracking mirrors or heliostats that focus the radiation to a fixed point. The principle is illustrated in Fig. 11.19 [14]. The redirected radiation absorbed by the receiver is converted to heat and used to vaporize a circulating heat transfer fluid. The resulting high temperature, high-pressure fluid (usually a gas) can be used to drive an electrical turbine-generator unit. Fluids such as water, gas, molten salt or liquid sodium have been considered.

Very high temperatures can be realized with the upper theoretical limit set by the source temperature (5800 K). In practice a working limit of about 4000°C can be achieved using a tracking system. But in order to produce steam for conventional electricity-generating plants, a temperature of 600°C is adequate for the concentrated collector fluid. The high temperature furnaces of tracking systems have been

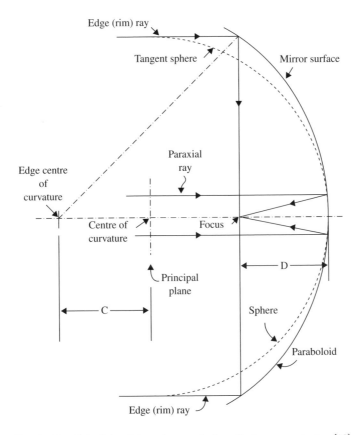

Fig. 11.18.   Principle of the solar parabolic reflector-concentrator [13].

used for the smelting of metals [14]. Solar thermal electricity generation can be realized in some locations by non-tracking systems.

Experimental heliostat power plants exist in several parts of the world. Felix Trombe of France was the pioneer in developing large solar furnaces [15]. The best-known of these is a 1MW unit at Odeillo in southern France, which uses 63 tracking heliostats covering an area of $2835\,\mathrm{m}^2$. The peak solar flux at the focus of the parabolic reflector is about $16 \times 10^6\,\mathrm{W/m^2}$, representing a concentration ratio of about 16,000 times the standard insolation of $1000\,\mathrm{W/m^2}$.

The principal heat loss from a heliostat power plant is the radiant losses from the absorber, which are given by

$$P = A\varepsilon\tau T^4 \tag{11.7}$$

where

$P =$ heat power loss, W
$A =$ absorber area, m^2
$\varepsilon =$ emissivity, which is dimensionless

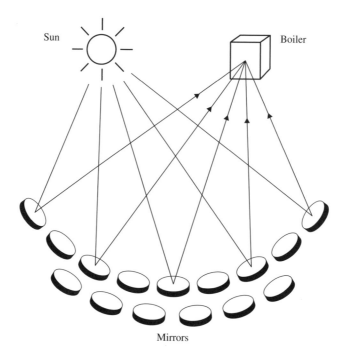

Fig. 11.19.   Array of solar tracking reflector mirrors [14].

T = absolute temperature, K

τ = Stefan–Boltzmann constant, of value $5.67 \times 10^{-8}\,\mathrm{W/m^2 \cdot K^4}$

The emissivity is the ratio of radiation emitted by a surface compared with the ideal value predicted by Planck's law. An ideal absorber is totally absorbent and has an emissivity of zero. An ideal reflector surface has an emissivity of unity since all the radiation is re-radiated. In practice, a good mirrored surface might reflect 98% of the input energy while absorbing 2%. A good absorber surface will, by comparison, absorb 98% of the input energy while reflecting only 2% [16].

Practical absorber surfaces have emissivity values of the order $\varepsilon = 0.05$ (i.e. 95% of the input energy is absorbed and 5% is re-radiated).

The emittance characteristics of Fig. 11.14 confirm the relationship in Eq. (11.7) that $P \propto T^4$ at constant wavelength.

From Eq. (11.6) the necessary absorber area A varies inversely with the concentration ratio, CR.

$$A \propto 1/CR \tag{11.8}$$

A high value of CR therefore permits a smaller absorber area and reduced re-radiation losses. Absorber design is a compromise between the concentration

Fig. 11.20.   10-MW solar power plant, Barstow, California, USA.

ratio, which is desired to be large, and absorber losses, which are desired to be a minimum. The working temperature of the absorber is seen in Eq. (11.7) to be the dominant variable in the re-radiation power loss.

Many experimental systems are under investigation in the USA. Flat (or nearly flat) heliostat tracking mirrors are arranged in a $360°$ array around a central collector tower. The best-known of these is a 10-MW_e electrical generating plant near Barstow in the Mojave Desert of southern California (Fig. 11.20). The 91-metre tower uses the concentrated heat to boil water, which is converted to high-pressure steam at $560°$C. Any surplus steam is stored for later use. The collector field consists of 1,818 slightly concave heliostats with a total reflecting area of $72538\,\text{m}^2$.

The amount of useful heat power from a solar tower can be expressed as

$$P = I_D \cdot A \cdot \varepsilon_H \cdot \eta_S \cdot \eta_{\text{real}}$$
$$= I_D \cdot A_H \cdot \eta_S \cdot \eta_{\text{real}} \tag{11.9}$$

where

$$I_D = \text{direct solar radiation (W/m}^2)$$

$$A_H = \text{area of heliostat collectors}$$

$$= \text{actual ground contour of} \times \text{fraction of the ground}$$

$$\text{heliostat site } (A) \quad \text{covered by heliostats } (\varepsilon_H)$$

or

$$A_H = A\varepsilon_H \tag{11.10}$$

$$\eta_S = \frac{\text{incident energy converted into heat}}{\text{total energy incident on heliostats}}$$

$$= \text{typical design value of } 0.53 \tag{11.11}$$

Now, from Section 1.7.4 of Chapter 1, it can be seen that

$$\eta_{Carn} = \text{Carnot efficiency of an ideal (i.e. lossless) heat–work cycle}$$

$$= \left[\frac{T_2 - T_1}{T_2}\right] \tag{11.12}$$

where T_2 is the working fluid temperature and T_1 is the dump or sink temperature, both in K.

$$\eta_{real} = \text{realistic proportion of the ideal Carnot efficiency}$$

$$\text{(increases with working fluid temperature)}$$

$$= \text{usually in the range } 0.4\text{--}0.7 \times \eta_{Carn}$$

$$= (0.4\text{--}0.7) \left[\frac{T_2 - T_1}{T_2}\right] \tag{11.13}$$

A numerical example incorporating Eq. (11.9) is given as Example 11.5 in Section 11.5.3, below.

The initial cost of a large-scale heliostat-power tower is very large in terms of money, raw materials and land. To generate $100\,MW_e$ would require about a square mile of land. The installation would use 30,000–40,000 tons of steel, 5,000 tons of glass and 200,000 tons of concrete; several times the raw material demand of a fossil fuel or nuclear station of the same capacity. Mining and manufacturing the steel, glass and concrete would produce air pollutants — notably sulphur and nitrogen oxides, carbon monoxide and particulates — equivalent to one year of effluent from an equivalent coal-burning plant [16]. Nevertheless, the longer-term accounting in financial and environmental terms is greatly in favour of solar power rather than in terms of fossil fuels.

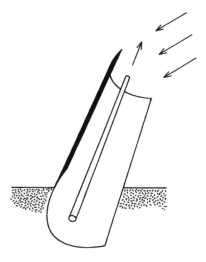

Fig. 11.21.   Parabolic trough collector [12].

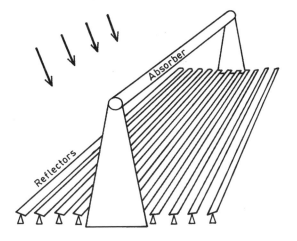

Fig. 11.22.   Linear focus concentrator [4].

11.5.1.3.   *Linear focus collectors*

There is a range of solar concentrator designs, known as distributed systems, in
which the radiation is concentrated not at a point but along a length of absorber.
In the parabolic trough collector (Fig. 11.21), the heated fluid, at a temperature
in the range 100–500°C, is pumped along the pipe that is placed at the focus of
the reflector [12]. Concentration ratios for this design of collector are much more
modest than for the point concentrators of the previous section and are typically in
the range 10 ≤ CR ≤ 100. An alternative form of linear focus concentrator, having
the same sort of performance, is shown in Fig. 11.22 [4].

If the axes of these collectors are horizontal in the east–west direction, they can be made to track continuously or be adjusted manually every few days. On the other hand, the axes may be oriented in the north–south direction at an optimum tilt for the particular latitude and continuous tracking is needed.

Adjustment of the parabolic mirror on the faceted mirror collectors of linear focus systems needs to take place only in the plane perpendicular to the collector tube. This represents only one degree of freedom compared with the two degrees of freedom of, for example, solar power tower collectors. An alternative form of description is to say that the collectors in Figs. 11.21 and 11.22 have only one axis of tracking rotation, whereas those in Figs. 11.19 and 11.20 have two axes of rotation.

The collector area of a linear focus type of collector is typically a few square metres. In order to provide a power output of $100\,\mathrm{MW_e}$, equivalent to a solar power tower, several thousand modules would be required and a distributed collector system is not then practicable. If thermal losses at high working temperatures are minimized, distributed systems operate at about 60% efficiency in the characteristics of Fig. 11.15.

11.5.2. *Solar non-tracking systems*

One of the features of modern research in solar thermal systems is the search for effective non-tracking techniques. This represents an obvious saving of first cost and reduction of system complexity. Elimination of the need for an electrical supply also makes non-tracking systems suitable for remote locations. Recently developed forms of non-tracking collector can achieve temperatures in excess of 100°C, which is suitable in industrial process heating and can be applied to solar cooling.

11.5.2.1. *Evacuated tube collectors*

In the evacuated tube collector an inner absorber tube with a selective surface coating is surrounded by an evacuated glass tube (Fig. 11.23) [13]. The vacuum cavity reduces conduction and convection losses. Like a flat-plate solar collector, the tube collector can absorb both direct and diffuse radiation. A liquid or gas may be used as the heat transfer medium, flowing inside the inner absorber tube. The focussing element of the mounting does not play a major role in the operation. Commercially available versions of this type of collector have typical values $0.8\,\mathrm{W/m^2 °C}$ for heat loss coefficient U and 0.84–0.86 for the transmittance–absorbance product $\tau\alpha$.

11.5.2.2. *Compound parabolic concentrator*

The compound parabolic concentrator developed by Winston in the USA has a cross-section like a parabola with its bottom end truncated [13, 18]. The two walls of the collector (Fig. 11.24) are each part of different but equal paraboloids. The

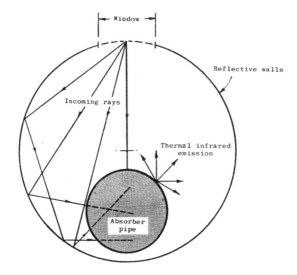

Fig. 11.23.   Optical cavity absorber [13].

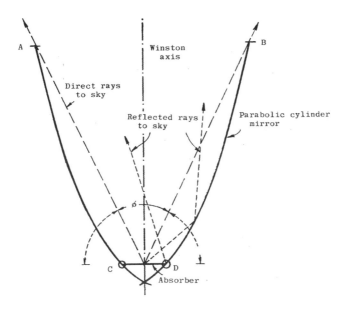

Fig. 11.24.   Basis of the compound parabolic concentrator [18].

left-hand wall of the collector is part of a parabola with its focus at point D, which is at the base of the right-hand collector wall. Similarly, the right-hand collector wall has its focus at point C. With appropriate dimensioning all the incoming radiation can be concentrated near the base and used to irradiate a thermionic or photovoltaic collector. If the design is such that points C and D are coincident, all of the radiation can be concentrated along the base line, with a concentration ratio of about 10:1.

11.5.3. *Worked examples involving solar thermionic concentrator systems*

Example 11.5. A flat-plate solar collector feeds a storage tank of capacity 50 imperial gallons (227 litres). The flow rate of the pumped water is $20 \times 1^{-6}\,\text{m}^3/\text{s}$. The ambient temperature is 15°C and the temperature rise of the collector water is 22°C for 6 hours. If the temperature rise in the storage tank is 16.5°C, what is the thermal efficiency? What is the power rating of the collector? Calculate the Carnot efficiency of the collector and hence estimate the overall efficiency of converting the collector heat gain into mechanical work.

The calculations referring to the thermal operation of this particular plate collector are given in Worked Example 11.2 in Section 11.4.4. Relevant values were calculated as

$$\text{Thermal efficiency } \eta_{\text{thermal}} = 39.4\%$$

$$\text{Power rating of the collector} = 1848\,\text{watts}$$

The Carnot efficiency for an ideal heat–work exchange cycle is defined by Eq. (11.12).

$$\eta_{\text{Carn}} = \frac{T_2 - T_1}{T_2} = \frac{22}{273 + 22} = \frac{22}{295} = 7.46\%$$

The low value for η_{Carn} is typical of lower-temperature systems. A multiplier is now used to represent the realistic proportion of the Carnot ideal efficiency that is obtained in practical energy transfer from heat to mechanical work systems, as in Eq. (11.13).

$$\eta_{\text{real}} = (0.4 - 0.7) \times \eta_{\text{Carn}}$$

For flat-plate solar systems the lower value of 0.4 is appropriate. The overall efficiency in using this flat-plate collector as a power source to produce mechanical work is therefore

$$\eta_{\text{overall}} = \eta_{\text{thermal}} \times 0.4 \times \eta_{\text{Carn}}$$
$$= 0.394 \times 0.4 \times 0.0746 = 1.18\%$$

The very low value of overall efficiency is typical and demonstrates that the flat-plate solar collector is completely unsuitable as a power source to produce mechanical work. This is largely due to the relatively low working temperature. In order to produce mechanical energy with an acceptable level of efficiency, it is necessary to use solar concentrator types of collector with much higher working fluid temperatures.

Example 11.6.   A solar power tower plant receives an average direct component of concentrated incident radiation of $950\,\text{W/m}^2$. The conversion efficiency of the heliostats into thermal energy is 53%. The operating temperature of the fluid is 560°C and the sink temperature is 100°C. Calculate the area of heliostats and the land area required to generate $100\,\text{MW}$ of thermal power.

   The Carnot efficiency in this case is

$$\eta_{\text{Carn}} = \frac{560 - 100}{560 + 273} = \frac{460}{833} = 55.2\%$$

For a thermal plant working at 560°C the realizable thermal efficiency is taken as 70% of the ideal Carnot value

$$\eta_{\text{real}} = 0.7\eta_{\text{Carn}} = 0.7 \times 0.552 = 0.386$$

The conversion efficiency of the heliostats, η_S, defined in Eq. (11.11), is specified as

$$\eta_S = 0.53$$

In Eq. (11.9), I_D is given as $0.95\,\text{kW/m}^2$ so that

$$P = 0.95 \times 0.386 \times 0.53 \times A_H = 0.19A_H$$

where P is in kW when the heliostat area A_H is in square kilometres. In terms of the units of the present problem

$$P = 194A_H$$

where P is in MW and A_H is in square kilometres.

   It is necessary to assume that the fractional value of ε_H in Eq. (11.10) is 1/4. In other words, the heliostats occupy one-quarter of the area of the ground contour A of the power plant site

$$A = \frac{A_H}{\varepsilon_H} = 4A_H$$

The thermal power generated by the heliostats in this case can therefore be expressed as

$$P = \frac{194}{4}A = 48.5\,A$$

where P is in MW and A is in square kilometres of land.

   To generate $100\,\text{MW}$ of thermal power therefore requires a heliostat land site covering an area

$$A = \frac{100}{48.5} = 2.075\,\text{km}^2$$

If the corresponding land site was square it would have a side length of 1,467 metres or about 1,600 yards.

Example 11.7. A modern conventional power station can generate up to $2000\,\mathrm{MW_e}$ of electrical power. If this amount of electrical power is to be obtained from a solar tower heliostat facility, what is the approximate land area that would be required?

The thermal energy from the heliostats would need to be processed through steam turbines or gas turbines that are then used to rotate electrical generators. The station full load rating of $2000\,\mathrm{MW_e}$ is likely to be achieved using four $500\,\mathrm{MW_e}$ generators, of very high efficiency. Assume that

$$\eta_{\mathrm{gen}} \approx 0.98$$

The associated turbines are likely to have an efficiency of the order

$$\eta_{\mathrm{turbine}} \approx 0.5$$

To provide an electrical output of $2000\,\mathrm{MW_e}$ the thermal power input is therefore required to be

$$P_{\mathrm{thermal\ input}} = \frac{2000}{0.5 \times 0.98} = 4082\,\mathrm{MW}$$

In Example 11.6 it was found that a thermal output power of $100\,\mathrm{MW}$ required a land area of $2.075\,\mathrm{km^2}$. With a solar furnace and (say) steam plant of the same thermal rating it is seen that the required land area A would be

$$A = \frac{2000}{100} \cdot 2.075 = 41.5\,\mathrm{km^2}$$

Now

$$1\,\mathrm{mile} = \frac{1760 \times 3}{3.281 \times 1000} = 1.609\,\mathrm{km}$$

and

$$1\,\mathrm{square\ mile} = 1.609 = 2.59\,\mathrm{km^2}$$

Therefore

$$A = \frac{41.5}{2.58} = 16.02\,\mathrm{square\ miles}$$

If the required site covered a square area the side length would be 4 miles. This is far more ambitious a scale of operation than anything that has so far been attempted, anywhere in the world.

11.6.   Passive Solar Space Heating of Buildings

Solar radiation can be used directly to heat the air in buildings. In effect the whole building is then being used as a solar collector. If a house design is tuned to the local climate, then nature does most of the work. The passive solar approach operates on the principles of direct admission of the solar radiation, thermal insulation and heat storage, control of the heat gain, and with all of the heat energy input distributed by natural conduction, convection and radiation. This has the advantages of simplicity, reliability, durability and economy.

In all passive solar systems the architecture and structure of the building must be designed so that much of the input energy can be stored. This can be achieved using large masses of materials that have a high heat-retaining capacity, such as masonry, concrete, adobe, stone and water.

11.6.1.   *Direct gain solar systems*

In so-called "direct gain" systems the incoming radiation directly heats the air space of the building and also contributes to the heat store. The process works best when the long axis of the building is oriented east–west and there is a large area of south-facing window (in the northern hemisphere). Only limited window area should be incorporated on the east and west walls.

The principle of one form of direct gain solar house is illustrated in Fig. 11.25 [19]. The wintertime sun is low in the sky and provides solar input through the south-facing windows. A rough estimate of the area of glazing required is to provide 1 square foot for each 4 square feet of floor area in the living accommodation.

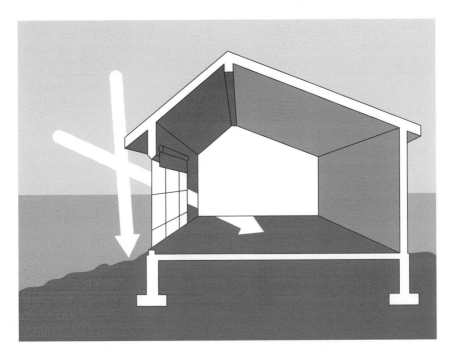

Fig. 11.25.   Solar house with direct gain heating [19].

During the summer periods, when the sun is high in the sky, the overhangs prevent excessive radiation from entering. The use of a roof overhang to prevent excessive summer radiation is very ancient. Cliff dwellings have been discovered in Mesa Verde, Colorado, USA, dated about 1200–1300 AD, that were built to incorporate this principle.

In addition to directly heating the air in the building space, a lot of energy is absorbed into the walls and floors, which might be up to 1 foot in thickness. After dark this absorbed heat is radiated back into the building space and it is advisable to insulate the south-facing glazing to prevent too much heat loss by re-radiation. When the storage material receives direct radiation, a US design criterion is that at least 40 lb (18 kg) of storage water or 150 lb (68 kg) of rock or masonry (not shown in Fig. 11.25) is required for each square foot of south-facing glazing. The thermal storage properties of some common materials are given in Table 11.8 [12].

One of the earliest and most successful direct gain passive solar buildings is St. George's County Secondary School in Wallasey, near Liverpool, England, which opened in 1965. This uses concrete floors and walls with an exterior wall insulation consisting of 13.5 cm of polystyrene. The south-facing wall is 8.2 metres high and 70 metres long, with mostly double glazing. All of the necessary heating is obtained from the solar input, lighting and the body heat of the students. Because of the massive thermal storage of the thick concrete, the temperature fluctuation is only 2–3°C. The structure of the building performs the equivalent function of the fly-wheel in rotating mechanical systems. Its thermal inertia tends to dampen any large temperature changes.

Table 11.8.   Heat storage capacities of materials at room temperature.

Materials	Specific heat Btu/lb.F° Cal/g C°	Density 1/ft^3	Kg/m^3	Btu/ft^{3}F	Heat storage capacity Kcal/m^3
Water	0.999	62.3	998	62.2	997
Mild steel	0.12	489	7830	58.7	940
Silica	0.316	140	2240	44.2	709
Paraffin	0.69	56	899	38.6	620
Rock	0.21	165	2640	34.7	550
Concrete	0.23	140	2240	32.2	516
Chalk	0.215	143	2290	30.8	493
Asphalt	0.22	132	2120	29	465
Glass	0.18	154	2470	27.7	444
White oak	0.57	47	750	26.8	429
Building brick	0.2	123	1974	24.6	395
Sand	0.191	94.6	1520	18.1	290
Clay	0.22	63	1010	13.9	222
Cork, granulated	0.485	5.4	87	2.6	42
Polyurethane	0.38	1.5	24	0.6	9

Ref.: *ASHRAE Handbook of Fundamentals*, American Society of Heating, Refrigerating and Air-Conditioning Engineers, Inc., New York, USA, 1977.

Direct gain houses of the type shown in Fig. 11.25 require adjustable window screens or blinds to prevent excessive solar input at certain times of the year. Even with the absorbtive capacity of the masonry, the air temperature can become uncomfortably warm. Two of the chief difficulties with direct gain systems are the glare from reflections inside the house and ultraviolet damage to the glass [12].

11.6.2.   *Indirect gain solar systems*

11.6.2.1.   *Thermal storage wall*

One alternative to direct gain systems is the use of a thermal storage wall. The best-known of these is the Trombe wall. A large concrete wall with a blackened outer surface is located immediately inside the south-facing double-glazed window (Fig. 11.26) [20]. The incoming radiation heats the air in the narrow cavity (i.e. 7–17 cm) between the wall and the window but is mostly absorbed by the wall. Heat absorbed by the wall causes the interior wall temperature to rise and to transfer heat energy into the house indirectly by conduction, convection and radiation. This

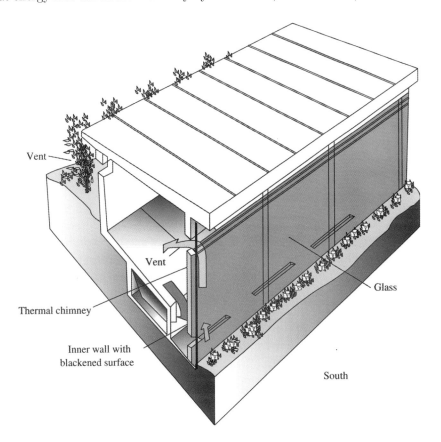

Fig. 11.26.   French solar house incorporating a Trombe wall [20].

Table 11.9.   Measured performance of a Trombe thermal storage wall.

Thickness (inches)	Temperature swing of the inside surface		Time delay to achieve peak temperature on inside surface (hours)
	°F	°C	
8	40	22	6.8
12	20	11	9.3
16	10	5.5	11.9
20	5	3	14.5
24	2	1	17.1

Ref.: *Engineers' Guide to Solar Energy*, by Y. Howell, J. A. Bereny, Solar Energy Information Services, CA, USA, 1979.

usually involves a time delay of several hours so that the heat stored from the morning sun becomes available to heat the house in the evening. Some information about the intrinsic time delay of operation of a Trombe wall is given in Table 11.9, which includes data on typical wall thicknesses.

If the wall is vented at the top and bottom, as in Fig. 11.26, the cavity-heated air is circulated round the house by natural convection or by the use of a fan blower. This facilitates the daytime transfer of about 30% of the total input heat throughout the living space. The dark outer surface of a vented wall can reach a temperature of 160°F and an unvented cavity can reach 200°F. Night-time re-radiation can be greatly reduced by the use of a screen across the wall window. The two vents should be closed at night to prevent warm air from the living space circulating through the wall-window cavity and being cooled by contact with the glazing.

Summertime cooling can be achieved by admitting air through north-facing windows and closing off the top vent of the Trombe wall. The cool air then passes through the house, up the cavity, and is expelled through an open window in the glazed south surface.

Trombe walls have the important disadvantages of high construction costs and a reduction of the available space within the house. The important design features of a Trombe wall are the area, the thickness and the thermal conductivity. A high conductivity together with a high thermal storage capacity is desired. Materials that have poor thermal conductivity, such as wood, paper, plasterboard (gypsum), cork, cellulose, etc., are to be avoided [12].

The fraction of the total winter heating load that is obtained from the solar system is sometimes defined as the "solar fraction". Variation of the solar fraction versus wall thickness with thermal conductivity as a parameter has the form shown in Fig. 11.27 [12]. The infinity line in Fig. 11.27 represents a water Trombe wall since the natural convection of heated water causes very rapid conductivity. It can be seen that in order to obtain maximum solar heating, a wall thickness between 8 and 16 inches is desired, using a material such as concrete, which combines good

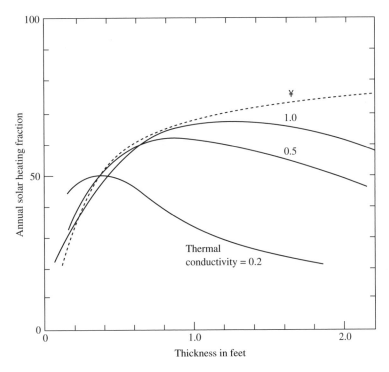

Fig. 11.27.   Performance of a thermal storage wall (with thermal conductivity as parameter) [12].

conductivity with high storage capacity, as seen in Table 11.8. Too thin a wall would lack thermal inertia and would permit too great an internal temperature variation.

11.6.2.2.   *Solar greenhouse (sunspace)*

The benefits of direct gain radiation and indirect gain energy storage and transfer may be combined by the use of the solar greenhouse effect (Fig. 11.28 [19] and Fig. 11.29 [14]). The physical principle of the greenhouse effect is discussed in Section 11.4.1. Any excess heat collected in the sunspace can be stored in the thermal mass of the wall and used in the manner described in the preceding section.

11.6.2.3.   *Roof pond*

A house with a flat roof can incorporate a roof pond of water contained in transparent plastic containers. In the winter months the water is exposed to the incoming daytime radiation and is covered with thermal insulation during night-times. The daytime heat energy input is then retained and used to supplement the house heating by radiation inwards from the metal roof. During summer months the opposite is true: in daylight hours the water pond is screened from incoming radiation. At night the insulation is rolled back and the water becomes cold by night-time evaporation.

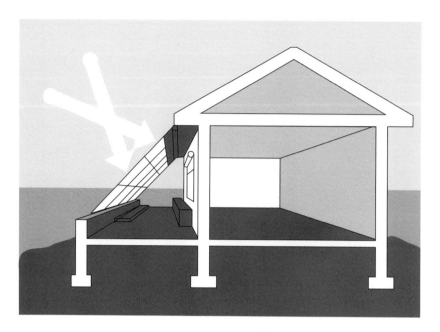

Fig. 11.28.   Use of an attached sunspace [19].

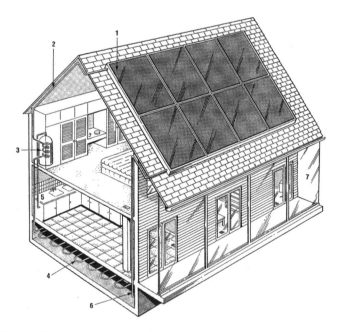

Fig. 11.29.   Energy-efficient house design [14]. 1. Solar-heated water. 2, 3. Solar energy transferred to domestic water system. 4. Underfloor heating. 5. Domestic hot water. 6. Solar water return pipe. 7. Built-in greenhouse.

This cold water absorbs heat from the building during the day and constitutes a form of daytime air-conditioning.

11.6.2.4. *Solar salt pond* [3]

In a pond of natural water, the water below the surface, heated by solar radiation, rises to the surface by convection currents. A solar pond contains concentrations of dissolved salts that gradually increase in concentration with depth. The bottom layer is highly saline, covering a blackened pond-bed. On top of the pond the water is less saline so that there is a gradual increase of water density with depth.

Solar radiation penetrates to the lowest (high density) region of the pond. Any rising convection currents are suppressed by the density gradient. Heat losses from the pond surface are lower than would be obtained with natural water. The bottom saline region acts as a solar collector and its temperature can rise to 200°F, which is sufficient to drive a turbine. Although there are daily fluctuations in the surface water temperature and the surrounding ambient temperature, the bottom saline basin temperature remains fairly constant, acting as a massive heat store.

Several experimental solar ponds have been developed in the Dead Sea region of Israel, the largest being of 5-MW rating. Much larger rated installations are planned in the USA. Solar ponds can operate continuously throughout the year and can provide peaks of "topping-up" power on demand.

11.7. Problems and Review Questions

Problems on solar flat-plate collectors

11.1. A solar collector is mounted at an inclination of 45° to the horizontal. If the sun rises to an inclination of 54° above the horizon, what proportion of the radiation is then falling normally onto the collector?

11.2. A solar collector in northern England is mounted with its axis inclined at the latitude angle 54° to the horizontal. If the total annual radiation energy on a horizontal is $1000\,kWh/m^2$, divided equally between direct and diffuse components, what is the radiation received by the collector?

11.3. A solar thermal collector is mounted at an angle $\alpha = 39.5°$ to the horizontal. If the direct and diffuse components of the solar radiation are both of value $320\,W/m^2$, then what is the total radiation?

11.4. What is a typical working efficiency for a flat-plate solar collector? Show mathematically how this efficiency is affected by an increase of the circulating fluid temperature.

11.5. Explain what is meant by the term "greenhouse effect" with regard to solar radiation. Is this effect utilized in the design of collectors for solar water heating?

What is a typical working efficiency of a flat-plate, solar water heating system? How could this efficiency be increased?

11.6. Describe, using a diagram, the essential features of an arrangement to heat indirectly the water of a domestic hot-water system using a flat-plate solar collector. In a typical system the flat-plate collector is housed in a thermally insulated container with a glass cover, exposed to the radiation. Explain the action of the glass-fronted container.

11.7. Sketch the form of a solar collector system for pre-heating the water in a domestic dwelling. Where would you prefer to locate the supplementary water tank? How would you check if the system is working?

11.8. Explain the basic principles of operation of a flat-plate solar collector. Sketch a scaled characteristic of efficiency versus operating temperature (above ambient) for a collector with a single glass window. Identify a typical working point. How would this characteristic be modified if double-glazing was introduced into the window?

11.9. Describe the main features of power loss in a flat-plate solar collector. How are these related to the terms of the general collector efficiency equation? Referring to this equation, explain how you would attempt to improve the collector performance.

11.10. A flat-plate solar collector, mounted at the latitude angle and south-facing, has an effective area of $2\,\mathrm{m}^2$. Water is pumped through this collector at the rate $20 \times 1^{-6}\,\mathrm{m}^3/\mathrm{s}$ and the mean temperature difference between the inflow and outflow is 18.4°C. The collector is used to heat indirectly the water in a storage tank of capacity 50 imperial gallons (227 litres) for 5 hours continuously. If the system operates at a typical efficiency, calculate the temperature rise in the storage tank. What is the power rating of the flat-plate collector in W/m^2?

11.11. A solar collector is to be mounted on the south-facing roof a dwelling house, feeding a storage tank with a capacity of 30 imperial gallons (136 litres). The circulating pump is to operate at the rate of $20 \times 10^{-6}\,\mathrm{m}^3/\mathrm{s}$. On a warm, sunny day the difference of water temperature between the inflow and outflow at the collector is typically 15.5°C and this difference exists for 6 hours. What is the operating efficiency of the system if the temperature of the water in the storage tank is increased by 20°C?

11.12. A solar collector is to be mounted on a south-facing roof, feeding a storage tank of capacity 200 litres. The circulating pump operates at $20 \times 10^{-6}\,\mathrm{m}^3/\mathrm{s}$. On a typical day the difference of water temperature between the collector output and input is 17.2°C and lasts for 6.8 hours. If the temperature of the water in the storage tank is increased by 13.75°, what is the thermal efficiency of the system?

11.13. It is proposed to use a roof-mounted solar water-heating system to supplement the energy input into a certain industrial process. The south-facing solar collector is to be used to heat indirectly the water in a storage tank of capacity 5,000 imperial gallons (22,700 litres). Water can be pumped through the collector by a range of available water pumps. On a typical summer day, there are 5.6 hours of sunshine, which causes an average temperature difference of 16.5°C between the inflow and outflow of the collector.

(a) If the anticipated efficiency of the system is 42.6%, what rate of water pump flow in m^3/s is needed to cause a temperature rise of 12°C in the storage tank?

(b) If this pump is used and the temperature of the storage tank becomes 14.5°C, what is the efficiency of the collector system?

11.14. A flat solar collector of area $2\,m^2$ has a water inflow temperature of 13.5°C and an outflow of 52°C, while the incident radiation is constant at $900\,W/m^2$. Use the empirical relation in Eq. (11.5) to calculate the approximate thermal efficiency.
What value of efficiency is indicated if $T_c = T_a$, and why is this efficiency indication wrong?

11.15. Give a broad specification for a solar water heating installation for a typical UK domestic dwelling (i.e. three-bedroom, semi-detached house) occupied by four persons. In particular, specify the necessary area of collector and the capacity of the supplementary water tank. The installed commercial cost is quoted at £3,200. Estimate the pay-back period if the household is (i) a heavy user of hot water and (ii) a light user of hot water.

Problems on solar concentrator systems

11.16. What are the main differences between the countries of northern Europe and those of the Middle East with regard to solar radiation? What are the chief obstacles to the widespread use of solar water heating in the two regions?

11.17. Industrial applications where heat is converted into mechanical work require working fluids of temperatures in the range 250–600°C (much higher temperatures than are realizable by solar flat-plate collectors). Why are such high temperature fluids needed?

11.18. Calculate the Carnot efficiency of a concentrated heat solar system in which the high temperature is 450°C while the low temperature is 40°C.

11.19. The difference between the inflow and outflow fluid temperatures in a flat-plate solar collector is 20°C. Explain why this is completely unsuitable as an energy supply for a thermodynamic energy converter (e.g. a steam engine).

11.20. A heat collector for an ideal heat–work (Carnot) system has a constant ambient temperature. Show, mathematically, that increase of the working fluid temperature will increase the Carnot efficiency.

11.21. A solar collector with constant incident radiation has an efficiency given by

$$\eta = A - B(T_2 - T_1)$$

where T_2 is the collector outlet temperature and T_1 is the ambient temperature. If the heated water is used as a source for an ideal heat engine (i.e. operating on a Carnot cycle), show that the maximum system efficiency occurs when

$$BT_2^2 = T_1(A + BT_1)$$

Calculate the efficiency when $T_1 = 25°C$ if $A = 0.6$ and $B = 0.001$ [3].

11.22. Define the term "concentration ratio" (CR) for solar collectors. How is the CR related to the working temperature of the solar absorber? What is the value of the CR for a flat-plate solar water heater?

11.23. How is the thermal power developed by a solar power tower system related to:

(a) the area of the heliostats,
(b) the solar radiation,
(c) the thermal efficiency of the heliostats, and
(d) the Carnot efficiency?

11.24. The absorber of a solar concentrator system operates at 55°C. The collector receives $200\,W/m^2$ of input power. If the concentration ratio is 50 and the absorber emissivity is 0.05, what proportion of the input power is re-radiated? [Hint: Use Eq. (11.7).]

11.25. A solar power tower plant receives an effective average radiation of $1000\,W/m^2$ from its concentrator collectors. The conversion efficiency of the collector heliostats into thermal energy is 53%. If the plant fluid operates at 600°C and the sink temperature is 100°C, calculate the area of heliostats and the land area required to generate 100 MW of thermal power.

11.26. Enumerate and briefly discuss the advantages and disadvantages of a proposed 100 MW heliostat power tower facility.

11.27. One of the largest electricity generation stations using fossil fuels is now rated at $2000\,MW_e$ and uses four 500-MW_e turbine-generator sets. Use the result of Problem 11.25 to estimate the ground area that would need to be occupied by a solar power tower heliostat installation delivering the same electrical power. If such an installation proved to be technically feasible, what considerations would arise in the choice of a suitable site?

11.28. Is access to the sun a legal right? Suppose, for example, that the owner of a small processing plant had spent US$50,000 (or equivalent) on building a solar water heating system that incorporated an array of roof-mounted solar collectors. The owner of an adjacent property then proposes to erect a taller building nearby such that it would effectively screen the solar collectors from the sun and render them useless. Does the owner of the processing plant have any legal right of protest or redress?

References

[1] Courtney, R.G. (1976). *Solar Energy Utilisation in the UK: Current Research and Future Prospects*, Building Research Establishment, London, Current Paper 64/76.

[2] Brinkworth, B.J. (1975). *Solar Energy for Man*, The Compton Press, Salisbury.

[3] McVeigh, J.C. (1984). *Energy around the World*, Pergamon Press, Oxford.

[4] Swinden, D.J. (1980). *Energy*, Lecture Support Notes, England.

[5] Mackay, D. (2009). *Sustainable Energy — Without the Hot Air*, Cambridge University Press, Cambridge.

[6] "Solar Energy: A UK Assessment", International Solar Energy Society (ISES), UK Section, London, 1976.

[7] "Renewable Energy Annual 2000 (Giving Data for 1999)", Energy Information Administration, US Department of Energy, Washington, DC, 2001, http://www.eia.gov/ renewable/.

[8] Dorf, R.C. (1978). *Energy Resources and Policy*, Addison Wesley Publishing Co., Reading, MA.

[9] "A Consumer's Guide: Heat your Water with the Sun", The National Renewable Energy Laboratory, US Department of Energy, Washington, DC, 2003.

[10] "Solar Energy for Heating and Cooling", US House of Representatives, Subcommittee on Energy of the Committee on Science and Astronautics, Washington, DC, June 1973.

[11] Shepherd, W. and Shepherd, D.W. (2008). *Problems and Solutions in Energy Studies*, Imperial College Press, London.

[12] McDaniels, D.K. (1984). *The Sun — Our Future Energy Source*, 2nd Edition, John Wiley and Sons, Inc., New York.

[13] Meinel, A.B. and Meinel, M.P. (1976). *Applied Solar Energy*, Addison Wesley Publishing Co., Reading, MA.

[14] "Renewable Energy — A Resource of Key Stages 3 and 4 of the UK National Curriculum", Renewable Energy Enquiries Bureau, Energy Technology Support Unit, Dept. of Trade and Industry, Harwell, 1995.

[15] Trombe, F., Gion, L., Royere, C. and Robert, J.-F. (1973). "First Results Obtained with the 1000 kW Solar Furnace", *Solar Energy*, **15**, 63–66.

[16] "An Emissivity Primer", Electro Optical Industries Inc., 2000, http://www.electro-optical.com/eoi_page.asp?h=What%20Is%20Emissivity?

[17] "Energy in Transition 1985–2010", National Research Council, Washington, DC, 1979.

[18] Winston, R. (1974). *Principles of Solar Collectors of a Novel Design*, Solar Energy, **16**, 89–95.

[19] "Guide to Passive Solar Home Design", Office of Energy Efficiency and Renewable Energy, US Department of energy, Washington, DC, 2010.

[20] Clarke, D. (ed.) (1978). *Energy*, Marshall Cavendish Books, Ltd., London.

Additional References

Duffie, J.A. and Beckman, W.A. (1980). *Solar Engineering of Thermal Processes*, John Wiley and Sons, New York.

The physical constants have been taken from Bishop, R.L. (ed.) (1987). *Handbook of Royal Astronomical Society of Canada*, University of Toronto Press, Toronto.

CHAPTER 12

SOLAR PHOTOVOLTAIC CONVERSION

12.1. Basic Features of Solar Cells and Solar Systems

Solar radiation can be converted into electrical energy directly at the atomic level, without any intermediate process at all, by the use of solar photovoltaic (PV) cells. The solar cells are made of semiconductor materials such as silicon. The principle of the operation is shown in Fig. 12.1, whereby direct current electricity is generated by photochemical action. Solar cells are usually fabricated as flat discs, up to a few inches in face dimension.

A number of solar cells can be electrically connected to each other and mounted in a support structure or frame, called a photovoltaic module. Modules are usually designed to supply electricity at a certain voltage level, such as a common 12 volts system. The modules themselves can be electrically connected into an array (Fig. 12.2) to supply electricity at predetermined voltage and current levels. The advantages of this form of static electricity generation, compared with thermal energy conversion methods, are considerable. Advantages of photovoltaic generation include [1, 2, 3]:

(a) there are no moving parts so that little maintenance is required;
(b) they utilize an infinitely renewable (compared with the human lifespan) and pollution-free power source;
(c) the cells are reliable and long-lasting, with no harmful waste products;
(d) there is no discernible health hazard;
(e) the cells are usually made of silicon, which is one of earth's most abundant and cheapest materials;
(f) the cells can be used on site in remote locations, such as buoys anchored at sea, or spacecraft in orbit;
(g) they have a high power-to-weight ratio, which is required in aerospace applications; and

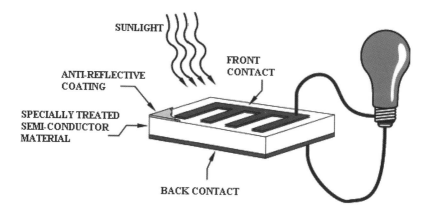

Fig. 12.1. Principle of the photovoltaic cell [3].

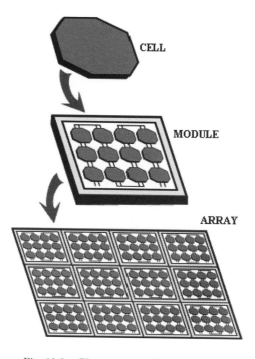

Fig. 12.2. Photovoltaic cell structures [3].

(h) they are manufactured and researched in a highly developed, scientifically based, well-funded industry, so that continual improvement of performance can be expected.

With this list of advantages one can understand the very large investment of time and effort that has been, and still is, devoted to solar cell development. If solar photovoltaic converters were available more cheaply they would completely change the whole scene of electricity generation for mass consumption. Unfortunately they still (2013) remain several times too expensive for mass use, but are viable for specialized applications such as spacecraft, isolated communication stations and certain defence needs.

In considering the cost of solar cells, the term "peak watt" of power is used. This means that the cell is required to generate 1 watt of power when the solar insolation (i.e. solar power per unit area) is $1000 \, W/m^2$. With a typical efficiency of 10%, $1 \, m^2$ of cell array area would then generate 100 peak watts.

12.2. Physical Nature of Semiconductor Materials [4]

Materials that are commonly used as semiconductors, such as silicon and germanium, are found to lie in the fourth column of the periodic table. The outer layer of electrons, sometimes called the valence shell, in the Bohr model of the atom consists of four electrons that enable a pure crystal of material to form tight covalent bonds. The nuclei consist of protons and neutrons summing to the respective atomic mass units. Quantum shells of electrons surround each nucleus so that the total number of electrons is equal to the atomic number and determines the placement in the periodic table.

The covalent bonding in the lattice of a pure crystal is tight, in that a relatively large amount of energy is necessary in order to free an electron from the bond and make it available for charge carrying. Pure semiconductor materials have such high resistivity that they are insulators at room temperature. The application of a voltage across a pure crystal of silicon or germanium, at room temperature, would cause only a small leakage current. For example, a potential difference of 100 volts across a cubic centimetre crystal of silicon would result in a leakage current of about 4 milliamperes.

To convert pure semiconductor material from a light-sensitive but grossly inefficient electricity converter into a form suitable for use in transistors or photovoltaic cells, capable of much higher current capacity, it is necessary to increase the conversion efficiency. This can be done by increasing the energy of the outer shell electrons. Such modification can be achieved by a very complex industrial process known as "doping", whereby other materials, lying in the third or fifth column of the periodic table, are combined with the pure semiconductor. This revolutionary breakthrough was pioneered at the Bell Telephone Laboratories in New Jersey, USA, in the 1950s.

12.2.1. *Group-3 (acceptor) impurities*

In the process of doping, atoms of a selected impurity material are substituted for some atoms of the pure semiconductor material within its semiconductor crystal lattice. Materials such as boron, aluminium, gallium and indium lie in the third column of the periodic table and are characterized by having three electrons in their

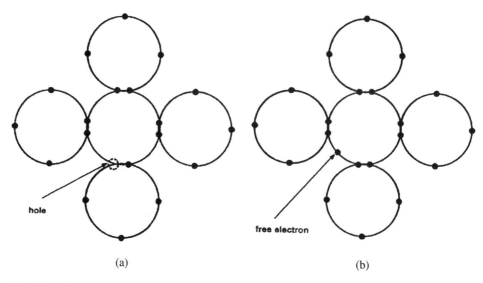

Fig. 12.3.   Planar representation of doped semiconductor lattice. (a) Acceptor impurity. (b) Donor impurity [4].

outer shells. Incorporation of (say) a boron atom into a lattice of silicon atoms results in the effect depicted in Fig. 12.3(a). The impurity atom forms three covalent bonds with adjacent atoms. In the fourth side, however, no impurity electron is available to bond with the semiconductor electron and a "hole" appears in the location of the missing electron. Such incomplete bonding can be arranged to occur throughout the crystal. Group-3 or trivalent impurities are called acceptor impurities because they have the capability of accepting free electrons. The application of a potential difference across the crystal will cause a migration of holes.

Doping a pure semiconductor material with acceptor impurity forms a p-type (positive) semiconductor in which the holes are called majority carriers and the free electrons are called minority carriers.

12.2.2.   *Group-5 (donor) impurities*

Materials such as arsenic, phosphorus and antimony lie in the fifth column of the periodic table and are characterized by having five electrons in their outer shells available for chemical reactions. Incorporation of (say) a phosphorus atom into a lattice of germanium atoms results in the effect depicted in Fig. 12.3(b). The impurity atom forms four covalent bonds with adjacent atoms but one electron per phosphorus atom lies in the valence shell. The free electron has relatively high energy and may be easily detached from its parent atom by the application of an electromotive force (emf) to the crystal.

Group-5 or pentavalent impurities are called donor impurities because they have the capability of donating free electrons. Doping a pure semiconductor material with donor impurity forms an *n*-type (negative) semiconductor in which the electrons are called majority carriers and the holes are called minority carriers, which is the opposite way round from acceptor-impurity-doped material.

Because of doping, the high intrinsic resistivity of pure semiconductor materials is markedly reduced and doped materials are adequate conductors of electric current. In this book the term "semiconductor" should always be taken to mean the doped, commercially available materials now so abundantly used in the electronics industry.

12.3.   Photovoltaic Materials [4, 5]

12.3.1.   *Crystalline silicon (c–Si)*

Early forms of silicon photovoltaic cells were very expensive because of difficulties in the industrial preparation of sufficiently pure high-grade silicon. Very pure single crystals of silicon needed to be grown as cylindrical ingots, about 10 cm in diameter, in order to maximize the cell exposure area. This is now known as "monocrystalline" silicon.

Processing and fabrication problems still exist in the preparation of single crystalline silicon cells, which remain very expensive. The wafers are typically 250–300 μm thick and need to be cut by diamond slitting discs of about the same thickness, which is both expensive and wasteful of the silicon material. Preparation of the pure crystals involves temperature control, within $\pm 0.1°C$ of a melt at 1420°C. After cutting, grinding and polishing, all labour-intensive operations, the silicon wafers have to undergo a gaseous diffusion process involving the bonding of another material. One development that has been responsible for a reduction in the cost of solar photovoltaic cell silicon is to grow the silicon crystal in the form of a ribbon rather than an ingot. By drawing the seed crystal upwards through a die in the molten silicon, a ribbon crystal several feet long may be obtained. Ribbon crystal material can now be made of thicknesses in the range 4–300 μm, with much less stringent temperature control. The costly process of cutting thin wafers from large single-crystal cylindrical ingots is then eliminated. The ribbon process results in less pure silicon than the traditional method and produces cells with efficiencies of the order 12%–15%.

Another development in the use of crystalline forms of silicon is the use of poly-crystalline silicon rather than single-crystal material. Many small silicon crystals are oriented randomly within thin layers of polycrystalline material. This is much cheaper to produce than single-crystal forms and uses much less silicon material. Reported efficiencies with polycrystalline solar cells are now up to 12%.

12.3.2.   *Amorphous (uncrystalline) silicon (a–Si)*

In amorphous silicon there is no regular crystal structure. The very expensive production techniques involving pure single-crystal forms are unnecessary. The absorption coefficient for amorphous silicon, in the visible light range, is more than ten times the value for single-crystal silicon. Amorphous silicon can be deposited onto backing material (sometimes called "substrates") in very thin films, of the order 1 μm thick. This greatly reduces the amount of silicon material used and, consequently, the cost of mass production.

Although amorphous silicon solar cells are presently available and relatively cheap, their maximum efficiency is low, of the order 6%. This raises interesting cost efficiency considerations in the design of any specific photovoltaic cell array. Amorphous silicon cells with an efficiency range 5–6% were used in a 50-MW photovoltaic power station built in the desert outside Los Angeles, California, USA [6].

Can it be anticipated that future generations of amorphous silicon cells will eventually achieve the same levels of operational efficiency as crystalline silicon cells? Amorphous silicon (a–Si) modules have dominated the consumer electronics market since the 1980s, but are only currently being launched into the power market [7].

12.3.3.   *Materials other than silicon*

In the search for more efficient solar photovoltaic cells many different semiconductor materials have been investigated. An historical perspective of the results is given in Fig. 12.4 [7]. The market is still dominated by crystalline silicon cells, with the most recent forms claiming efficiencies approaching 25%, which is higher than the historical maximum theoretical value. There is a 7–8% gap between the realized efficiencies

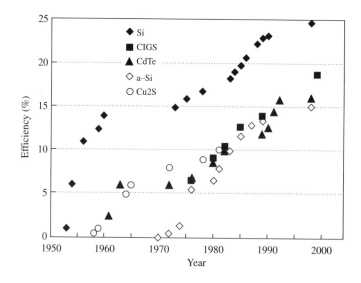

Fig. 12.4.   Development of thin-film, photovoltaic cell efficiencies [7]. ◆ Crystalline silicon. ■ Copper indium gallium sulphide (CIGS). ▲ Cadmium tellurium. ◇ Amorphous silicon. ○ Copper sulphide.

of crystalline silicon cells and thin-film versions (the most economic) of other options (Fig. 12.4). Some of this loss is due to inhomogeneities in the polycrystalline materials, but there is also an inherent mechanism, not fully identified, that reduces the efficiencies of CdTe and CIS materials [7]. Another way of interpreting Fig. 12.4 is that CdTe, CIS and other thin-film technologies (not shown) are about 20 years later in development than Si. But Si cells are now approaching the theoretical limit of their efficiency; will thin-film cells eventually catch up? A thin-film competitor, not shown in Fig. 12.4, is the gallium arsenide (GaAs) cell, attractive because of its high efficiency. It will operate at temperatures up to 200°C, whereas the maximum working temperature for silicon, about 100°C, is already accompanied by a serious loss of efficiency. Because of the higher temperature tolerance, gallium arsenide cells can be used in focussing concentrator systems. Gallium arsenide cells, however, cost several times the price of silicon cells and it is questionable if the world supply of gallium would be adequate to facilitate its widespread mass use.

A recent development is to use multilayer cells, with different energy gaps, so that sunlight first strikes the material with the largest band gap. One recent design stacks layers of GaAs onto layers of silicon. A GaAs cell on top absorbs solar energy from the blue (short-wavelength, high-energy) end of the spectrum, passing redder light to a silicon cell beneath. Higher conversion efficiency is realized by capturing a larger portion of the solar spectrum. The laboratory prototype of this new cell was reported to have an efficiency of 31% [8].

Interesting research is underway to attempt to make solar cells from plastic materials rather than from crystalline silicon. This is, in effect, an attempt to produce artificial photosynthesis using polymer (organic) materials to provide a self-assembling, two-phase, photovoltaic material [6].

The most remarkable form of photovoltaic conversion occurs in nature in the operation of photosynthesis in plants, described in the following chapter. Green leaves convert and store solar energy at a phenomenal rate in a chemical process that separates the functions of light absorption and charge transportation, a process that has been going on naturally for millions of years. Much research is underway to develop thermoelectric materials for electricity generation and energy harvesting, such as by the use of Si/SiGe based materials.

12.4.   Cost of Solar Photovoltaic Electricity

The quoted cost of electricity from solar photovoltaic cells varies widely. In 1973 the cost was US $300/peak watt. By 1977, due to vastly increased research, the cost dropped to US $15/peak watt, falling further to $10/peak watt in 1983. In the late 1980s the cost had fallen to US $2–3/peak watt [2]. During 2000 the global installed base of PV generation exceeded 1 GW for the first time. The costs of PV systems in the USA in 2002 varied between US$6 and US$12 per watt, resulting in costs for PV-generated electricity from US$0.25/kWh to over US$1.0/kWh.

In order to be competitive with conventional generation the cost of electricity delivered to consumers must now be around US$0.07 per kWh in the USA and £0.07 per kWh in Great Britain. This energy cost corresponds to an installed cost for solar cell arrays of US $0.15–0.3 per peak watt. When maintenance and interest upon capital costs are added (probably more than doubling the price per peak watt) the cost of solar-photovoltaic-generated electricity is found to be about four times the cost of conventional generation using fossil fuels. A relevant numerical example is given as Example 12.9 in Section 12.8.

In the UK an installed photovoltaic array on a dwelling house would cost about four times the price of a solar thermal system of the same rating.

12.5. Operation of the Semiconductor Diode and Solar Photovoltaic Cell

Let a crystal of p-type material be brought into conjunction with a crystal of n-type material such that the junction represents perfect continuation of the lattice. Even in the unexcited state all of the charged sub-atomic particles are in constant motion due to thermal energy. Any increase of temperature causes the release of further electrons and holes due to the breaking of covalent bonds.

The most common form of p–n junction device is the semiconductor diode rectifier, which is probably the most frequently used component in electronic circuits. The application of an externally applied reverse voltage across a p–n junction device causes a small reverse leakage or thermal current. This condition is known as reverse bias or reverse blocking in electronic circuits. If a forward bias voltage is applied across a p–n junction device, the low resistivity of the semiconductor material results in the possibility of high forward current, which depends upon the external circuit impedance. The diode rectifier therefore has the nature of a terminal voltage-actuated on–off switch.

A semiconductor p–n device can also be switched on by irradiating the p–n junction with photons of sufficient energy, and this is the basis of the solar photovoltaic cell. In a solar photovoltaic cell the incident solar radiation passes through the p-type material into the junction. Some photons of the incident radiation collide with the valence electrons of the silicon (for example) and are absorbed, releasing electrons and holes into the crystal lattice. If the silicon cell is electrically isolated on an open circuit a direct emf or voltage will then appear across its terminals. If the cell has an external electrical circuit connected to its terminals, then a direct electric current (DC) will flow.

A p–n junction photovoltaic cell therefore performs two functions simultaneously: it harvests sunlight by converting photons to electric charges and it also conducts the charge carriers to the device terminals to become direct voltages. If a load is connected the charges go into motion and can be collected as DC electric current.

12.6.   Physical Properties of the Solar Photovoltaic Cell

The energy content W of the incoming radiation is in discrete packets that depend on its frequency, according to the relation

$$W = hf \tag{12.1}$$

where f is the frequency in Hz or s^{-1} and h is the Planck constant (6.626×1^{-34} Js or 4.136×1^{-15} eVs).

The frequency f of the radiation is related to its wavelength λ by the relation

$$f\lambda = c \tag{12.2}$$

where c is the velocity of light ($c = 2.998 \times 10^8$ m/s).

Combining Eqs. (12.1) and (12.2) expresses the radiation energy in terms of wavelength:

$$W = \frac{hc}{\lambda} = \frac{1.986 \times 10^{-25}}{\lambda} \tag{12.3}$$

where the energy W is in joules when the wavelength λ is in metres. Alternatively, if the Planck constant is expressed in electron volt seconds (eVs),

$$W = \frac{hc}{\lambda} = \frac{1.24 \times 10^{-6}}{\lambda} \tag{12.4}$$

where now the energy W is in electron volts when the wavelength λ is in metres. The energy per photon at various parts of the solar spectrum (Fig. 11.1 of Chapter 11) is given in Table 12.1.

It is significant to note that only part of the incident solar radiation can produce a photovoltaic effect. The minimum amount of input energy per photon needed to liberate electrons into a lattice of crystalline silicon (sometimes called the "energy gap") is found to be almost 1.08 eV or 1.73×1^{-19} J. From Eq. (12.3) or Eq. (12.4) this occurs at a wavelength $\lambda = 1.15\,\mu$m. The infrared portion of the solar spectrum, with $\lambda > 1.15\,\mu$m, is useful for heating purposes but will not photovoltaically energize silicon [9].

Table 12.1.   Photon energy, frequency and wavelength for solar radiation.

Wavelength (μm)	Frequency (Hz or s^{-1})	Energy per photon	
		joules	electron volts
0.3 (ultraviolet)	9.99×10^{14}	6.62×10^{-19}	4.133
0.5 (visible light)	5.996×10^{14}	3.972×10^{-19}	2.48
1.0 (near infrared)	2.998×10^{14}	1.986×10^{-19}	1.24
1.15 (near infrared)	2.607×10^{14}	1.727×10^{-19}	1.0783
2.0 (infrared)	1.499×10^{14}	0.993×10^{-19}	0.62
3.0 (infrared)	0.9993×10^{14}	0.662×10^{-19}	0.413

When the wavelength is less than 1150 nm (1.15 μm) its energy content is greater than the critical energy value of 1.08 eV. Electrons are released but the excess energy above 1.08 eV is absorbed by the silicon and re-appears as heat without contributing to the electric current flow. In the wavelength range $0.3 \leq \lambda \leq 1.15\,\mu$m, at any particular frequency, a proportion $1.8/W^t$ of the associated energy W is wasted as heat. Over this spectral range a 33% proportion of the input energy is lost. This reduces the maximum theoretical conversion efficiency to 44% for a perfectly constructed crystalline silicon cell.

In addition to the intrinsic physical limitations above, any photovoltaic cell is subjected to further losses due to its electrical operation. The passage of current through the semiconductor material causes heating (I^2R) losses and this may be exacerbated by the contact resistance between the terminals and the cell active material. There is also power loss at the cell junction, which must be strictly con-trolled to avoid overheating. These various forms of loss contribute additional com-ponents to the limitation of working efficiency. The final result is that an operational photovoltaic cell constructed of crystalline silicon has a maximum theoretical work-ing efficiency of about 23%, although this is seldom attained or even approached in commercial practice [10]. Typical commercial photovoltaic cells operate at an effi-ciency of about 10%. More efficient designs, for example using multiple junctions, realize efficiencies between 10% and 20% [5, 11].

Maximum realizable efficiencies for various materials are shown in Fig. 12.5, which is a smoothed approximate characteristic [12]. Each material has its own characteristically critical value of energy gap, corresponding to the 1.08 eV for silicon. As the junction temperature of a semiconductor material increases, its realizable efficiency is reduced. Typical efficiency loss is reported to be 0.38% loss per °C. At 100°C, for example, the characteristic of Fig. 12.5 would be lowered,

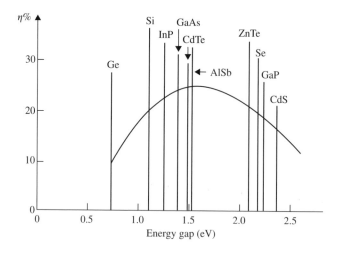

Fig. 12.5.   Maximum realizable solar cell efficiencies for various materials at 0°C [12].

roughly uniformly, by about 40% and the maximum efficiency for silicon would drop to 14%. It can be seen that the materials gallium arsenide (GaAs), cadmium tellurium (CdTe) and aluminium antimony (AlSb), for example, have potentially higher achievable efficiencies than silicon, but at present these are more expensive.

12.7. Electrical Output Properties of the Solar Photovoltaic Cell

The external characteristic of a solar cell is the property of current versus voltage. An ideal characteristic would be rectangular in shape. Practical characteristics are roughly rectangular (Fig. 12.6). Each different level of incident radiation results in a different characteristic. The intercept of a characteristic on the current axis represents a zero voltage drop across the cell terminals and is the short circuit current I_{SC} that is (almost) directly proportional to the incident light intensity. This is true for individual silicon cells (Fig. 12.7) and also for solar arrays of many identical cells (Fig. 12.8) [12]. The intercept of an I–V characteristic on the voltage axis in Figs. 12.6 and 12.7 is the open circuit voltage V_{OC}, which varies logarithmically with the incident light intensity (Fig. 12.8). Most solar cells operate with a working level of direct voltage of less than 1 volt.

12.7.1. *Maximum power delivery*

In DC circuits with a solar cell source the power generated is the product of the cell voltage and current. For typical characteristics like Fig. 12.6, the maximum power delivery point lies in the region of the knee of the curve. The current and voltage

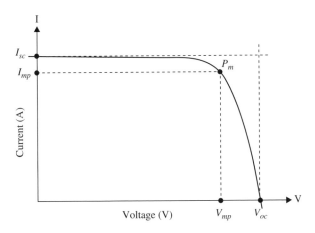

Fig. 12.6.   External characteristic (current versus voltage) for typical photovoltaic cell.

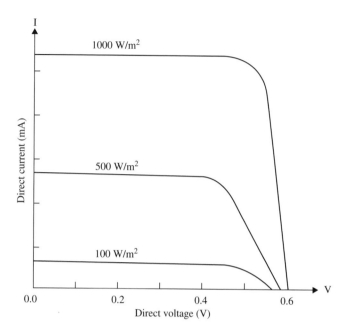

Fig. 12.7.　Typical current–voltage characteristics of a solar photovoltaic cell [10].

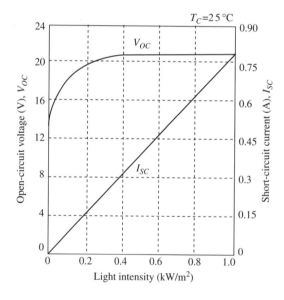

Fig. 12.8.　Variation of V_{oc} and I_{sc} with solar insolation for a 36-cell solar module [12].

at the maximum power point P_m are defined here as I_{mp} and V_{mp} respectively, and the maximum power deliverable, for that particular value of insolation, is

$$P_m = I_{mp}V_{mp} \qquad (12.5)$$

The cells manufactured by different companies have slightly different I–V shaped characteristics but they all follow the same pattern.

Most solar cell loads are resistive in nature. A load resistor R_L can be represented in the I–V plane by a straight line through the origin. Load resistance R_L can vary from zero for short circuit operation to infinity for open circuit operation. In order to deliver the maximum possible power for a specified level of insolation, R_L must satisfy the relationship

$$R_L = R_{mp} = \frac{V_{mp}}{I_{mp}} \qquad (12.6)$$

12.7.2. *Equivalent circuits*

The electrical performance of a photovoltaic cell can be approximately represented by the equivalent circuits of Fig. 12.9, for the constant current regions of Figs. 12.6 and 12.7. A constant current generator that delivers its short circuit current I_{SC} is the power source and this is shunted by a nonlinear (i.e. does not obey Ohm's law) resistor R_j, representing the variable junction resistance. For most circuit calculations the equivalent circuit of Fig. 12.9(b) is satisfactory. A more sophisticated, but more appropriate, representation is given in Fig. 12.9(a), where the internal series resistor R_S is much smaller than R_L, while the internal shunt resistor R_{sh} is much larger than R_L, for maximum power delivery.

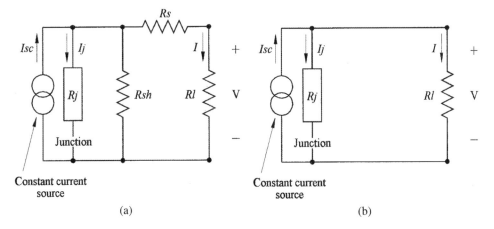

Fig. 12.9. Equivalent circuits of an irradiated solar cell [12]. I_{SC} = short circuit current. I = load current. R_j = junction resistance. R_L = load resistance. V = load voltage.

The equations of circuit operation for Fig. 12.9(b) are

$$I = \frac{V}{R_{\mathrm{L}}} \tag{12.7}$$

$$I_{\mathrm{j}} = \frac{V}{R_{\mathrm{j}}} \tag{12.8}$$

$$I_{\mathrm{j}} = I_{\mathrm{SC}} - I \tag{12.9}$$

$$R_{\mathrm{j}} = \frac{V}{I_{\mathrm{j}}} = \frac{V}{I_{\mathrm{SC}} - \frac{V}{R_{\mathrm{L}}}}$$

$$\therefore \ R_{\mathrm{j}} = \frac{V R_{\mathrm{L}}}{I_{\mathrm{SC}} R_{\mathrm{L}} - V} = \frac{I R_{\mathrm{L}}}{I_{\mathrm{SC}} - I} \tag{12.10}$$

Both forms of Eq. (12.10) for R_{j} depend on the accurate reading of the current–voltage characteristic. Also, the calculation of R_{j} involves taking the small difference between two relatively large values in the denominator term. This leads to inaccuracy, so R_{j} should be considered an approximate value.

The output power P_{o} of a solar cell is given by the product of two DC values:

$$P_0 = VI = \frac{V^2}{R_{\mathrm{L}}} = I^2 R_{\mathrm{L}} \tag{12.11}$$

12.7.3.   *Load lines in the current–voltage plane*

A more detailed set of performance characteristics for a typical cell is given in Fig. 12.10. The slope of a load resistance line is defined by Ohm's law, Eq. (12.7). With a load resistance of $10\,\Omega$, for example, the load line passes through the co-ordinates 0.1 V and 10 mA, 0.2 V and 20 mA, etc. while the 100-Ω load line passes through 0.1 V and 1 mA, 0.2 V and 2 mA, 0.3 V and 3 mA, etc.

If the load on a cell was, for example, a small electric motor, its load impedance characteristic would not be linear but roughly parabolic. In such a case an equivalent circuit would not be appropriate and analysis using Eqs. (12.7)–(12.11) would not apply.

The current–voltage characteristics of a solar cell cannot be expressed in analytical form but only by a general expression:

$$I = f(V) \tag{12.12}$$

Any particular graphical characteristic could be approximated by a power series of many terms, but this would need to be obtained by a computer-based curve-fitting process. Usually there is no need for this level of calculation.

There is no way of calculating a maximum power point mathematically unless an analytic expression for the current–voltage characteristic is known. A maximum power point can, however, be calculated by iteration: making an initial guess as to

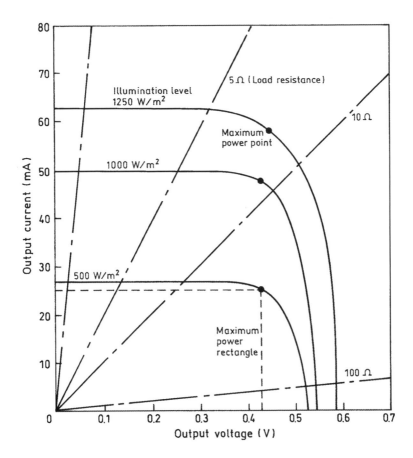

Fig. 12.10. Cell and load characteristic for typical cell (adapted from Swinden [13]).

the location and then taking successive co-ordinates and calculating the current–voltage product until a maximum value is reached.

Maximum power points for different levels of radiation usually occur at about the same voltage level and are therefore roughly proportional to the current. With the International Standard insolation of $1000\,\text{W/m}^2$, for example, the maximum power is delivered at $0.43\,\text{V}$ and $48\,\text{mA}$ in Fig. 12.10, which gives $0.43 \times 48/1000$ or $20.64\,\text{mW}$.

12.7.4. *Arrays of solar photovoltaic cells*

In order to deliver increased current to a load the appropriate number of solar cells has to be connected in parallel, whereas to deliver increased voltage the cells must be connected in series. Some load voltage and current specifications require values such that a series-parallel combination is needed. Clusters of cells are often referred to as solar arrays. The electrical output characteristics of simple series or

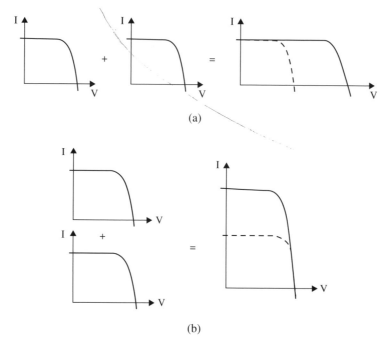

Fig. 12.11.   Output characteristics of solar cell arrays. (a) Series combination. (b) Parallel combination [14].

parallel combinations of two identical cells are shown diagrammatically in Fig. 12.11 [14]. A common load application is for the full charging of a 12-V battery, which requires a level of 14–15 V from the battery charger. At a radiation of 1000 W/m^2 (Fig. 12.10) it would be necessary to use 15/0.54 or 28 cells in series of that particular manufacture. There would probably need to be several parallel-connected modules, each consisting of 28 cells in series, to supply the specified charging current.

An application requiring the direct connection of the array to provide a DC power supply of 120 V would require, effectively, 120/0.54 or 222 cells in series. Systems producing 600-V DC and hundreds of amperes of current are in operation worldwide.

12.7.5.   *Effect of temperature on solar cell operation*

When a solar cell is working, delivering current to a load, its temperature rises above the ambient. Temperature rises of 25–30°C are typical and this rise varies directly with the solar insolation. A rise of about 25°C can be expected at 900 W/m^2:

$$\text{Cell temperature} = \text{Ambient temperature} + \text{Temperature rise} \qquad (12.13)$$

In a high ambient temperature of (say) 25°C the cell working temperature could become $25 + 25 = 50$°C. With a low ambient temperature of (say) 5°C the temperature rise is the same, at the same insolation level, so that the cell working temperature will then be $5 + 25 = 30°C$.

A common effect of temperature rise is to slightly increase the short circuit current but more significantly to decrease the open circuit voltage. Overall, the effect is to reduce the maximum power available for a fixed value of radiation. Temperature effects are calibrated against an internationally agreed standard temperature of 25°C. These vary slightly for different designs of silicon photovoltaic cells. Data for several different cells are in the following ranges (with respect to 25°C) [14].

Effect of cell temperature on the short circuit current:

$+ (0.004 \text{ to } 0.013) \text{ mA/cm}^2/°C$

Effect of cell temperature on the open circuit voltage

(voltage reduction coefficient):

$- (0.0023 \text{ to } 0.0028) \text{ V/cell}/°C$

Effect of cell temperature on the maximum power:

$- (0.3 \text{ to } 0.5)\%/°C$

The negative signs for V_{oc} and P_m above imply a reduction of value if the temperature rises above 25°C. Of course, the cell working temperature cannot naturally fall below the ambient level. The reduction of voltage of a module of series-connected cells can therefore be calculated from the relationship

$$\text{Voltage } (V_{oc}) \text{ reduction} = \text{Number of cells in series} \times \text{Temp rise}$$
$$\times \text{Voltage reduction coefficient} \qquad (12.14)$$

The overall temperature effect on the operating characteristic of a 36-cell solar array is given in Fig. 12.12 [12]. Increase of temperature has a greater effect on the voltage reduction than on current increase.

In locations where the ambient temperature is high the working temperature of a cell can become so large as to seriously impair its performance. It may then be necessary to employ some form of forced cooling, as shown diagrammatically in Fig. 12.13. It should be noted from Eq. (12.13) that forced reduction of the ambient temperature, below the standard value of 25°C, causes a reduction of the cell working temperature. But the reduction of working voltage due to cell temperature rise still satisfies Eq. (12.14).

12.8.   Applications of Photovoltaic Cells

The only mass use of the silicon photovoltaic cell is as the exposure meter or "electric eye" in cameras. In this application the low efficiency of 3–4% caused by the use of low-grade silicon is not significant.

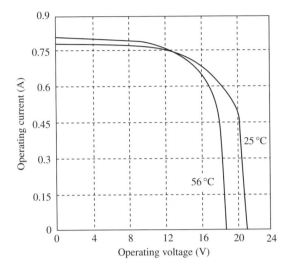

Fig. 12.12. Effect of temperature change on the operating characteristic for a 36-cell solar module [12].

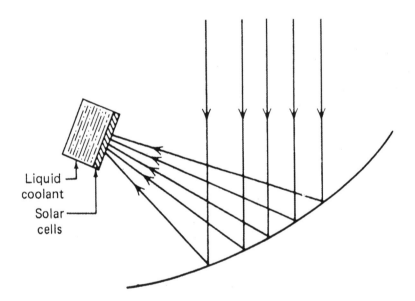

Fig. 12.13.   Forced cooling of photovoltaic cells.

Low power applications that incorporate photovoltaic cells include portable apparatus for remote locations. An array with an area of about $1\,m^2$ will provide several tens of watts of power, which is suitable for lighting purposes or for portable radio transmitter/receiver systems. With about $2\,m^2$ of cells, sufficient power is available to drive a small water pump for irrigation purposes.

Fig. 12.14.   100 kW photovoltaic array [15].

In a number of locations in the world solar cell arrays are used to power street lighting. During daylight hours the solar-generated energy is stored in a battery. After dark, the battery is electronically connected to power the lighting and then is automatically switched off again soon after dawn.

Large arrays have been used in certain commercial applications. For example, Fig. 12.14 shows a 100-kW system at Beverley High School, Beverley, Massachusetts, USA, that uses 3,200 modules, each containing 36 solar cells. The system provides about 10% of the electricity demand [15]. There appears to be no electrical limitation of array size. Limitations may be posed by difficulties of fabrication, space available and the weight of the structure.

Large-scale use of solar photocells is also evident in weather and telecommunication satellites and in space vehicles. The cells are constructed in sails, which are folded away during launching and are unfurled by radio control after the satellite is in stable orbit. In space the cell assembly and support structure is weightless and can be directed towards the sun almost continuously, without intervening clouds. The failure of one of the solar arrays on the Hubble orbiting telescope was one of the problems that had to be solved by the US National Aeronautics and Space Administration (NASA) in the 1993 rescue mission.

The vast scale of modern space stations is illustrated in Fig. 12.15, which is a newspaper photograph showing the space shuttle docked at the International Space Station (ISS). Power is provided in the main (American) section by 16 solar arrays, each 58 metres long and having an area of 375 m^2. While the sun is eclipsed by the earth, power is delivered by nickel-hydrogen rechargeable batteries.

The US part of the ISS operated at 130–180 V DC from the photovoltaic arrays arranged as four wing-pairs, each producing 32.8 kW [16].

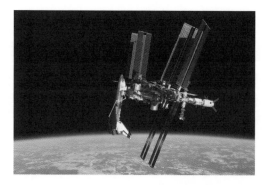

Fig. 12.15.   Space shuttle docked at the International Space Station, June 2011 (reproduced by permission of the "Times", London, 9 June 2011).

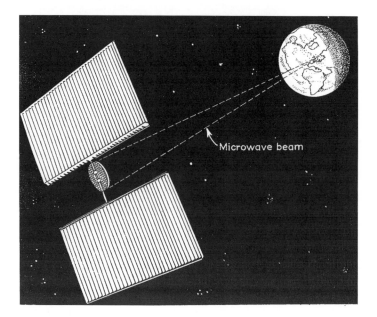

Fig. 12.16.   Electricity power plant in space [17]. (Artistic impression by P. E. Glaser of Arthur D. Little, Inc.)

An intriguing possibility is the idea of power stations in space. A satellite in synchronous (i.e. geostationary) orbit could focus a beam of microwave power continuously onto a receiver antenna at a fixed location on earth, as illustrated by the artistic drawing of Fig. 12.16 [17].

Photovoltaic energy may be collected in space, using concentrator systems, and converted from DC power to microwave power. Energy storage is not a problem because a single satellite could view the sun for 23 out of every 24 hours and the solar flux is almost constant at its maximum (m = 0) value. To provide a power of $1000\,MW_e$, the equivalent of a medium size power plant on earth, would require about $5\,km^2$ of cells, at an efficiency of 15%.

The basic scientific and engineering knowledge needed to mount such a massive project already exists. It would involve some major problems of satellite tracking and very precise guidance of the microwave beam. Waste heat from the microwave converters and generators would need to be radiated into space. The intensity of radiation of the microwave beam could have destructive effects on plant and animal life if it was misdirected, rather like the "death ray" that features in some films and novels. Building and maintaining a project of this size would involve continuous manning by operational personnel and frequent trips to the space shuttle. It therefore raises the questions of human life and health in space. The massive cost could only be borne by governmental funding, or with governmental participation.

Solar photovoltaic electricity power stations on earth are already a reality. They comprise a solar cell array, usually combined with some form of concentrator system, a storage battery facility capable of accepting most or all of the solar power output and a suitable control system. Terrestrial or earth-bound solar power stations are only viable in regions of intense radiation, such as north Africa, the Middle East, and countries with a Mediterranean climate, like the southern parts of the USA, and Mexico.

The Winston compound parabolic concentrator, described in Section 11.5.2 of Chapter 11, can also be used for solar photovoltaic energy capture. Because the efficiency of conversion is reduced at high temperatures it is usually necessary to use liquid cooling for silicon cells, as illustrated conceptually in Fig. 12.13.

A commonly used form of concentrator in photovoltaic systems is the Fresnel lens optical system. This is a refracting lens system that uses glass or plastic lens material in which concentric (or parallel) grooves are cut or moulded. The grooves are shaped and arranged to make all the light converge to a point or along a line (Fig. 12.17). Common uses of Fresnel lens systems include theatrical spotlights. Lighthouse rotating warning lights invariably use large Fresnel lens systems. Lower power applications include solar cookers and solar furnaces. The Fresnel lens concentrates the input radiation like a magnifying glass. Use of Fresnel lens systems can provide the equivalent of 50 times the normal solar radiation, sometimes referred to as "50 suns". (Note that this is not the same as a concentration ratio of 50, as defined in Eq. (11.6) of Chapter 11, which refers to a ratio of two areas.)

A 350-kW solar voltaic system is in operation, supplying two villages outside Riyadh, the capital city of Saudi Arabia. An assembly of 160 photovoltaic arrays (Fig. 12.18) using Fresnel lens concentrators delivers power to the 1100 kWh lead-acid storage batteries and also provides input power to the local 60 Hz electricity grid through a DC-to-AC three-phase inverter. One of the local environmental features is desert sandstorms and the Fresnel lens surfaces are cleaned manually by washing with water [18].

Most of the serious research work on solar photovoltaic electric power generation is now taking place in Germany, Japan and the USA. Various installations, with

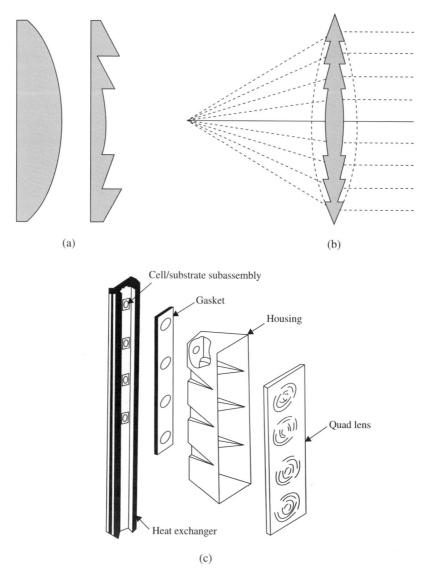

Fig. 12.17.   Fresnel lens. (a) Basic structure of three-ring lens. (b) Light distribution in double-sided, four-ring lens. (c) Assembly structure of four-cell array.

ratings of the order of a few hundred kW up to a few tens of MW, are planned, under development or are now in commission. This includes a design for a community college in Arkansas, USA, with an electrical output of 320 kW from 59,400 single-crystal silicon cells incorporating parabolic trough tracking concentrator collectors with an average concentration of 30 suns.

In Arizona, USA, an experimental 20 kW photovoltaic array has been built using innovative cells made of crystalline silicon. These are mounted under acrylic

Fig. 12.18.   Concentrator photovoltaic array in Saudi Arabia, using four-cell units [18].

concentrating lenses and are reported to have conversion efficiencies greater than 20%. The global capacity of photovoltaic power installations increases every year and in 2012 surpassed 100 GW [19]. Annual US shipments of photovoltaic cells increased from almost zero in the early 1970s to greater than 75 MW in 1996 and to 77 peak MW in 1999. Crystalline silicon cells continue to dominate the US market, accounting for 96% of the total shipments in 1999 [20]. In the USA in 2002, the industrial sector replaced the residential sector as the largest market for PV cells and modules.

12.9.   The Future Challenge for Photovoltaics

The challenge in the development of photovoltaic materials is demonstrated in Fig. 12.19, which compares the spectral profile of the sun (at $m = 0$) with the corresponding profile for commercial grade silicon. In a radio or telecommunications system these two curves would be described as a bad mismatch. The sun profile is not going to change. What is needed is a cheap and abundant material with a spectral response that is a better fit to the solar characteristic and has a high energy conversion efficiency. Alternatively, the need could be satisfied with some type of radiation-matching device (which is, in effect, a frequency changer) between the input solar radiation and the spectral profile of the solar cell material. Enormous research effort is presently devoted to this task all over the world. The highest scientific accolade and great wealth will go to the successful inventor/discoverer.

Most of the work now underway involves government-sponsored or government-subsidized research prototypes. The big commercial future for photovoltaic electric power generation still awaits the development of a much cheaper photocell.

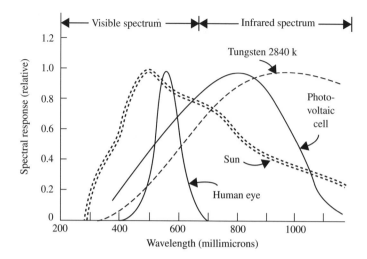

Fig. 12.19.   Spectral responses of sunlight and a photovoltaic cell (unknown origin).

12.10.   Worked Examples

Example 12.1.   A solar cell array has a current–voltage characteristic at the standard insolation of $1000\,\text{W/m}^2$, in which $V_{oc} = 11.5\,\text{V}$ and $I_{sc} = 1.75\,\text{A}$. For a particular resistive load $V_{mp} = 7.75\,\text{V}$ and $I_{mp} = 1.15\,\text{A}$. Calculate the maximum power deliverable and the load resistance that will result in maximum power delivery From Eq. (12.5)

$$P_m = V_{mp}I_{mp} = 7.75 \times 1.5 = 11.625\,\text{W}$$

From Eq. (12.6)

$$R_{mp} = \frac{V_{mp}}{I_{mp}} = \frac{7.75}{1.5} = 5.17\,\Omega$$

Example 12.2.   Two modules of solar cells were measured on different days with different weather conditions to give the following data:

	I_{sc} (A)	Insolation (W/m^2)
Module 1	1.25	400
Module 2	1.05	325

Which module has the larger I_{sc} measured at the standard insolation of $1000\,\text{A/m}^2$?

It can be assumed that I_{sc} is proportional to the insolation

$$\text{For module 1} \quad I_{sc} = \frac{1000}{400} \times 1.25 = 3.125 \text{ A}$$

$$\text{For module 2} \quad I_{sc} = \frac{1000}{400} \times 1.05 = 3.23 \text{ A}$$

The calculation shows that module 2 has the higher standard value of I_{sc}.

Example 12.3. A 28-cell module of a solar array has $V_{oc} = 17$ V at 25°C. What will be the value V_{oc} at 60°C if the cell data specifies 0.0024 V/°C/cell?

$$\text{Temperature difference} = 35°\text{C}$$

From Eq. (12.14) the reduction of open circuit voltage is given by:

$$\text{Voltage reduction} = 28 \times 35 \times 0.0024 = 2.35 \text{ V}$$
$$\text{Working value of } V_{oc} \text{ is } V_{oc} = 17 - 2.35 = 14.65 \text{ V}$$

This represents a reduction of 2.35/17 or 13.8%.

If the short circuit current is only slightly increased by the temperature rise, the implication here is that the maximum power will decrease by the order 10–12%.

Example 12.4. In the current–voltage characteristics of Fig. 12.10 for a typical photocell, what values of load resistance would result in maximum power output at the three specified radiation levels?

At 1250 W/m^2, maximum output power P_o occurs when $V = 0.447$ V and $I = 58.1$ mA so that $P_o = 0.447 \times 58.2/1000 = 25$ mW.

Similarly at 1000 W/m^2, for maximum output power it is seen that $V = 0.428$ V and $I = 48$ mA, giving $P_o = 20.54$ mW.

Also, at 500 W/m^2 input radiation, the maximum output power occurs at $V = 0.428$ V and $I = 25.2$ mA, so that $P_o = 10.8$ mW.

It can be deduced from these results that the maximum output power is proportional to the input radiation power, within about 5% error.

The load resistances for the three cases are obtained from Eq. (12.6).

P_{in} (W/m^2)	(P_o max) (mW)	R$_{load}$ (ohms)
1250	26	447/58.2 = 7.68
1000	20.54	428/48 = 8.92
500	10.8	428/25.2 = 16.9

Example 12.5. The typical photocell with characteristics depicted in Fig. 12.20 is delivering power to the load resistance $R_L = 7.5\,\Omega$ with an input radiation of 1000 W/m^2. What is the value of the junction resistor R_j in the equivalent circuit?

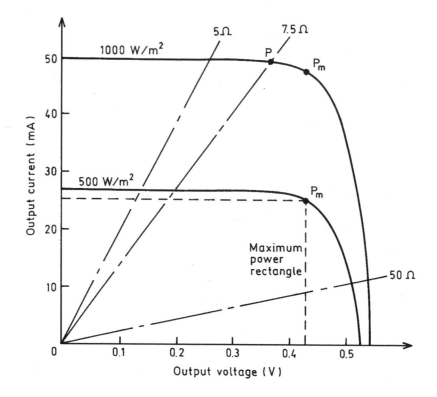

Fig. 12.20.   Specimen photovoltaic cell characteristics.

With $R_{\mathrm{L}} = 7.5\,\Omega$ the resistance line intersects the $1000\,\mathrm{W/m^2}$ characteristic at a point P (Fig. 12.20), where the terminal voltage $V = 0.364\,\mathrm{V}$.

If the simplified equivalent circuit of Fig. 12.9(b) is used, then the load current is

$$I = \frac{V}{R_{\mathrm{L}}} = \frac{0.364}{7.5} = 0.0485\,\mathrm{A} = 48.5\,\mathrm{mA}$$

The constant current delivered by the constant current generator is the short circuit value of $50\,\mathrm{mA}$. With a 7.5-Ω load the junction resistor current is therefore, from Eq. (12.9),

$$I_{\mathrm{j}} = I_{\mathrm{s}} - I = 50 - 48.5 = 1.5\,\mathrm{mA}$$

Junction resistor R_{j} therefore has the value

$$R_{\mathrm{j}} = \frac{V}{I_{\mathrm{j}}} = \frac{0.364}{\frac{15}{1000}} = \frac{364}{1.5} = 242.7\,\Omega$$

Example 12.6.   A certain type of commercial solar photovoltaic cell has $V_{\mathrm{mp}} = 0.48\,\mathrm{V}$ and $I_{\mathrm{mp}} = 20\,\mathrm{mA/cm^2}$ under standard insolation conditions. What combination of cells would be required to fully charge a nickel–cadmium battery requiring 4.2 V and 70 mA?

$$\text{Number of cells in series to supply the voltage} = \frac{4.2}{0.48} = 8.7 \quad (\text{say } 9)$$

$$\text{Area of solar cell material to generate the required current} = \frac{70}{29} = 2.4 \, \text{cm}^2$$

The number of parallel-connected cells to generate the required current will depend on the individual cell areas. A standard size of cell is $1 \, \text{cm}^2$, which would require 2.4 cells in parallel. This is obviously not possible and the choice might be 3 cells of standard size.

With 3 cells of $1 \, \text{cm}^2$ in parallel,

$$I_{\text{mp}} = 3 \times 29 \times = 87 \, \text{mA}$$

One possible design choice is therefore to use 3 parallel branches, each containing 9 cells in series. A parallel-connected resistor across the load would divert the excess $87 - 79 = 8 \, \text{mA}$ of current if required, although this would be a wasteful solution.

Example 12.7.   A number of identical solar cells, having the characteristics of Fig. 12.20, are connected in parallel. How many cells would be needed to deliver 1 A of current to a 10-Ω load resistor with an input radiation level of $1000 \, \text{W/m}^2$?

With a load resistance of $10 \, \Omega$ the load line intersects the current–voltage characteristic for $1000 \, \text{W/m}^2$ at the co-ordinate $V = 0.46 \, \text{V}$ and $I = 45 \, \text{mA}$.

At $1000 \, \text{W/m}^2$ the short circuit current is 50 mA. Each constant current generator is therefore delivering 5 mA of current to its identical junction resistor, and 45 mA to the load resistor (Fig. 12.21). To deliver 1 A, therefore,

$$\text{Necessary number of cells } n = \frac{1}{\frac{45}{1000}} = 22.22$$

The design choice is to use either 22 or 23 cells.

$$\text{With 22 cells,} \quad I = 22 \times 0.045 = 0.99 \, \text{A}$$
$$\text{With 23 cells,} \quad I = 23 \times 0.045 = 1.035 \, \text{A}$$

Example 12.8.   A number of identical solar cells, having the characteristics of Fig. 12.20, are connected in series. How many series connected cells are needed to deliver an output of 10 V to a 10-Ω load resistor with an insolation level of $1000 \, \text{W/m}^2$?

With a load resistance of $10 \, \Omega$ the load line intersects the $1000 \, \text{W/m}^2$ current–voltage characteristic at the co-ordinate $V = 0.46 \, \text{V}$ and $I = 45 \, \text{mA}$.

At $1000 \, \text{W/m}^2$ the short circuit current is 50 mA. Each constant current generator therefore delivers 5 mA of current to its junction resistor R_{j}, and 45 mA to the load resistor (Fig. 12.21(b)).

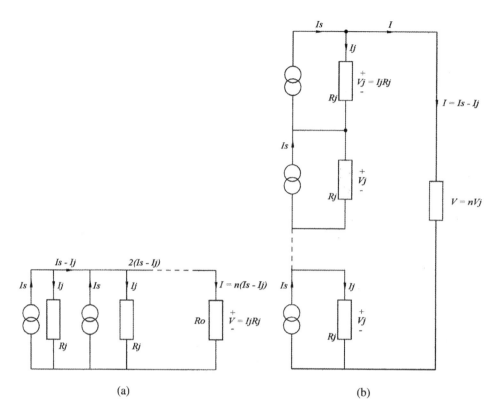

Fig. 12.21.   Equivalent circuits for the connection of identical solar cells. (a) Parallel. (b) Series.

The load current is equal to the constant current-source value of 45 mA. The load voltage is the sum of n series-connected identical voltage drops V_j.

$$V = nV_j = n \times 0.46$$

To deliver a load value of 10 V requires

$$n = \frac{10}{0.46} = 21.74$$

The design choice is to connect either 21 or 22 cells in series.

$$\text{With 21 cells,}\quad V = 21 \times 0.46 = 9.66\,\text{V}$$
$$\text{With 22 cells,}\quad V = 22 \times 0.46 = 10.12\,\text{V}$$

Example 12.9.   In a particular location in the USA the solar insolation has a power density of 1000 W/m² for an average of 112 hours/month. The energy demand is for 1500 kWh/month. This can be provided by a solar cell array with an estimated lifetime of 20 years. The cost of the solar cells is about US$3 per peak watt. Fabrication, maintenance and interest on the capital cost effectively (at least) doubles the price per peak watt. Calculate the cost of the energy per kWh.

$$\text{Required power} = \frac{1500}{112} = 13.4\,\text{kW average}$$
$$\text{Cost} = 13400 \times 3 \times 2 = \$80400$$
$$\text{Energy consumed in 20 years} = 1500 \times 12 \times 20$$
$$= 360000\,\text{kWh}$$
$$\text{Cost per kWh} = 80400/360000$$
$$= \$0.223/\text{kWh}$$

This should be regarded as a minimum and rather optimistic figure. The present (2013) tariff for electricity in the USA is about US$0.07/kWh. The cost of solar-generated electricity therefore needs to be reduced by a factor of three (and probably more) in order to be competitive. The fabrication, maintenance and financing charges on an installation are not likely to undergo dramatic change. A major reduction (i.e. at least five-fold) in the cost of manufacturing the photovoltaic material, improving its efficiency, or both, is required.

12.11.   Problems and Review Questions

Problems on the basic physics of solar photovoltaic cells

12.1. List the advantages of photovoltaic cells compared with solar thermionic applications.

12.2. Describe the atomic structure of silicon and explain the term "covalent bond".

12.3. Describe the differences between the types of silicon semiconductors known as (a) monocrystalline, (b) polycrystalline and (c) amorphous.

12.4. Explain the terms "n-type silicon" and "p-type silicon".

12.5. What is the minimum energy of the incoming radiation that will cause electrons to flow across the energy gap in silicon? How is this energy related to the frequency, wavelength and velocity of the radiation?

12.6. Why does the efficiency characteristic of Fig. 12.5 fall off at high values of the energy gap?

12.7. What is the effect of temperature on the efficiency of a photovoltaic cell operation?

Problems on photovoltaic materials

12.8. What are the principal difficulties in manufacturing single-crystal silicon for use in photovoltaic cells?

12.9. What is the principal advantage of producing single-crystal silicon by growing the crystal in the form of a ribbon rather than a cylindrical ingot?

12.10. Gallium arsenide is capable of converting energy at higher theoretical working efficiencies than silicon in photovoltaic cells. What are the objections to its widespread use?

12.11. What is the main advantage of polycrystalline silicon solar cells compared with single-crystal cells?

12.12. Specify the reasons why solar cells have such low values of conversion efficiency.

12.13. Why are solar concentrators often used in photovoltaic cell applications?

12.14. What is the effect of increased temperature on the photo-conversion efficiency of a photovoltaic cell?

12.15. In space the area of photocells needed to produce a specified power output is only a fraction of the area that would be required at the earth's surface. Why is this?

Problems on the electrical output properties of photovoltaic cells

12.16. Sketch the current–voltage characteristic of a typical solar photovoltaic cell. Identify the points of operation for (i) open circuit voltage, (ii) short circuit current and (iii) maximum power delivery.

12.17. Sketch the current–voltage characteristic of a typical solar photovoltaic cell. Identify the point of operation for maximum power delivery. How would you determine this location?

12.18. For a silicon photovoltaic cell, what are the approximate values of the ratios (i) V_{mp}/V_{oc} and (ii) I_{mp}/I_{sc}?

12.19. A module of solar cells has $I_{sc} = 1.5\,\text{A}$ when the radiation is $1000\,\text{W/m}^2$. What will be the value of I_{sc} when the radiation is (i) $850\,\text{W/m}^2$ and (ii) $300\,\text{W/m}^2$?

12.20. For the solar cell characteristics of Fig. 12.10 identify the operating voltage and current values, with load resistances of $8\,\Omega$ and $20\,\Omega$, for radiation levels of (i) $1000\,\text{W/m}^2$ and (ii) $500\,\text{W/m}^2$.

12.21. The temperature effects on a certain solar cell are specified as $-0.0024\,\text{V/}^\circ\text{C/cell}$ and $+0.006\,\text{mA/}^\circ\text{C/cm}^2$. A modular array of 30 cells generates $V_{oc} = 19\,\text{V}$ at 25°C. What is the change of V_{oc} for each 10°C rise of cell temperature? Estimate the percentage change in the value of maximum power P_m at the same insolation level.

12.22. Sketch an equivalent circuit for a silicon solar photovoltaic cell. How is the source current of your equivalent circuit related to the solar radiation falling on the cell?

12.23. For the solar cell characteristic representing $500\,\text{W/m}^2$ of Fig. 12.20, develop an equivalent circuit for operation at the maximum power point P_m. Calculate values for the load resistance, junction resistance and branch currents.

12.24. In a silicon solar cell the effective junction resistance R_j varies with current. For a constant insolation of $1000\,\text{W/m}^2$, using the characteristic in Fig. 12.20, calculate values of R_j with load resistances of $5\,\Omega$, $10\,\Omega$ and $50\,\Omega$. Sketch the form of variation of R_j with current over the whole operating range.

12.25. The solar cell that is characterized in Fig. 12.20 operates with a fixed load resistance of $5\,\Omega$. How does the value of the junction resistance R_j vary as the radiation level falls from $1000\,\mathrm{W/m^2}$ to $500\,\mathrm{W/m^2}$?

12.26. For the solar cell characteristics of Fig. 12.10, what value of load resistor would result in a load line passing through the maximum power point P_m at $1250\,\mathrm{W/m^2}$?

12.27. For the solar cell characteristics of Fig. 12.20, what value of load resistor would result in a load line passing through the maximum power point P_m at $1000\,\mathrm{W/m^2}$?

Problems on combinations of solar cells

12.28. An array of solar cells consists of two parallel branches, each of three cells connected in series. Sketch a simple diagram, of the form of Fig. 12.11, to illustrate the overall $I-V$ characteristic.

12.29. A solar array consists of three parallel branches, each containing two identical cells in series. Sketch a diagram of the overall $I-V$ characteristic, compared with the $I-V$ characteristic of each individual cell.

12.30. Twenty solar cells of the type characterized in Fig. 12.20 are connected in parallel to a 10-Ω load. Calculate the load current, voltage and power when the radiation is (i) $1000\,\mathrm{W/m^2}$ and (ii) $500\,\mathrm{W/m^2}$.

12.31. How would you realize a terminal voltage of $10\,\mathrm{V}$ using solar cells of the type characterised in Fig. 12.20?

12.32. At a radiation level of $1000\,\mathrm{W/m^2}$, with a load resistance of $10\,\Omega$, 100 solar cells of the type characterized in Fig. 12.20 are connected in series. Calculate the current, voltage and power at the load terminals.

References

[1] McVeigh, J.C. (1983). *Sun Power*, 2nd Edition, Pergamon Press, Oxford.

[2] Zweibel, K. (1990). *Harnessing Solar Power*, Plenum Press, New York.

[3] Knier, G. (2011). *How do Photovoltaics Work?*, NASA Science: Science News, April 2011, http://science.nasa.gov/science-news/science-at-nasa/2002/solarcells.

[4] Shepherd, W. (1975). *Thyristor Control of AC Circuits*, Crosby, Lockwood, Staples Ltd., St. Albans.

[5] Mackay, D. (2009). *Sustainable Energy — Without the Hot Air*, Cambridge University Press, Cambridge.

[6] Gross, M. (2001). *Organic Light Harvesting*, Chemistry in Britain, 22–23.

[7] Schock H.-W. and Pfisterer, F. (2001). *Thin-Film Solar Cells-Past, Present and Future*, Renewable Energy World, **4**, 75–87.

[8] Gratzel, M. (1983). *Low Cost Solar Cells*, The World and I, 228–234.

[9] Strong, S.J. (1996). *Power Windows*, IEEE Spectrum, **33**, 49–55.

[10] McDaniels, D.K. (1984). *The Sun — Our Future Energy Source*, 2nd Edition, John Wiley and Sons, New York.

[11] Borenstein, S. (2008). *The Market Value and Cost of Solar Photovoltaic Electricity Production*, University of California Report CSEM WP176, USA.

[12] "Panasonic Solar Cells Technical Handbook 2000", Matsushita Battery Industrial Co. Ltd., Janurary 2000.

[13] Swinden, D.J. (1980). *Energy*, Lecture Support Notes, England.

[14] "Photovoltaic Technology and System Design", *Output Curves*, Siemens Solar Industries, USA, undated, Chapter 4.

[15] Vogt, G. (1986). *Generating Electricity*, Franklin Watts, New York.

[16] "International Space Station" Wikipedia, 2011, http://en.wikipedia.org/wiki/International_Space_Station.

[17] Denton, J.D., Glanville, R., Glidden, B.J., Harrison, P.L., Hotchkiss, R.C., Hughes, E.M., Swift-Hook, D.T. and Wright, J.K. (1975). *The Potential of Natural Energy Resources*, CEGB Research, **2**, 29.

[18] "Solar Village", National Center for Science and Technology, Associated with King Saud University, Riyadh, Saudi Arabia, 1979.

[19] "Global Solar PV Installed Capacity Crosses 100 GW Mark", *EPIA Solar Projections for 2013*, European Photovoltaic Industry Association, Brussels, Belgium, 2013.

[20] "Renewable Energy Annual 2000", Energy Information Administration, US Department of Energy, Washington, DC, 2000.

CHAPTER 13

BIOLOGICAL ENERGY
AND CHEMICAL ENERGY

13.1. Biomass and Biofuels

Any type of animal or plant material that can be converted into energy is called biomass. This includes trees and shrubs, crops and grasses, algae, aquatic plants, agricultural and forest residues plus all forms of human, animal and plant waste [1, 2]. When the material is used for energy production it becomes a biofuel. There are many forms of biofuel, existing in solid, liquid or gaseous categories.

The main sources of raw material that constitute biofuels include:

(a) natural vegetation,
(b) energy tree plantations,
(c) specific energy crops (industrial biomass),
(d) wastes, and
(e) water-based biomass.

Each of the above five categories is briefly discussed below.

13.1.1. *Natural vegetation*

In harvesting a natural vegetation site no energy costs are involved in clearing or replanting. Where an area may be unsuitable for agriculture the use of its vegetation for biomass would constitute a useful bonus biofuel source. There still remain the necessary costs of harvesting the vegetation and transporting it to a user site. A disadvantage is that even in fertile locations the yields are low: about one-half of the value that might be obtained from customized energy plantations.

13.1.2. *Energy tree plantations*

Trees and other types of lignocellulose materials may be grown specifically for burning as biofuels. By the choice of appropriate species, sites, planting densities and harvesting schedules, biofuels can be grown at competitive costs. This process is

Photo credit: USDA–Soil Conservation Service

Fig. 13.1.   Commercial forestry [10].

sometimes referred to as "short rotation forestry". An example of coniferous forestry in the USA is given in Fig. 13.1.

13.1.3.   *Specific energy crops*

Certain crops have high energy conversion efficiency. In appropriate locations, crops such as eucalyptus trees, rubber plants or sunflowers might be used because of their rapid growth and high-energy content [3]. Such crops do not have to be consumable by humans or animals, and the entire crop can be used, including leaves, stalks and roots. The stored chemical energy can be converted directly to heat by combustion or processed into liquid or gaseous fuels. Liquid biofuels are discussed in detail in Section 13.3.4, below.

Energy farming, like other agricultural operations, requires large quantities of water. In the developed countries of the Western world there is competition for the use of land, water and nutrients with various forms of food production, including animal farming. An interesting economic issue is whether crops in growth command higher prices as fuel rather than food or fibre. Crop surpluses would provide low-cost biomass, but is surplus food production the best overall use of the land and water?

There is a further potential disadvantage to the use of land-based energy farming. The hardy and fast-growing species required for energy use could become widespread nuisances if they escaped from the confines of the farm. They may then displace native plant species and impact on animal and insect life. Also, the methods necessarily used to control infestations and disease in the energy plants, especially if they are monocultures, could have adverse effects on neighbouring food production crops.

The same soil plantation may be used for two species of plants with out-of-phase growth patterns. If both species are in leaf together the leaf canopy cover profile is increased. Non-nitrogen-fixing crops can be grown side by side with legumes to reduce the need for nitrogenous fertilizers. Mixed cropping of this kind is less susceptible to damage by external parasites or predators than monoculture plantations [4].

There are some plants that produce high concentrations of "petroleum-like" products. These can be grown en masse and the "oils" extracted or squeezed out. Pilot schemes have been carried out in Mexico.

Widespread energy farming on marginal land implies the need for some level of overall control. A fast-growing plant that thrived in marginal conditions would create incentives to adapt the plant for food production. This is not necessarily bad but would create further pressure on land use for purely biomass reasons. In other words, the food versus biomass competition would increase.

It would seem sensible to think in terms of integrated growth mixed cropping, in which energy crops are developed alongside crops for food, fertilizers and chemicals [4].

13.1.4. *Use of wastes*

The burning and biodegradation of wastes as a source of energy is so significant an issue that a complete section below (Section 13.5) is devoted to this.

13.1.5. *Water-based biomass*

Aquatic plants do not require irrigation or appreciable rainfall. Water temperatures fluctuate less than land temperatures and light absorption is high, giving rise to high photosynthetic efficiencies.

Algae, seaweed and other aquatic lower plants can be intensively grown in certain areas of the sea or in inland lakes or ponds. This would eliminate the competition for land with conventional agriculture. On average, the oceans are low in plant nutrients. Surface waters have low natural productivity but sometimes act as nutrient repositories due to runoff from the land. Deep ocean waters are rich in nutrients. If such deep water can be pumped to the surface a biomass crop such as giant kelp (large brown seaweed) could be grown [1, 3]. Yields of up to several hundred tons per acre per year are possible.

A typical alga farm might contain an acre of land, excavated to a depth of 18 inches, flooded, and alga plants introduced. The pond would be covered with plastic, injected with carbon dioxide and maintained at a temperature of 100°F using (say) power station cooling water, which is presently dumped into rivers. As the algae are harvested the nutrient-rich water used for growth is returned to the pond and re-seeded to recycle the operation [5].

The aquatic weed known as water hyacinth has been studied as a tropical water source of biogases, particularly by NASA [6]. On a dry weight base, 1 kg of water hyacinth can produce $0.4\,m^3$ of biogas with a calorific value of $22\,MJ/m^3$. Aquatic weeds are a hazard in some waterways and have to be harvested. The biofuel value is then a useful by-product [1].

13.2.   Photosynthesis

Most living plants obtain their energy from sunlight. The basic metabolic processes are photosynthesis and respiration. In photosynthesis solar energy is converted into chemical energy in the plant and stored naturally as carbohydrates, including starch and cellulose. Solar energy is absorbed by the chlorophyll (i.e. the green material) and other plant chloroplast pigments, removing electrons from water molecules and liberating molecular oxygen. In generalized terms the photosynthetic reaction is

$$\text{sunlight absorbed by plant} + CO_2 + H_2O \rightarrow Cx(H_2O)y + O_2 \qquad (13.1)$$

The integer values x and y in Eq. (13.1) differ for different plants. Each reaction converts 114 kcal/mole (477 kJ/mole) of solar energy into stored chemical energy.

The carbohydrate may take the form of cane or beet sugar, $(C_{12}H_{22}O_{11})_n$, more complex starches or cellulose, $(C_6H_1O_5)_n$. Cellulose and starch are the major end products of photosynthesis and most of it remains unutilized, forming a large renewable source of chemicals and stored energy.

The biochemical reaction of photosynthesis, in Eq. (13.1), requires at least eight quanta of light energy. This usable input energy is equivalent to that of monochromatic light of wavelength 575 nm ($0.575\,\mu m$). Eight quanta of $0.575\,\mu m$ radiation have an energy content of 1665 kJ, giving a maximum theoretical photosynthetic efficiency of absorption of 477/1665 or 0.286 (28.6%). The actual theoretical efficiency is likely to be lower than this because experiments indicate that 8–10 quanta of energy may be needed to initiate the reaction.

Only the visible part of the solar spectrum (Fig. 11.1) occupying wavelengths in the range 400–700 nm (0.4–0.7 μm), is used in photosynthesis. The photosynthetically active radiation (PAR) constitutes about 43% of the total incident radiation so that the maximum theoretical conversion efficiency from input radiation to plant energy content, under ideal conditions, is $0.43 \times 0.286 = 0.123$ or 12.3% [4].

When carbohydrates are ingested by animals the energy is released to do muscular work, to generate nerve impulses and to create proteins for the building of new cells [7]. Photosynthesis is therefore a process in which solar energy is used to convert inorganic raw materials into organic compounds.

Respiration is the reverse of the photosynthetic reaction. In the respiration of a plant or animal the combustion of carbohydrates and oxygen yields energy, carbon dioxide and water vapour. The respiration process of a plant uses typically about one-third of the energy stored by photosynthesis. The basic processes of plant photosynthesis and respiration are depicted in Fig. 13.2 [7].

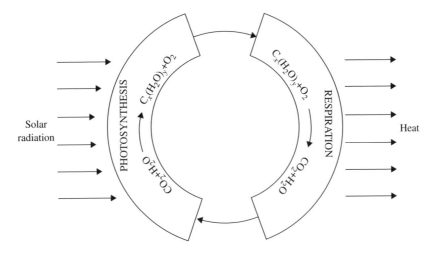

Fig. 13.2.   Basic processes of photosynthesis and respiration [7].

Table 13.1.   Maximum photosynthetic efficiency of land plants [4].

Parameter	Factor
Photosynthetically active radiation/total radiation	0.43
Maximum leaf absorption	0.8
Maximum efficiency of absorbed light conversion	0.286
(Photosynthesis–respiration)/photosynthesis	0.667
Overall efficiency = 0.43 × 0.8 × 0.286 × 0.667 = 0.066	

The value 0.43 for the PAR could only be realized if there was complete and continuous radiation of a plant foliage. Such an optimally arranged leaf canopy is not normally realizable. In practice a plant can absorb, at best, about 80% of the PAR due to shading, shielding, overlap of growth, etc. Combining the photosynthetic efficiency with the absorption and respiration factors gives a maximum value of overall efficiency for the conversion of solar energy into stored chemical energy $0.123 \times 0.8 \times 0.667 = 0.066$ or 6.6%. The various stage efficiencies are summarized in Table 13.1 [4].

Many figures of photosynthetic efficiency have been suggested, based on experimental evidence. These are all lower than the theoretical overall maximum value above of 6.6%. For optimum field conditions efficiencies in the range 3–5% are possible for limited periods. Typical annual conversion efficiencies are 0.5–1.3% for temperate crops and 0.5–2.3% for tropical and subtropical plants [4]. For large scale applications an average overall photosynthetic efficiency value of 1% has been suggested [7]. Figures for the photosynthetic efficiencies of selected agricultural crops are given in Table 13.2 [4].

Table 13.2.   Annual production and photosynthetic efficiencies of selected agricultural crops [4].

Crop	Location	Yield (t ha^{-1} y^{-1} dry wt)	Photosynthetic efficiency (%)
Temperate			
Sugar beet	Washington, USA	32	1.1
Wheat	Washington, USA	30 (grain)	0.1
Barley	UK	7 (grain)	0.3
Maize (C$_4$)	Japan	26	1.1
Subtropical			
Alfalfa	California, USA	33	1
Sugar beet	California, USA	42	1.2
Maize (C$_4$)	Egypt	29	0.6
Tropical			
Sugar beet	Hawaii (2 crops)	31	0.9
Cassava	Malaysia	38	1.1
Rice and sorghum (C$_4$) (multiple cropping)	Philippines	23 (grain)	0.7
Sugarcane (C$_4$)	Hawaii	64	1.8
Maize (C$_4$)	Peru	26	0.8
Napier grass (C$_4$)	El Salvador	85	2.4

13.3.   Methods of Industrial Biomass Conversion

The principal processes for converting biomass material into usable forms of energy are shown in Fig. 13.3 [4]. Some liquid or gaseous biofuels such as methane can be obtained by several different processes, including biodegradation.

13.3.1.   *Combustion*

Combustion is the best-known and probably most efficient form of converting solid biomass to steam or electricity. Water pipes in incinerator walls can produce heat for combined heat and power (CHP) district heating or other co-generation purposes from forestry residues and industrial or urban wastes. In principle, the energy system of biomass combustion is the same as that for fossil fuel combustion. A diagram of the appropriate stages, applied for the combustion of wood, is shown in Fig. 13.4. Fluidized-bed boilers used in the UK coal industry and discussed in Section 4.6 of Chapter 4 have proven to be particularly appropriate for wood and wood wastes.

Industries that produce solid biomass residues such as rice husks, bagasse (cane sugar residue), corn cobs, cotton stalks and nut shells often burn their own residues as supplementary fuels.

Solid-biomass-fuelled generation systems are usually most competitive in relatively small sizes. As system size increases, the conventional alternative becomes cheaper, transport costs rise disproportionately and the amount of land needed

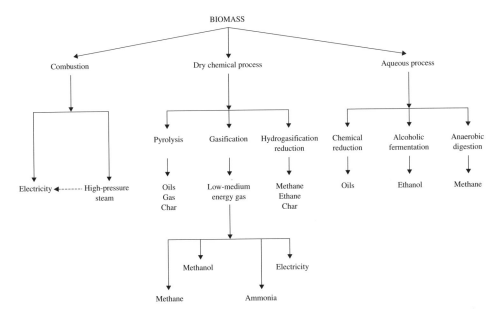

Fig. 13.3.   Biomass energy conversion processes and products [4].

to supply the biomass becomes very large. For example, with an average annual yield of 7 tons/hectare, almost 1,500 hectares (1500 × 2.471 = 3706 acres = 5.8 square miles) would be needed to produce 1 MW of power. This would constitute a very poor optional use of the land. In addition, industrial size solid-fuel boilers for wood (or coal) burning are of low efficiency and relatively high cost. If steam is not required as part of the process, then power requirements of below about 10 MW$_e$ are better met by the use of pyrolytic gasification processes [8, 9].

13.3.2.   *Pyrolysis*

If organic matter is heated with insufficient oxygen to support combustion, the resulting action is called pyrolysis. It is common to heat in a non-oxygen atmosphere, such as nitrogen. For example, if wood is heated to about 250°C (480°F) the moisture and volatile materials evaporate, leaving carbon and inert materials in the form of charcoal (char). Utilization of the gaseous and liquid products as well as the charcoal can result in an overall efficiency up to about 80%.

Charcoal can be transported and handled much more economically than an equivalent amount of wood. In many countries charcoal is the most widely used household fuel in urban (though not rural) areas. Charcoal stoves are more efficient than wood fires. There are many industrial applications that use charcoal in chemical reactions, including steel processing, cement manufacture, drying crops and metal smelting.

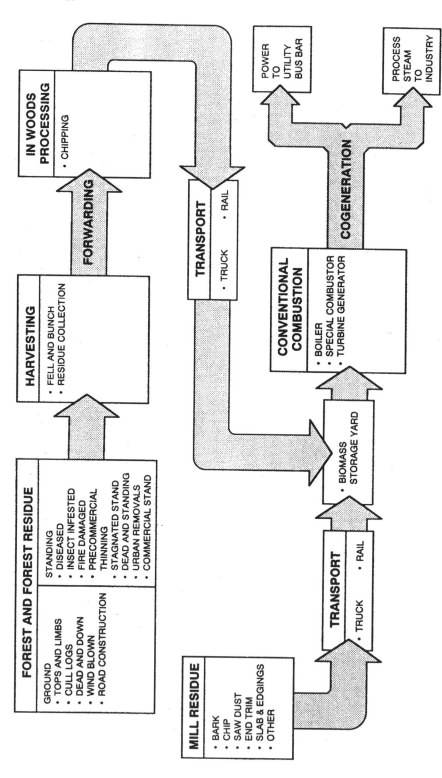

Fig. 13.4.   Network for wood fuel electricity generation [9].

If organic material is heated between 50°C and 90°C at ordinary pressures in the absence of oxygen, then methanol is produced. This is widely used as an automobile fuel additive [1].

13.3.3.   *Gasification of biomass*

The gasification of solid biomass is carried out in the presence of limited air or oxygen but at higher temperatures and/or pressures than for pyrolysis. The low-energy "producer" gas can be intermediate in several production processes and contains carbon monoxide, hydrogen, carbon dioxide and methane. If the gaseous phase is condensed, followed by distillation to 98% purity, then methanol is formed via a catalytic reforming process. Another product realizable is ammonia, used in the manufacture of chemical fertilizers. Producer gas can also be used in combined cycle gas turbine-steam cycle electricity generation systems [4]. Wood and charcoal gasifiers were used in Europe during the 1930s and 1940s (i.e. including the gasoline-starved years of the Second World War) to fuel motor vehicle engines. Such systems were very demanding of maintenance, even under the most favourable conditions [10]. Producer gas typically contains about 25% carbon dioxide and 15% hydrogen, so its fuel value is about one-sixth that of natural gas.

Biomass can be converted to methane and ethane by reduction with hydrogen at 540°C and 6.9×10^6 Pa pressure. This process is limited in usefulness by the fact that the hydrogen itself is a premium fuel [4].

Gasification technologies for solid biomass fuels, including solid wastes, have a number of potential environmental problems. The quantity and mix of resulting air pollutants depend on the gas-producing technology and on the chemical content of the initial feedstock. There is likely to be formation of ammonia (NH_3), hydrogen sulphide (H_2S), hydrogen cyanide (HCN) and dicarbon sulphide (C_2S), as well as phenols. Leaks of raw gas would be a hazard in the immediate locality.

Most biomass feedstocks used in gasification processes have concentrations of trace elements such as potassium, magnesium, sodium, iron, boron, barium, cadmium, chromium, copper, lead, strontium and zinc. These levels of chemical concentration, plus fly-ash and sulphur content, are much lower than in coal combustion. Feedstock derived from farm waste may be contaminated with pesticides, fertilizers and soil [10].

In biomass gasification water is present in the initial feedstock and is also formed during combustion. Biomass storage sites can also leach polluted rain water. The safe disposal of effluent water is part of the overall biomass safety problem.

13.3.4.   *Liquid and gaseous fuels from biomass*

13.3.4.1.   *Chemical reduction*

Fuel oils of varying compositions can be obtained by the chemical and physical reduction of aqueous biomass (Fig. 13.3). For example, carbon monoxide, steam and a slurry of cellulosic waste react together at 250–400°C and $13.8-27.6 \times 10^6$ Pa pressure, with an alkaline catalyst, to yield a fuel oil of approximate formula $C_{11}H_{19}$ O with an energy content of 40 MJ/kg [4]. This compares favourably with all of the types of coal in Table 4.1 of Chapter 4.

13.3.4.2.   *Alcoholic fermentation (ethanol)*

The alcoholic fermentation process uses biomass materials containing starches and simple sugars. Starch feedstocks are mainly grain crops and cereals but also include root plants like potatoes. Sugar feedstocks include cane and beet sugar, sorghum and artichokes. Fermentation produces ethyl alcohol (ethanol), which is the intoxicant ingredient in beer, wine and spirits [10].

Ethanol is a relatively clean burning fuel that can be used as a substitute for gasoline (petrol) in modified internal combustion engines or as a non-lead, octane-enhancing gasoline additive. The major producers of ethanol include Brazil and the USA. Production figures for 2000–2010 are given in Table 13.3 [11]. In Brazil many car use ethanol, made from sugar cane, in place of petrol. An alcohol powered car uses a smaller engine, with smaller cooling and exhaust systems, than a petrol car [9]. The energy equivalence is that 1 barrel of ethanol = 0.57 barrels of oil [11]. Ethanol consumption in an engine is higher than for petrol, since the energy per unit volume of fuel is lower.

The USA is desperate to reduce its dependence on imported oil, which is mostly used as automobile fuel. Running a car on a gasoline (petrol)-ethanol mix partly achieves the objective and many cars can run on blends of up to 10% ethanol. In several US states it is currently mandatory to use blended fuel.

In the UK many formerly green fields now grow crops of bright yellow oilseed rape or blue linseed. Rape grain can be processed into the biodegradable chemical called rape methylester, which is a biodiesel fuel similar to diesel fuel but much more expensive to produce. Valuable by-products of the process include protein-rich animal feed and glycerine [12].

13.3.4.3.   *Anaerobic digestion to produce biogas*

In the anaerobic ("without oxygen") decomposition of organic materials, bacteria consume the material in an airtight container, called a digester. The bacteria may be in the original charge material, such as animal manure, or it may be intentionally introduced to facilitate the process. The digester operates at a working temperature in the range 95–140°F and yields a mixture of gases often referred to as "biogas". When this gas occurs naturally due to rotting vegetation it is called "swamp gas". The same biochemical process is used in some forms of sewage treatment.

Biogas contains mainly methane, similar to natural gas, with the remainder made up of mostly carbon dioxide. It has the heat value of its methane component, 39 MJ/m^3, as shown in Table 13.4 [4], and can be used directly for heating or in

Table 13.3.   World biofuels production, 2000–2010 [11].

Thousand tonnes oil equivalent	2000	2001	2002	2003	2004	2005	2006	2007	2008	2009	**2010**	Change 2010 over 2009	2010 share of total
US	2991	3288	3987	5226	6357	7478	9746	13456	19096	21670	**25351**	17.0%	42.8%
Canada	105	111	113	113	113	133	160	461	536	721	**996**	38.1%	1.7%
Total North America	3096	3399	4100	5339	6470	7612	9906	13922	19637	22399	**26355**	17.7%	44.5%
Argentina	4	9	9	9	9	9	29	228	632	1054	**1687**	60.0%	2.8%
Brazil	5212	5600	6149	7068	7135	7835	8729	11323	14132	13962	**15573**	11.5%	26.3%
Colombia	—	—	—	—	—	14	131	141	239	326	**351**	7.8%	0.6%
Jamaica	—	—	54	74	56	62	147	138	182	196	**196**	—	0.3%
Other S. & Cent. America	31	30	69	78	93	171	369	472	741	457	**457**	—	0.8%
Total S. & Cent. America	5248	5639	6281	7228	7292	8091	9405	12302	15927	15994	**18264**	14.2%	30.8%
Austria	18	18	22	26	48	70	105	220	263	354	**383**	8.3%	0.6%
Belgium	—	—	—	—	—	1	21	140	278	473	**454**	−4.0%	0.8%
France	315	315	337	368	385	439	798	1121	2012	2312	**2312**	—	3.9%
Germany	215	298	473	688	909	1788	2561	3181	2727	2728	**2930**	7.4%	4.9%
Italy	70	123	180	232	272	340	482	443	617	758	**670**	−11.5%	1.1%
Netherlands	—	—	—	—	6	3	22	80	77	241	**283**	17.6%	0.5%
Poland	—	—	—	27	23	84	158	116	279	393	**338**	−14.0%	0.6%
Portugal	—	—	—	—	—	1	79	153	136	202	**275**	36.3%	0.5%
Spain	70	70	134	184	221	288	248	320	356	958	**1179**	23.1%	2.0%
Sweden	—	14	31	32	43	48	54	99	118	173	**212**	22.8%	0.4%

(*Continued*)

Table 13.3.   (*Continued*)

Thousand tonnes oil equivalent	2000	2001	2002	2003	2004	2005	2006	2007	2008	2009	2010	Change 2010 over 2009	2010 share of total
United Kingdom	—	—	3	9	9	39	166	136	196	180	**180**	—	0.3%
Other Europe & Eurasia	57	113	126	138	166	301	406	536	1031	1825	**2135**	17.0%	3.6%
Total Europe & Eurasia	744	951	1305	1704	2081	3401	5103	6546	8091	10597	**11354**	7.1%	19.2%
Total Middle East	—	—	—	—	—	—	—	—	—	—	—	—	—
Total Africa	6	6	6	6	6	6	6	6	10	14	**14**	—	◆
Australia	—	—	—	—	4	20	54	70	110	174	246	41.8%	0.4%
China	—	4	146	396	492	622	858	1076	1323	1399	1399	—	2.4%
India	82	85	91	94	99	114	134	92	148	82	151	84.5%	0.3%
Malaysia	—	—	—	—	—	—	48	110	197	250	97	−61.2%	0.2%
South Korea	—	—	1	2	4	9	39	74	140	217	287	31.9%	0.5%
Thailand	—	—	—	—	3	52	80	138	495	618	647	4.6%	1.1%
Other Asia Pacific	—	—	—	—	—	18	109	176	215	353	448	26.7%	0.8%
Total Asia Pacific	82	89	238	491	603	833	1323	1736	2628	3094	3275	5.9%	5.5%
Total World	**9176**	**10084**	**11930**	**14767**	**16452**	**19944**	**25743**	**34512**	**46294**	**52098**	**59261**	**13.8%**	**100.0%**
of which: OECD	3841	4350	5406	7045	8549	11013	15054	20494	27728	32569	**37130**	14.0%	62.7%
Non-OECD	5336	5734	6523	7723	7903	8930	10688	14018	18566	19528	22131	13.3%	37.3%
European Union	744	951	1305	1704	2073	3378	5052	6469	7944	9970	**10447**	4.8%	17.6%
Former Soviet Union	—	—	—	—	11	22	28	49	129	645	**913**	41.5%	1.5%

◆ Less than 0.05%.

Source: Includes data from F.O. Licht; US Energy Information Administration.

Notes: Consumption of fuel ethanol and biodiesel is included in oil consumption.

Table 13.4.   Bioconversion processes and products [4].

Process		Initial product	Final product
	Anaerobic digestion	Biogas: $CH_4 2 : 1CO_2$ (22–28 $MJ\ m^{-3}$)	Methane (38 $MJ\ m^{-3}$)
Aqueous	Alcohol fermenta- tion		Ethanol (19 $MJ\ l^{-1}$)
	Chemical reduction		Oils (35–40 $MJ\ kg^{-1}$)
Dry Thermochemical	Pyrolysis		Pyrolytic oils (23–30 $MJ\ kg^{-1}$) Gas (8–15 $MJ\ m^{-3}$) Char (19–31.5 $MJ\ kg^{-1}$)
	Gasification	Low-medium energy gas (7–15 $MJ\ m^{-3}$)	Methane (38 $MJ\ m^{-3}$) Methanol (16.9 $MJ\ l^{-1}$) Ammonia Electricity (3.6 $MJ\ kW\ h^{-1}$)
	Hydrogasification		Methane (38 $MJ\ m^{-3}$) Ethane (70.5 $MJ\ m^{-3}$) Char (19–31.5 $MJ\ kg^{-1}$)
Direct combustion of:		High-pressure steam	High-pressure steam
Wood chips (18.6–20.9 $MJ\ kg^{-1}$ dry wt)			Electricity (3.6 $MJ\ kW\ h^{-1}$)
Sugar-cane bagasse (9.5 $MJ\ kg^{-1}$)			
Cereal straw (16–17 $MJ\ kg^{-1}$)			
Organic refuse (13.2 $MJ\ kg^{-1}$)			
Biophotolysis (see Chapter 7)			Hydrogen (12.7 $MJ\ m^{-3}$)

internal combustion engines. In the developing countries there is a lot of interest in biogas units for household fuel, for the improvement of sanitation and to improve the fertilizer value of animal dung and other organic wastes [9].

The anaerobic microbiological digestion process is especially well adapted to slurry-type wastes. Such processes have been used for many years to treat sewage and its by-products as fertilizers. They can be used as waste treatment operations to reduce pollution hazards and nuisance odours. An additional benefit is that a fertilizer rich in nitrogen is retrieved from the digested slurry as a by-product. Furthermore, the residual from the process can be returned to the land, unlike other biomass conversion processes, which almost totally destroy the input material [10].

There are many forms of anaerobic digester design, ranging from the small domestic size to large industrial systems. In China, for example, over seven million small digestion systems have been installed, mostly family-sized units, to partially meet the cooking, space heating and lighting demands of small rural communities [8, 9].

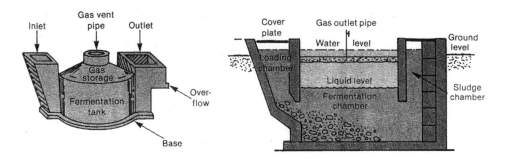

Fig. 13.5. Chinese design of biogas plant [10].

A diagram of the Chinese design is given in Fig. 13.5 [10]. In the USA the majority of biogas installations are used to collect gaseous emissions from landfill waste deposits. Because biogas units operate at low pressure, the pipeline distribution is generally limited to a restricted area. The small hydrogen sulphide component is corrosive to metal. However, with suitable adjustments to the burners, most appliances made for natural gas or bottled gas can be adapted for use with biogas [9].

13.4. Wood as a Fuel

Most poor people in the developing countries use wood as a cooking fuel. In the early 1980s it was estimated that "more than 90% of the wood cut in Africa — five million acres a year — is burnt as fuel" [13]. The World Bank has estimated that between 500,000 and one billion people use agricultural or animal wastes as fire fuel [14, 15]. Moreover, a lot of this fuel is used inefficiently in open fires with cooking utensils sitting on it or suspended over it. The efficiency of use would be increased by a factor of 4–5 by enclosing the fire, regulating the flow of air into the stove and venting through a chimney. Many governments and aid agencies all over the world are addressing the need for better and different forms of wood-burning stoves. For fuel wood to satisfy this need, greatly increased planting levels would be required all over the developing world [9]. Moreover, deforestation and desertification is widespread and increasing in Africa. For example, the southern edge of the Sahara desert has moved 100 km south in 17 years [1].

In medieval Europe wood was the main fuel and the whole continent, including Great Britain, was heavily forested. Between 1550 and 1700, the population of Britain doubled and the demand for timber resulted in massive deforestation. A transition took place from wood cutting to coal mining as the main source of heat. This changed the economic history of Britain, then of the rest of Europe and finally of the world. It led to the Industrial Revolution, which started in Britain around 1780, with the expansion of manufacturing industries and the exploitation of untapped natural resources [16].

By the middle of the 19th century, wood was the principal fuel in the USA. By 1900 it provided 25% of the country's annual energy, but this usage had fallen to 1.5% in 1976. Most of the 1.5% came from the forest products industry, which burns tree wastes. The use of biomass, typically in wood-burning stoves, is increasing in North America and now represents about 4% of energy use. This still lags far behind wood heating in Sweden and Finland, where, in 1981, 8% and 15%, respectively, of the energy needs were met by wood [17]. More than 300 US power plants use biomass to produce $5000\,MW_e$ of electricity.

The use of wood and waste in the USA is of a lower order of magnitude compared with the use of conventional hydroelectric power for electricity generation (Table 13.5). It is significant that in 2009 wood and waste provided about one-half the electricity generation capability of wind energy, which was about 50 times the contribution of solar energy [18].

Various options for the use of wood as a biomass fuel are included in Table 13.6. The direct combustion of wood to produce steam or oil represents one of the cheapest options available [4]. In comparison with coal, wood fuel is bulkier and contains more volatile constituents, leading to the need for less expensive combustion plants.

Table 13.5.   US electric net summer capacity (MW) [18].

Source	(Megawatts)				
	2005	2006	2007	2008	2009
Total	978,020	986,215	994,888	1,010,171	1,027,584
Renewable Total	98,746	101,934	107,954	116,423	125,800
Biomass	9,802	10,100	10,839	11,050	11,353
Waste	3,609	3,727	4,134	4,186	4,405
Landfill Gas	887	978	1,319	1,429	1,514
MSW[1]	2,167	2,188	2,218	2,215	2,215
Other Biomass[2]	554	561	598	542	676
Wood and Derived Fuels[3]	6,193	6,372	6,704	6,864	6,948
Geothermal	2,285	2,274	2,214	2,256	2,351
Hydroelectric Conventional	77,541	77,821	77,885	77,930	77,951
Solar/PV	411	411	502	536	603
Wind	8,706	11,329	16,515	24,651	33,542
Nonrenewable Total	879,274	884,281	886,934	893,747	901,785

[1] Includes total capacity whose primary energy source is MSW.
[2] Agriculture by-products/crops, sludge waste and other biomass solids, liquids and gases. Does not include tyres.
[3] Black liquor, and wood/wood waste solids and liquids.
MSW = Municipal Solid Waste.
PV = Photovoltaic.
Notes: Totals may not equal sum of components due to independent rounding.
Revisions to biomass capacity removed tyres from renewable waste energy.
Data for 2009 is preliminary.
Source: U.S. Energy Information Administration, Form EIA-860, "Annual Electric Generator Report".

Table 13.6.   Comparative costs for various biomass conversions per unit of energy output [4].

Process	Comparative cost rating
Wood to char and oil for direct combustion	1.0–1.3
Steam production from wood via direct combustion	1.1–1.2
Medium-energy gas production from cattle manure via anaerobic digestion	1.5–3.3
Wood to oil via catalytic liquefaction	2
Substitute natural gas production from cattle manure via anaerobic digestion	2.2–5.3
Substitute natural gas production from wood gasification (oxygen-blown reactor)	2.4–2.9
Wood to methanol via gasification (oxygen-blown reactor)	2.9–3.7
Ammonia from wood via gasification (oxygen-blown reactor)	3.3–7.0
Wheat straw to medium-energy gas via anaerobic digestion	4.9–8.8
Electricity generation from wood via direct combustion	6.1–6.9
Algae to ethanol via acid hydrolysis and fermentation	7.0–9.9
Corn straw to ethanol via enzymatic hydrolysis and fermentation	7.4–16.2
Kelp to substitute natural gas via anaerobic digestion	7.7–8.3
Sugarcane to ethanol via fermentation	11.9
Wheat straw to ethanol via enzymatic hydrolysis and fermentation	19.5

If combusted properly, wood is far less polluting than either coal or oil, particularly with regard to sulphur emissions [19]. It is significant to note that the carbon dioxide (CO_2) released on wood combustion is the same as the carbon dioxide taken up during the plant growth. Wood can therefore be described as carbon dioxide-neutral. This means that when wood is burned in preference to a fossil fuel, there is a net reduction of atmospheric carbon dioxide and therefore a lower contribution to the production of greenhouse gases. But note that wood (trees) has taken in CO_2 over many, many years whereas burning wood releases CO_2 quickly; this alters the carbon dioxide equilibrium of the atmosphere.

In the UK, research is now under way to look into the possibility of building wood-fuelled electricity-generating stations. Year-old trees, such as willows, are cut down near to the ground, causing them to throw up around six new shoots, which grow rapidly over the next 3–5 years. This is known as coppicing. The harvested coppice wood can be dried, chipped and used as fuel.

The successful and economic use of wood as a primary fuel, whether from conventional forestry or from energy forestry, depends on a chain consisting of resource, harvesting, storage and transport [20]. A detailed flow graph of the processing of wood fuel to steam or electricity generation is given in Fig. 13.4.

13.5.   Energy from Wastes

All human and industrial processes produce waste. In the industrial countries the amount of municipal solid waste derived from domestic, commercial and industrial sources increases every year. Data for the USA, from 1960 to 2009, is shown in Fig. 13.6 [21].

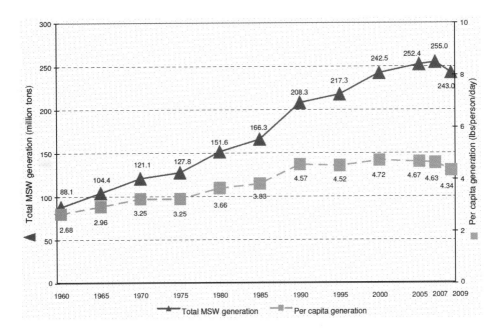

Fig. 13.6.   US municipal solid waste generation, 1960–2009 [21].

Wastes that can form sources of biofuels include domestic refuse, industrial wastes, agricultural wastes, forestry residues, sewage and industrial effluents. It is highly desirable that the recovery of energy from waste should form part of an integrated approach to waste management, designed to maximize waste recycling and reclamation. Recycling prevents the emission of many greenhouse gases and water pollutants, saves energy, supplies valuable raw materials to industry, creates jobs, stimulates the development of greener technologies, conserves resources for the future and reduces the need for new landfills and combustors (incinerators). In 1996, recycling of solid waste in the USA prevented the release of 33 million tons of carbon into the air, roughly the amount emitted annually by 25 million cars [21]. The US recycling rates of selected materials for 2009 are shown in Fig. 13.7.

Using wastes as a direct source of energy can be highly cost-effective, especially if the alternative disposal cost is discounted.

Wastes do have certain disadvantages when used as fuel sources. They may be difficult to handle and process and generally they have low energy density. Some wastes may be contaminated with non-fuel materials. Also, it may be necessary to transport the waste from its source to a conversion site [19].

13.5.1.   *Solid waste disposal in landfill sites*

Large municipal or industrial landfills produce gas that can be tapped to generate electricity. Microorganisms that live in organic materials such as food wastes, paper

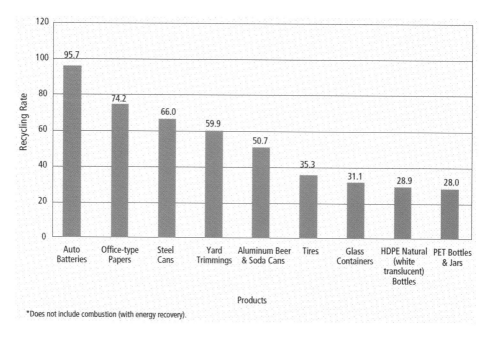

Fig. 13.7.   US recycling rates of selected materials, 2009 [21].

or yard clippings cause these materials to decompose. This produces landfill gas, typically comprised of roughly 60% methane and 40% carbon dioxide, similar to the biogas described in Section 13.3.4.

Landfill gas is collected from landfills by drilling "wells" into the landfills, and collecting the gases through pipes. Once the landfill gas is processed, it can be combined with natural gas to fuel conventional combustion turbines or used to fuel small combustion or combined cycle turbines. Landfill gas may also be used in fuel cell technologies, which use chemical reactions to create electricity, and are much more efficient than combustion turbines [21].

In a highly developed society like Great Britain or the USA, solid waste from all sources is produced at the rate of about one tonne per person per year [5, 18]. Most domestic waste is solid in form. The constituent proportions typical of UK municipal waste are shown in Table 13.7 [12]. Corresponding data for the USA in 2009 is given in Fig. 13.8 [21].

Each year the UK produces 28 million tonnes of solid domestic waste plus a similar amount of industrial waste. Solid wastes are usually disposed of either by burning or by burial in landfill sites. The majority of the domestic waste produced in the UK is not burned but is buried in about 1,500 landfill sites. At the end of 1990 there were 33 landfill gas schemes in operation in the UK, including 18 MW^e of electricity generation. By 1995 there were 50 gas-producing sites in operation and an estimated further 400 sites that could be used. In 2009 the number of controlled gas-producing sites had increased to 75, generating 110 MW of electricity. The largest

Table 13.7. Composition of typical UK municipal solid waste [12].

Category	Per cent	Calorific value (MJ/kg)
Paper and cardboard	30.5	14.6
Food waste	24	6.7
Metal	7.8	—
Glass	11.2	—
Dust and cinders	9.0	9.6
Textiles	4.9	16
Leather, rubber	8.3	17.6–37
Miscellaneous	4.3	17.6
Total	100%	

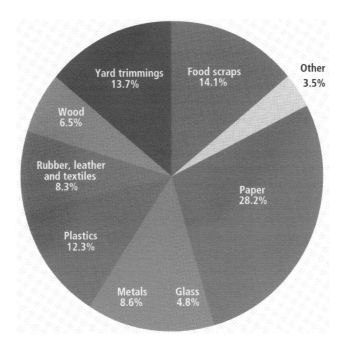

Fig. 13.8. US municipal solid waste, 2009 [21].

current site in the UK produces $3,500 \, m^3$ of methane per hour to make steam for use in paper production. A diagrammatic representation of a landfill gas system is given in Fig. 13.9 [12].

As of 2009, over 3,000 landfill gas recovery and utilization projects have been operational in the USA. The US Environmental Protection Agency (EPA) estimates that about 700 other landfill sites present attractive opportunities for project development [23].

The venting of landfill gases poses some environmental problems. Within the site vicinity there may be objectionable odours. Uncontrolled discharges from landfill

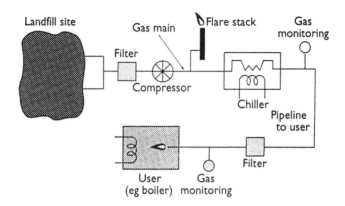

Fig. 13.9.   Production of landfill gas [12].

sites account for an estimated 21% of UK methane emissions. If the landfill gas can be used as a fuel its own methane discharge is minimal and it also reduces the carbon dioxide that would be emitted by the equivalent fossil fuels [19].

In the USA the EPA requires that all large landfills install collection systems at landfill sites to minimize the release of methane, a major contributor to global climate change. While new EPA regulations require gathering and flaring of methane from large landfill operations, small landfills, which fall outside the federal agency's jurisdiction, may amount to as much as 40% of the methane generated by landfills nationwide.

Landfill gas generators produce nitrogen oxide emissions that vary widely from one site to another, depending on the type of generator and the extent to which steps have been taken to minimize such emissions. Combustion of landfill gas can also result in the release of organic compounds and trace amounts of toxic materials, including mercury and dioxins, although such releases are at levels lower than if the landfill gas is flared [23].

There are few water impacts associated with landfill gas power plants. Unlike other power plants that rely upon water for cooling, landfill gas power plants are usually very small, and therefore pollution discharges into local lakes or streams are typically quite small [23].

13.5.2.   *Solid waste disposal using municipal incinerators (combustors)*

The burning of municipal solid waste (MSW) can generate energy while reducing the amount of waste by up to 90% by volume and 75% by weight. In the USA in 1999, there were 102 incinerators with energy recovery capability, having the capability to burn 96,000 tons of waste per day [20].

"In 2009, Americans recovered about 61 million tons of MSW (excluding composting) through recycling. Composting recovered about 21 million tons of waste. We combusted about 29 million tons for energy recovery (about 12 percent). Subtracting out what we recycled and composted we combusted (with energy recovery) or discarded 2.9 pounds per person per day.

"In 2009, office-type paper recovery rose to about 74 percent (4 million tons), and about 60 percent of yard trimmings were recovered [...]. Metals were recycled at a rate of about 34.5 percent [...]. By recycling more than 7 million tons of metals (which includes aluminum, steel, and mixed metals), we eliminated greenhouse gas (GHS) emissions totaling about 25 million metric tons of carbon dioxide equivalent (MMTCO$_2$E). This is equivalent to removing almost 5 million cars from the road for one year.

"About 132 million tons of MSW (54.3 percent) were discarded in landfills in 2009 in the USA." [21]

For combustion in a modern waste incinerator the refuse can be first sorted to separate out materials such as glass for recycling. The remainder is shredded to convert the burnable component into refuse-derived fuel pellets [18].

The pellets burn like wood but produce less heat than coal and leave more ash. Modern municipal waste incinerators produce no smoke and reduce the bulk of the waste to about one-eighth. Some wastes, such as synthetic polymers (e.g. plastics produced from petrochemicals), produce toxic gases on combustion and these have to be safely contained within the incineration process.

General industrial waste consists mainly of paper, cardboard, wood and plastics. It contains a lower moisture and lower ash content than municipal waste, making combustion easier to control, and is less contaminated with non-fuel ingredients. Special forms of industrial waste such as batteries, motor tyres, poultry litter and hospital wastes are potential hazards and require special forms of treatment but are all useful biofuels for combustion systems [18].

Refuse incineration in municipal plants for energy recovery is a relatively new but growing technology in the UK, as illustrated in Tables 13.8 [12] and 13.9 [24].

Table 13.8. Refuse incineration in the European Union [12].

	Number of plants	Total waste processed (tonnes/day)	Energy recovered		
			Hot water	Steam	Electricity
Belgium	6	3650	3	1	2
Denmark	63	9100	58	5	0
Eire	0	0	0	0	0
France	33	13700	10	21	9
Germany	43	21780	13	27	29
Greece	0	0	0	0	0
Italy	6	3250	6	0	1
Luxembourg	2	800	0	0	2
The Netherlands	5	5670	1	1	4
Portugal	0	0	0	0	0
Spain	4	1350	0	4	2
UK	7	3500	5	0	3
Total	169	62800	96	60	52

Table 13.9.   UK renewable energy sources [24].

Renewable energy sources, 2009

Geothermal and active solar heating 1.0%
Small scale hydro 0.8%

Other 1.8%

Wind 11.6%

Hydro
(Large-scale)
5.8%

Biomass
80.7%

Landfill gas 23.8%

Sewage gas 4.0%
Domestic wood 5.5%
Industrial wood 2.4%
Co-firing 8.6%

Waste combustion 9.5%
Animal biomass 4.0%
Plant biomass 8.2%

Liquid biofuels 14.7%

Total renewables used= 6,875 thousand tonnes of oil equivalent (ktoe)

Total use of renewables	Thousand tonnes of oil equivalent				
	1990	2000	2007	2008	2009
Geothermal and active solar heating	7.2	12.0	46.9	58.0	72.0
Wind and wave	0.8	81.3	453.5	610.3	800.0
Hydro (small- and large-scale)	447.7	437.3	437.5	444.4	452.4
Landfill gas	79.8	731.1	1,547.5	1,573.9	1,637.8
Sewage gas	138.2	168.7	215.2	231.7	277.3
Wood (domestic and industrial)	174.1	458.4	433.2	520.8	539.7
Municipal waste combustion	100.8	374.8	520.5	538.2	655.8
Liquid biofuels	-	-	361.7	825.5	1,008.6
Other biomass	71.9	265.0	1,154.7	1,203.8	1,431.2
Total	**1,020.5**	**2,528.5**	**5,170.6**	**6,001.8**	**6,874.9**

In the UK in 2009, biofuels accounted for 81% of renewable energy sources (Table 13.9) with most of the remainder coming from wind energy. The largest and growing components are seen to be landfill gas, liquid biofuels and municipal solid waste incineration.

There has been much consideration of the relative merits of landfill gas extraction and municipal solid waste incineration. For example:

"A Royal Commission on Environmental Pollution in Great Britain studied greenhouse emissions from both landfills and incinerators (Royal Commission, 1993). They reported that incinerating one million tonnes of municipal garbage produces net emissions of 15,000 tonnes of carbon in the form of carbon dioxide, whereas landfilling it with energy recovery produces emissions of greenhouse gases equivalent to 50,000 tonnes of carbon as carbon dioxide. Thus, even with gas collection and burning, landfills have a worse impact on global warming than incinerators." [25]

Nevertheless there is widespread opposition to the siting and installation of municipal incinerators, both in Europe and in the USA. The misgivings are chiefly related to environmental safety and the release of contaminated gases and fluids.

13.5.3. *Worked examples on solid waste incineration*

Example 13.1. A ton of municipal refuse contains the proportions of material indicated in Table 13.6. Calculate the calorific value if all the energy is recoverable. What is the equivalent energy in (a) kWh and (b) tonnes of oil?

$$1 \text{ ton} = 2240 \text{ lb} = \frac{2240}{2.2046} = 1016 \text{ kg} = 1.016 \text{ tonnes}$$

Applying the percentage proportions in Table 13.6 and multiplying by the appropriate calorific value gives the data of the following table:

Material	Weight in kilogrammes	Calorific value (MJ)
Paper	$0.305 \times 1016 = 309.9$	$309.9 \times 14.6 = 4524.5$
Food waste	$0.24 \times 1016 = 243.8$	$243.8 \times 6.7 = 1633.5$
Metal	$0.078 \times 1016 = 79.25$	$79.25 \times 0 = 0$
Glass	$0.112 \times 1016 = 113.8$	$113.8 \times 0 = 0$
Dust, cinders	$0.09 \times 1016 = 91.44$	$91.44 \times 9.6 = 877.8$
Textiles	$0.049 \times 1016 = 49.8$	$49.8 \times 16 = 796.5$
Leather, rubber, plastic	$0.083 \times 1016 = 84.3$	$84.3 \times 25^* = 2108.2$
Miscellaneous	$0.043 \times 1016 = 43.7$	$43.7 \times 17.6 = 768.9$
		Total 10709 MJ

*Estimated value

From Table 1.5 of Chapter 1,

$$MJ \equiv 1 \text{ kWh}$$
$$\text{(a)} \therefore \ 1 \text{ ton refuse} \equiv 10709/3.6 \equiv 2974.7 \text{ kWh}$$
$$1 \text{ tonne oil} \equiv 4000 \text{ kWh}$$
$$\text{(b)} \therefore \ 1 \text{ ton refuse} = 2974.7/4000 \equiv 0.744 \text{ tonne of oil}$$

In practical terms, perhaps 20–50% of the energy in the refuse might be reclaimable. Then:
1 ton of refuse yields the equivalent of about 750 kWh or 0.19 tonne of oil minimum to 1500 kWh or 0.38 tonne of oil maximum.

The estimated "energy cost" of collecting and disposing of municipal waste is about 5000 MJ/tonne [5]. Does the energy value of 1 ton of waste justify the collection in energy terms?

The total energy value of 1 ton of waste in Example 13.1 is 10709 MJ. This is more than twice the specified energy cost of collection and disposal.

The actual energy profit will depend on what proportion of the maximum potential value (10709 MJ) is extractable. If 50% energy is available there will be a small energy profit. But part of the collection and disposal cost is for public health and social reasons and has to be undertaken anyway. Any energy profit is a bonus.

13.5.4. *Liquid and gaseous wastes*

Large-scale liquid and gaseous wastes from industry are often processed at the producer sites. This permits the retrieval of salvageable materials and contributes to energy costs when the residue is burned. Liquid wastes from domestic sources are usually poured down the drain.

Sewage disposal is an issue of concern in the UK. Many communities adjacent to rivers, lakes or the sea eject raw sewage directly into the water. This poses problems of water pollution. It also eliminates the possibility of sewage treatment to minimize the effluent and to obtain useful sewage gases, such as methane.

13.6. The Fuel Cell

A fuel cell uses the gases hydrogen and oxygen as energy sources to produce electricity and water. It consists of two electrodes which enclose an electrolyte (Fig. 13.10).

Hydrogen fuel is fed into the anode of the fuel cell. Oxygen (or air) enters the fuel cell through the cathode. Encouraged by a catalyst, the hydrogen atom splits into a proton and an electron, which take different paths to the cathode. The proton passes through the electrolyte. The electrons create a separate electric current that can be utilized before they return to the cathode, to be reunited with the hydrogen and oxygen in a molecule of water [26, 27].

A fuel cell system usually includes a "fuel reformer" that can utilize the hydrogen from any hydrocarbon fuel, including methanol, ethanol, natural gas, liquid propane, gasified coal, gasoline and diesel fuel. Input energy can also be supplied by hydrogen derived, via methane, from biomass, wind and solar renewable sources, including gas from landfills. Since the fuel cell relies on chemistry and not combustion, emissions from this type of system are much smaller than emissions from the cleanest fuel combustion processes.

Serious interest in the fuel cell as a practical and reliable source of electricity began in the 1960s in the US space programme. Fuel cells furnished power for the Gemini and Apollo spacecraft and provided electricity and water for the space shuttle.

Fuel cells are ideal for home power generation, either connected to the electricity grid to provide supplemental power and back-up or installed as a grid-independent generator for on-site service in areas that are inaccessible by power lines. Since fuel

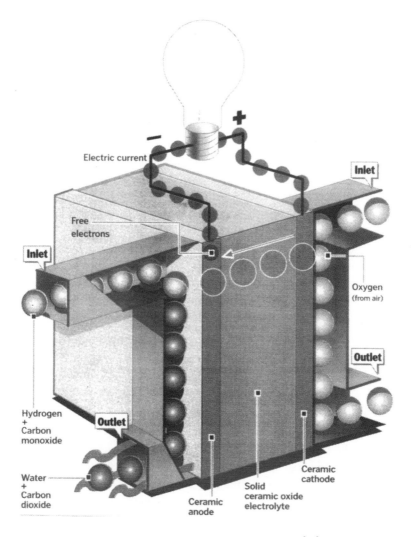

Fig. 13.10.   Basic principle of the fuel cell [26].

cells operate silently, they reduce noise pollution as well as air pollution and the waste heat from a fuel cell can be used to provide hot water or space heating.

Fuel cell automobiles are an attractive advance from electric battery-powered and hybrid (i.e. battery plus gasoline power) vehicles. They offer the advantages of battery-powered vehicles but can also be refuelled quickly and could go longer between refuelling.

A vehicle using fuel cells that utilize pure hydrogen as a fuel would be a zero emission vehicle. Those using other fuels would produce near zero emissions. They are also more efficient than "grid"-powered battery vehicles. In addition, fuel cell cars could produce fewer "system-wide" releases of greenhouse gases, taking into account all emissions associated with resource recovery, fuel processing and use. Studies in the USA by General Motors and by Ford noted that fuel cell car engines could be built for about the same price as an internal combustion engine. The energy efficiency of fuel cells also makes them an attractive alternative for automakers. Automakers and component suppliers are spending billions of dollars to drive fuel cell technology towards commercialization. Some are concentrating on using pure hydrogen, while others are trying to find new ways to use gasoline-like hydrocarbons [26].

At present (2013) hydrogen is a more expensive fuel than conventional fossil fuels. A US company offers commercial fuel cell power plants for about US$3,000 per kilowatt. At that price the units are competitive in high value, "niche" markets and in areas where electricity prices are high and natural gas prices are low. Several companies are now selling small units for research purposes. Fuel cells will have to be much cheaper to become commercial in passenger vehicles. Conventional car engines cost about US$3,000 to manufacture in the USA. More research is needed to bring the cost of fuel cell systems down to that level [26]. The widespread use of fuel-cell-driven automobiles would dramatically reduce the world oil consumption and the emissions from its use [28].

13.7.   Problems and Review Questions

13.1. What are the principal disadvantages of the use of land for intensive biofuel growth?

13.2. Describe the process of photosynthesis in plants. In particular, what are the input forms of energy of the photosynthetic process?

13.3. Enumerate the stages of the photosynthetic process to show that the maximum efficiency under ideal conditions from input radiation to plant energy storage is 6.6%.

13.4. What are the practical values of photosynthetic efficiency in (a) temperate locations and (b) tropical locations?

13.5. What is the nature and purpose of the bright yellow crop now seen in many English fields?

13.6. What is the chief constituent of the biogas in anaerobic digester systems? Why are such systems so abundant in the developing countries?

13.7. What are the chief advantages of wood as a source of fuel?

13.8. In what sense can the burning of wood be described as "carbon dioxide-neutral"?

13.9. What are the advantages and disadvantages of burning plastics as a biofuel?

13.10. How would the wider use of biofuels affect the appearance of the British countryside?

13.11. How would the contents of refuse collection in Victorian England or 19th century USA compare with the present day?

13.12. In the period 1990–2009, how did the proportions of (a) landfill gas, (b) wood and (c) municipal solid waste of the renewable energy sources change in the UK?

13.13. In the period 2005–2009, how did the consumption of (a) wood, (b) conventional hydropower, (c) solar energy and (d) wind energy of the renewable energy consumption change in the USA?

13.14. Compare the recent compositions of municipal solid waste in the UK and USA.

13.15. In the UK there are about 28 million tonnes of domestic waste each year. If the distribution of the contents is the same as in Table 13.5 and the energy efficiency of collection is 50%, how much energy in kWh is available?

13.16. The estimated energy cost of waste collection is about 5000 MJ/tonne. What is the overall annual energy cost in kWh of collecting 28 million tonnes/year of domestic waste?

13.17. Why is the hydrogen-powered fuel cell such an attractive option as a future energy source?

References

[1] McVeigh, J.C. (1984). *Energy around the World*, Pergamon Press, Oxford.
[2] Hall, D.O. (1979). "Solar Energy use through Biology", *Solar Energy*, **22**, 307–329.
[3] "Energy in Transition, 1985–2010", Report of the Committee on Nuclear and Alternative Energy Systems, National Academy of Sciences, Washington, DC, 1979.
[4] Lewis, C. (1983). *Biological Fuels*, Edward Arnold, London.
[5] Dorf, R.C. (1978). *Energy Resources and Policy*, Addison-Wesley Publishing Co., Inc., Reading, MA.
[6] "Making Aquatic Weeds Useful: Some Perspectives for Developing Countries", National Academy of Sciences, Washington, DC, 1976.
[7] Gates, D.M. (1971). "The Flow of Energy in the Biosphere", in Flanagan, D. (ed.) *Energy and Power*, Scientific American, New York, pp. 43–52.
[8] Siddayao, C.M. and Griffin, L.A. (eds.) (1990). *Energy Investments and the Environment*, Economic Development Institute of the World Bank, Washington, DC.
[9] "Renewable Energy Resources in Developing Countries", World Bank, Washington, DC, November 1980.
[10] "Energy from Biological Processes" (2 vols.), Office of Technology Assessment, Document, OTA-E-123, Congress of the United States, Washington, DC, July 1980.
[11] "BP Statistical Review of World Energy", British Petroleum plc, London, June 2011.
[12] "Renewable Energy — A Resource for Key Stages 3 and 4 of the UK National Curriculum", Renewable Energy Enquiries Bureau, Energy Technology Support Unit, Department of Trade and Industry, Harwell, 1995.
[13] Hancock, G. (1981). *Premiers to Discuss Use of Renewable Energy*, The Guardian, August 10 1981.
[14] "Energy in the Developing Countries", The World Bank, Washington, DC, 1980.

[15] Imran, M. and Barnes, P. (1990). *Energy Demand in the Developing Countries — Prospects for the Future*, World Bank Staff Commodity Working Paper No. 23, The World Bank, Washington, DC.

[16] Nef, J.U. (1977). *An Early Energy Crisis and Its Consequences*, Scientific American, **237**, 97–110.

[17] Stobaugh, R. and Yergin, D. (eds.) (1983). *Energy Future, Report of the Energy Project at the Harvard Business School*, Vintage Books, New York.

[18] "Annual Electric Generation Report", US Energy Information Administration (EIA), Form EIA-860, 2010.

[19] "Making Fuels from Wastes and Crops", Renewable Energy Enquiries Bureau, Energy Technology Support Unit, Harwell, 1991.

[20] "Wood as a Fuel", wall poster on "Biofuels — A Renewable Energy", Renewable Energy Enquiries Bureau, Energy Technology Support Unit, Harwell, undated.

[21] "Municipal Solid Waste Generation, Recycling, and Disposal in the United States: Facts and Figures for 2009", Office of Solid Waste, US Environmental Protection Agency (EPA), Washington, DC, June 2010.

[22] "Electricity from Landfill Gas", Power Scorecard, Pace University, White Plains, New York, February 2002.

[23] "Landfill Methane Technology", The Greenpower Market Development Group, World Resources Institute, Washington, DC, USA, 2000.

[24] "UK Energy in Brief", Department of Energy and Climate Change (DECC), London, July 2010.

[25] Tammemagi, H. Y. (1999). *The Waste Crisis — Landfills, Incinerators and the Search for a Sustainable Future*, Oakhill Environmental, Loudon, NH.

[26] "Fuel Cells & Hydrogen", *Fuel Cells 2000*, Online Fuel Cell Information Center, 2002, http://www.fuelcells.org/base.cgim?template=fuel_cells_and_hydrogen.

[27] Funk, J. (2001). *Generating Jobs, Revenue*, The Plain Dealer, Cleveland, OH. Based on a report from the Solid Oxide Fuel Cell Co., McDermott Technology Inc., Alliance, OH.

[28] Polymer Fuel Cells — Cost Reduction and Market Potential", Report by the Carbon Trust, London, England, September 2012.

CHAPTER 14

THE ENERGY FUTURE

14.1. The Energy Problems

All the indicators and projections suggest that there will be increased world use and demand for energy into the foreseeable future (Fig. 14.1). Energy demand is likely to grow faster than the increase of world population, due to increased industrialization and higher living standards. Moreover, the rate of energy demand is rising fastest in the developing countries of the non-OECD grouping. The US Department of Energy projection (Fig. 14.1) shows that the use of fossil fuels and of renewable sources is expected to increase for another 25 years.

A further indication of the anticipated increase of energy use is given by the projected increase of net electricity generation (Fig. 14.2). Most of the increased generation will be provided by fossil fuels.

"Of the 4.5 trillion kilowatt hours of increased renewable generation over the projection period, 2.4 trillion kilowatt hours (54 percent) is attributed to hydroelectric power and 1.2 trillion kilowatt hours (26 percent) to wind. Except for those two sources, most renewable generation technologies are not expected to be economically competitive with fossil fuels over the projection period, outside a limited number of niche markets. Typically, government incentives or policies provide the primary support for construction of renewable generation facilities. Although they remain a small part of total renewable generation, renewables other than hydroelectricity and wind — including solar, geothermal, biomass, waste, and total/wave/oceanic energy — do increase at a rapid rate over the projection period." [1]

Estimates of world energy supplies for the remainder of the 21st century are given in Fig. 14.3, from an oil industry source [2]. The trends are seen to be entirely consistent with the forecasts of the US Energy Information Administration in Fig. 14.1. Crude oil supplies will go into steep decline after about 2025, partly ameliorated by supplies of shale oil and the use of tar sands. The decline of natural gas supplies will commence in about the middle of the century. By the turn of the 21st century it is anticipated that liquid fuels and natural gas will contribute about 15/125 or 12% of the total energy demand of the increased population.

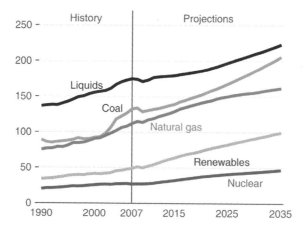

Fig. 14.1.   World marketed energy use, by fuel type [1].

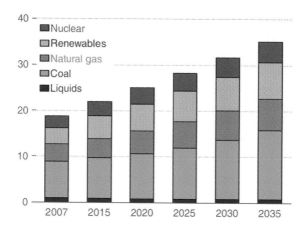

Fig. 14.2.   World electricity generation, by fuel [1].

For the greater part of the 21st century the supply of energy will be taken up, in roughly equal proportions, by coal, nuclear power and renewables (mainly wind power).

Concurrent with the increased energy use is anticipated to be an increase in the creation of energy related carbon dioxide gas emissions. The US Energy Information Administration figures anticipate that:

"World energy-related carbon dioxide emissions will rise from 29.7 billion metric tons in 2007 to 33.8 billion metric tons in 2020 and 42.4 billion metric tons in 2035 — an increase of 43 percent over the projection period. With strong economic growth and continued heavy reliance on fossil fuels expected for most non-OECD nations. In 2007, non-OECD emissions exceeded OECD emissions by 17 percent; in 2035, they are projected to be double OECD emissions, Fig. 14.4." [1]

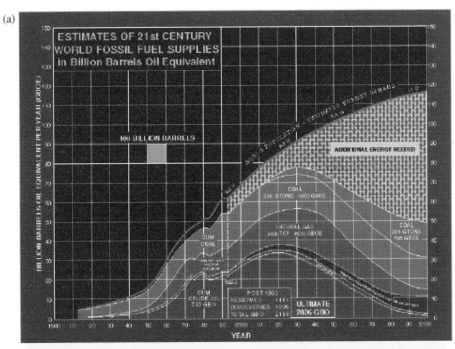

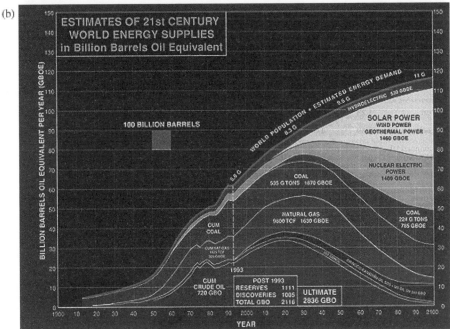

Fig. 14.3. Estimates of world energy supplies [2]. (a) Fossil fuel supplies. (b) World energy supplies.

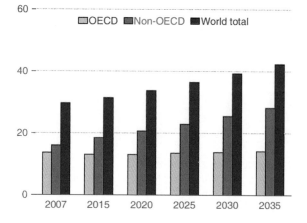

Fig. 14.4. Anticipated increase of energy related carbon dioxide emissions [1].

The scenarios described by Figs. 14.1–14.4 predict that carbon emissions from the use of fossil fuels will peak in about 100 years, stabilizing the atmospheric carbon dioxide at about 60% above present levels with significant climatic effects.

The many uncertainties make it impossible to accurately predict detailed forward demand for energy. But if there is no further major dislocation due to, for example, large-scale and lengthy global wars, the main energy issues discussed in Section 2.7 of Chapter 2 will arise. These are repeated here.

Problem 1   A need for continuing sources of gasoline and diesel fuel for motor vehicles and aircraft.

Problem 2   A need for continuing sources of prime fuel for use in the generation of electricity (on the assumption that oil and natural gas will ultimately be unavailable).

14.2.   An Energy Strategy [3]

Problem 1, concerning oil supply, can be addressed in terms of:

- new oilfield discoveries;
- enhanced recovery rates of natural crude oil;
- the development of an economic synthetic fuels industry, such as tar sands;
- major social adjustments in the pattern of private motoring;
- some alternative form of road transportation (such as the electric car); and
- further development of mass transportation systems using electricity-powered rail vehicles.

Problem 2, concerning electricity generation, can be addressed simultaneously on both the short-term and long-term levels.

In the short term:

- the use of coal and coal products, and
- the increased use of nuclear fission power, using breeder reactors.

In the longer term, the future seems to lie with renewable energy sources:

- Solar energy:
  - photovoltaic conversion,
  - solar-thermal systems,
  - wind turbine systems,
  - ocean thermal systems,
  - water wave energy,
  - hydropower, and
  - biomass and photosynthesis;
- geothermal energy;
- gravitational energy:
  - tidal energy; and
- thermonuclear fusion.

14.3. The Long-Term Energy Future

In the opinion of the authors there appear to be four sources of energy that are adequate to sustain a greatly increased world population. The energy sources need to supply heat and electrical energy to homes, commercial premises and manufacturing industry, to fuel transportation services and to contribute to food production.

Supplies of the major fossil fuels oil and natural gas will probably cease to be available in bulk at economic prices from the middle of this century onwards. A fuel mainstay of the future will be coal, especially for the economies of the eastern countries of China, India and Indonesia. What other options are available?

Use of the term "alternative energy" should be avoided on the grounds that, ultimately, there will not be alternatives [4]. Also, rather than the use of the now common tern "renewable energy", it might be better to classify the ultimate sources below as "indefinitely sustainable energy sources". Each of the four energy sources listed below could supply up to ten times our present energy requirements and for thousands of years.

1. Nuclear fission, using breeder reactors.
2. Solar energy, of various forms.
3. Controlled thermonuclear fusion.
4. Geothermal energy.

These four forms differ widely in their readiness for use, in their likely side effects, and in their economics. Moreover, present knowledge is not sufficient to make meaningful economic comparisons and permits only limited comparisons in respect of environmental and safety risks or of the likelihood of successful technical development. In engineering design the risks involved in compromises between safety and economics often cannot be fully seen until full-scale operation is realized and operational experience is gained [5].

It would be wise to continue research and development in all four of the areas listed above, even though the investment required is massive. Choices and priorities should not be made at this stage. A combination of long-term options is more flexible and more reliable than dependence on a single option.

A useful and helpful survey of future opportunities, especially with regard to the generation of electricity, is given in Ref. 6, which contains an extensive bibliography.

14.3.1.   *Nuclear fission using breeder reactors* [5]

Since uranium, like the fossil fuels, is finite and non-renewable, the long-term future of nuclear fission as an electricity generation source depends on the use of breeder reactors, discussed in Section 8.5.4 of Chapter 8. The liquid-metal fast breeder reactor is the choice of those countries operating or planning to operate fast breeder reactor (FBR) stations, namely France, Germany, Japan, Russia and the UK.

Early enthusiasm for FBR programmes, in the 1970s and 1980s, has significantly waned. Of the 430 nuclear reactors operating worldwide in 1993, only four were breeder systems: one in Russia, two in France and one in the UK. A further FBR system started in Japan in 1995, rated at $246\,MW_e$. In 2001, the world total capacity of the breeder stations was $2473\,MW_e$ out of a total world nuclear capacity of $351746\,MW_e$, which is 0.70%. By 2011 there were 442 uranium reactors in operation, generating $374.958\,MW_e$ of electricity, as shown in Table 8.3 of Chapter 8. About 20 FBRs have in the past been in operation worldwide in the USA, the UK, France, Germany, India, Japan, Kazakhstan, Russia and China. In 2011, FBRs (sometimes called fast neutron reactors) were in operation in China, India, Japan and Russia. No breeder reactors are presently (2013) in commission in France, the UK or the USA.

The loss of public confidence in the nuclear industry, now recovering, that has arisen in recent years in Britain and the USA has greatly inhibited the funding and development of breeder reactors. In particular, the USA, which is energy-rich and a world leader in so many fields of endeavour, has no FBR programme in operation or planned.

14.3.2.   *Solar energy*

Solar energy is, in several ways, the most appealing option to provide an indefinitely sustainable energy source. The risks associated with solar heating are small and public confidence is high. Controversies that arise in the proposed development of nuclear energy, and even wind energy, are avoided.

At present certain solar heating applications are economical. Domestic space heating, domestic hot water heating, the industrial production of hot water or low-pressure steam and some agricultural heating applications are well developed.

Photovoltaic arrays for the direct production of electricity are an established technology of high appeal but, at present, are costly. Fundamental research could yield dramatic returns and progress to date is encouraging.

In the UK the form of solar energy that is likely to be the most economically viable is that which leads to the creation of wind to power large wind turbines. The amount of land per unit of electrical capacity is, in this form, larger than for other forms of solar energy [5]. There are environmental objections to land-based wind farms such that the use of offshore sites is now (2013) established and is moving forwards.

A long-term potential option is to use solar energy for the production of fluid fuels by direct photochemical conversion. For example, the decomposition of water can be used to produce hydrogen. This could be used directly as a fuel or in the synthesis of hydrocarbon fuels. It is possible to realize photochemical conversion efficiencies of 20–30%, based on incident solar energy, compared with an average photosynthetic efficiency of 0.1% for natural ecosystems and up to 1% for intensive biomass systems [5].

14.3.3.   *Controlled thermonuclear fusion*

Despite the enormous sums spent on research all over the world, large-scale nuclear fusion has yet to be demonstrated as technically feasible. The successful development of fusion-derived energy would solve mankind's energy problems permanently. Radioactivity produced in fusion devices could be from ten to several hundred times smaller than from fission reactors. The problems associated with commercial traffic in weapons-usable fissile materials are largely absent [5]. Compared with the use of fast breeder fission reactors, the fusion option has fewer environmental, proliferation or safety problems. Moreover, its raw material source base is very large.

Thermonuclear fusion will be (hopefully) part of tomorrow's technology. It represents so immense a potential that the momentum of international research must be maintained, even though progress is slow. If and when the basic scientific problem of controlled and sustained fusion is realized, there may need to follow something like 20 years of engineering development.

14.3.4.   *Geothermal energy* [5]

At present the only usable geothermal sources are deposits of hot water or natural steam. In the long-term future it may be possible to extract heat from the natural thermal gradient in the earth's crust and from unusually hot rock formations lying close to the surface. The possible use of hot dry rock as a heat source depends on developing a fracture system large enough to be economical. The technical and environmental implications of this are speculative.

A potentially large source of low temperature geothermal energy is geopressurized brines, for example those off the Mexican Gulf coast. If the heat and dissolved natural gas can be exploited simultaneously this could become a major energy source. Such exploitation is still speculative and the technical implications are not fully known.

14.4.   What Shall We Do When the Oil Runs Out?

After the significant depletion of oil and natural gas, about the middle of the 21st century (less than two human generations from now), what options remain for the provision of energy on the scale demanded by the present and projected future populations?

The remaining fossil fuel is coal, which is abundant and widely distributed worldwide. But should we be using coal as a source of chemicals rather than the wasteful process of direct combustion?

In the absence of the fossil fuels there remain only four sources. These are listed in the preceding section and are repeated here:

1. Nuclear fission, using breeder reactors.
2. Solar energy, of various forms.
3. Controlled thermonuclear fusion.
4. Geothermal energy.

The last two options are not likely to be available on the required scale for 20–25 years, and perhaps longer.

There seems to be a strong case for a big upsurge of investment in both solar and nuclear fission systems [7]. Political misgivings in Britain and the USA about nuclear electric power are being reconsidered [8]. Smaller contributions such as energy efficiency [9], reduction of wastes, use of biomass, hydroelectric schemes, tidal barrages, etc. are important and welcome, but they will not close the energy gap.

References

[1] "International Energy Outlook 2010", Energy Information Administration, US Department of Energy, Washington, DC, September 2010.
[2] Salvador, A. (2005). *Energy Consumption and Probable Energy Sources during the 21st Century*, AAPG Studies in Geology, **54**, 131–147.
[3] Shepherd, W. and Shepherd D. W. (2003). *Energy Studies*, 2nd Edition, Imperial College Press, London.
[4] Bondi, H. (1978). *Who Needs the World's Energy?*, The Fourth Mountbatten Lecture, National Electronics Council, London.
[5] "Energy in Transition 1985–2010", National Academy of Sciences, Washington, DC, 1980.

[6] Chu, S. and Majumdar, A. (2012). *Opportunities and Challenges for a Sustainable Energy Future*, Nature, **488**, pp. 294–303.

[7] Douglas, J. (1995). *Reopening the Nuclear Option*, IEEE Power Engineering Review, **15**, 5–9.

[8] "Nuclear Plants to be Built in the UK", Sunday Telegraph, London, 4 March 2001.

[9] Eastop, T.D. and Croft, D.R. (1990). *Energy Efficiency for Engineers and Technologists*, Longman Science and Technology, London.

ANSWERS

Chapter 1

1.1. Use Eq. (1.6).

1.2. (a) PE $= 100 \times 9.81 \times 25 = 24525\,\text{J}$

(b) PE $= 0$, KE $= 0$

(c) $mgh = mgh/2 + 1/2\,\text{mv}^2$

$$v = \sqrt{gh} = 15.66\,\text{m}$$

(d) PE $= 12262\,\text{J}$, KE $= 12262\,\text{J}$

1.3. (a) $a = 100/100 = 1\,\text{m/s}^2$

(b) KE $= 1/2\,\text{mv}^2 = 1/2(100)(100) = 5000\,\text{J}$ Momentum $= \text{mv} = (100)\,(10) = 1000\,\text{kg/s}$

1.4. $v = wr = 10 \times 0.5 = 5\,\text{m/s}$

$T = Fr = 10 \times 0.5 = 5\,\text{Nm}$

$\alpha = T/mr^2 = 5/(1 \times 0.52) = 5/0.25 = 20\,\text{rad/s}^2$

1.5. $J = mr^2 = 1 \times (0.5)2 = 0.25\,\text{kg}^2$

1.6. $W_{\text{KE}} = 1/2\,J\omega^2 = 1/2 \times 1 \times 0.52 \times 102 = 12.5\,\text{J}$

1.7. $1800\,\text{rpm} \equiv 6\pi\,\text{rad/s}$

KE $= 1/2mr^2\omega^2 = 1/2 \times 10 \times 1^2 \times (6\pi)^2 = 1776\,\text{J}$

1.8. 1 UK gallon $= 4.55$ litres $= 4500\,\text{cm}^3 = 4.55\,\text{kg}$

$Q = 4550 \times 1 \times 20 = 91000\,\text{cal}$

$W = 4.2 \times 91000 = 382.2\,\text{kJ}$

1.9. (a) Final temperature $T_{\text{f}} =$ initial temperature T_{in}.

(b) From the formula in Example 1.4 the mass and specific heat terms cancel out:

$$T_{\text{f}} = \frac{T_{\text{in}} + 2T_{\text{in}}}{2} = \frac{3}{2}T_{\text{in}}$$

1.10. $10°C = 283\,\mathrm{K}$, $100°C = 373\,\mathrm{K}$

$\eta_{\mathrm{Carn}} = 1 - 283/373 = 1 - 0.759 \equiv 24.1\%$

1.11. (a) $\eta_{\mathrm{Carn}} = 1 - 300/900 = 1 - 0.333 = 0.667$

$\eta_{\mathrm{working}} = 0.667/2 = 0.333$

(b) $W = 0.333 \times 500 = 166.5\,\mathrm{MJ}$

1.12. $\eta 1 = 1 - 373/673 = 0.45$, $W_1 = 0.45Q$

$\eta 2 = 1 - 373/473 = 0.211$, $W_2 = 2 \times 0.211Q = 0.422Q$

Better to use a single source.

1.13. $35°\mathrm{F} = 1.67°C = 274.8\,\mathrm{K}$

$68°\mathrm{F} = 20°C = 293\,\mathrm{K}$

$\eta_{\mathrm{Carn}} = 1 - 274.67/293 = 1 - 0.937 = 6.3\%$

$\eta_{\mathrm{pract}} = 0.65 \times 6.3\% = 4.1\%$

1.13. $35°\mathrm{F} = 1.67°C = 274.8\,\mathrm{K}$

$68°\mathrm{F} = 20°C = 293\,\mathrm{K}$

$\eta_{\mathrm{Carn}} = 1 - \frac{274.67}{293} = 1 - 0.937 = 6.3\%$

$\eta_{\mathrm{pract}} = 0.65 \times 6.3 = 4.1\%$

1.14. (a) $Q_{\mathrm{H}} = 11.194\,\mathrm{MBTU}$, $Q_{\mathrm{L}} = 5.933\,\mathrm{MBTU}$, $W = 5.261\,\mathrm{MBTU}$,

$Q_{\mathrm{H}} - Q_{\mathrm{L}} = W$

(b) $\eta_{\mathrm{turbine}} = 5.261/11.194 = 47\%$

(c) It is necessary to determine the energy content of 1000 pounds of solid fuel and deduct $12.72\,\mathrm{MBTU}$.

(d) $\eta_{\mathrm{gen}} = 5.209/5.261 = 99\%$

1.15. $T_{\mathrm{L}} = 10°C = 283\,\mathrm{K}$

initially $0.38 = 1 - 283/T_{\mathrm{H}}$ or $T_{\mathrm{H}} = 456.5\,\mathrm{K}$

finally $0.5 = 1 - 283/T_{\mathrm{H}}$ or $T_{\mathrm{H}} = 566\,\mathrm{K}$

$$\text{Temperature rise} = 566 - 456.5 = 109.5°C$$

1.16. The refrigerator mechanism tries to refrigerate the room. It works continuously at full load drawing its maximum input power.

1.17. $T_{\mathrm{H}} = 873\,\mathrm{K}$

(a) $0.3 = 1 - T_{\mathrm{L}}/873$ so that $T_{\mathrm{L}} = 611\,\mathrm{K}$

(b) From Eq. (1.33), $0.3 = 1 - Q_{\mathrm{L}}/1000$, giving $Q_{\mathrm{L}} = 700\,\mathrm{MJ}$

From Eq. (1.31), $\Delta S = 700/611 - 1000/873 = 1.1457 - 1.1455 = +0.0002\,\mathrm{J/K}$

1.18. (a) $100°C$, (b) $37.8°C$, (c) $0°C$, (d) $-17.8°C$

1.19. (a) $413.6°\mathrm{F}$, (b) $212°\mathrm{F}$, (c) $89.6°\mathrm{F}$, (d) $32°\mathrm{F}$

1.20. $-40°C = -40°\mathrm{F}$

1.21. $24.33°C$

1.22. $67.2°\mathrm{F}$

1.23. (a) $(2 \times 10^9)/746 = 2.68 \times 10^6\,\mathrm{HP}$, (b) $2000\,\mathrm{MJ/s}$, (c) $2 \times 10^6\,\mathrm{kW}$, (d) $1474 \times 10^6\,\mathrm{ft\,lb/s}$

1.24. $50/3.6 = 13.9\,\mathrm{kWh}$

1.25. $(3 \times 10^6)/747 = 4021.5\,\mathrm{HP}$

Chapter 2

2.1. $1.256 \times 10^{14}\,\mathrm{m}^2$

2.2. $6.36 \times 10^6\,\mathrm{m}$ or 3961 miles

2.3. Two different interpretations possible:

(i) if the radiation intercepted is $1.73 \times 10^{17}\,\mathrm{W}$ (Sec. 2.1.1), then

$$\text{fraction} = \frac{1.73 \times 10^{17} \times \text{number of seconds in 24 hours}}{3 \times 10^{32}}$$

$$= 5 \times 10^{-11}$$

(ii) $\text{earth interception} = \dfrac{\text{earth interception area}}{\text{sun radiation area at earth distance}}$

$$= 4.48 \times 10^{-10}$$

Why are the two values in (i) and (ii) different?

2.4. $1.733 \times 10^{17}\,\mathrm{J/s}$

2.5. (i) 47%, (ii) 0.21%, (iii) 23%, (iv) 0.0023%

2.6. The Industrial Revolution in England. Mass migration of population from country to towns. Building of factories. Invention and use of steam engines.

2.7. (a) 40 years   (b) 10 years   (c) 20 years ⎫

2.8. (a) 60 years   (b) 12 years   (c) 80 years ⎬ Precise reading of Fig. 2.3

2.9. (a) 220 years (b) 215 years (c) 660 years ⎭ is not feasible.

2.10. Small population (4.46 M), abundant hydroelectric power and 50% stake in the North Sea oil and gas fields; by implication, a massive fuel exporter.

2.11. To move towards higher GNP/capita.

To move towards lower energy/capita.

2.12. They perceive that economic success is based on industrial development (not on agriculture). But this is energy-intensive and requires a vast capital endowment of funding (Japan was rebuilt after the Second World War with the aid of American money).

2.13. Australia, Canada and New Zealand have small populations in large areas of agricultural land.

2.14. Large country with mobile population. Much air travel. Traditional users of large size motor vehicles. A tradition of cheap gasoline. Major fuel deposits of coal, gas and oil. Higher level of general prosperity. Labour-saving (i.e. energy-burning) homes.

2.15. See answer to 2.12.

There are no available sources of foreign capital to increase their rate of industrialization. Further demands on fossil fuel sources would reduce the world reserves at an increased rate, assuming that they could afford to buy the fuel. The possibility that some countries might take military action against others to appropriate their oil.

2.16. Since 1999, in the UK, certain important changes have occurred.

  (i) The coal industry has steadily declined. Production (Table 2.5) in 2009 reduced to 10.9/22.4 or 48.7% while consumption (Table 2.6) reduced to 29.7/34.3 or 87% of the 1999 figure. In 1973, UK coal production and consumption were in balance. By 2009 (29.7–10.9)/29.7 or 63.3% of the coal consumed was imported.

  (ii) In 1999 the UK production and consumption of natural gas were almost in balance. Production then exceeded consumption until 2004. Since 2005 the UK has needed to import natural gas. In 2009, (77.9–53.7)/77.9 or 31% of the natural gas consumed was imported.

  (iii) From 1999–2005 oil production exceeded consumption. Since 2006, the UK has become a net oil importer.

2.17. Norway, Sweden, France (nuclear), Iceland.

2.18. Should there be legislation to limit the numbers of automobiles? Should vehicle taxation levels be adjusted to favour low carbon dioxide-emitter level vehicles? Should there be further incentives to encourage the use of electric and hybrid vehicles?

2.19. The big emission producers are the energy industry and big manufacturers. Further taxation would reduce profits and possibly result in job losses. Is this an acceptable situation?

2.20. The authors see no way in which the Kyoto Protocol can be levied other than by voluntary action.

2.21.

Largest Emissions:	Largest Per Capita Emissions:
China	Netherlands
USA	USA
India	Australia
Russia	Canada
Japan	Saudi Arabia
Germany	Kazakhstan

2.22. Small populations generating little carbon dioxide. Large agricultural sector with the use of nitrogenous fertilizers.

2.23. Very energy-conscious. Widespread use of nuclear-powered electricity. No oil industry.

Chapter 3

3.1. 0.2 A, 40 W

3.2. 20 ohms, 2880 W

3.3. 24 ohms, 0.3125

3.4. Pout $= 2 \times 746 = 1492$ W
Pin $= 200 \times 10 = 2000$ W
$\eta = 74.6\%$

3.5. 74.6%, 0.805

3.6. The input power is unchanged.

$I = P/V = 2000/120 = 16.67\,A$

3.7. With balanced sinusoidal currents it is possible to supply three separate phase circuits using only three conductors. (Whereas six conductors would be needed with three single-phase generators.) The saving on copper wire and transmission system construction costs is very large.

3.8. High voltage permits low current and smaller size transmission line conductors.

3.9. See Section 3.4.

3.10. From Table 3.1, USA, China, Japan, Russia.

3.11. From Table 3.1, in MWh/year/person:

Norway (24.7)	Taiwan (9.65)
Canada (16.6)	South Korea (8.08)
USA (13)	Japan (7.86)

3.12. From Fig. 3.6, the proportions of prime fuels used for electricity generation will change little up to 2035. The most significant change is expected to be increased use of renewables.

3.13. From Fig. 3.8, it is seen that there has been a declining use of coal but increased use of natural gas and nuclear.

3.14. The figures for 2009 are shown in Fig. 3.10. From Table 3.4, the significant growth fuel is natural gas, which is expected to grow by (903–683)/903 or 24.4% of the 2008 value. Renewable sources are expected to grow by (569–347)/569 = 39% of the 2008 value. Coal and nuclear indicate only small growth.

3.15. From Table 3.2. In 2009 electricity imports were 2.9/357.2, or less than 1%. This was imported from France via the undersea cable.

3.16. From Table 3.5 and Table 3.2.

	2000	2009
CHP	25.246	27.777
Total	371.4	357.2
CHP	6.8%	7.78%

3.17. From Table 3.4. In 2009, 161/3814 = 4.22%

3.18. See Section 3.8 and Fig. 3.11. Figures quoted for the improvement of overall efficiency range between 35% and 70%.

3.19. More expensive first cost. Motor operates below its rated load, with reduced efficiency.

3.20. Its light appears to be yellow in colour and casts people in an unflattering aspect; they look pale and peculiar.

3.21. Lamps operate for $8 \times 5 \times 40 = 1600$ hours/year.
   (a) $20 \times 100 \times 8 \times 5 \times 40 = 3200\,\text{kWh}$
      $3200 \times 0.07 = £224$
   (b) Lamp replacement time $= 1000/1600 = 0.625$/year
   Annual labour cost $= 20 \times 5/0.625 = £160$
   Lamp replacement cost $= 20 \times 0.5/0.625 = £16$
   (c) Annual cost $= £224 + £160 + £16 = £400$
3.22. Lamps operate for 1600 hours/year as in Eq. (3.21).
   (a) Annual consumption $= 30 \times 70 \times 1600 = 3360\,\text{kWh}$
   Annual cost $= 3360 \times 0.70 = £235.2$
   (b) Lamp replacement times $= 10,000/1600 = 6.25$ years
   Annual labour cost $= 30 \times 5/6.25 = £24$
   Tube replacement cost $= 30 \times 4/6.25 = £19.2$
   (c) Total annual electricity costs $= £235.2 + £24 + £19.2 = £278.4$
3.23. (a) Annual cost saving $= £400 - £278.4 = £121.6$
   (b) Payback period $= 660/121.6 = 5.43$ years
   (c) Levels of illumination/unit area

Incandescent	Fluorescent
$20 \times 100 \times 12/150$ $= 1600\,\text{fm/m}^2$	$30 \times 70 \times 64/150$ $= 896\,\text{fm/m}^2$

Although the fluorescent option represents an energy saving, the reduction of illuminance could be a serious disadvantage. The Illuminating Energy Society of North America recommended lighting level for classrooms/lecture rooms is category F, which is 1000–2000 lux (lumens/m^2) [16].

Chapter 4

4.1

Coal producers (2010 mtoes)		Coal consumers (2010 mtoes)	
China	1800.4	China	1713.5
USA	552.2	USA	524.6
Australia	235.4	India	277.6
India	216.1	Japan	123.7
Indonesia	188.1	Russian Fed.	93.8
Russian Fed.	148.8	South Africa	88.7
South Africa	143	South Korea	76
Kazakhstan	56.2	Germany	76.5
Poland	55.5	Poland	54
Columbia	48.3	Australia	43.4

4.2. Big importers: Japan, South Korea, Germany, UK, France
   Big exporters: Australia, South Africa, Indonesia

4.3. (a) Increases: China, India Indonesia
   (b) Decreases: Germany, Czech Republic, Poland

4.4. (a) Increases: China, India, Indonesia, Japan
   (b) Decreases: Czech Republic, Russian Fed., UK, Germany.

4.5. In the UK coal production decreased by $(130.1–19.9)/130.1$ or 86.2% of the 1980 figure.

4.6. In the USA coal production decreased by 3.1% but coal consumption decreased by 8.8% in the period. Production still exceeds consumption, making the USA a potential coal exporter.

4.7. From Table 4.6, in the period 1980–2009, deep mined coal production fell by $(112.4–7.5)/112.4 = 93.3\%$ of the 1980 figure while open-cast (surface) mined production fell by $(15.8 – 9.9)/15.8 = 37.3\%$. In the same period coal imports increased by $(38.2 – 7.3) = 30.9$ million tonnes to form 68.1% of the total.

4.8. The UK domestic sector accounted for $8.9/123.5 = 7.21\%$ of the total consumption in 1980, reduced to $0.7/48.8 = 1.43\%$ in 2009 (Table 4.7). The reduction represents the UK preference for gas-fired domestic heating.

4.9. In 1980, $89.6/123.5 = 72.6\%$ of coal consumption was accounted for by electricity generation. In 2009, this proportion had changed to $40.1/48.8$, which is 82.2% (Table 4.7)

4.10. European coal has to be deep mined, whereas US coal is surface mined and the cost/ton of extraction is cheaper.

4.11. Mostly as fuel for electricity generation.

4.12. (a) Cheap, readily available, easily transportable
   (b) Dirty, unhealthy, inefficient fire grate systems, smoke pollution

4.13. (a) Cheap, readily available, easily transportable
   (b) Pollutant gases and particulates, creation of solid ash, acid precipitation, greenhouse gases

4.14. Mainly by rail, but by waterway if feasible.

4.15. See Section 4.3.2.

4.16. Sulphur oxides, nitrogen oxides, carbon monoxide, carbon dioxide, particulates

4.17. See Section 4.4.2.2.

4.18. Ionization of solid carbon particles by passing them through an electric field.

4.19. See Section 4.6.

4.20. Domestic consumers in Europe prefer central heating systems to open coal fires. The price of coal has doubled since 1990. Coal from surface mine sites in Australia and South Africa is cheaper than deep mined coal in Europe.

Chapter 5

5.1. The oil producer countries, OPEC, realized that they had the power to fix their own oil prices and not be dependent on western oil companies.

5.2.

Consumption (mtoes)		Production (2010 mtoes)	
USA	580	Russian Fed.	505.1
China	428.6	Saudi Arabia	467.8
Japan	201.6	USA	339.1
India	155.5	Iran	203.2
Russian Fed.	147.6	China	203
Saudi Arabia	125.5	Canada	162.8
Brazil	116.9	Mexico	146.3
Germany	115.1	UAE	130.8
South Korea	105.6	Venezuela	126.6
Canada	102.3		

5.3. Increasing industrialization of China, India, Japan and the Pacific Rim countries.

5.4. Use Table 5.4 to compile data on 2000 consumption minus 2010 consumption and divide by 11 for the average increase/year. The two biggest changes are China and India.

5.5. Use Table 5.3 to calculate the 2000 production minus the 2010. The result is Russian Fed. 182, Angola, 53.8, Brazil 42.5, China 40.4.

5.6. The Sultan of Brunei owns all of Brunei's oil (8.4 million tonnes in 2010).

5.7. Atlantic Ocean, into USA; Indian and Pacific Oceans, to the Far East Asia.

5.8. (a) The output of Iran, the Gulf States and Saudi Arabia totalling roughly one-quarter of world oil production in 2010.

(b) Large-scale curtailment of industrial production in Western Europe and Japan.

Severe restrictions in North America for any period of closure.

5.9. $\frac{850-339.1}{200\times10^3} \times 10^6 = 2550$ tanker journeys

5.10. See Table 5.5.

5.11. See Table 5.5.

5.12. See Table 5.5.

In 1980, the UK made a net loss of £0.3 billion.

In 2010, the UK made a net profit of £3.2 billion.

5.13. From Fig. 5.6, Venezuela and Saudi Arabia.

5.14. In 1995, consumption 807.7 mtoes, production 384 mtoes, deficit $= 807.7 - 384 = 423.7$ mtoes.

5.15. Large country with widespread population. Long car journeys are routine. Large internal air services greatly used. Car ownership universal, from teenage onwards. Historical tradition of cheap gasoline. High levels of affluence. Labour-saving (i.e. energy-using) homes.

5.16. Fuel prices and fuel tax levies are a political "hot potato". In the UK in 2000, farmers and self-employed transport workers, enraged by yet another fuel tax increase, blockaded fuel depots, causing a considerable crisis. Within three days fuel pumps were dry. Within five days supermarket shelves were becoming empty and reports of theft of fuel and groceries were rife. The government's opinion poll rating dropped like a stone as everyday life became seriously disrupted within just one week.

Despite the relatively tiny tax levy on US fuel, sharp increases in gas pump prices in 1999 caused threats of industrial action and an expedient partial climb-down by federal authorities.

The extremely sensitive link between public opinion and fuel prices gives sobering food for thought about the consequences that could ultimately occur if the world supply of oil brought about shortages and/or large price increases. A future "oil war" is not at all beyond the realm of possibility; indeed, a regional oil war (the UN–Iraq war over Kuwait) has already taken place.

If the shortages were not sudden, but a gradual squeeze on supply forced prices very high, diplomatic pressure to increase supply would be exerted on the producer countries, which rely heavily on global markets for prosperity. Secondly, consumer countries would move to utilize more of their less economically viable resources (e.g. coal, shale oil) to buffer the shortfall.

5.17. New oilfield discoveries. Improved extraction rates.

5.18. Recognition of the problem. Addressing of the problem at government and international level. An assessment of overall energy use and supply.

5.19. Move out of oil and natural gas into renewables or nuclear or coal or some combination.

5.20. Without the military and economic leadership of the USA there could be oil anarchy. A possibility of major wars to possess and control the world oil supply.

5.21. Deep mining. Large amounts of rock waste for surface disposal. Large requirement of processing water that becomes contaminated. Gas or dust emissions. Despoliation of mining sites. Very expensive compared with imported crude oil.

5.22. The oil companies are immensely rich and realize the value of good public relations. Everybody needs oil for transportation. The need for coal to generate electricity (universally sought) is less well perceived. Carbon dioxide emissions are mostly invisible. Power station effluent is highly visible. Coal burning results in acid rain.

Chapter 6

6.1.

Producers (mtoes)		Consumers (mtoes)	
USA	556.8	USA	621
Russian Fed.	530.8	Russian Fed.	372.7
Canada	143.8	Iran	123.2
Iraq	127.4	China	98.1
Qatar	105	Japan	85.1
Norway	95.7	Canada	84.5
China	87.1	UK	4.5
Saudi Arabia	75.5	Saudi Arabia	75.5
Indonesia	73.8	Germany	73.2
Algeria	72.4	Italy	68.5

6.2. Political breakup of the former Soviet Union. Abandonment of central control (from Moscow). Autonomous decision making by the countries of the former Eastern Bloc.

6.3. In the period 2000–2010 the consumption of natural gas increased most significantly in China, from 22.1 to 98.1 mtoes. Consumption more than doubled in India (23.7 → 55.7) and South Korea (17 → 38.6). There were significant increased in Algeria, Japan, Iran, Mexico, Saudi Arabia and the United Arab Emirates.

6.4. From Fig. 6.2 the trade routes are (i) Eastern Europe to Western Europe, (ii) Canada to the USA, (iii) Sout East Asia to Japan, (iv) north Africa (Algeria) to Europe.

6.5. The information is given in Fig. 6.4. The UK was an exporter from 1980–1995. Since 1995 consumption has exceeded production and the UK has imported natural gas.

6.6. The USA has imported natural gas every year in the period 2000 to 2010.

6.7. From Table 6.3 the average consumption was $9285/11 = 84.4$ mtoes.

6.8. From Table 6.3: $\frac{621-600.4}{11} = 1.87$ mtoes/year

6.9. From Table 6.3, the OECD group of countries consumed 48.9% of the world's natural gas in 2010.

6.10. From Table 6.1, in 2010:
(a) 5.3% (b) 33.7% (c) 23.9%
(d) 40.5% (e) 7.9% (f) 8.7%

6.11. New discoveries, revised estimates of reserves and improved extraction rates.

6.12. Most likely by producing liquid natural gas (LNG) and using refrigerator ships. The alternative, by use of long pipelines, is too uncertain for political reasons.

6.13. Coal-bed methane (see Section 6.5).

6.14. low BTU 3–6 MJ/m^3
medium BTU 10–22 MJ/m^3
high BTU 37 MJ/m^3

6.15. Natural gas is abundant, easy to transport and cheaper to extract than coal.

6.16. Natural gas and coal-bed methane and coal gas all consist mainly of methane. For natural gas and coal-bed methane the fuel is obtained directly in its gaseous state. Coal gas has to be manufactured by burning coal.

6.17. Until recent years (i.e. since the turn of the 21st century) the price of oil was too low for other options to be viable.

Chapter 7

7.1. Section 7.3.

7.2. Due to high pressure.

7.3. In the seismic areas described in Fig. 7.3.

7.4.

Cotopaxi — Brazil	Mauna Loa — Hawaii
Fujiyama — Japan	Mount Etna — Italy
Hekla — Iceland	Mount St. Helens — USA
Katmai — Alaska	Ngauruhoe — New Zealand
Kilauea — Hawaii	Osorno — Chile
Krakatoa — Indonesia	Paricutin — Mexico
Lassen Peak — USA	Popocatépetl — Mexico
Semeru — Indonesia	Stromboli — Italy

Reference to Fig. 7.3 shows that they are all in seismic regions.

7.5. Geothermally heated greenhouses.

7.6. Larderello — Italy

The Geysers — west coast of USA

Otake — Japan

Matsukawa — Japan

Wairakei — New Zealand

Stromboli — Italy

Pauzhetsk — eastern Russia

Cerro Prieto — Mexico

Niland — USA

Ahuachapan — Central America

Hveragerdhi — Iceland

Reykjanes — Iceland

Namafjall — Iceland

7.7. The aquifers are too low in temperature and of inadequate flow rate.

7.8. Dry steam, wet steam, hot brine, dry rocks, molten magma.

7.9. Necessity to fracture the rock underground in a controlled manner.

7.10. See Fig. 7.6.

7.11. Geothermal energy is largely unused. The risks and problems of exploration are similar to those involved in oil exploration. Environmental problems might include [2]: land use, noise and damage during drilling, visual impact of power and heat extraction plant, need for suitable heat distribution system, release of gases, liquids and chemicals, physical effects of the local area geological structure.

7.12. Uneconomical due to low flow rate and lower-than-expected temperature.
7.13. See Table 7.3.

UK does not lie in a seismic zone.

7.14. 1295 tonnes

Chapter 8

8.1. 2667 tons of coal
8.2. (i) $3.2 \times 10 - 11$ J, (ii) $8.91 \times 10 - 18$ kWh
8.3. 0.33 ton ore (compared with 1.8 ton without enrichment, see Example 8.2).
8.4. From Eq. (8.10),

(i) 3.32 T1/2, (ii) 6.64 T1/2

8.5. See Section 8.3.1.
8.6. 2337 million years
8.7. $\lambda = 0.639/30 = 0.0231$
8.8. From Eq. (8.9), $T_{1/2} = 0.693/0.131 = 5.29$ years
8.9. 10 years
8.10. 1 rad dose $= 1$ rem

1 mrem $= 1/1000$ rem

100 rems $= 1$ sievert

100 rads $= 1$ gray

8.11. $W = 1$ joule/kg $= 1$ gray at 100 rads dose

For a person weighing 150 lb ($150 \times 2.2 = 330$ kg) the whole body radiation energy input is 330 J or 330 watt seconds.

8.12. See Subsection 8.4.3.1; 200 mrem/year
8.13. See Subsection 8.4.3.2;

0.3 mrem/year from the nuclear industry

500 mrems/year from all sources

8.14. See Section 8.5.1.
8.15. Business is not convinced of the commercial viability, especially with regard to decommissioning costs. The public is not convinced regarding the safety of nuclear reactor operation.
8.16. The natural uranium supply is being depleted and is not renewable. Breeder reactors simultaneously generate heat and breed plutonium, using smaller reactor core sizes.
8.17. See answer to 8.15.
8.18. See Section 8.6.
8.19. See Sections 8.2.1 and 8.8.1.
8.20. See Section 8.8.1.
8.21. (i) 35.84×10^{-13} J, (ii) 9.98×10^{-19} kWh

8.22. Deuterium and lithium are cheap, abundant and non-radioactive materials.

8.23. No. It is a complex, multifaceted situation. Nuclear fission is a mature industry that is making great contributions to electricity generation worldwide but is in decline in the USA, which is the richest, most powerful and most influential country in world energy. Nuclear fusion is still at the scientific experiment stage. If it works on a commercial scale, which is probable but not certain, such a development is at least 20 years hence.

8.24.

Advantages	Disadvantages
Cheap, abundant, non-fossil fuel.	Still experimental and far from maturity.
Level of associated radioactivity much lower than for nuclear fission.	Requires major advances in several scientific areas.
International collaboration (not competition as with nuclear weapons).	Requires massive investment in an unproven technology.
Another energy option.	Still a generation of time from realization.
Does not create any "greenhouse" gases.	Other options are starting to look more attractive.

8.25. See Section 8.8.5.

Chapter 9

9.1. Values in TWh:
Canada 369.5 China 652.05
Brazil 363.8 Russian Fed. 167
USA 250.6 Norway 140.5

9.2. (a) China, Brazil
(b) There are no overall significant decreases.

9.3. In the year 2009, $740.3/11164.3 = 6.63\%$ of the world total primary energy consumption was due to hydroelectricity.

9.4. $74.6 \text{ m}^3/\text{s}$

9.5. See Section 9.1.1.

9.6. 424.7 m

9.7. (a) Axial-flow (propeller) turbine (e.g. Kaplan)
(b) Reaction turbine (e.g. Francis)
(c) Impulse turbine (e.g. Pelton)

9.8. $365.2 \text{ m}^3/\text{s}$
6.47×10^6 tons

9.9. (v) Not all rain is available as surface run-off.
Not all run-off appears in streams that are worth damming.
If run-off descent is at too small a slope, piping difficulties limit the available head.
(vi) 10–30% in general

9.10. See Subsection 9.3.3.4.

9.11. From Eq. (9.20) with $h = 1$, $R = 6$

(a) $W = \frac{1}{2}pgAR^2[\frac{2}{6}]$; net gain is 33.33%

(b) From Eq. (9.23), with $h = 1$, $R = 6$, $k = 0.8$ net gain is $0.333 - 0.007 \equiv$ 32.6%

9.12. Use Eq. (9.22) and differentiate w.r.t. h/R

9.13. 21.7%

9.14. 213 MW

9.15. Geometric solution, showing P is proportional to R^3

9.16. 4491 MW

9.17. 0.1176 Hz, 112.8 m, 13.27 m/s

9.18. 74.8 kW/m

But the extractable power is likely to be about 30% of this.

9.19. $T = 9.47$ s, $P = 46.9$ kW/m

9.20. 20.1 kW/m

9.21. See Section 9.3.6.

9.22. Approx. 1660 m if wave converter is 30% efficient.

Chapter 10

10.1. Differentiate Eq. (10.12) w.r.t. (V_2/V_1) and equate the derivative to zero.

10.2. Since P ∞V^3 doubling V causes a $2^3 = 8$ times increase of P.

10.3. In Fig. 10.14 the vertical projection is about 40% of the distance from 10^4 to 10^5.

Estimate $A = 30{,}000$ m², $D = 195$ m

10.4. See Section 10.2.4.

r = 90 ft = 27.43 m

V = 20 mph = 8.94 m/s

$\omega = 1.96$ rad/s = 18.7 rpm

10.5. TSR = 7.61

10.6. (a) $0.26 - 0.45$ per unit

(b) $0.084 - 0.26$ per unit

10.7. If $\eta = 0.25$, $D = 4.45$ m.

10.8. If $\eta_g = 0.75$, $C_p = 0.35$ and there is no gearbox, $D = 12.12$ m.

10.9. $D = 3.65$ m (12 ft), when $\eta = 0.25$

10.10. $V = 8.61$ m/s = 19.3 mph

10.11. Let the overall efficiency be 25%

(a) $V = 6.26$ m/s = 14 mph

(b) TSR = 1.34

(c) $D = 3.54$ cm = 1.39 in

10.12. (a) $T = 3.54 \times 10^6$ Nm

$D = 68.4$ cm = 26.93 in

(b) See Section 10.2.4

(c) TSR $= 6.46$

$D = 84.2\,\text{m}$ (276.3 ft)

(d) $\eta_g = 0.95$, $\eta_{gb} = 0.9$, $C_p = 0.351$

10.13. $d = 12\,\text{in} = 0.304\,\text{m}$

$T = 0.305 \times 10^6\,\text{Nm}$

10.14. $N_{\text{max}} = 35\,\text{rpm}$

Propeller is feathered (turned into the wind) to limit rotational speed. Excessive speed would cause large centrifugal forces on the blades plus possible bearing damage.

A vertical (horizontal axis) rotor acts as a rudder or stabilizer.

10.15. (i) Air speed is not affected.

(ii) Ground speed is increased (America to Europe) by 100 mph.

Ground speed is decreased (Europe to America) by 100 mph.

10.16. Airspeed $= 445\,\text{mph}$

Ground speed $= 445 - 95 = 350\,\text{mph}$

Time $= 3400/350 = 9.71$ hours

10.17. (a) 24.12 rpm

(b) By "feathering", as in Problem 10.14

(c) 398 kW

(d) Power increased by $(1.05)^3$ to 461 kW

10.18. (a) TSR $= \frac{v}{V} = \frac{rw}{V}$, see Section 10.2.4

(b) 42 rpm

(c) $P_{\text{in}} = 0.83\,\text{MW}$

$V = 27.1\,\text{mph}$.

10.19. (a) $P_{\text{extract}} = 2\,\text{MW}$

(b) 25.21 rpm

(c) $0.76 \times 10^6\,\text{Nm}$, 16.93 ins

(d) 1.62 MW

Chapter 11

11.1. $\cos 9° = 0.988$

11.2. From Eq. (11.1). Assume that the inclination of the collector compensates the altitude angle, then the radiation falls normally onto the collector: $G = 500 + 500 = 1000\,\text{kWh/m}^2$.

11.3. $G = D(1 + \sin \alpha) = 320(1 + 0.636) = 523.5\,\text{W}$

11.4. In Eq. (11.4), if $T_c \uparrow$, $(T_c - T_a) \uparrow$ and $\eta \downarrow$

11.5. The greenhouse effect is discussed in Subsection 2.6.6.2 of Chapter 2.

This greenhouse effect is not utilized directly in the design and operation of solar flat-plate collectors.

Typical working efficiency $\approx 40\%$

Efficiency features are summarized in Table 11.6.

11.6. Typical efficiency 30–40%. See Section 11.4.1.

11.7. See Fig. 11.16. Locate supplementary tank near to main tank for minimum pipe loss. System is working if supplementary tank is delivering water greater than cold tap temperature.

11.8. See Fig. 11.13. Depending on the working temperature, the use of double glazing could increase the thermal efficiency by up to 10% (Fig. 11.15).

11.9. See the latter part of Section 11.4.2.

11.10. Mass transferred in 5 hours $= 360\,\text{kg}$
Temperature rise $= 11.68°\text{C}$. Power $= 773\,\text{W/m}^2$

11.11. Heat collected in 6 hours $= 28.12 \times 10^6$ joules
Heat required to raise the tank temperature $= 11.424 \times 10^6$ joules
$\eta = 41\%$

11.12. $Q = 35.36\,\text{MJ}$, $\eta = 32.7\%$

11.13. (a) $1.92 \times 10^{-3}\,\text{m}^3/\text{s}$, (b) 51.5%

11.14. From Eq. (11.5), $\eta_c = 45.1\%$
Now if $T_c = T_a$ the apparent efficiency is 78%
But if $T_c = T_a$ there is no temperature rise and the actual efficiency is zero.

11.15. S facing roof, inclined at $54°$ to horizontal, $1\,\text{m}^2/\text{person}$ of collector $\equiv 4\,\text{m}^2$
$10\,\text{gal/person}$ storage $\equiv 40$ gallon tank
Cost $\approx £3200 + 20\text{--}25\%$ if borrowed
or $\approx £3200 + 10\text{--}15\%$ if loss of existing capital
Electricity cost in UK (2002) is $7\,\text{p/kWh}$.
Lifetime of system (assuming proper maintenance) is about 20 years.

(i) For a heavy user of water, annual saving on electricity bill might be £300. Payback period ≈ 10 years.

(ii) For a light user of water, annual saving on electricity bill might be £100. Payback period $\approx £3200/100 \approx 30$ years, which exceeds the expected plant lifetime.

11.16. In northern Europe: average insolation $= 100\,\text{W/m}^2$, with 3:1 energy input split between summer and winter. Cold ambient temperature in winter. Solar energy input is in time antiphase with the energy demand.
    In the Middle East: average insolation $= 300\,\text{W/m}^2$ with 1.5:1 split between summer and winter. Warm ambient temperature is winter. Solar energy input is in time phase with the refrigeration and air conditioning load. The chief obstacle to solar energy use is the cheap price of oil, especially in the Arab countries.

11.17. Carnot (heat $\rightarrow$ work) efficiency is greatly dependent on (and varies directly with) the working fluid temperature.
The working fluid is often required to be gaseous (e.g. steam).

11.18. $\eta_{\text{Carn}} = 410/723 = 56.7\%$

11.19. Low Carnot ideal efficiency $= 20/293 = 6.83\%$

11.20.

$$\eta_{\text{Carn}} = \frac{T_{\text{fluid}} - T_{\text{amb}}}{T_{\text{fluid}}}$$

Let T_{fluid} increase to $(T + \Delta)_{\text{fluid}}$. Is, then,

$$\frac{(T + \Delta)_{\text{fluid}} - T_{\text{amb}}}{(T + \Delta)_{\text{fluid}}} > \frac{T_{\text{fluid}} - T_{\text{amb}}}{T_{\text{fluid}}}?$$

11.21. Combine η_{c} with η_{Carn}. Differentiate $\eta_{\text{c}}\eta_{\text{Carn}}$ and equate to zero to give QED solution.

$\eta_{\text{Carn}} = 0.424$, $\eta_{\text{c}} = 0.381$, $\eta_{\text{syst}} = 16.15\%$

11.22. See Eq. (11.6).

Heat losses in absorber $\propto T^4$

For flat-plate collector $CR = 1$

11.23. From Eq. (11.9),

$P \propto A$, $P \propto I_D$, $P \propto \eta_{\text{t}}$, $\text{P} \propto \eta_{\text{Carn}}$

11.24. Re-radiated power $= 26 \,\text{W/m}^2 = 13\%$

11.25. $A_H = 0.471 \,\text{km}^2$, $A = 1.88 \,\text{km}^2$

11.26.

Advantages:	Disadvantages:
• free fuel • large construction project: jobs • diversifies the sources of energy supply • encourages new technologies	• large land area required • large first cost (materials) • pollution of manufacturing the materials • destroys several square miles of animal habitat • modifications of hydrological cycle due to heliostat canopies • modification of wind and water erosion due to site plus access roads

11.27. $37.65 \,\text{km}^2$ or 14.54 square miles

11.28. One can devise any number of examples that incorporate the same arguments. For example, can a homeowner permit one of his trees to grow such that it will gradually shade the solar collector of a neighbour? Would it be reasonable or unreasonable to seek to go to law over such an issue? What is reasonable?

If a case arises such that the actions of one person prevent access to sunlight by another person, is this an infringement of legal right, moral right, good neighbourliness, reasonable behaviour, professional conduct, etc.? The issue is not merely academic. In order to reach solar collectors the radiation often has to pass through air space not owned or controlled by the solar collector site owner. It would seem prudent on the part of someone intending to install solar collecting equipment to ensure that the necessary intervening air space would not be subsequently blocked by the actions of other people.

There is no law in Western Europe or North America that at present that covers the above eventualities.

Chapter 12

12.1. See the listing in Section 12.1.

12.2. Covalent bonding (Fig. 12.4)

12.3. (a) Monocrystalline (single crystal) (Section 12.4.1)
(b) Polycrystalline (Section 12.4.1)
(c) Amorphous (Section 12.4.2)

12.4. "*n*-type" silicon is doped with an element containing five electrons. This increases the density of free electrons in the conduction band.
"*p*-type" silicon is doped with an element containing three electrons. This decreases the density of free electrons (increases the "holes") in the conduction band.

12.5. $1.08\,\text{eV}$ or 2.63×10^{-19} joules
This energy corresponds to $f = 2.59 \times 10^{-14}\,\text{Hz}$,
$\lambda = 1.15\,\mu\text{m}$, $c = 2.98 \times 10^{8}\,\text{m/s}$.

12.6. The fraction of the solar spectrum that causes electrons to cross the energy gap decreases as the gap energy increases.

12.7. Increased working temperature causes increased thermal agitation of the lattice electrons. External radiation has fewer free electrons to dislodge and harvest, causing reduced current and reduced output power.

12.8. High purity material;
Slow growth rate of crystal formation;
High waste factor by diamond slitting process;
Labour-intensive fabrication.

12.9. No need to slice it into thin wafers (minimizes waste).

12.10. More expensive than silicon;
Raw material stock inadequate for mass production.

12.11. Much easier (and cheaper) to form into cell wafers.

12.12. Upper wavelengths of the solar spectrum will not produce free electrons and low wavelengths have only limited capacity to produce free electrons; junction temperature losses; cell material temperature losses.

12.13. Reduces the necessary amount of (expensive) solar cell material.

12.14. Increase of temperature causes significant decrease of efficiency. Forced cooling is needed in some applications.

12.15. (a) In space the exposure is for 24 hours, compared with (say) 12 hours on earth.

(b) In space the insolation is four times the value at the earth surface.
In combination, (a) and (b) reduce the necessary area in space, compared with earth, by a factor of $2 \times 4 = 8$.

12.16. See Fig. 12.6.

12.17. See Figs. 12.7 and 12.10.
By iteration (trial and error) taking the current–voltage product for each co-ordinate.

12.18. $V_{mp}/V_{oc} \approx 0.8$, $I_{mp}/I_{sc} \approx 0.9$

12.19. (i) $I_{sc} = \frac{850}{1000} \times 1.5 = 1.275\,\mathrm{A}$, (ii) $I_{sc} = \frac{3000}{1000} \times 1.5 = 0.45\,\mathrm{A}$

12.20.

	1000 W/m²		500 W/m²	
Load resistance	8 Ω	20 Ω	8 Ω	20 Ω
voltage current (mA)	0.386	0.525	0.21	0.46
	49	26	26	23

12.21. $-0.72\,\mathrm{V}/10°\mathrm{C}$, 4%

12.22. The short circuit current I_{sc} is almost directly proportional to the incident radiation.
The equivalent circuit is shown in Fig. 12.9(b).

12.23. $2.75\,\mathrm{mA}$ for $500\,W/m^2$
At the maximum power point P_n, $V = 0.43\,\mathrm{V}$ and $I = 25\,\mathrm{mA}$, $R_0 = (0.43 \times 1000)/25 = 17.2\,\Omega$. $I_j = 27.5 - 25 = 2.5\,\mathrm{mA}$, $R_j = (0.43 \times 1000)/2.5 = 172\,\Omega$.

12.24. At $1000\,\mathrm{W/m^2}$, $I_s = 50\,\mathrm{mA}$.

R_0 (ohms)	V (V)	I (mA)	I_j (mA)	$V/I_j = \mathrm{Rj}$
5	0.26	49.5	0.5	520
10	0.46	45	5	92
50	0.54	11.5	38.5	14

Variation of R_j with current is shown in Fig. P.12.24, in which the bend of the curve follows the knee of the current–voltage characteristic.

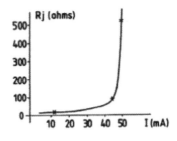

Fig. P.12.24. Solution of problem 12.24.

12.25. For the $5\,\Omega$ characteristic in Fig. 12.24:

Insolation	V (V)	I (mA)	I_s (mA)	I_j (mA)	$R_j\,(\Omega)$
$1000\,\text{W/m}^2$	0.26	49.5	50	0.5	520
$500\,\text{W/m}^2$	0.14	37	27.5	22.5	6.22

12.26. $R_L = V_m/I_m = 0.44/57.5/1000 = 7.65\,\Omega$

12.27. $R_L = 0.425/36/1000 = 16.25\,\Omega$

12.28. I_{sc} is double and V_{oc} is tripled, compared with a single cell.

12.29. I_{sc} is tripled and V_{oc} is doubled, compared with a single cell.

12.30. (i) At $1000\,\text{W/m}^2$, $I_s = 50\,\text{mA}$

With $R_0 = 10\,\Omega$, $V = 0.46\,V$ and $I = 45\,\text{mA/cell}$.

With n cells in parallel:

$$I = n(I_s - I_j) = nI = 20 \times 45/100 = 0.9\,\text{A}$$
$$P_0 = VI = 0.46 \times 0.9 = 0.414\,\text{W}$$

(ii) At $500\,\text{W/m}^2$, $I_s = 27.5\,\text{mA}$

With $R_0 = 10\,\Omega$, $V = 0.27\,\text{V}$, $I = 27\,\text{mA/cell}$

With 20 cells in parallel:

$$V = 0.27\,\text{V}, \quad I = 20 \times 27/1000 = 0.54\,\text{A}$$
$$P_0 = 0.27 \times 0.54 = 0.15\,\text{W}$$

By connecting 20–25 cells in series.

At $1000\,\text{W/m}^2$ and $R_0 = 10\,\Omega$, $V = 0.46\,\text{V/cell}$

$I = 45\,\text{mA/cell}$

12.31. With 100 cells in series $V = 100 \times 0.46 = 46\,\text{V}$

12.32. $I = 45\,\text{mA}$, $P = 46 \times 45/1000 = 2.07\,\text{W}$

Chapter 13

13.1. Competition with food production;

Change of water and nutrient demand;

Effect on wildlife;

Contamination of local food crops.

13.2. See Section 13.2.

Solar radiation $\rightarrow$ stored chemical energy

13.3. See Section 13.2 and Table 13.1.

Temperate: 0.5–1.3%

Tropical: 0.5–2.3%

Average $\approx 1\%$

13.4. See Section 13.2.

Temperate 0.5–1.3%

Tropical 0.5–2.3%

13.5. Rapeseed: a biofuel plant.

13.6. Methane;

Abundance of human and animal dung for fuel;

Need for cooking fuel.

13.7. Widespread growth, ease of intensive farming, combustibility of forestry products, less polluting than coal or oil, carbon dioxide-neutral.

13.8. Carbon dioxide given up on combustion is the same amount as the carbon dioxide absorbed in photosynthetic growth.

13.9. High calorific value (37 MJ/kg) but releases toxic fumes.

13.10. More rapeseed (bright yellow) and linseed (blue). Fewer conventional food crops. More arable coppice plantation of small (up to 3 ft) trees.

13.11. No plastic. Less paper and packaging. Less food waste. More glass, more ash and cinders.

13.12. from Table 13.8

	1990	2000
(a)	7.82%	23.80%
(b)	17.10%	7.85%
(c)	9.90%	9.54%

13.13.

(a) Wood increased by $6948/6193 = 12\%$
(b) Hydropower increased by $77951/77541 = 1\%$
(c) Solar power increased by $603/411 = 47\%$
(d) Wind power increased by $33542/8706 = 385\%$

$\left.\right\}$ from Table 13.4

13.14. Comparing Table 13.6 and Fig. 13.8.

	UK	USA
Paper and cardboard	30.5%	28.2%
Food waste	24%	14.1%
Glass	11.2%	4.8%
Rubber leather & textiles	13.2%	8.3%

13.15. 28 mtoes yields a gross 3×10^{12} MJ.

$$\text{At 50\% efficiency} = 1.5 \times 10^{12} \text{ MJ}$$
$$= 1.5 \times 1012/3.6 = 0.417 \times 10^{12} \text{ kWh}$$

13.16. $28 \times 10^6 \times 5000/3.6 = 38.9 \times 10^9$ kWh

13.17. Reduced dependence on oil;

Reduced emissions.

INDEX